Engineering Surveying

Theory and Examination Problems for Students

Fifth Edition

W. Schofield
Principal Lecturer, Kingston University

ELSEVIER
BUTTERWORTH
HEINEMANN

AMSTERDAM BOSTON HEIDELBERG LONDON NEW YORK OXFORD
PARIS SAN DIEGO SAN FRANCISCO SINGAPORE SYDNEY TOKYO

Elsevier Butterworth-Heinemann
Linacre House, Jordan Hill, Oxford OX2 8DP
30 Corporate Drive, Burlington MA 01803

First published 1972
Second edition 1978
Third edition 1984
Fourth edition 1993
Reprinted 1995, 1997, 1998
Fifth edition 2001
Reprinted 2002, 2004

British Library Cataloguing in Publication Data
Schofield, W. (Wilfred)
    Engineering surveying: theory and examination problems for students. - 5th ed.
    1 Surveying
    1 Title
    526.9'024624

Library of Congress Cataloguing in Publication Data
Schofield, W. (Wilfred)
    Engineering surveying: theory and examination problems for students/W. Schofield.-5th ed.
    p. cm.
    ISBN 0 7506 4987 9 (pbk.)
    1 Surveying   I Title

    TA545.S263 2001
    526.9'024'62-dc21

For information on all Elsevier Butterworth-Heinemann
publications visit our website at www.bh.com

ISBN 0 7506 4987 9

Typeset in Replika Press Pvt Ltd. 100% EOU, Delhi 110 040, (India)
Printed and bound in Great Britain by Martins the Printers

Preface to the fourth edition

This book was originally intended to combine volumes 1 and 2 of *Engineering Surveying,* 3rd and 2nd editions respectively. However, the technological developments since the last publication date (1984) have been so far-reaching as to warrant the complete rewriting, modernizing and production of an entirely new book.

Foremost among these developments are the modern total stations, including the automatic self-seeking instruments; completely automated, 'field to finish' survey systems; digital levels; land/ geographic information systems (L/GIS) for the managing of any spatially based information or activity; inertial survey systems (ISS); and three-dimensional position fixing by satellites (GPS).

In order to include all this new material and still limit the size of the book a conscious decision was made to delete those topics, namely photogrammetry, hydrography and field astronomy, more adequately covered by specialist texts.

In spite of the very impressive developments which render engineering surveying one of the most technologically advanced subjects, the material is arranged to introduce the reader to elementary procedures and instrumentation, giving a clear understanding of the basic concept of measurement as applied to the capture, processing and presentation of spatial data. Chapters 1 and 4 deal with the basic principles of surveying, vertical control, and linear and angular measurement, in order to permit the student early access to the associated equipment. Chapter 5 deals with coordinate systems and reference datums necessary for an understanding of satellite position fixing and an appreciation of the various forms in which spatial data can be presented to an L/GIS. Chapter 6 deals with control surveys, paying particular attention to GPS, which even in its present incomplete stage has had a revolutionary impact on all aspects of surveying. Chapter 7 deals with elementary, least squares data processing and provides an introduction to more advanced texts on this topic. Chapters 8 to 10 cover in detail those areas (curves, earthworks and general setting out on site) of specific interest to the engineer and engineering surveyor. Each chapter contains a section of 'Worked Examples', carefully chosen to clearly illustrate the concepts involved. Student exercises, complete with answers, are supplied for private study. The book is aimed specifically at students of surveying, civil, mining and municipal engineering and should also prove valuable for the continuing education of professionals in these fields.

W. Schofield

Preface to the fifth edition

Since the publication of the fourth edition of this book, major changes have occurred in the following areas:

- surveying instrumentation, particularly Robotic Total Stations with Automatic Target Recognition, reflectorless distance measurement, etc., resulting in turnkey packages for machine guidance and deformation monitoring. In addition there has been the development of a new instrument and technique known as laser scanning
- GIS, making it a very prominent and important part of geomatic engineering
- satellite positioning, with major improvements to the GPS system, the continuance of the GLONASS system, and a proposal for a European system called GALILEO
- national and international co-ordinate systems and datums as a result of the increasing use of satellite systems.

All these changes have been dealt with in detail, the importance of satellite systems being evidenced by a new chapter devoted entirely to this topic.

In order to include all this new material and still retain a economical size for the book, it was necessary but regrettable to delete the chapter on Least Squares Estimation. This decision was based on a survey by the publishers that showed this important topic was not included in the majority of engineering courses. It can, however, still be referred to in the fourth edition or in specialised texts, if required.

All the above new material has been fully expounded in the text, while still retaining the many worked examples which have always been a feature of the book. It is hoped that this new edition will still be of benefit to all students and practitioners of those branches of engineering which contain a study and application of engineering surveying.

W. Schofield
February 2001

Acknowledgements

The author wishes to acknowledge and thank all those bodies and individuals who contributed in any way to the formation of this book.

For much of the illustrative material thanks are due to Intergraph (UK) Ltd, Leica (UK) Ltd, Trimble (UK) Ltd, Spectra-Precision Ltd, Sokkisha (UK) Ltd, and the Ordnance Survey of Great Britain (OSGB).

I am also indebted to OSGB for their truly excellent papers, particularly 'A Guide to Co-ordinate Systems in Great Britain', which formed the basis of much of the information in chapter 7.

I must also acknowledge the help received from the many papers, seminars, conferences, and continued quality research produced by the IESSG of the University of Nottingham.

Finally, may I say thank you to Pat Affleck of the Faculty of Technology, Kingston University, who freely and unstintingly typed all this new material.

1

Basic concepts of surveying

The aim of this chapter is to introduce the reader to the basic concepts of surveying. It is therefore the most important chapter and worthy of careful study and consideration.

1.1 DEFINITION

Surveying may be defined as the science of determining the position, in three dimensions, of natural and man-made features on or beneath the surface of the Earth. These features may then be represented in analog form as a contoured map, plan or chart, or in digital form as a three-dimensional mathematical model stored in the computer. This latter format is referred to as a *digital ground model* (DGM).

In engineering surveying, either or both of the above formats may be utilized in the planning, design and construction of works, both on the surface and underground. At a later stage, surveying techniques are used in the dimensional control or setting out of the designed constructional elements and also in the monitoring of deformation movements.

In the first instance, surveying requires management and decision making in deciding the appropriate methods and instrumentation required to satisfactorily complete the task to the specified accuracy and within the time limits available. This initial process can only be properly executed after very careful and detailed reconnaissance of the area to be surveyed.

When the above logistics are complete, the field work – involving the capture and storage of field data – is carried out using instruments and techniques appropriate to the task in hand.

The next step in the operation is that of data processing. The majority, if not all, of the computation will be carried out by computer, ranging in size from pocket calculator to mainframe. The methods adopted will depend upon the size and precision of the survey and the manner of its recording; whether in a field book or a data logger. Data representation in analog or digital form may now be carried out by conventional cartographic plotting or through a totally automated system using a computer-driven flat-bed plotter. In engineering, the plan or DGM is used for the planning and design of a construction project. This project may comprise a railroad, highway, dam, bridge, or even a new town complex. No matter what the work is, or how complicated, it must be set out on the ground in its correct place and to its correct dimensions, within the tolerances specified. To this end, surveying procedures and instrumentation are used, of varying precision and complexity, depending on the project in hand.

Surveying is indispensable to the engineer in the planning, design and construction of a project, so all engineers should have a thorough understanding of the limits of accuracy possible in the construction and manufacturing processes. This knowledge, combined with an equal understanding of the limits and capabilities of surveying instrumentation and techniques, will enable the engineer to successfully complete his project in the most economical manner and shortest time possible.

1.2 BASIC MEASUREMENTS

Surveying is concerned with the fixing of position whether it be control points or points of topographic detail and, as such, requires some form of reference system.

The physical surface of the Earth, on which the actual survey measurements are carried out, is mathematically non-definable. It cannot therefore be used as a reference datum on which to compute position.

An alternative consideration is a level surface, at all points normal to the direction of gravity. Such a surface would be formed by the mean position of the oceans, assuming them free from all external forces, such as tides, currents, winds, etc. This surface is called the geoid and is the equipotential surface at mean sea level. The most significant aspect of this surface is that survey instruments are set up relative to it. That is, their vertical axes, which are normal to the plate bubble axes used in the setting-up process, are in the direction of the force of gravity at that point. Indeed, the points surveyed on the physical surface of the Earth are frequently reduced to their equivalent position on the geoid by projection along their gravity vectors. The reduced level or elevation of a point is its height above or below the geoid as measured in the direction of its gravity vector (or plumb line) and is most commonly referred to as its height above or below mean sea level (MSL). However, due to variations in the mass distribution within the Earth, the geoid is also an irregular surface which cannot be used for the mathematical location of position.

The mathematically definable shape which best fits the shape of the geoid is an ellipsoid formed by rotating an ellipse about its minor axis. Where this shape is used by a country as the surface for its mapping system, it is termed the reference ellipsoid. *Figure 1.1* illustrates the relationship of the above surfaces.

The majority of engineering surveys are carried out in areas of limited extent, in which case the reference surface may be taken as a tangent plane to the geoid and the rules of plane surveying used. In other words, the curvature of the Earth is ignored and all points on the physical surface are orthogonally projected onto a flat plane as illustrated in *Figure 1.2*. For areas less than 10 km square the assumption of a flat Earth is perfectly acceptable when one considers that in a triangle of approximately 200 km^2, the difference between the sum of the spherical angles and the plane angles would be 1 second of arc, or that the difference in length of an arc of approximately 20 km on the Earth's surface and its equivalent chord length is a mere 10 mm.

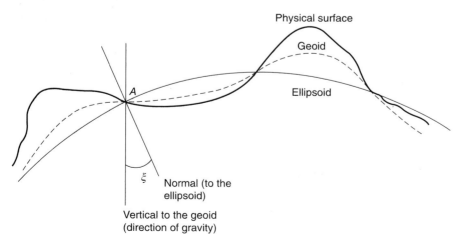

Fig. 1.1

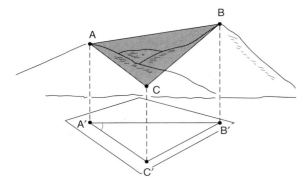

Fig. 1.2 *Projection onto a plain surface*

The above assumptions of a flat Earth are, however, not acceptable for elevations as the geoid would deviate from the tangent plane by about 80 mm at 1 km or 8 m at 10 km. Elevations are therefore referred to the geoid or MSL as it is more commonly termed. Also, from the engineering point of view, it is frequently useful in the case of inshore or offshore works to have the elevations related to the physical component with which the engineer is concerned.

An examination of *Figure 1.2* clearly shows the basic surveying measurements needed to locate points *A*, *B* and *C* and plot them orthogonally as *A′*, *B′* and *C′*. In the first instance the measured *slant distance AB* will fix the position of *B* relative to *A*. However, it will then require the *vertical angle* to *B* from *A*, in order to reduce *AB* to its equivalent horizontal distance *A′B′* for the purposes of plotting. Whilst similar measurements will fix *C* relative to *A*, it requires the *horizontal angle BAC* (*B′A′C′*) to fix *C* relative to *B*. The *vertical distances* defining the relative elevation of the three points may also be obtained from the slant distance and vertical angle (trigonometrical levelling) or by direct levelling (Chapter 2) relative to a specific reference datum. The five measurements mentioned above comprise the basis of plane surveying and are illustrated in *Figure 1.3*, i.e. *AB* is the slant distance, *AA′* the horizontal distance, *A′B* the vertical distance, *BAA′* the vertical angle (α) and *A′AC* the horizontal angle (θ).

It can be seen from the above that the only measurements needed in plane surveying are angle and distance. Nevertheless, the full impact of modern technology has been brought to bear in the acquisition and processing of this simple data. Angles are now easily resolved to single-second accuracy using optical and electronic theodolites; electromagnetic distance measuring (EDM)

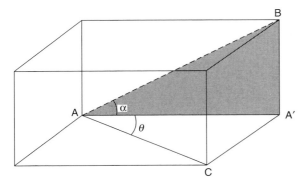

Fig. 1.3 *Basic measurements*

equipment can obtain distances of several kilometres to sub-millimetre precision; lasers and north-seeking gyroscopes are virtually standard equipment for tunnel surveys; orbiting satellites and inertial survey systems, spin-offs from the space programme, are being used for position fixing off shore as well as on; continued improvement in aerial and terrestrial photogrammetric equipment and remote sensors makes photogrammetry an invaluable surveying tool; finally, data loggers and computers enable the most sophisticated procedures to be adopted in the processing and automatic plotting of field data.

1.3 CONTROL NETWORKS

The establishment of two- or three-dimensional control networks is the most fundamental operation in the surveying of an area of large or small extent. The concept can best be illustrated by considering the survey of a relatively small area of land as shown in *Figure 1.4*.

The processes involved in carrying out the survey can be itemized as follows:

(1) A careful reconnaissance of the area is first carried out in order to establish the most suitable positions for the survey stations (or control points) *A*, *B*, *C*, *D*, *E* and *F*. The stations should be intervisible and so positioned to afford easy and accurate measurement of the distances between them. They should form 'well-conditioned' triangles with all angles greater than 45°, whilst the sides of the triangles should lie close to the topographic detail to be surveyed. If this procedure is adopted, the problems of measuring up, over or around obstacles, is eliminated.

The survey stations themselves may be stout wooden pegs driven well down into the ground, with a fine nail in the top accurately depicting the survey position. Alternatively, for longer life,

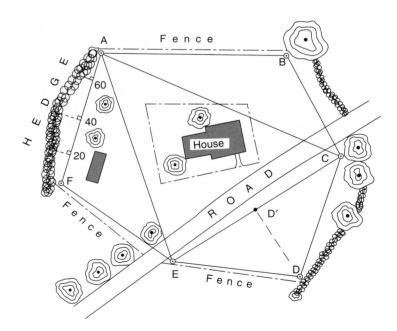

Fig. 1.4   *Linear survey*

concrete blocks may be set into the ground with some form of fine mark to pinpoint the survey position.

(2) The distances between the survey stations are now obtained to the required accuracy. Steel tapes may be laid along the ground to measure the slant lengths, whilst vertical angles may be measured using hand-held clinometers or Abney levels to reduce the lengths to their horizontal equivalents. Alternatively, the distances may be measured in horizontal steps as shown in *Figure 1.5*. The steps are short enough to prevent sag in the tape and their end positions at 1, 2 and B fixed using a plumb-bob and an additional assistant. The steps are then summed to give the horizontal distances.

Thus by measuring all the distances, relative positions of the survey stations are located at the intersections of the straight lines and the network possesses shape and scale. The surveyor has thus established in the field a two-dimensional horizontal control network whose nodal points are positioned relative to each other. It must be remembered, however, that all measurements, no matter how carefully carried out, contain error. Thus, as the three sides of a triangle will always plot to give a triangle, regardless of the error in the sides, some form of *independent check* should be introduced to reveal the presence of error. In this case the horizontal distance from D to a known position D' on the line EC is measured. If this distance will not plot correctly within triangle CDE, then error is present in one or all of the sides. Similar checks should be introduced throughout the network to prove its reliability.

(3) The proven network can now be used as a reference framework or huge template from which further measurements can now be taken to the topographic detail. For instance, in the case of line FA, its position may be physically established in the field by aligning a tape between the two survey stations. Now, offset measurements taken at right angles to this line at known distances from F, say 20 m, 40 m and 60 m, will locate the position of the hedge. Similar measurements from the remaining lines will locate the position of the remaining detail.

The method of booking the data for this form of survey is illustrated in *Figure 1.6*. The centre column of the book is regarded as the survey line FA with distances along it and offsets to the topographic detail drawn in their relative positions as shown in *Figure 1.4*.

Note the use of oblique offsets to more accurately fix the position of the trees by intersection, thereby eliminating the error of estimating the right angle in the other offset measurements.

The network is now plotted to the required scale, the offsets plotted from the network and the relative position of all the topographic detail established to form a plan of the area.

(4) As the aim of this particular survey was the production of a plan, the accuracy of the survey is governed largely by the scale of the plan. For instance, if the scale was, say, 1 part in 1000, then a plotting accuracy of 0.1 mm would be equivalent to 100 mm on the ground and it would not be economical or necessary to take the offset measurements to any greater accuracy than this. However, as the network forms the reference base from which the measurements are taken, its position would need to be fixed to a much greater accuracy.

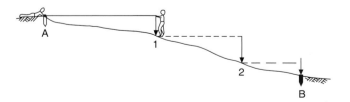

Fig. 1.5 *Stepped measurement*

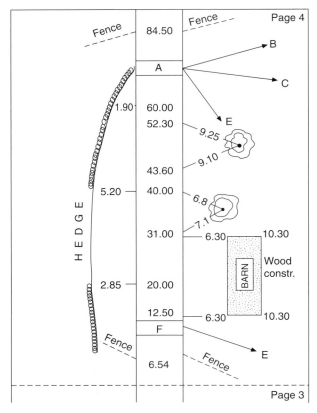

Fig. 1.6   *Field book*

The above comprises the steps necessary in carrying out this particular form of survey, generally referred to as a linear survey. It is naturally limited to quite small areas, due to the difficulties of measuring with tapes and the rapid accumulation of error involved in the process. For this reason it is not a widely used surveying technique. It does, however, serve to illustrate the basic concepts of all surveying in a simple, easy to understand manner.

Had the area been much greater in extent, the distances could have been measured by EDM equipment; such a network is called a *trilateration*. A further examination of *Figure 1.4* shows that the shape of the network could be established by measuring all the horizontal angles, whilst its scale or size could be fixed by a measurement of one side. In this case the network would be called a *triangulation*. If all the sides and horizontal angles are measured, the network is a *triangulateration*. Finally, if the survey stations are located by measuring the adjacent angles and lengths shown in *Figure 1.7*, thereby constituting a polygon A, B, C, D, E, F, the network is a *traverse*. These then constitute all the basic methods of establishing a horizontal control network, and are dealt with in more detail in Chapter 6.

1.4   LOCATING POSITION

The method of locating the position of topographic detail by right-angled offsets from the sides of the control network has been mentioned above. However, this method would have errors in establishing

Contents

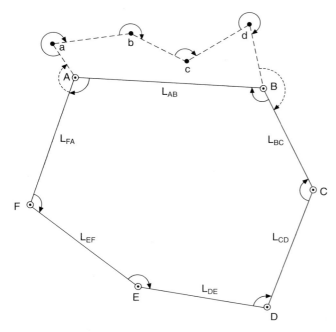

Fig. 1.7 *Traverse*

the line *FA*, in setting out the right angle (usually by eye) and in measuring the offset. It would therefore be more accurate to locate position directly from the survey stations. The most popular method of doing this is by polar coordinates as shown in *Figure 1.8*. *A* and *B* are survey stations of known position in a control network, from which the measured horizontal angle *BAP* and the horizontal distance *AP* will fix the position of point *P*. There is no doubt that this is the most popular method of fixing position, particularly since the advent of EDM equipment. Indeed, the method of traversing is a repeated application of this process.

An alternative method is by intersection where *P* is fixed by measuring the horizontal angles *BAP* and *ABP* as shown in *Figure 1.9*. This method forms the basis of triangulation. Similarly, *P* may be fixed by the measurement of horizontal distances *AP* and *BP* and forms the basis of the method of

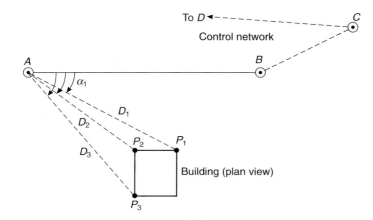

Fig. 1.8 *Polar coordinates*

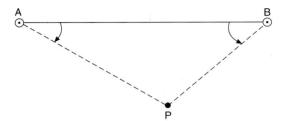

Fig. 1.9 *Intersection*

trilateration. In both these instances there is no independent check as a position for *P* (not necessarily the correct one) will always be obtained. Thus at least one additional measurement is required either by combining the angles and distances (triangulateration) by measuring the angle at *P* as a check on the angular intersection, or by producing a trisection from an extra control station.

The final method of position fixing is by resection (*Figure 1.10*). This is done by observing the horizontal angles at *P* to at least three control stations of known position. The position of *P* may be obtained by a mathematical solution as illustrated in Chapter 6.

Once again, it can be seen that all the above procedures simply involve the measurement of angle and distance.

1.5 LOCATING TOPOGRAPHIC DETAIL

Topographic surveying of detail is, in the first instance, based on the established control network. The accurate relative positioning of the control points would generally be by the method of traversing or a combination of triangulation and trilateration (Chapter 6). The mean measured angles and distances would be processed, to provide the plane rectangular coordinates of each control point. Each point would then be carefully plotted on a precisely constructed rectangular grid. The grid would be drawn with the aid of a metal template (*Figure 1.11*), containing fine drill holes in an exact grid arrangement. The position of the holes is then pricked through onto the drawing material using the precisely fitting punch shown. Alternatively, the grid would be drawn using a computer-driven coordinatorgraph on a flat-bed or drum plotter. The topographic detail is then drawn in from the plotted control points which were utilized in the field.

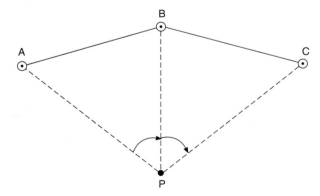

Fig. 1.10 *Resection*

Fig. 1.11   *Metal template and punch*

1.5.1 Field survey

In the previous section, the method of locating detail by offsets was illustrated. In engineering surveys the more likely method is by polar coordinates, i.e. direction relative to a pair of selected control points, plus the horizontal distance from one of the known points, as shown in *Figure 1.8*.

The directions would be measured by theodolite and the distance by EDM, to a detail pole held vertically on the detail (*Figure 1.12*); hence the ideal instrument would be the electronic tacheometer or total station.

The accuracy required in the location of detail is a function of the scale of the plan. For instance, if the proposed scale is 1 in 1000, then 1 mm on the plan would represent 1000 mm on the ground. If the plotting accuracy was, say, 0.2 mm, then the equivalent field accuracy would be 200 mm and distance need be measured to no greater accuracy than this. The equivalent angular accuracy for a length of sight at 200 m would be about 3′ 20″. From this it can be seen that the accuracy required to fix the position of detail is much less than that required to establish the position of control points. It may be, depending on the scale of the plan and the type of detail to be located, that stadia tacheometry could be used for the process, in the event of there being no other alternative.

The accuracy of distance measurement in stadia tacheometer ($D = 100 \times S \cos^2 \theta$), as shown in Chapter 2, is in the region of 1 in 300, equivalent to 300 mm in an observation distance of 100 m. Thus before this method can be considered, the scale of the plan must be analysed as above, the average observation distance should be considered and the type of detail, hard or soft, reconnoitred. Even if all these considerations are met, it must be remembered that the method is cumbersome and uneconomical unless a direct reading tacheometer is available.

1.5.2 Plotting the detail

The purpose of the plan usually defines the scale to which it is plotted. The most common scale for construction plans is 1 in 500, with variations above or below that, from 1 in 2500 to 1 in 250.

The most common material used is plastic film with such trade names as 'Permatrace'. This is an

Fig. 1.12 *'Detail pole' locating topographic detail*

extremely durable material, virtually indestructible with excellent dimensional stability. When the plot is complete, paper prints are easily obtained.

Although the topographic detail could be plotted using a protractor for the direction and a scale for the distances, in a manner analogous to the field process, it is a trivial matter to produce 'in-house' software to carry out this task. Using the arrangement shown in *Figure 1.13*, the directions and distances are input to the computer, changed to two-dimensional coordinates and plotted direct. A simple question asks the operator if he wishes the plotted point to be joined to the previous one and in this way the plot is rapidly progressed. This elementary 'in-house' software simply plots points and lines and the reduced level of the points, where the vertical angle is included. However, there is now an abundance of computer plotting software available that will not only produce a contoured plot, but also supply three-dimensional views, digital ground models, earthwork volumes, road design, drainage design, digital mapping, etc.

1.5.3 Computer systems

To be economically viable, practically all major engineering/surveying organizations use an automated plotting system. Very often the total station and data logger are purchased along with the computer hardware and software, as a total operating system. In this way interface and adaptation problems are precluded. *Figure 1.14* shows such an arrangement including a 'mouse' for use on the digitizing tablet. An AO flat-bed plotter is networked to the system and located separately.

The essential characteristics of such a system are:

(1) Capability to accept, store, transfer, process and manage field data that is input manually or directly from an interfaced data logger (*Figure 1.15*).
(2) Software and hardware to be in modular form for easy accessing.
(3) Software to use all modern facilities, such as 'windows', different colour and interactive screen graphics, to make the process user friendly.
(4) Continuous data flow from field data to finished plan.

Fig. 1.13 *Computer driven plotter*

Fig. 1.14 *Computer system with digitizing tablet*

(5) Appropriate data-base facility, for the storage and management of coordinate and cartographic data necessary for the production of digital ground models and land/geographic information systems.
(6) Extensive computer storage facility.
(7) High-speed precision flat-bed or drum plotter.

Fig. 1.15 *Data logger*

To be truly economical, the field data, including appropriate coding of the various types of detail, should be captured and stored by single-key operation, on a data logger interfaced to a total station. The computer system should then permit automatic transfer of this data by direct interface between the logger and the system. The modular software should then: store and administer the data; carry out the mathematical processing, such as network adjustment, production of coordinates and elevations; generate data storage banks; and finally plot the data on completion of the data verification process.

Prior to plotting, the data can be viewed on the screen for editing purposes. This can be done from the keyboard or by light pen on the screen using interactive graphics routines. The plotted detail can be examined, moved, erased or changed, as desired. When the examination is complete, the command to plot may then be activated. *Figure 1.16* shows an example of a computer plot.

1.5.4 Digital ground model (DGM)

A DGM is a three-dimensional, mathematical representation of the landform and all its features, stored in a computer data base. Such a model is extremely useful in the design and construction process, as it permits quick and accurate determination of the coordinates and elevation of any point.

The DGM is formed by sampling points over the land surface and using appropriate algorithms to process these points to represent the surface being modelled. The methods in common use are modelling by 'strings', 'regular grids' or 'triangular facets'. Regardless of the methods used, they will all reflect the quality of the field data.

A 'string' comprises a series of points along a feature and so such a system stores the position of features surveyed. It is widely used for mapping purposes due to its flexibility, its accuracy along the string and its ability to process large amounts of data very quickly. However, as it does not store the relationship between strings, a searching process is essential when the levels of points not

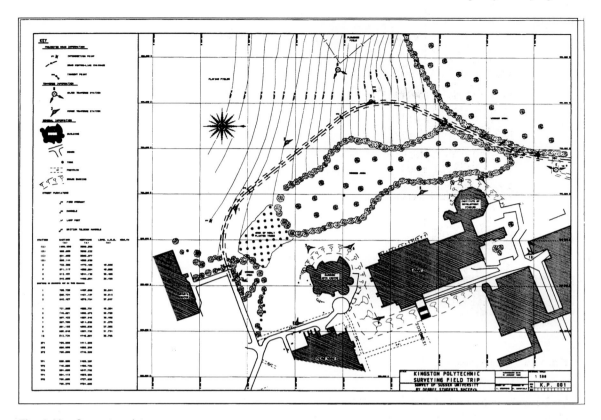

Fig. 1.16 *Computer plot*

included in a string are required. Thus its weakness lies in the generation of accurate contours and volumes.

The 'regular grid' method uses appropriate algorithms to convert the sampled data to a regular grid of levels. If the field data permit, the smaller the grid interval, the more representative of landform it becomes. Although a simple technique, it only provides a very general shape of the landform, due to its tendency to ignore vertical breaks of slope. Volumes generated also tend to be rather inaccurate.

In the 'triangular grid' method, 'best fit' triangles are formed between the points surveyed. The ground surface therefore comprises a network of triangular planes at various angles (*Figure 1.17(a)*). Computer shading of the model (*Figure 1.17(b)*) provides an excellent indication of the landform. In this method vertical breaks are forced to form the sides of triangles, thereby maintaining correct ground shape. Contours, sections and levels may be obtained by linear interpolation through the triangles. It is thus ideal for contour generation (*Figure 1.18*) and highly accurate volumes. The volumes are obtained by treating each triangle as a prism to the depth required; hence the smaller the triangle, the more accurate the final result.

1.5.5 Computer-aided design (CAD)

In addition to the production of DGMs and contoured plans, the modern computer surveying system permits the easy application of the designed structure to the finished plan. The three-

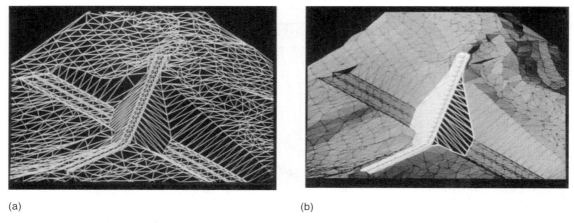

(a)   (b)

Fig. 1.17 *(a) Triangular grid model, and (b) Triangular grid model with computer shading*

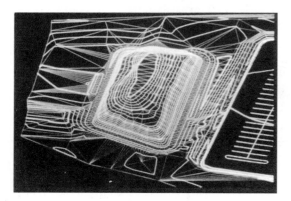

Fig. 1.18 *Computer generated contour model*

dimensional information held in the data base supplies all the ground data necessary to facilitate the finished design. *Figure 1.19* illustrates its use in road design.

The environmental impact of the design can now be more readily assessed by producing perspective views as shown in *Figures 1.20(a)* and (*b*). The new environmental impact laws make this latter tool extremely valuable.

1.5.6 Land/geographic information systems (LIS/GIS)

Prior to the advent of computers, land-related information was illustrated by means of overlay tracings on the basic topographic map or plan. For instance, consider a plan of an urban area on which it is also required to show the public utilities, i.e. the gas mains, electrical cables, substations, drainage system, manholes, etc. As adding all this information to the base plan would render it completely unreadable, each system was drawn on separate sheets of tracing paper. Each tracing could then be overlain over the base plan, as and when required (*Figure 1.21*). In addition, a large ledger was kept, as part of the arrangement, itemizing the dimensions of the pipes, the material used, the ownership, the condition, the ownership of the land under which it passed, etc. All this information, used with the base plan and overlays, comprised a cumbersome land information system.

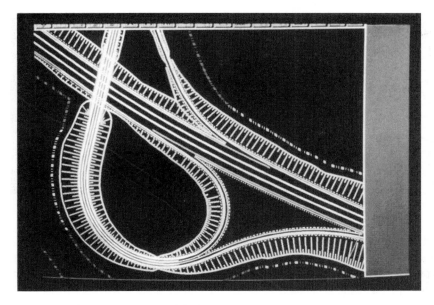

Fig. 1.19 *Computer-aided road design*

(a)   (b)

Fig. 1.20 *Perspectives with computer shading*

All this information and more can now be stored in a computer to form the basis of the modern-day L/GIS. Thus a L/GIS is a land-related data base held in a highly structured form within the computer, in order to make it easier to manage, update, access, interrogate and retrieve. Although many sophisticated commercial packages are available, the process is still in a state of evolution. The ultimate GIS is one which could supply all the information relating to land from, say, 10 km above its surface to 100 km below; the amount of information to be stored is almost incomprehensible. It may be necessary to consider land boundaries, areas of land, type of soil, erosion characteristics, type of property, ownership, street names, rateable values, landslip data, past and future land use, agricultural areas, flood protection, mineral resources, public utilities; the list is inexhaustible. In addition, all this information must be related to good-quality large-scale maps or plans. Further to this, there is the problem of different individuals wishing to access the system for their own

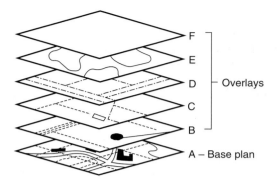

Fig. 1. 21   *The concept of a L/GIS: B, gas pipes; C, electric cables; D, drainage system, etc.*

requirements. There is the private landowner wishing to know about future land use, the planners, the local authority administrators, the civil engineer, the mineral operator, the lawyer, all requiring rapid and easy access to the information specific to their needs. The system would thereby improve the administration of all legal matters appertaining to land, furnish data for the better administration of the land, facilitate resource management and environmental planning, etc.

The problems of producing an efficient L/GIS are complex and numerous. The information must be efficiently filed, uniquely coded, conveniently stored, easily accessed, interrogated and retrieved, and highly flexible in its applications.

The first problem is the availability of good-quality large-scale plans on an approved coordinate system. This can be achieved by surveying the areas concerned or, where acceptable plans are available, digitizing them. A system of quality control is necessary to ensure a common standard from all the sources.

A system of identifying and indexing the various land parcels is then necessary, based in the first instance on the coordinate system used.

When the topographic structure is in place for on-screen analysis and hard copy availability, the massive problem of finding, checking, proving and storing the large volume of land-related data follows. It may be necessary to layer this information in files within the data base and combine this with powerful data-base management software to ensure its efficient manipulation. The coding process is far more complex than the surveyor is normally used to. In surveying an area, for instance, the surveyor is concerned essentially with the shape, size and position of a feature. Therefore, if surveying a number of buildings, a simple code of B1, B2, etc. may be used, i.e. B for Building, the number denoting the number of buildings. In a L/GIS system, not only is the above information required, but it is necessary to know the type of building (office, residential, industrial, etc.), the mode of construction (brick or concrete), the number of storeys, the ownership, the present occupancy, the specific use, the rateable value, etc. Thus it can be seen that the coding is an extremely complex issue. The situation may be further complicated by the problem of confidentiality, for whilst the system should be user friendly, it should not be possible to access confidential data.

Integration of all sources of data may be rendered extremely difficult, if not impossible, by the attitudes of the various institutions holding the information.

It can be seen that the problems of producing a multi-purpose land information system are complex. In the case of a geographic information system these problems are magnified. The GIS is similarly concerned with the storage, management and analysis of spatially related data, but on a much greater scale. The ultimate GIS would be a global information system. The geographic information could be necessary for such processes as weather forecasting, flood forecasting from rainfall records, stream and river location, drainage patterns and systems, position and size of dams and reservoirs, land use and transportation patterns over very wide areas; once again the list is inexhaustible.

Thus, although the formation of a L/GIS is a formidable problem, the necessity for an efficient and accurate source of land-related data makes it mandatory as a powerful land management tool. As good-quality plans form the basis of such a system, it is feasible that surveyors, who are the experts in measurement and position, should play a prominent part in the design and management of such systems.

1.5.6.1 GIS data

From the broad introduction given to GIS it can be seen that a GIS is a computer-based system for handling not only physical location but also the attributes associated with that location. It thus possesses a graphical display in two or three dimensions of the *spatial data*, combined with a database for the *non-spatial data*, i.e. the attribute information. The prime aspect in the construction of a GIS is the acquisition, from many different sources, conversion and entry of the data.

The spatial data may be acquired from a variety of sources: from digitized maps and plans, from aerial photographs, from satellite imagery, or directly from GPS surveys. However, in order to represent this complex, three-dimensional reality in a spatial database, it is modelled using *points*, *lines*, *areas*, *surfaces* and *networks*. For instance, if we consider an underground drainage system, the pipes would be represented by lines; the manhole positions by points; the parcels of land ownership forming closed boundaries whose polygon shape is defined by coordinates would be represented by areas; whilst the three-dimensional land surface through which the pipes pass would be represented by a surface. Such a GIS would probably incorporate a network, which represents the whole branching system of pipes (line segments), and is used to simulate flow through the pipes or indicate the buildings affected by a break in the pipe network at a specific point. The attributes attached to this network, such as type and size of pipe, depth below ground, rate of flow, gradients, etc., would be stored in the associated database. The linking of the spatially referenced data with their attributes is the basis of GIS.

The above features can be represented within a GIS in either *vector* or *raster* format; their relative spatial relationships are given by their *topology* (*Figure 1.22*).

Vector data uses dots and lines, similar to the plotting of x, y coordinates on a plan, and the joining up of those coordinated points with lines and curves to give shape and position. The vector format provides an accurate representation of the spatially referenced data incorporating the topology and other spatial relationships between the individual entries.

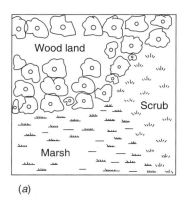
(a)

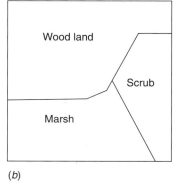
(b)

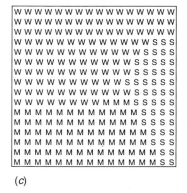
(c)

Fig. 1.22 *(a) Shows standard topographic plan. (b) shows vector representation of (a). (c) Shows raster representation of (a)*

The GIS vector model differs from that of CAD or simple drawing packages as each dot (called vertices in GIS), line segment, area or polygon is uniquely identified and their relationships stored in the database. Computer data storage is very economical, but certain analytical processes have high computational requirements resulting in slow operations or the use of high specification hardware. The vector data model is ideally suited to the representation of linear networks such as roads, railways and pipelines. It also provides accurate measurement of areas and lengths. It is the obvious format for inputting digital data obtained by conventional survey procedures or by digitizing existing plans or maps.

The raster format uses pixels (derived from '*pic*ture *el*ements') or grid cells. It is not as accurate or flexible as vector format as each coordinate may be represented by a cell and each line by an array of cells. Thus data can be positioned only to the nearest grid cell. Examples of data in the raster format are aerial photographs, satellite imagery and scanned maps or plans. An example from reality would be moorland comprising areas of marsh and scrub, etc., where the vague boundaries would not be unduly affected by the inaccuracy of the format presentation. In addition to producing a coarse resolution of the data, each cell contains a single value representing the attribute contained within the area of the cell. The resolution of the data may be improved using a smaller cell size, but this would increase the computer storage, which tends, in any case, to be uneconomical using the raster format. The computer finds it easier to collect, store and manage raster data using such techniques as overlay, buffering and network analysis.

The above are the two main data models, but a third *object-based* model is available which represents the data as it appears in the real world, thereby making it easier to understand (*Figure 1.20*). It does, however, result in very high processing requirements.

Initially, GIS systems used one format or the other. However, modern GIS software permits conversion between the two and can display vector data over the top of raster data.

In all GIS systems, the data is layered. For instance, one layer would contain all the houses in the area, another layer all the water pipes, and so on, as shown in *Figure 1.21*. This allows data to be shown separately but still retains cross-referencing between the layers for analyses or interrogation. All the layers are interrelated and to a common scale so that they can be accurately overlain.

1.5.6.2 Topology

Topology is a branch of mathematics dealing with the relative relationships between individual entities. It is a method of informing the computer how to arrange the data input into its correct relative position. Important topological concepts are:

- *Adjacency:* consider a line defining the edge of a road: on which side of that line does the road lie?
- *Connectivity:* which points must be connected to show each side of the road?
- *Orientation:* defines the starting point and ending point in a chain of points describing the road?
- *Nestedness:* what spatial objects, such houses, lie within a given polygon, such as its property boundaries?

Once these concepts are placed in the computer data files the relative relationships, or topology, of the spatial data can be realized.

1.5.6.3 Functionality

The GIS is not just a simple graphic display of spatial data or of attribute data, but a system combining both to provide sophisticated functions that assist management and decision making.

The first and most important step is the acquisition and input of data. It is important because the

GIS is only as good as the data provided. The data may be obtained from many sources already mentioned, such as the digitization of existing graphic material; the scanning of topographic maps/plans; aerial photographs (or the photographs of satellite imagery); keyboard entry of survey data, attribute data or direct interface of GPS data; all of which must be transformed, where necessary, into digital form. In addition, it may be possible to use existing digital data sets.

The data is not only sorted within the computer, but is indexed and managed to ensure controlled and co-ordinated access. The data must be structured in such a way as to ensure the reliability, security and integrity of the data.

The GIS provides links between spatial and non-spatial data, allowing sophisticated analysis of the total data set. Interrogation may be graphics-driven or data-driven and require the selective display of spatial and non-spatial data. Examples of the more common spatial analysis and computational functions are illustrated below.

- *Buffering* involves the creation of new polygons or buffer zones around existing nodes or points at set intervals. An example may be a break in a water pipe: a buffer zone may be created around that point showing the area which may be flooded. Similarly, the creation of buffers around a source of contamination, indicating the various areas of intensity of contamination.
- *Overlay* is the process of overlaying spatial data of one type onto another type. For instance, the overlaying of soil type data on drainage patterns may indicate the best positions to site land drains.
- *Network analysis* may be used to simulate traffic flows through a network of streets in a busy urban complex in order to optimize and improve traffic conditions.
- *Terrain analysis* could involve the creation of a three-dimensional ground model in order to investigate the environmental impact of a proposed construction, for instance.
- *Contouring* is the connection of points of equal value to form lines. These could be points of elevation to give ground contours, or points of a particular attribute to, perhaps, give population density contour lines.
- *Area and length calculations* is largely self-explanatory and could involve the area of derelict land for future housing development, or lengths of highway to be widened.

All these functions can be viewed on the screen, or output in the form of plans, graphs, tables or reports.

The use of GIS, therefore, removes the need for paper plans and associated documents and greatly speeds up operations as the data, both spatial and non-spatial, can be rapidly updated, edited and transferred to other computers networked to the central GIS. It thus has the advantages of transferring data between multiple users, thereby minimizing duplication and increasing security and reliability of the data. Specific scenarios can be modelled to test possible outcomes and create better-informed decision making. For instance, using various layers of data such as drainage patterns, surface and sub-soil data, ground slopes, and rainfall values, areas of potential erosion or landslip can be identified. Thus the GIS not only provides effective data management and analysis, but also allows spatial features and their relationships to be visualized. In this way planning and investment decisions can be made with confidence.

1.5.6.4 Applications of GIS

GIS can be applied in any situation where spatially referenced data requires modelling, analysis and management. Some examples are:

Facilities management   Organizations such as those dealing with gas, water, electricity or sewerage are responsible for vast amounts of pipelines, cables, tunnels, buildings and land, all of which require monitoring, maintenance and management in order to give an efficient and effective service to customers.

Highways maintenance   This situation is very similar to the above but deals with roads, motorways, bridges, road furniture, etc., all of which is spatially referenced and requires maintenance and management. Three-dimensional ground models can be used for design and environmental impact studies.

Housing associations   These organizations are responsible for the building, maintenance, leasing, renting or sale of houses on a massive scale. Not only is the geographic distribution of the properties required, but full details of the properties are also vital. To assist in operational management and strategic planning such information as rent arrears and the geographic clustering; housing types; properties sold, leased or rented; conditions/repairs; population trends; development sites; bad debt hotspots – the list is endless. Thus paper-based land terriers are replaced, there is high-quality visual representation of spatial data, improved productivity and more efficient management tools.

The above examples clearly illustrate the importance of GIS and the manner of its application. Other areas which would benefit from its use are environmental management; transportation; market analysis using, say, socio-economic population distribution patterns; and land use patterns. Indeed, wherever the relationship and interaction of various spatially referenced data is required, GIS provides a powerful analytical tool.

1.5.7 Laser scanner

Laser scanning, in a terrestrial or airborne form, is a relatively new and powerful surveying technique. The system provides 3-D location of features and surfaces quickly and accurately, in real time if necessary.

The system is a combined hardware and software package. The hardware consists of a tripod-mounted pulsed laser range finder and a mechanical scanner. The time taken by the laser pulse to hit the target and return is measured by the picosecond timing circuitry of the unit's signal detector, and the range calculated. The amount of energy reflected by the target surface is a function of the target's characteristics, such as roughness, colour, etc. The amplitude of the returned pulse gives an intensity or brightness value. A Class 1, eye safe laser, operating in the near-infrared region at $0.9\,\mu m$ is used, with an operating range of 0.1–$350\,m$ and a beam width of about $300\,mm$ at $100\,m$ distance. The scanning density can be altered and set in increments of $0.25°$, $0.5°$ and $1°$. A rotating polygonal mirror directs the laser beam in the horizontal and vertical directions. Angle encoders record the orientation of the mirror. Thus, each point within the raster image of range and intensity is accurately positioned in 3-D and illustrated via the controlling laptop PC. Data can be acquired at rates as high as 6000 measurements per second using a laser pulsing at 20 kHz, with accuracies of ± 5 mm. In some systems, using special targets other than the actual ground or structure surfaces, accuracies of ± 2 mm are achievable. If the tripod is set over a point of known coordinates and orientated into the coordinate system in use, then the spatial position of the points scanned can be defined in that system. At the present time the laser scanning device can vary in weight from 13.5 kg to 30 kg, depending on the make of the unit. One particular unit incorporates a colour CCD camera to capture scenes for later analysis. This latter point indicates the many and varied ways in which modern technology is being utilized in spatial data capture.

The laser device is controlled and the data processed by means of a PC connected to it through serial and parallel cables. The scanner parameters are set by the operator and the data downloaded in real time for 3-D screen viewing. The raster style 3-D picture can be rotated in space for viewing from any angle as scanning takes place. The range to points can be queried and inter-distances between points measured. The screen image enables the operator to evaluate the quality of the data and, if necessary, change the parameter settings or move the scanner to a better site position. If the survey area is extensive, reflectors may be used in the scanned portions to allow the co-ordination and merging of various scans. The intensities of the laser signals, which in effect describe the characteristics of the points in question, may be illustrated on the screen using different colours, thereby highlighting variations in the data. The data files are naturally quite large, and a figure quoted for the survey of a room area of 30 m^2 with pillars and windows, was 2 Mb. For best results the field data can be transferred to a more powerful graphics workstation for further processing, editing and analysis. Precise 2-D drawings with elevations, or 3-D models can be generated.

Applications of this revolutionary system occur in all aspects of surveying, mining and civil engineering. It is particularly useful in inaccessible locations such as building facades, mine and quarry faces, and areas which are unsafe such as cliff faces, airport runways, busy highways and hazardous areas in chemical and nuclear installations. The applications mentioned are those that are particularly difficult for conventional surveying procedures. However, this does not preclude its use in all those areas of conventional survey, including tunnelling.

The principles outlined above can also be used in airborne situations where the aircraft equipped with GPS is positioned in space by a single ground-based GPS station and an inertial navigation unit is used for the determination of roll, pitch and yaw. In this way the position and attitude of the scanner is fixed in the GPS coordinate system (WGS84), and so also are the terrain positions. Transformation to a local reference system will also require a geoid model.

The flying height varies from 300–1000 m, with the laser beam scanning at a rate as high as 25 000 pulses per second across a swath beneath the aircraft.

At the present time, ground-based systems are large, heavy and expensive, but there is no doubt that, within a very short period by time, they will become smaller, more sophisticated, and a major method of 3-D detailing.

1.6 SUMMARY

In the preceding sections an attempt has been made to outline the basic concepts of surveying. Because of their importance they will now be summarized as follows:

(1) *Reconnaissance* is the first and most important step in the surveying process. Only after a careful and detailed reconnaissance of the area can the surveyor decide upon the techniques and instrumentation required to economically complete the work and meet the accuracy specifications.

(2) *Control networks* not only form a reference framework for locating the position of topographic detail and setting out constructions, but may also be used as a base for minor control networks containing a greater number of control stations at shorter distances apart and to a lower order of accuracy, i.e. *a, b, c, d* in *Figure 1.7*. These minor control stations may be better placed for the purpose of locating the topographic detail.

This process of establishing the major control first to the highest order of accuracy, as a framework on which to connect the minor control, which is in turn used as a reference framework for detailing, is known as *working from the whole to the part* and forms the basis of all good surveying procedure.

(3) *Errors* are contained in all measurement procedures and a constant battle must be waged by the surveyor to minimize their effect.

It follows from this that the greater the accuracy specifications the greater the cost of the survey for it results in more observations, taken with greater care, over a longer period of time, using more precise (and therefore more expensive) equipment. It is for this reason that major control networks contain the minimum number of stations necessary and surveyors adhere to the economic principle of working to an accuracy neither greater than nor less than that required.

(4) *Independent checks* should be introduced not only into the field work, but also into the subsequent computation and reduction of field data. In this way, errors can be quickly recognized and dealt with.

Data should always be measured more than once. Examination of several measurements will generally indicate the presence of blunders in the measuring process. Alternatively, close agreement of the measurements is indicative of high precision and generally acceptable field data, although, as shown later, high precision does not necessarily mean high accuracy, and further data processing may be necessary to remove any systematic error that may be present.

(5) *Commensurate accuracy* is advised in the measuring process, i.e. the angles should be measured to the same degree of accuracy as the distances and vice versa. The following rule is advocated by most authorities for guidance: $1''$ of arc subtends 1 mm at 200 m. This means that if distance is measured to, say, 1 in 200 000, the angles should be measured to $1''$ of arc, and so on.

(6) The model used to illustrate the concepts of surveying is limited in its application and for most engineering surveys may be considered obsolete. Nevertheless it does serve to illustrate those basic concepts in simple, easily understood terms, to which the beginner can more easily relate.

In the majority of engineering projects, sophisticated instrumentation such as 'total stations' interfaced with electronic data loggers is the norm. In some cases the data loggers can directly drive plotters, thereby producing plots in real time.

Further developments are in the use of satellites to fix three-dimensional position. Such is the accuracy and speed of positioning using the latest GPS satellites that they may be used to establish control points, fix topographic detail, set out position on site and carry out continuous deformation monitoring. Indeed, in the very near future, the use of networks may be of purely historical interest.

Also, inertial positioning systems (IPS) provide a continuous output of position from a known starting point, independent of any external agency, environmental conditions or location. Integration of GPS and IPS may provide a formidable positioning process in the future.

However, regardless of the technological advances in surveying, attention must always be given to instrument calibration, carefully designed projects and meticulous observation. As surveying is essentially the science of measurement, it is necessary to examine the measured data in more detail, as follows.

1.7 UNITS OF MEASUREMENT

The system most commonly used in the measurement of distance and angle is the 'Systeme Internationale', abbreviated to SI. The basic units of prime interest are:

Length in metres (m)

from which we have:

1 m = 10^3 millimetres (mm)

1 m = 10^{-3} kilometres (km)

Thus a distance measured to the nearest millimetre would be written as, say, 142.356 m.
Similarly for areas we have:

1 m^2 = 10^6 mm^2

10^4 m^2 = 1 hectare (ha)

10^6 m^2 = 1 square kilometre (km^2)

and for volumes, m^3 and mm^3.

There are three systems used for plane angles, namely the sexagesimal, the centesimal and radiants (arc units).

The sexagesimal units are used in many parts of the world, including the UK, and measure angles in degrees (°), minutes (′) and seconds (″) of arc, i.e.

$1° = 60′$

$1′ = 60″$

and an angle is written as, say, 125° 46′ 35″.

The centesimal system is quite common in Europe and measures angles in gons (g), i.e.

1 gon = 100 cgon (centigon)

1 cgon = 10 mgon (milligon)

A radian is that angle subtended at the centre of a circle by an arc on the circumference equal in length to the radius of the circle, i.e.

2π rad = 360° = 400 gon

Thus to transform degrees to radians, multiply by $\pi/180°$, and to transform radians to degrees, multiply by $180°/\pi$. It can be seen that:

1 rad = 57.2957795° = 63.6619972 gon

A factor commonly used in surveying to change angles from seconds of arc to radians is:

α rad = $\alpha″/206\,265$

where 206 265 is the number of seconds in a radian.

Other units of interest will be dealt with where they occur in the text.

1.8 SIGNIFICANT FIGURES

Engineers and surveyors communicate a great deal of their professional information using numbers. It is important, therefore, that the number of digits used, correctly indicates the accuracy with which the field data were measured. This is particularly important since the advent of pocket calculators, which tend to present numbers to as many as eight places of decimals, calculated from data containing, at the most, only three places of decimals, whilst some eliminate all trailing zeros. This latter point is important, as 2.00 m is an entirely different value to 2.000 m. The latter number implies estimation to the nearest millimetre as opposed to the nearest 10 mm implied by the former. Thus in the capture of field data, the correct number of significant figures should be used.

By definition, the number of significant figures in a value is the number of digits one is certain of plus one, usually the last, which is estimated. The number of significant figures should not be confused with the number of decimal places. A further rule in significant figures is that in all numbers less than unity, the number of zeros directly after the decimal point and up to the first non-zero digit are not counted. For example:

Two significant figures: 40, 42, 4.2, 0.43, 0.0042, 0.040

Three significant figures: 836, 83.6, 80.6, 0.806, 0.0806, 0.00800

Difficulties can occur with zeros at the end of a number such as 83 600, which may have three, four or five significant figures. This problem is overcome by expressing the value in powers of ten, i.e. 8.36×10^4 implies three significant figures, 8.360×10^4 implies four significant figures and 8.3600×10^4 implies five significant figures.

It is important to remember that the accuracy of field data cannot and should not be improved in the computational processes to which it is subjected.

Consider the addition of the following numbers:

155.486
7.08
2183.0
42.0058

If added on a pocket calculator the answer is 2387.5718; however, the correct answer with due regard to significant figures is 2387.6. It is rounded off to the most extreme right-hand column containing all the significant figures, which in the example is the column immediately after the decimal point. In the case of $155.486 + 7.08 + 2183 + 42.0058$ the answer is 2388. This rule also applies to subtraction.

In multiplication and division, the answer should be rounded off to the number of significant figures contained in that number having the least number of significant figures in the computational process. For instance, $214.8432 \times 3.05 = 655.27176$, when computed on a pocket calculator; however, as 3.05 contains only three significant figures, the correct answer is 655. Consider $428.4 \times 621.8 = 266\,379.12$, which should now be rounded to $266\,400 = 2.664 \times 10^5$, which has four significant figures. Similarly, $41.8 \div 2.1316 = 19.609682$ on a pocket calculator and should be rounded to 19.6.

When dealing with the powers of numbers the following rule is useful. If x is the value of the first significant figure in a number having n significant figures, its pth power is rounded to:

$n - 1$ significant figures if $p \leq x$

$n - 2$ significant figures if $p \leq 10x$

For example, $1.5831^4 = 8.97679$ when computed on a pocket calculator. In this case $x = 1$, $p = 4$ and $p \leq 10x$; therefore, the answer should be quoted to $n - 2 = 3$ significant figures = 8.98.

Similarly, with roots of numbers, let x equal the first significant figure and r the root; the answer should be rounded to:

n significant figures when $rx \geq 10$

$n - 1$ significant figures when $rx < 10$

For example:

$36^{\frac{1}{2}} = 6$, because $r = 2$, $x = 3$, $n = 2$, thus $rx < 10$, and answer is to $n - 1 = 1$ significant figure.

$415.36^{\frac{1}{4}} = 4.5144637$ on a pocket calculator; however, $r = 4$, $x = 4$, $n = 5$, and as $rx > 10$, the answer is rounded to $n = 5$ significant figures, giving 4.5145.

As a general rule, when field data are undergoing computational processing which involves several intermediate stages, one extra digit may be carried throughout the process, provided the final answer is rounded to the correct number of significant figures.

1.9 ROUNDING OFF NUMBERS

It is well understood that in rounding off numbers, 54.334 would be rounded to 54.33, whilst 54.336 would become 54.34. However, with 54.335, some individuals always round up, giving 54.34, whilst others always round down to 54.33. This process creats a systematic bias and should be avoided. The process which creates a more random bias, thereby producing a more representative mean value from a set of data, is to round up when the preceding digit is odd but not when it is even. Using this approach, 54.335 becomes 54.34, whilst 54.345 is 54.34 also.

1.10 ERRORS IN MEASUREMENT

It should now be apparent that position fixing simply involves the measurement of angles and distance. However, all measurements, no matter how carefully executed, will contain error, and so the true value of a measurement is never known. It follows from this that if the true value is never known, the true error can never be known and the position of a point known only within certain error bounds.

The sources of error fall into three broad categories, namely:

(1) Natural errors caused by variation in or adverse weather conditions, refraction, gravity effects, etc.
(2) Instrumental errors caused by imperfect construction and adjustment of the surveying instruments used.
(3) Personal errors caused by the inability of the individual to make exact observations due to the limitations of human sight, touch and hearing.

1.10.1 Classification of errors

(1) Mistakes are sometimes called gross errors, but should not be classified as errors at all. They are blunders, often resulting from fatigue or the inexperience of the surveyor. Typical examples are omitting a whole tape length when measuring distance, sighting the wrong target in a round of angles, reading '6' on a levelling staff as '9' and vice versa. Mistakes are the largest of the errors likely to arise, and therefore great care must be taken to obviate them.
(2) Systematic errors can be constant or variable throughout an operation and are generally attributable to known circumstances. The value of these errors can be calculated and applied as a correction to the measured quantity. They can be the result of natural conditions, examples of which are: refraction of light rays, variation in the speed of electromagnetic waves through the atmosphere, expansion or contraction of steel tapes due to temperature variations. In all these cases, corrections can be applied to reduce their effect. Such errors may also be produced by instruments, e.g. maladjustment of the theodolite or level, index error in spring balances, ageing of the crystals in EDM equipment.

There is the personal error of the observer who may have a bias against setting a micrometer or in bisecting a target, etc. Such errors can frequently be self-compensating; for instance, a person setting a micrometer too low when obtaining a direction will most likely set it too low when obtaining the second direction, and the resulting angle will be correct.

Systematic errors, in the main, conform to mathematical and physical laws; thus it is argued that appropriate corrections can be computed and applied to reduce their effect. It is doubtful, however, whether the effect of systematic errors is ever entirely eliminated, largely due to the inability to obtain an exact measurement of the quantities involved. Typical examples are: the difficulty of obtaining group refractive index throughout the measuring path of EDM distances; and the difficulty of obtaining the temperature of the steel tape, based on air temperature measurements with thermometers. Thus, systematic errors are the most difficult to deal with and therefore they require very careful consideration prior to, during, and after the survey. Careful calibration of all equipment is an essential part of controlling systematic error.

(3) Random errors are those variates which remain after all other errors have been removed. They are beyond the control of the observer and result from the human inability of the observer to make exact measurements, for reasons already indicated above.

Random variates are assumed to have a continuous frequency distribution called normal distribution and obey the law of probability. A random variate x, which is normally distributed with a mean and standard deviation, is written in symbol form as $N(\mu, \sigma^2)$. It should be fully understood that it is random errors alone which are treated by statistical processes.

1.10.2 Basic concept of errors

The basic concept of errors in the data captured by the surveyor may be likened to target shooting.

In the first instance, let us assume that a skilled marksman used a rifle with a bent sight, which resulted in his shooting producing a scatter of shots as at A in *Figure 1.23*.

That the marksman is skilled (or reliable) is evidenced by the very small scatter, which illustrates excellent precision. However, as the shots are far from the centre, caused by the bent sight (systematic error), they are completely inaccurate. Such a situation can arise in practice when a piece of EDM equipment produces a set of measurements all agreeing to within a few millimetres (high precision) but, due to an operating fault and lack of calibration, the measurements are all incorrect by several metres (low accuracy). If the bent sight is now corrected, i.e. systematic errors are minimized, the result is a scatter of shots as at B. In this case, the shots are clustered near the centre of the target and thus high precision, due to the small scatter, can be related directly to accuracy. The scatter is, of course, due to the unavoidable random errors.

If the target was now placed face down, the surveyors' task would be to locate the most probable position of the centre based on an analysis of the position of the shots at B. From this analogy several important facts emerge, as follows.

(1) Scatter is an 'indicator of precision'. The wider the scatter of a set of results about the mean, the less reliable they will be compared with results having a small scatter.

(2) Precision must not be confused with accuracy; the former is a relative grouping without regard to nearness to the truth, whilst the latter denotes absolute nearness to the truth.

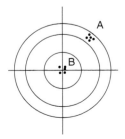

Fig. 1.23

(3) Precision may be regarded as an index of accuracy only when all sources of error, other than random errors, have been eliminated.

(4) Accuracy may be defined only by specifying the bounds between which the accidental error of a measured quantity may lie. The reason for defining accuracy thus is that the absolute error of the quantity is generally not known. If it were, it could simply be applied to the measured quantity to give its true value. The error bound is usually specified as symmetrical about zero. Thus the accuracy of measured quantity x is $x \pm \varepsilon_x$ where ε_x is greater than or equal to the true but unknown error of x.

(5) Position fixing by the surveyor, whether it be the coordinate position of points in a control network, or the position of topographic detail, is simply an assessment of the most probable position and, as such, requires a statistical evaluation of its reliability.

1.10.3 Further definitions

(1) The true value of a measurement can never be found, even though such a value exists. This is evident when observing an angle with a one-second theodolite; no matter how many times the angle is read, a slightly different value will always be obtained.

(2) True error (ε_x) similarly can never be found, for it consists of the true value (X) minus the observed value (x), i.e.

$$X - x = \varepsilon_x$$

(3) Relative error is a measure of the error in relation to the size of the measurement. For instance, a distance of 10 m may be measured with an error of ± 1 mm, whilst a distance of 100 m may also be measured to an accuracy of ± 1 mm. Although the error is the same in both cases, the second measurement may clearly be regarded as more accurate. To allow for this, the term relative error (R_x) may be used, where

$$R_x = \varepsilon_x / x$$

Thus, in the first case $x = 10$ m, $\varepsilon_x = \pm 1$ mm, and therefore $R_x = 1/10\ 000$; in the second case, $R_x = 1/100\ 000$, clearly illustrating the distinction. Multiplying the relative error by 100 gives the percentage error. 'Relative error' is an extremely useful definition, and is commonly used in expressing the accuracy of linear measurement. For example, the relative closing error of a traverse is usually expressed in this way. The definition is clearly not applicable to expressing the accuracy to which an angle is measured, however.

(4) Most probable value (MPV) is the closest approximation to the true value that can be achieved from a set of data. This value is generally taken as the arithmetic mean of a set, ignoring at this stage the frequency or weight of the data. For instance, if A is the arithmetic mean, X the true value, and ε_n the errors of a set of n measurements, then

$$A = X - \frac{[\varepsilon_n]}{n}$$

where $[\varepsilon_n]$ is the sum of the errors. As the errors are equally as likely to be positive or negative, then for a finite number of observations $[\varepsilon_n]/n$ will be very small and $A \approx X$. For an infinite number of measurements, it could be argued that $A = X$. (*N.B.* The square bracket is Gaussian notation for 'sum of'.)

(5) Residual is the closest approximation to the true error and is the difference between the MPV of a set, i.e. the arithmetic mean, and the observed values. Using the same argument as before, it can be shown that for a finite number of measurements, the residual r is approximately equal to the true error ε.

1.10.4 Probability

Consider a length of 29.42 m measured with a tape and correct to ±0.05 m. The range of these measurements would therefore be from 29.37 m to 29.47 m, giving 11 possibilities to 0.01 m for the answer. If the next bay was measured in the same way, there would again be 11 possibilities. Thus the correct value for the sum of the two bays would lie between 11 × 11 = 121 possibilities, and the range of the sum would be 2 × ±0.05 m, i.e. between −0.10 m and +0.10 m. Now, the error of −0.10 m can occur only once, i.e. when both bays have an error of −0.05 m; similarly with +0.10. Consider an error of −0.08; this can occur in three ways: (−0.05 and −0.03), (−0.04 and −0.04) and (−0.03 and −0.05). Applying this procedure through the whole range can produce *Table 1.1*, the lower half of which is simply a repeat of the upper half. If the decimal probabilities are added together they equal 1.0000. If the above results are plotted as error against probability the histogram of *Figure 1.24* is obtained, the errors being represented by rectangles. Then, in the limit, as the error interval gets smaller, the histogram approximates to the superimposed curve. This curve is called the normal probability curve. The area under it represents the probability that the error must lie between ±0.10 m, and is thus equal to 1.0000 (certainty) as shown in *Table 1.1*.

More typical bell-shaped probability curves are shown in *Figure 1.25*; the tall thin curve indicates small scatter and thus high precision, whilst the flatter curve represents large scatter and low precision. Inspection of the curve reveals:

(1) Positive and negative errors are equal in size and frequency; they are equally probable.
(2) Small errors are more frequent than large; they are more probable.
(3) Very large errors seldom occur; they are less probable and may be mistakes or untreated systematic errors.

The equation of the normal probability distribution curve is

$$y = h\pi^{-\frac{1}{2}} e^{-h^2\varepsilon^2}$$

where y = probability of an occurrence of an error ε, h = index of precision, and e = exponential function.

As already illustrated, the area under the curve represents the limit of relative frequency, i.e. probability, and is equal to unity. Thus tables of standard normal curve areas can be used to calculate probabilities provided that the distribution is the standard normal distribution, i.e.

Table 1.1

Error	Occurrence	Probability
−0.10	1	1/121 = 0.0083
−0.09	2	2/121 = 0.0165
−0.08	3	3/121 = 0.0248
−0.07	4	4/121 = 0.0331
−0.06	5	5/121 = 0.0413
−0.05	6	6/121 = 0.0496
−0.04	7	7/121 = 0.0579
−0.03	8	8/121 = 0.0661
−0.02	9	9/121 = 0.0744
−0.01	10	10/121 = 0.0826
0	11	11/121 = 0.0909
0.01	10	10/121 = 0.0826

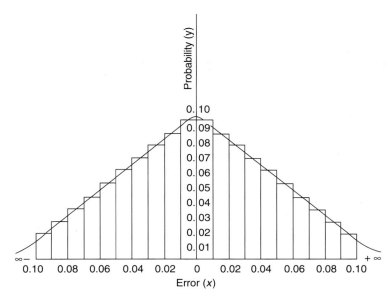

Fig. 1.24

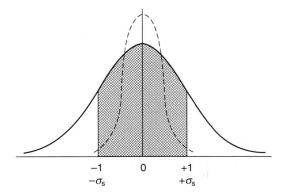

Fig. 1.25

$N(0, 1^2)$. If the variable x is $N(\mu, \sigma^2)$, then it must be transformed to the standard normal distribution using $Z = (x - \mu)/\sigma$, where Z has a probability density function equal to $(2\pi)^{-\frac{1}{2}} e^{-Z^2/2}$

> when $x = N(5, 2^2)$ then $Z = (x - 5)/2$
>
> When $x = 9$ then $Z = 2$

Thus the curve can be used to assess the probability or certainty that a variable x will fall between certain values. For example, the probability that x will fall between 0.5 and 2.4 is represented by area A on the normal curve (*Figure 1.26(a)*). This statement can be written as:

> $P(0.5 < x < 2.4) = \text{area } A$
>
> Now Area A = Area B – Area C (*Figures 1.26(b)* and *(c)*)
>
> where Area B represents $P(x < 2.4)$

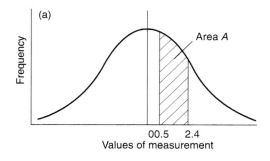

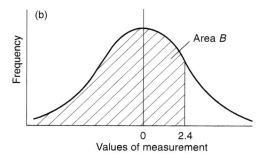

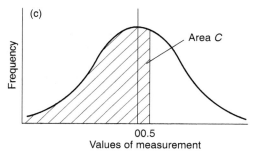

Fig. 1.26

and Area *C* represents $P(x < 0.5)$

i.e. $P(0.5 < x < 2.4) = P(X < 2.4) - P(X < 0.5)$

From the table of Standard Normal Curve Areas

When $x = 2.4$, Area $= 0.9916$

When $x = 0.5$, Area $= 0.6915$

∴ $P(0.5 < x < 2.4) = 0.9916 - 0.6195 = 0.3001$

That is, there is a 30.01% probability that x will lie between 0.5 and 2.4.

If verticals are drawn from the points of inflexion of the normal distribution curve (*Figure 1.27*) they will cut that base at $- \sigma_x$ and $+ \sigma_x$, where σ_x is the standard deviation. The area shown indicates the probability that x will lie between $\pm\sigma_x$ and equals 0.683 or 68.3%. This is a very important statement.

Standard deviation (σ_x), if used to assess the precision of a set of data, implies that 68% of the time, the arithmetic mean ($\bar{x}$) of that set should lie between ($\bar{x} \pm \sigma_x$). Put another way, if the

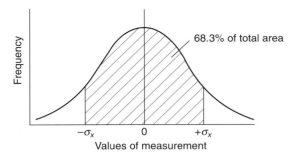

Fig. 1.27

sample is normally distributed and contains only random variates, then 7 out of 10 should lie between $(\bar{x} \pm \sigma_x)$. It is for this reason that two-sigma or three-sigma limits are preferred in statistical analysis:

$$\pm 2\sigma_x = 0.955 = 95.5\% \text{ probability}$$
$$\text{and } \pm 3\sigma_x = 0.997 = 99.7\% \text{ probability}$$

Thus using two-sigma, we can be 95% certain that a sample mean $(\bar{x})$ will not differ from the population mean μ by more than $\pm 2\sigma_x$. These are called 'confidence limits', where $\bar{x}$ is a point estimate of μ and $(\bar{x} \pm 2\sigma_x)$ is the interval estimate.

If a sample mean lies outside the limits of $\pm 2\sigma_x$ we say that the difference between $\bar{x}$ and μ is statistically significant at the 5% level. There is, therefore, reasonable evidence of a real difference and the original null hypothesis $(H_0. \bar{x} = \mu)$ should be rejected.

It may be necessary at this stage to more clearly define 'population' and 'sample'. The 'population' is the whole set of data about which we require information. The 'sample' is any set of data from the population, the statistics of which can be used to describe the population.

1.11 INDICES OF PRECISION

It is important to be able to assess the precision of a set of observations, and several standards exist for doing this. The most popular is standard deviation (σ), a numerical value indicating the amount of variation about a central value.

In order to appreciate the concept upon which indices of precision devolve, one must consider a measure which takes into account *all* the values in a set of data. Such a measure is the deviation from the mean $(\bar{x})$ of each observed value (x_i), i.e. $(x_i - \bar{x})$, and one obvious consideration would be the mean of these values. However, in a normal distribution the sum of the deviations would be zero; thus the 'mean' of the squares of the deviations may be used, and this is called the *variance* (σ^2).

$$\sigma^2 = \sum_{i=1}^{n} (x_i - \bar{x})^2 / n \tag{1.1}$$

Theoretically σ is obtained from an infinite number of variates known as the *population*. In practice, however, only a *sample* of variates is available and S is used as an unbiased estimator. Account is taken of the small number of variates in the sample by using $(n-1)$ as the divisor, which is referred to in statistics as the *Bessel correction*; hence, variance is

$$S^2 = \sum_{i=1}^{n} (x_i - \bar{x})^2 / n - 1 \tag{1.2}$$

As the deviations are squared, the units in which variance is expressed will be the original units squared. To obtain an index of precision in the same units as the original data, therefore, the square root of the variance is used, and this is called *standard deviation (S)*, thus

$$\text{Standard deviation} = S = \pm \left\{ \sum_{i=1}^{n} (x_i - \bar{x})^2 / n - 1 \right\}^{\frac{1}{2}} \tag{1.3}$$

Standard deviation is represented by the shaded area under the curve in *Figure 1.27* and so establishes the limits of the error bound within which 68.3% of the values of the set should lie, i.e. seven out of a sample of ten.

Similarly, a measure of the precision of the mean $(\bar{x})$ of the set is obtained using the *standard error* $(S_{\bar{x}})$, thus

$$\text{Standard error} = S_{\bar{x}} = \pm \left\{ \sum_{i=1}^{n} (x_i - \bar{x})^2 / n(n - 1) \right\}^{\frac{1}{2}} = S / n^{\frac{1}{2}} \tag{1.4}$$

Standard error therefore indicates the limits of the error bound within which the 'true' value of the mean lies, with a 68.3% certainty of being correct.

It should be noted that S and $S_{\bar{x}}$ are entirely different parameters. The value of S will *not* alter significantly with an increase in the number (n) of observations; the value of $S_{\bar{x}}$, however, will alter significantly as the number of observations increases. It is important therefore that to describe measured data both values should be used.

Although the weighting of data has not yet been discussed, it is appropriate here to mention several other indices of precision applicable to weighted (w_i) data

Standard deviation (of weighted data)

$$= S_w = \pm \left\{ \sum_{i=1}^{n} w_i (x_i - \bar{x})^2 / n - 1 \right\}^{\frac{1}{2}} \tag{1.5}$$

Standard deviation of a single measure of weight w_i

$$= S_{wi} = \pm \left\{ \sum_{i=1}^{n} w_i (x_i - \bar{x})^2 / w_i (n - 1) \right\}^{\frac{1}{2}} = S_w / (w_i)^{\frac{1}{2}} \tag{1.6}$$

Standard error (the weighted mean)

$$= S_{\bar{w}} = \pm \left\{ \sum_{i=1}^{n} w_i (x_i - \bar{x})^2 \bigg/ \sum_{i=1}^{n} (w_i)(n - 1) \right\}^{\frac{1}{2}} = S_w \bigg/ \left(\sum_{i=1}^{n} w_i \right)^{\frac{1}{2}} \tag{1.7}$$

N.B. The conventional method of expressing *sum of* has been used for the various indices of precision, as this is the format used in texts on statistics, and is therefore more easily recognizable. However, for the majority of the expressions the neater Gaussian square bracket format has been used.

1.12 WEIGHT

Weights are expressed numerically and indicate the relative precision of quantities within a set.

The greater the weight, the greater the precision of the observation to which it relates. Thus an observation with a weight of two may be regarded as twice as reliable as an observation with a weight of one. Consider two mean measures of the same angle: $A = 50° 50' 50''$ of weight one, and $B = 50° 50' 47''$ of weight two. This is equivalent to three observations, $50''$, $47''$, $47''$, all of equal weight, and having a mean value of

$$(50'' + 47'' + 47'')/3 = 48''$$

Therefore the mean value of the angle $= 50° 50' 48''$.

Inspection of this exercise shows it to be identical to multiplying each observation a by its weight, w, and dividing by the sum of the weights $[w]$, i.e.

$$\text{Weighted mean} = A_m = \frac{a_1 w_1 + a_2 w_2 + \ldots + a_n w_n}{w_1 + w_2 + \ldots + w_n} = \frac{[aw]}{[w]} \tag{1.8}$$

Weights can be allocated in a variety of ways, such as: (a) by personal judgement of the prevailing conditions at the time of measurement; (b) by direct proportion to the number of measurements of the quantity, i.e. $w \propto n$; (c) by the use of variance and co-variance factors. This last method is recommended and in the case of the variance factor is easily applied as follows. Equation (1.4) shows

$$S_{\bar{x}} = S/n^{\frac{1}{2}}$$

That is, error is inversely proportional to the square root of the number of measures. However, as $w \propto n$, then

$$w \propto 1/S_{\bar{x}}^2$$

i.e. weight is proportional to the inverse of the variance.

1.13 REJECTION OF OUTLIERS

It is not unusual, when taking repeated measurements of the same quantity, to find at least one which appears very different from the rest. Such a measurement is called an *outlier*, which the observer intuitively feels should be rejected from the sample. However, intuition is hardly a scientific argument for the rejection of data and a more statistically viable approach is required.

As already indicated, standard deviation S represents 68.3% of the area under the normal curve and is therefore representative of 68.3% confidence limits. It follows from this that

$\pm 3.29S$ represents 99.9% confidence limits (0.999 probability)

Thus, any random variate x_i, whose residual error $(x_i - \bar{x})$ is greater than $\pm 3.29\, S$, must lie in the extreme tail ends of the normal curve and should therefore be ignored, i.e. rejected from the sample. In practice, this has not proved a satisfactory rejection criterion due to the limited size of the samples. Logan (*Survey Review*, No. 97, July 1955) has shown that the appropriate rejection criteria are relative to sample size, as follows:

Sample size	Rejection criteria
4	1.5 S
6	2.0 S
8	2.3 S
10	2.5 S
20	3.0 S

A similar approach to rejection is credited to Chauvenet. If a random variate x_i, in a sample size n has a deviation from the mean $\bar{x}$ greater than a $1/2n$ probability, it should be rejected. For example, if $n = 8$, then $1/2n = 0.06$ (94% or 0.94) and the probability of the deviate is $1.86S$ Thus, an outlier whose residual error or deviation from the mean was greater than $1.86S$. would be rejected. This approach produces the following table:

Sample size	Rejection criteria
4	1.53 S
6	1.73 S
8	1.86 S
10	1.96 S
20	2.24 S

It should be noted that *successive* rejection procedures should not be applied to the sample.

1.14 COMBINATION OF ERRORS

Much data in surveying is obtained indirectly from various combinations of observed data, for instance the coordinates of a line are a function of its length and bearing. As each measurement contains an error, it is necessary to consider the combined effect of these errors on the derived quantity.

The general procedure is to differentiate with respect to each of the observed quantities in turn and sum them to obtain their total effect. Thus if $a = f(x, y, z, \ldots)$, and each independent variable changes by a small amount (an error) $\delta x, \delta y, \delta z \ldots$, then a will change by a small amount equal to δa, obtained from the following expression:

$$\delta a = \frac{\partial a}{\partial x} \cdot \delta x + \frac{\partial a}{\partial y} \cdot \delta y + \frac{\partial a}{\partial z} \cdot \delta z + \ldots \tag{1.9}$$

in which $\partial a / \partial x$ is the partial derivative of a with respect to x, etc.

Consider now a set of measurements and let $x_i = \delta x_i$, $y_i = \delta y_i$, $z_i = \delta z_i$, equals a set of residual errors of the measured quantities and $a_i = \delta a_i$:

$$a_1 = \frac{\partial a}{\partial x} \cdot x_1 + \frac{\partial a}{\partial y} \cdot y_1 + \frac{\partial a}{\partial z} \cdot z_1 + \ldots$$

$$a_2 = \frac{\partial a}{\partial x} \cdot x_2 + \frac{\partial a}{\partial y} \cdot y_2 + \frac{\partial a}{\partial z} \cdot z_2 + \ldots$$

$$\vdots \qquad \vdots \qquad \vdots \qquad \vdots$$

$$a_n = \frac{\partial a}{\partial x} \cdot x_n + \frac{\partial a}{\partial y} \cdot y_n + \frac{\partial a}{\partial z} \cdot z_n + \ldots$$

Now squaring both sides gives

$$a_1^2 = \left(\frac{\partial a}{\partial x}\right)^2 \cdot x_1^2 + 2\left(\frac{\partial a}{\partial x}\right)\left(\frac{\partial a}{\partial y}\right)x_1 y_1 + \ldots \left(\frac{\partial a}{\partial y}\right)^2 y_1^2 + \ldots$$

$$a_2^2 = \left(\frac{\partial a}{\partial x}\right)^2 \cdot x_2^2 + 2\left(\frac{\partial a}{\partial x}\right)\left(\frac{\partial a}{\partial y}\right)x_2 y_2 + \ldots \left(\frac{\partial a}{\partial y}\right)^2 y_2^2 + \ldots$$

$$a_n^2 = \left(\frac{\partial a}{\partial x}\right)^2 \cdot x_n^2 + 2\left(\frac{\partial a}{\partial x}\right)\left(\frac{\partial a}{\partial y}\right)x_n y_n + \ldots \left(\frac{\partial a}{\partial y}\right)^2 y_n^2 + \ldots$$

In the above process many of the square and cross-multiplied terms have been omitted for simplicity. Summing the results gives

$$[a^2] = \left(\frac{\partial a}{\partial x}\right)[x^2] + 2\left(\frac{\partial a}{\partial x}\right)\left(\frac{\partial a}{\partial y}\right)[xy] + \ldots \left(\frac{\partial a}{\partial y}\right)[y^2] + \ldots$$

As the measured quantities may be considered independent and uncorrelated, the cross-products tend to zero and may be ignored.

Now dividing throughout by $(n - 1)$:

$$\frac{[a^2]}{n-1} = \left(\frac{\partial a}{\partial x}\right)\frac{[x^2]}{n-1} + \left(\frac{\partial a}{\partial y}\right)\frac{[y^2]}{n-1} + \left(\frac{\partial a}{\partial z}\right)\frac{[z^2]}{n-1} + \ldots$$

The sum of the residuals squared divided by $(n - 1)$, is in effect the variance σ^2, and therefore

$$\sigma_a^2 = \left(\frac{\partial a}{\partial x}\right)\sigma_x^2 + \left(\frac{\partial a}{\partial y}\right)\sigma_y^2 + \left(\frac{\partial a}{\partial z}\right)\sigma_z^2 + \ldots \qquad (1.10)$$

which is the general equation for the variance of any function. This equation is very important and is used extensively in surveying for error analysis, as illustrated in the following examples.

1.14.1 Errors affecting addition or subtraction

Consider a quantity $A(f) = a + b$ where a and b are affected by standard errors σ_a and σ_b, then

$$\sigma_A^2 = \left\{\frac{\partial(a+b)}{\partial a}\sigma_a\right\}^2 + \left\{\frac{\partial(a+b)}{\partial b}\sigma_b\right\}^2 = \sigma_a^2 + \sigma_b^2 \quad \therefore \sigma_A = \pm(\sigma_a^2 + \sigma_b^2)^{\frac{1}{2}} \qquad (1.11)$$

As subtraction is simply addition with the signs changed, the above holds for the error in a *difference*:

$$\text{If } \sigma_a = \sigma_b = \sigma, \text{ then} \quad \sigma_A = \pm\sigma(n)^{\frac{1}{2}} \qquad (1.12)$$

Equation (1.12) should not be confused with equation (1.4) which refers to the *mean*, not the *sum* as above.

Worked examples

Example 1.1 If three angles of a triangle each have a standard error of $\pm 2''$, what is the total error (σ_T) in the triangle?

$$\sigma_T = \pm(2^2 + 2^2 + 2^2)^{\frac{1}{2}} = \pm 2(3)^{\frac{1}{2}} = \pm 3.5''$$

Example 1.2 In measuring a round of angles at a station, the third angle c closing the horizon is obtained by subtracting the two measured angles a and b from 360°. If angle a has a standard error of ±2″ and angle b a standard error of ±3″, what is the standard error of angle c?

$$c \pm \sigma_c = 360° - (a \pm \sigma_a) - (b \pm \sigma_b)$$
$$= 360° - (a \pm 2″) - (b \pm 3″)$$

since $c = 360° - a - b$

then $\pm \sigma_c = \pm \sigma_a \pm \sigma_b = \pm 2″ \pm 3″$

and $\sigma_c = \pm(2^2 + 3^2)^{\frac{1}{2}} = \pm 3.6″$

Example 1.3 The standard error of a *mean* angle derived from four measurements is ±3″; how many measurements would be required, using the same equipment, to halve this error?

From equation (1.4) $\sigma_m = \pm \dfrac{\sigma_s}{n^{\frac{1}{2}}}$ $\therefore \sigma_s = 3 \times 4^{\frac{1}{2}} = \pm 6″$

i.e. the instrument used had a standard error of ±6″ for a single observation; thus for $\sigma_m = \pm 1.5″$, when $\sigma_s = \pm 6″$

$$n = \left(\frac{6}{1.5}\right)^2 = 16$$

Example 1.4 If the standard error of a single triangle in a triangulation scheme is ±6.0″, what is the permissible standard error per angle?

From equation (1.12) $\sigma_T = \sigma_p(n)^{\frac{1}{2}}$

where σ_T is the triangular error, σ_p the error per angle, and n the number of angles.

$$\therefore \sigma_p = \frac{\sigma_T}{(n)^{\frac{1}{2}}} = \frac{\pm 6.0″}{(3)^{\frac{1}{2}}} = \pm 3.5″$$

1.14.2 Errors affecting a product

Consider $A(f) = (a \times b \times c)$ where a, b and c are affected by standard errors. The variance

$$\sigma_A^2 = \left\{\frac{\partial(abc)}{\partial a} \sigma_a\right\}^2 + \left\{\frac{\partial(abc)}{\partial b} \sigma_b\right\}^2 + \left\{\frac{\partial(abc)}{\partial c} \sigma_c\right\}^2$$
$$= (bc\sigma_a)^2 + (ac\sigma_b)^2 + (ab\sigma_c)^2$$

$$\therefore \sigma_A = \pm abc\left\{\left(\frac{\sigma_a}{a}\right)^2 + \left(\frac{\sigma_b}{b}\right)^2 + \left(\frac{\sigma_c}{c}\right)^2\right\}^{\frac{1}{2}} \tag{1.13a}$$

The terms in brackets may be regarded as the relative errors R_a, R_b, R_c giving

$$\sigma_A = \pm abc \, (R_a^2 + R_b^2 + R_c^2)^{\frac{1}{2}} \tag{1.13b}$$

1.14.3 Errors affecting a quotient

Consider $A(f) = a/b$, then the variance

$$\sigma_A^2 = \left\{ \frac{\partial(ab^{-1})}{\partial a} \sigma_a \right\}^2 + \left\{ \frac{\partial(ab^{-1})}{\partial b} \sigma_b \right\}^2 = \left(\frac{\sigma_a}{b} \right)^2 + \left(\frac{\sigma_b a}{b^2} \right)^2$$

$$\therefore \ \sigma_A = \pm \frac{a}{b} \left\{ \left(\frac{\sigma_a}{a} \right)^2 + \left(\frac{\sigma_b}{b} \right)^2 \right\}^{\frac{1}{2}} \tag{1.14a}$$

$$= \pm \frac{a}{b} (R_a^2 + R_b^2)^{\frac{1}{2}} \tag{1.14b}$$

1.14.4 Errors affecting powers and roots

The case for the power of a number must not be confused with multiplication, since $a^3 = a \times a \times a$, with each term being exactly the same.

Thus if $A(f) = a^n$, then the variance

$$\sigma_A^2 = \left(\frac{\partial a^n}{\partial a} \sigma_a \right)^2 = (na^{n-1}\sigma_a)^2 \quad \therefore \ \sigma_A = \pm(na^{n-1} \sigma_a) \tag{1.15a}$$

Alternatively $\quad R_A = \dfrac{\sigma_A}{a^n} = \dfrac{na^{n-1}\sigma_a}{a^n} = \dfrac{n\sigma_a}{a} = nR_a \tag{1.15b}$

Similarly for roots, if the function is $A(f) = a^{1/n}$, then the variance

$$\sigma_A^2 = \left(\frac{\partial a^{1/n}}{\partial a} \sigma_a \right)^2 = \left(\frac{1}{n} a^{1/n-1}\sigma_a \right)^2 = \left(\frac{1}{n} a^{1/n} a^{-1}\sigma_a \right)^2$$

$$= \left(\frac{a^{1/n}}{n} \frac{\sigma_a}{a} \right)^2 \quad \therefore \ \sigma_A = \pm \left(\frac{a^{1/n}}{n} \frac{\sigma_a}{a} \right) \tag{1.16}$$

The same approach is adopted to general forms which are combinations of the above.

Worked examples

Example 1.5 The same angle was measured by two different observers using the same instrument, as follows:

	Observer A			Observer B	
°	′	″	°	′	″
86	34	10	86	34	05
	33	50		34	00
	33	40		33	55
	34	00		33	50
	33	50		34	00
	34	10		33	55
	34	00		34	15
	34	20		33	44

Calculate: (a) The standard deviation of each set.
(b) The standard error of the arithmetic means.
(c) The most probable value (MPV) of the angle.   (KU)

Observer A			r	r^2	Observer B			r	r^2
$\circ$	$'$	$''$	$''$	$''$	$\circ$	$'$	$''$	$''$	$''$
86	34	10	10	100	86	34	05	7	49
	33	50	-10	100		34	00	2	4
	33	40	-20	400		33	55	-3	9
	34	00	0	0		33	50	-8	64
	33	50	-10	100		34	00	2	4
	34	10	10	100		33	55	-3	9
	34	00	0	0		34	15	17	289
	34	20	20	400		33	44	-14	196
Mean = 86	34	00	0	1200 = $[r^2]$	86	33	58	0	624 = $[r^2]$

(a) (i) Standard deviation $([r^2] = \Sigma(x_i - \bar{x})^2)$

$$S_A = \pm\left(\frac{[r^2]}{n-1}\right)^{\frac{1}{2}} = \pm\left(\frac{1200}{7}\right)^{\frac{1}{2}} = \pm 13.1''$$

(b) (i) Standard error $S_{\bar{x}_A} = \pm\frac{S_A}{n^{\frac{1}{2}}} = \pm\frac{13.1}{8^{\frac{1}{2}}} = \pm 4.6''$

(a) (ii) Standard deviation $S_B = \pm\left(\frac{624''}{7}\right)^{1/2} = \pm 9.4''$

(b) (ii) Standard error $S_{\bar{x}_B} = \pm\frac{9.4}{8^{\frac{1}{2}}} = \pm 3.3''$

(c) As each arithmetic mean has a different precision exhibited by its $S_{\bar{x}}$ value, they must be weighted accordingly before they can be meaned to give the MPV of the angle:

Weight of $A \propto \dfrac{1}{S^2_{\bar{x}_A}} = \dfrac{1}{21.2} = 0.047$

Weight of $B \propto \dfrac{1}{10.9} = 0.092$

The ratio of the weight of A to the weight of B is $1 : 2$

$$\therefore \text{MPV of the angle} = \frac{(86°34'00'' + 86°33'58'' \times 2)}{3}$$

$$= 86°33'59''$$

As a matter of interest, the following point could be made here: any observation whose residual is greater than $2.3S$ should be rejected (see *Section 1.13*). As $2.3S_A = 30.2''$ and $2.3S_B = 21.6''$, all the observations should be included in the set. This test should normally be carried out at the start of the problem.

Example 1.6 Discuss the classification of errors in surveying operations, giving appropriate examples.

In a triangulation scheme, the three angles of a triangle were measured and their mean values recorded as 50°48′18″, 64°20′36″ and 64°51′00″. Analysis of each set gave a standard deviation of ±4″ for each of these means. At a later date, the angles were re-measured under better conditions, yielding mean values of 50°48′20″, 64°20′39″ and 64°50′58″. The standard deviation of each value was ±2″. Calculate the most probable values of the angles. (KU)

The angles are first adjusted to 180°. Since the angles within each triangle are of equal weight, then the angular adjustment within each triangle is equal.

$$
\begin{array}{llll}
50°48′18″ +2″ = & 50°48′20″ & 50°48′20″ + 1″ = & 50°48′21″ \\
64°20′36″ + 2″ = & 64°20′38″ & 64°20′39″ + 1″ = & 64°20′40″ \\
\underline{64°51′00″ + 2″ =} & \underline{64°51′02″} & \underline{64°50′58″ + 1″ =} & \underline{64°50′59″} \\
\underline{179°59′54″} & \underline{180°00′00″} & \underline{179°59′57″} & \underline{180°00′00″}
\end{array}
$$

Weight of the first set $= w_1 = 1/4^2 = \dfrac{1}{16}$

Weight of the second set $= w_2 = 1/2^2 = \dfrac{1}{4}$

Thus $w_1 = 1$, when $w_2 = 4$.

$$\therefore \text{MPV} = \frac{(50°48′20″) + (50°48′21″ \times 4)}{5} = 50°48′20.8″$$

Similarly, the MPVs of the remaining angles are

$$64°20′ 39.6″ \qquad 64°50′ 59.6″$$

These values may now be rounded off to single seconds.

Example 1.7 A base line of ten bays was measured by a tape resting on measuring heads. One observer read one end while the other observer read the other – the difference in readings giving the observed length of the bay. Bays 1, 2 and 5 were measured six times, bays 3, 6 and 9 were measured five times and the remaining bays were measured four times, the means being calculated in each case. If the standard errors of single readings by the two observers were known to be 1 mm and 1.2 mm, what will be the standard error in the whole line due only to reading errors? (LU)

Standard error in reading a bay $\qquad S_s = (1^2 + 1.2^2)^{\frac{1}{2}} = ±1.6 \text{ mm}$

Consider bay 1. This was measured six times and the mean taken; thus the standard error of the mean is

$$S_{\bar{x}} = \frac{S_s}{n^{\frac{1}{2}}} = \frac{1.6}{6^{\frac{1}{2}}} = ±0.6 \text{ mm}$$

This value applies to bays 2 and 5 also. Similarly for bays 3, 6 and 9

$$S_{\bar{x}} = \frac{1.6}{5^{\frac{1}{2}}} = ±0.7 \text{ mm}$$

For bays 4, 7, 8 and 10 $S_{\bar{x}} = \dfrac{1.6}{4^{\frac{1}{2}}} = \pm 0.8$ mm

These bays are now summed to obtain the total length. Therefore the standard error of the whole line is

$$(0.6^2 + 0.6^2 + 0.6^2 + 0.7^2 + 0.7^2 + 0.7^2 + 0.8^2 + 0.8^2 + 0.8^2 + 0.8^2)^{\frac{1}{2}} = \pm 2.3 \text{ mm}$$

Example 1.8

(a) A base line was measured using electronic distance-measuring (EDM) equipment and a mean distance of 6835.417 m recorded. The instrument used has a manufacturer's quoted accuracy of 1/400 000 of the length measured ± 20 mm. As a check the line was re-measured using a different type of EDM equipment having an accuracy of 1/600 000 ± 30 mm; the mean distance obtained was 6835.398 m. Determine the most probable value of the line.

(b) An angle was measured by three different observers, A, B and C. The mean of each set and its standard error is shown below.

Observer	Mean angle °	′	″	$S_{\bar{x}}$ ″
A	89	54	36	± 0.7
B	89	54	42	± 1.2
C	89	54	33	± 1.0

Determine the most probable value of the angle.   (KU)

(a) Standard error, 1st instrument $S_{\bar{x}_1} = \pm \left\{ \left(\dfrac{6835}{400\,000} \right)^2 + (0.020)^2 \right\}^{\frac{1}{2}}$

$$= \pm 0.026 \text{ m}$$

Standard error, 2nd instrument $S_{\bar{x}_2} = \pm \left\{ \left(\dfrac{6835}{600\,000} \right)^2 + (0.030)^2 \right\}^{\frac{1}{2}}$

$$= \pm 0.032 \text{ m}$$

These values can now be used to weight the lengths and find their weighted means as shown below.

	Length, L (m)	$S_{\bar{x}}$	Weight ratio	Weight, W	L × W
1st instrument	0.417	± 0.026	$1/0.026^2 = 1479$	1.5	0.626
2nd instrument	0.398	± 0.032	$1/0.032^2 = 977$	1	0.398
			[W] =	2.5	1.024 = [LW]

$\therefore$ MPV $= 6835 + \dfrac{1.024}{2.5} = 6835.410$ m

(b)

Observer	Mean angle			$S_{\bar{x}}$	Weight ratio	Weight, W	$L \times W$
	°	′	″	″			
A	89	54	36	±0.7	$1/0.7^2 = 2.04$	2.96	$6'' \times 2.96 = 17.8''$
B	89	54	42	±1.2	$1/1.2^2 = 0.69$	1	$12'' \times 1 = 12''$
C	89	54	33	±1.0	$1/1^2 = 1$	1.45	$3'' \times 1.45 = 4.35''$
					$[W] =$	5.41	$34.15 = [LW]$

$$\therefore \text{MPV} = 89°54'30'' + \frac{34.15''}{5.41} = 89°54'36''$$

The student's attention is drawn to the method of finding the weighted mean in both these examples, although since the advent of the pocket calculator there is no need to refine the weights down from the weight ratio, particularly in (b).

Example 1.9 In an underground correlation survey, the sides of a Weisbach triangle were measured as follows:

$$W_1W_2 = 5.435 \text{ m} \qquad W_1W = 2.844 \text{ m} \qquad W_2W = 8.274 \text{ m}$$

Using the above measurements in the cosine rule, the calculated angle $WW_1W_2 = 175°48'24''$. If the standard error of each of the measured sides is 1/20 000, calculate the standard error of the calculated angle in seconds of arc. (KU)

From *Figure 1.28*, by the cosine rule $\qquad c^2 = a^2 + b^2 - 2ab \cos W_1.$

Using equation (1.14) and differentiating with respect to each variable in turn

$$2c\delta c = 2ab \sin W_1 \delta W_1 \qquad \text{thus} \qquad \delta W_1 = \pm \frac{c\delta c}{ab \sin W_1}$$

Similarly $\qquad a^2 = c^2 - b^2 + 2ab \cos W_1$

$$2a\delta a = 2b \cos W_1 \delta a - 2ab \sin W_1 \delta W_1$$

$$\therefore \delta W_1 = \frac{2a\delta a - 2b \cos W_1 \delta a}{2ab \sin W_1} = \frac{\delta a (a - b \cos W_1)}{ab \sin W_1}$$

but, since angle $W_1 \approx 180°$, $\cos W_1 \approx -1$ and $(a + b) \approx c$

$$\therefore \delta W_1 = \pm \frac{\delta ac}{ab \sin W_1}$$

now $\qquad b^2 = a^2 - c^2 + 2ab \cos W_1$

and $\qquad 2b\delta b = 2a \cos W_1 \delta b - 2ab \sin W_1 \delta W_1$

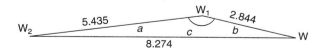

Fig. 1.28

$$\therefore \; \delta W_1 = \frac{\delta b(b - a \cos W_1)}{ab \sin W_1} = \pm \frac{\delta bc}{ab \sin W_1}$$

Making δW_1, δa, δb and δc equal to the standard deviations gives

$$\sigma_{w1} = \pm \frac{c}{ab \sin W_1} (\sigma_a^2 + \sigma_b^2 + \sigma_c^2)^{\frac{1}{2}}$$

where $\sigma_a = \dfrac{5.435}{20\,000} = \pm 2.7 \times 10^{-4}$

$\sigma_b = \dfrac{2.844}{20\,000} = \pm 1.4 \times 10^{-4}$

$\sigma_c = \dfrac{8.274}{20\,000} = \pm 4.1 \times 10^{-4}$

$$\therefore \; \sigma_{w1} = \pm \frac{8.274 \times 206\,265 \times 10^{-4}}{5.435 \times 2.844 \sin 175°48'24''} (2.7^2 + 1.4^2 + 4.1^2)^{\frac{1}{2}}$$

$$= \pm 770''$$
$$= \pm 0°12'50''$$

This is a standard treatment for small errors, and nothing is to be gained by further examples of this type here. The student can find numerous examples of its application throughout the remainder of this book.

Exercises

(*1.1.*) Explain the meaning of the terms *random error* and *systematic error*, and show by example how each can occur in normal surveying work.

A certain angle was measured ten times by observer *A* with the following results, all measurements being equally reliable:

74°38'18", 20", 15", 21", 24", 16", 22", 17", 19", 13"

(The degrees and minutes remained constant for each observation.)

The same angle was measured under the same conditions by observer *B* with the following results:

74°36'10", 21", 25", 08", 15" 20", 28", 11", 18" 24"

Determine the standard deviation for each observer and relative weightings. (ICE)

(*Answer*: ±3.4"; ±6.5". *A*:*B* is 9:2)

(*1.2.*) Derive from first principles an expression for the standard error in the computed angle W_1 of a Weisbach triangle, assuming a standard error of σ_w in the Weisbach angle W, and equal proportional standard errors in the measurement of the sides. What facts, relevant to the technique of correlation using this method, may be deduced from the reduced error equation? (KU)

(*Answer*: see Chapter 10)

2
Vertical control

2.1 INTRODUCTION

Vertical control refers to the various heighting procedures used to obtain the elevation of points of interest above or below a reference datum. The most commonly used reference datum is mean sea level (MSL). There is no such thing as a common global MSL, as it will vary from place to place depending on the effect of local conditions. It is important therefore that MSL is clearly defined wherever it is utilized.

The engineer is, in the main, more concerned with the relative height of one point above or below another, in order to ascertain the difference in height of the two points, rather than any relationship to MSL. It is not unusual, therefore, on small local schemes, to adopt a purely arbitrary reference datum. This could take the form of a permanent, stable position or mark, allocated such a value that the level of any point on the site would not be negative. For example, if the reference mark was allocated a value of 0.000 m, then a ground point 10 m lower would have a negative value of 10.000 m. However, if the reference value was 100.000 m, then the level of the ground point in question would be 90.000 m. As minus signs in front of a number can be misinterpreted, erased or simply forgotten about, they should, wherever possible, be avoided.

The vertical heights of points above or below a reference datum are referred to as the reduced level or simply the level of a point. Reduced levels are used in practically all aspects of construction: to produce ground contours on the plan; to enable the optimum design of road, railway or canal gradients; to facilitate ground modelling for accurate volumetric calculations. Indeed, there is scarcely any aspect of construction that is not dependent on the relative levels of ground points.

2.2 LEVELLING

Levelling is the most widely used method of obtaining the elevations of ground points relative to a reference datum and is usually carried out as a separate procedure to those used in fixing planimetric position.

The basic concept of levelling involves the measurement of vertical distance relative to a horizontal line of sight. Hence it requires a graduated staff for the vertical measurements and an instrument that will provide a horizontal line of sight.

2.3 DEFINITIONS

2.3.1 Level line

A level line or level surface is one which at all points is normal to the direction of the force of gravity as defined by a freely suspended plumb-bob. As already indicated in *Chapter 1* during the discussion of the geoid, such surfaces are ellipsoidal in shape. Thus in *Figure 2.1* the difference in level between A and B is the distance $A'B$.

2.3.2 Horizontal line

A horizontal line or surface is one which is normal to the direction of the force of gravity at a particular point. *Figure 2.1* shows a horizontal line through point C.

2.3.3 Datum

A datum is any reference surface to which the elevations of points are referred. The most commonly used datum is that of mean sea level (MSL).

In the UK the MSL datum was fixed by the Ordnance Survey (OS) of Great Britain, and hence it is often referred to as Ordnance Datum (OD). It is the mean level of the sea at Newlyn in Cornwall calculated from hourly readings of the sea level, taken by an automatic tide gauge over a six-year period from 1 May 1915 to 30 April 1921. The readings are related to the Observatory Bench Mark, which is 4.751 m above the datum. Other countries have different datums; for instance, Australia used 30 tidal observatories, interconnected by 200 000 km of levelling, to produce their national datum, whilst just across the English Channel, France uses a different datum, rendering their levels incompatible with those in the UK.

2.3.4 Bench mark (BM)

In order to make OD accessible to all users throughout the country, a series of permanent marks were established, called bench marks. The height of these marks relative to OD has been established by differential levelling and is regularly checked for any change in elevation.

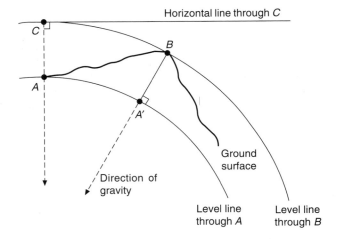

Fig. 2.1

(1) Fundamental bench marks (FBM)

In the UK, FBMs are established by precise geodetic levelling, at intervals of about 40 km. Each mark consists of a buried chamber containing two reference points, whilst the published elevation is to a brass bolt on the top of a concrete pillar (*Figure 2.2*).

(2) Flush brackets

These are metal plates, 90×175 mm, cemented into the face of buildings at intervals of about 1.5 km (*Figure* 2.3).

(3) Cut bench marks

These are the most common type of BM cut into the vertical surface of stable structures (*Figure 2.4*).

(4) Bolt bench marks

These are 60-mm-diameter brass bolts set in horizontal surfaces and engraved with an arrow and the letters OSBM (*Figure 2.5*).

Rivet and pivot BMs are also to be found in horizontal surfaces.

Details of BMs within the individual's area of interest may be obtained in the form of a Bench Mark List from the OS. Their location and value are also shown on OS plans at the 1/2500 and 1/1250 scales. Their values are quoted and guaranteed to the nearest 10 mm only.

Bench marks established by individuals, other than the OS, such as engineers for construction work, are called temporary bench marks (TBM).

2.3.5 Reduced level (RL)

The RL of a point is its height above or below a reference datum.

2.4 CURVATURE AND REFRACTION

Figure 2.6 shows two points A and B at exactly the same level. An instrument set up at X would give a horizontal line of sight through X'. If a graduated levelling staff is held vertically on A the horizontal line would give the reading A'. Theoretically, as B is at the same level as A, the staff reading should be identical (B'). This would require a level line of sight; the instrument, however, gives a horizontal line and a reading at B'' (ignoring refraction). Subtracting vertical height AA' from BB'' indicates that point B is lower than point A by the amount B'B''. This error (c) is caused by the curvature of the Earth and its value may be calculated as follows:

With reference to *Figure 2.7*, in which the instrument heights are ignored:

$$(XB'')^2 = (OB'')^2 - (OX)^2 = (R + c)^2 - R^2 = R^2 + 2Rc + c^2 - R^2 = (2Rc + c^2)$$

As c is a relatively small value, distance XB'' may be assumed equal to the arc distance XB = D. Therefore

FUNDAMENTAL BENCH MARK

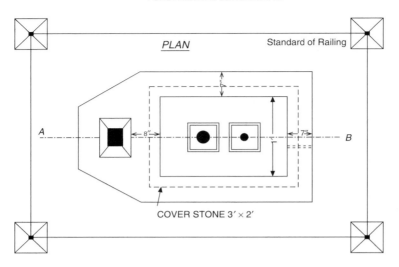

PLAN

Standard of Railing

COVER STONE 3′ × 2′

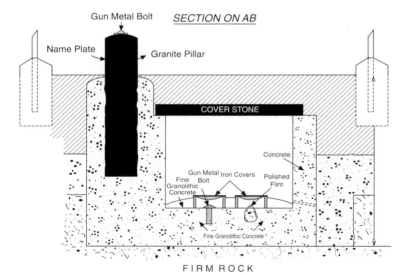

Gun Metal Bolt   SECTION ON AB

Name Plate

Granite Pillar

COVER STONE

Concrete

Gun Metal Iron Covers Polished
Fine   Bolt   Flint
Granolithic
Concrete

Fine Granolithic Concrete

FIRM ROCK

DESCRIPTION

The sites are specially selected with reference to the geological structure, so that they may be placed on sound strata clear of areas liable to subsidence. They are established along the Geodetic lines of levels throughout Great Britain at approximately 30 mile intervals. They have three reference points, two of which, a gun metal bolt and a flint are contained in a buried chamber. The third point is a gun metal bolt set in the top of a pillar projecting about l foot above ground level.

The pillar bolt is the reference point to be used by Tertiary Levellers and other users.
The buried chamber is only opened on instructions from Headquarters.

Some Fundamental Bench Marks are enclosed by iron railings, this was done where necessary, as a protective measure.

These marks are generally referred to as F.B.M's.

Fig. 2.2   *Fundamental bench mark*

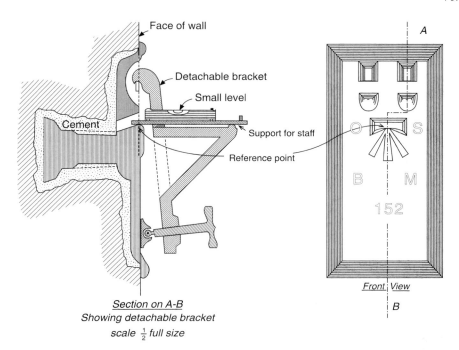

Face of wall
Detachable bracket
Small level
Cement
Support for staff
Reference point

Section on A-B
Showing detachable bracket
scale $\frac{1}{2}$ full size

Front View

DESCRIPTION

These are normally emplaced on vertical walls of buildings and in the sides of triangulation pillars.

They are cast in brass and rectangular in shape ($7'' \times 3\frac{1}{2}''$) with a large boss at the rear of the plate. The boss is cemented into a prepared cavity, so that the face of the bracket is vertical and in line with the face of the object on which it is placed. These marks in precise levelling necessitate the use of a special fitting as above.

Each flush bracket has a unique serial number and is referred to in descriptions as Fl. Br. No............

They are sited at approximately I mile intervals along Geodetic lines of levels and at 3 to 4 mile intervals on Secondary lines of levels.

Fig. 2.3 *Flush bracket (front and side view)*

$$D = (2Rc + c^2)^{\frac{1}{2}}$$

Now as c is very small compared with R, c^2 may be ignored, giving

$$c = D^2/2R \tag{2.1}$$

Taking the distance D in kilometres and an average value for R equal to 6370 km, we have

$$c = (D \times 1000)^2/2 \times 6370 \times 1000$$
$$c = 0.0785D^2 \tag{2.2}$$

with the value of c in metres, when D is in kilometres.

In practice the staff reading in *Figure 2.6* would not be at B'' but at Y due to refraction of the line of sight through the atmosphere. In general it is considered that the effect is to bend the line of sight down, reducing the effect of curvature by 1/7th. Thus the combined effect of curvature and refraction $(c - r)$ is $(6/7)(0.0785D^2)$, i.e.

$$(c - r) = 0.0673D^2 \tag{2.3}$$

*Cut bench marks
(vertical surfaces)*

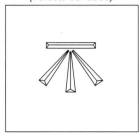

These are to be found on the
vertical faces of buildings, bridges,
walls, milestones, gate posts, etc.

The mark is approximately
4×4 inches, and cut to a depth of
$\frac{1}{4}''$ about 18 inches above ground
level.

Some very old marks may be
considerably larger than this and
Initial Levelling marks 1840–60
may have a copper bolt set in the
middle of the horizontal bar or
offset to one side of the mark.

The exact point of reference is
the centre of the V shaped horizontal bar.

*Cut bench marks
(horizontal surfaces)*

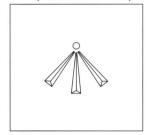

These are found on horizontal
surfaces such as parapets,
culverts, ledges, steps, etc. In place
of the horizontal bar, the reference
point is usually a small brass or
steel round headed rivet which is
inserted at the point of the arrow.

Some of these marks may have
instead of the rivet, a small pivot
hole at the point of the arrow. When
use is made of this type of mark, a
$\frac{5}{8}''$ ball bearing is placed in the
pivot hole.

Stable bench marks

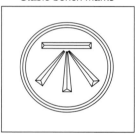

These marks are sometimes
found in the vicinity of coal mines.
They were sited on positions
which were considered to be
stable.

Since 1944 the use of the circle
has been discontinued, but a
number of stable marks bearing
this symbol are still in existence.
They cannot now be definitely
quoted as stable and enquiries in
this respect are referred to the
N.C.B.

Cancelled bench marks

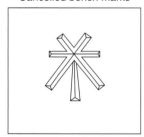

Bench marks found with the
arrow head extending across the
reference point indicates that
the mark has been considered
unsuitable and cancelled. Such
marks are not included in the re-
survey.

Fig. 2.4

Substituting a value of 1 mm in $(c - r)$ gives a value for D equal to 122 m. Thus in tertiary levelling,
where the length of sights is generally 25–30 m, the effect may be ignored.

It should be noted that although the effect of refraction has been shown to bend the line of sight
down by an amount equal to 1/7th that of the effect of curvature, this is a most unreliable assumption.

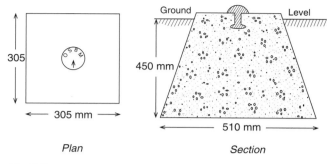

Plan   Section

O.S.B.M.bolts are established on horizontal surfaces where no suitable site exists for the emplacement of a flush bracket or cut bench mark.

They are made of brass and have a mushroom shaped head. The letters O.S.B.M. and an arrow pointing to the centre are engraved on the head of the bolt. Typical sites for the bolts are:—

(a) Living rock.
(b) Foundation abutments to buildings, etc.
(c) Steps, ledges, etc.
(d) Concrete blocks (as in fig.)

Fig. 2.5

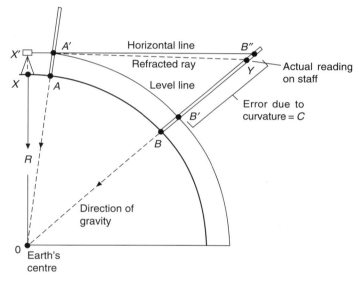

Fig. 2.6

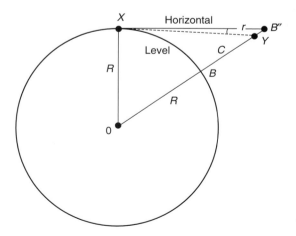

Fig. 2.7

Refraction is largely a function of atmospheric pressure and temperature gradients, which may casue the bending to be up or down by extremely variable amounts.

There are basically three types of temperature gradient (dT/dh):

(1) *Absorption*: occurs mainly at night when the colder ground absorbs heat from the atmosphere. This causes the atmospheric temperature to increase with distance from the ground and $dT/dh > 0$.
(2) *Emission*: occurs mainly during the day when the warmer ground emits heat into the atmosphere, resulting in a negative temperature gradient, i.e. $dT/dh < 0$.
(3) *Equilibrium*: no heat transfer takes place ($dT/dh = 0$) and occurs only briefly in the evening and morning.

The result of $dT/dh < 0$ is to cause the light ray to be convex to the ground rather than concave as generally shown. This effect increases the closer to the ground the light ray gets and errors in the region of 5 mm/km have resulted.

Thus, wherever possible, staff readings should be kept at least 0.5 m above the ground, using short observation distances (25 m) equalized for backsight and foresight.

2.5 EQUIPMENT

The equipment used in the levelling process comprises optical levels and graduated staffs. Basically, the optical level consists of a telescope fitted with a spirit bubble or automatic compensator to ensure long horizontal sights onto the vertically held graduated staff (*Figure 2.8*).

2.5.1 Levelling staff

Levelling staffs are made of wood, metal or glass fibre and graduated in metres and decimals. The alternate metre lengths are in black and red on a white background. The majority of staffs are telescopic or socketed in three sections for easy carrying. Although the graduations can take various forms, the type adopted in the UK is the British Standard (BS 4484) E-pattern type as shown in *Figure 2.9*. The smallest graduation on the staff is 0.01 m, with readings estimated to the nearest millimetre. As the staff must be held vertical during observation it should be fitted with a circular bubble.

2.5.2 Optical levels

The types of level found in general use are the tilting, the automatic level, and digital levels.

(1) Tilting level

Figure 2.10 shows the telescope of the tilting level pivoted at the centre of the tribrach. The footscrews are used to centre the circular bubble, thereby approximately setting the telescope in a horizontal plane. When the telescope has been focused on the staff, the line of sight is set more precisely horizontal using the highly sensitive tubular bubble and the tilting screw which raises or lowers one end of the telescope.

The double concave internal focusing lens is moved along the telescope tube by its focusing screw until the image of the staff is brought into focus on the cross-hairs. The Ramsden eyepiece,

Fig. 2.8 *Levelling procedure — using a Kern GKO-A automatic level and taking a horizontal sight onto a levelling staff held vertically on a levelling plate*

with a magnification of about 35 diameters, is then used to view the image in the plane of the cross-hairs.

The cross-hairs are etched onto a circle of fine glass plate called a reticule and must be brought into sharp focus by the eyepiece focusing screw prior to commencing observations. This process is necessary to remove any cross-hair parallax caused by the image of the staff being brought to a focus in front of or behind the cross-hair. The presence of parallax can be checked by moving the head from side to side or up and down when looking through the telescope. If the image of the staff does not coincide with the cross-hair, movement of the head will cause the cross-hair to move relative to the staff image. The adjusting procedure is therefore:

(1) Using the eyepiece focusing screw, bring the cross-hair into very sharp focus against a light background such as a sheet of blank paper held in front of the object lens.
(2) Now focus on the staff using the main focusing screw until a sharp image is obtained without losing the clear image of the cross-hair.
(3) Check by moving your head from side to side several times. Repeat the whole process if necessary.

Different types of cross-hair are shown in *Figure 2.11*. A line from the centre of the cross-hair and passing through the centre of the object lens is the line of sight or line of collimation of the telescope.

The sensitivity of the tubular spirit bubble is determined by its radius of curvature (R) (*Figure 2.12*); the larger the radius, the more sensitive the bubble. It is filled with sufficient synthetic alcohol to leave a small air bubble in the tube. The tube is graduated generally in intervals of 2 mm.

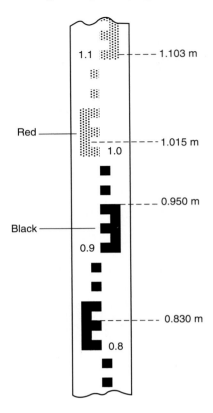

Red ――――――――――― 1.015 m

Black ――――――――

1.1 - - - - 1.103 m

1.0

0.950 m

0.9

0.830 m

0.8

Fig. 2.9

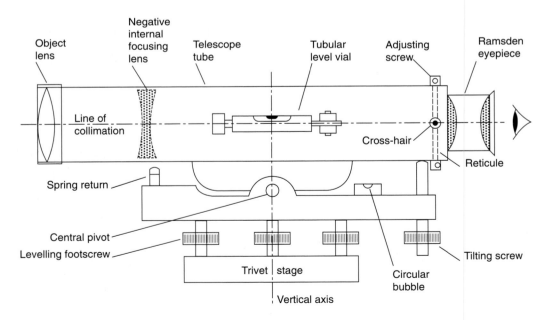

Fig. 2.10   *Tilting level*

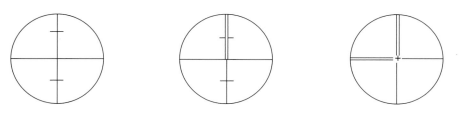

Fig. 2.11

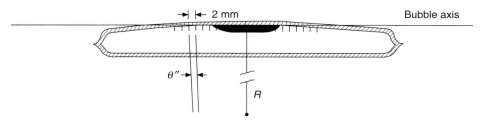

Fig. 2.12 *Tubular bubble*

If the bubble moves off centre by one such interval it represents an angular tilt of the line of sight of 20 seconds of arc. Thus if 2 mm subtends $\theta = 20''$, then:

$$R = (2\ \text{mm} \times 206\,265)/20'' = 20.63\ \text{m}$$

As attached to the tilting level it may be viewed directly or by means of a coincidence reading system (*Figure 2.13*). In this latter system the two ends of the bubble are viewed and appear as shown at (a) and (b). (a) shows the image viewed when the bubble is off centre (b) when the bubble is centred by means of the tilting screw. This method of viewing the bubble is four or five times more accurate than direct viewing.

The main characteristics defining the quality of the telescope are its powers of magnification, the size of its field of view, the brightness of the image formed and the resolution quality when reading the staff. All these are a function of the lens systems used and vary accordingly from low-order builders' levels to very precise geodetic levels.

Magnification is the ratio of the size of the object viewed through the telescope to its apparent size when viewed by the naked eye. Surveying telescopes are limited in their magnification in order to retain their powers of resolution and field of view. Also, the greater the magnification, the greater the effect of heat shimmer, on-site vibration and air turbulence. Telescope magnification lies between 15 and 50 times.

The field of view is a function of the angle of the emerging rays from the eye through the telescope, and varies from 1° to 2°. Image brightness is the ratio of the brightness of the image when viewed through the telescope to the brightness when viewed by the naked eye. It is argued that the lens system, including the reticule, of an internal focusing telescope loses about 40% of the light. If reflex-reducing 'T-film' is used to coat the lens, the light loss is reduced by 10%.

The resolution quality or resolving power of the telescope is the ability to define detail and is independent of magnification. It is a function of the effective aperture of the object lens and the wavelength (λ) of light and is represented in angular units. It can be computed from $P.$ rad $= 1.2\lambda$/(effective aperture).

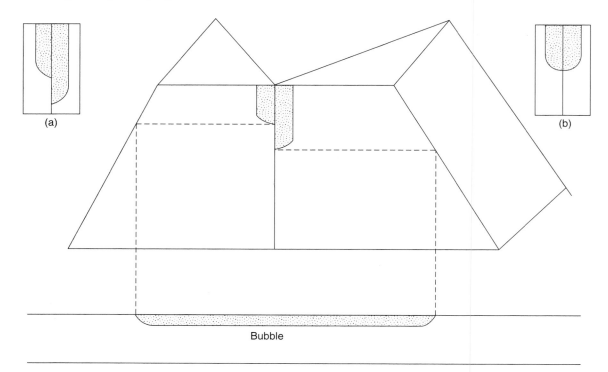

Bubble

Fig. 2.13 *Bubble coincidence reading system*

(2) Using a tilting level

(1) Set up the instrument on a firm, secure tripod base.
(2) Centralize the circular bubble using the footscrews or ball and socket arrangement.
(3) Eliminate parallax.
(4) Centre the vertical cross-hair on the levelling staff and clamp the telescope. Use the slow-motion screw if necessary to ensure exact alignment.
(5) Focus onto the staff.
(6) Carefully centre the tubular bubble using the tilting screw.
(7) With the staff in the field of view as shown in *Figure 2.14* note the staff reading (1.045) and record it.

Operations (4) to (7) must be repeated for each new staff reading.

(3) Automatic levels

The automatic level is easily recognized by its clean, uncluttered appearance. It does not have a tilting screw or a tubular bubble as the telescope is rigidly fixed to the tribrach and the line of sight is horizontalized by a compensator inside the telescope.

The basic concept of the automatic level can be likened to a telescope rigidly fixed at right angles to a pendulum. Under the influence of gravity, the pendulum will swing into the vertical, as defined by a suspended plumb-bob and the telescope will move into a horizontal plane.

As the automatic level is only approximately levelled by means of its low-sensitivity circular bubble, the collimation axis of the instrument will be inclined to the horizontal by a small angle α

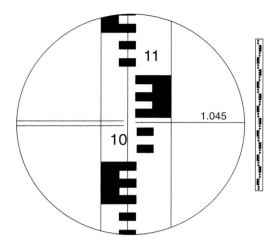

Fig. 2.14

(*Figure 2.15*) so the entering ray would strike the reticule at *a* with a displacement of *ab* equal to *fα*. The compensator situated at *P*, would need to redirect the ray to pass through the cross-hair at *b*. Thus

$$f\alpha = ab = s\beta$$

and $\quad \beta = \dfrac{f\alpha}{s} = n\alpha$

It can be seen from this that the positioning of the compensator is a significant aspect of the compensation process. For instance, if the compensator is fixed halfway along the telescope, then $s \approx f/2$ and $n = 2$, giving $\beta = 2\alpha$. There is a limit to the working range of the compensator, about 20′; hence the need of a circular bubble.

In order, therefore, to compensate for the slight residual tilts of the telescope, the compensator requires a reflecting surface fixed to the telescope, movable surfaces influenced by the force of gravity and a dampening device (air or magnetic) to swiftly bring the moving surfaces to rest and permit rapid viewing of the staff. Such an arrangement is illustrated in *Figure 2.16*.

The advantages of the automatic level over the tilting level are:

(1) Much easier to use, as it gives an erect image of the staff.
(2) Rapid operation, giving greater economy.

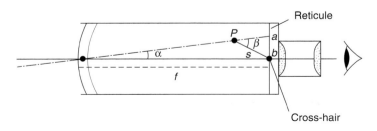

Fig. 2.15 *Basic principle of compensator*

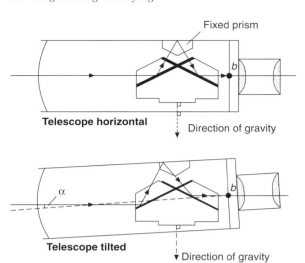

Fig. 2.16 *Suspended compensation*

(3) No chance of reading the staff without setting the bubble central, as can occur with a tilting
     level.
(4) No bubble setting error.

A disadvantage is that it is difficult to use where there is vibration caused by wind, traffic or, say,
piling operations on site, resulting in oscillation of the compensator. Improved damping systems
have, however, greatly reduced this effect.

(4) Using an automatic level

The operations are identical to those for the tilting level with the omission of operation (6). Some
automatic levels have a button, which when pressed moves the compensator to prevent it sticking.
This should be done just prior to reading the staff, when the cross-hair will be seen to move.
Another approach to ensure that the compensator is working is to move it slightly off level and note
if the reading on the staff is unaltered, thereby proving the compensator is working.

2.6 INSTRUMENT ADJUSTMENT

For equipment to give the best possible results it should be frequently tested and, if necessary,
adjusted. Surveying equipment receives continuous and often brutal use on construction sites.
In all such cases a calibration base should be established to permit weekly checks on the
equipment.

2.6.1 Tilting level

The tilting level requires adjustment for collimation error only.

(1) Collimation error

Collimation error occurs if the line of sight is not truly horizontal when the tubular bubble is centred, i.e. the line of sight is inclined up or down from the horizontal. A check known as the 'Two-Peg Test' is used, the procedure being as follows (*Figure 2.17*):

(a) Set up the instrument midway between two pegs *A* and *B* set, say, 20 m apart and note the staff readings, a_1 and b_1, equal to, say, 1.500 m and 0.500 m respectively.
  Let us assume that the line of sight is inclined up by an angle of α; as the lengths of the sights are equal (10 m), the error in each staff reading will be equal and so cancel out, resulting in a 'true' difference in level between *A* and *B*.

$$\Delta H_{TRUE} = (a_1 - b_1) = (1.500 - 0.500) = 1.000 \text{ m}$$

Thus we know that *A* is truly lower than B by 1.000 m. We do not at this stage know that collimation error is present.

(b) Move the instrument to *C*, which is 10 m from *B* and in the line *AB* and observe the staff readings a_2 and b_2 equal to, say, 3.500 m and 2.000 m respectively. Then

$$\Delta H = (a_2 - b_2) = (3.500 - 2.000) = 1.500 \text{ m}$$

Now as 1.500 ≠ the 'true' value of 1.000, it must be 'false'.

$$\Delta H_{FALSE} = 1.500 \text{ m}$$

and it is obvious that the instrument possesses a collimation error the amount and direction of which is as yet still unknown, but which has been revealed by the use of unequal sight lengths *CB*

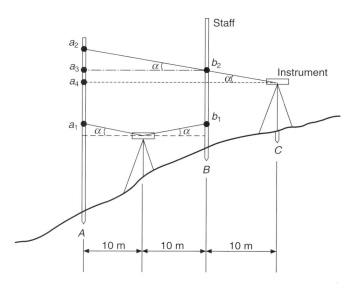

Fig. 2.17 *Two-peg test*

(10 m) and *CA* (30 m). Had the two values for ΔH been equal, then there is no collimation error present in the instrument.

(c) Imagine a horizontal line from reading b_2 (2.000 m) cutting the staff at *A* at reading a_3, because *A* is truly 1.000 m below *B*, the reading at a_3 must be 2.000 + 1.000 = 3.000 m. However, the actual reading was 3.500 m, and therefore the line of sight of the instrument is too high by 0.500 m in 20 m (the distance between the two pegs). This is the amount and direction of collimation error.

(d) Without moving the instrument from *C*, the line of sight must be adjusted down until it is horizontal. To do this one must compute the reading (a_4) on staff *A* that a horizontal sight from *C*, distance 30 m away, would give.
  By simple proportion, as the error in 20 m is 0.500, the error in 30 m = (0.500 × 30)/20 = 0.750 m. Therefore the required reading at a_4 is 3.500 – 0.750 = 2.750 m.

(e) (i) Using the 'tilting screw', tilt the telescope until it reads 2.750 m on the staff. (ii) This movement will cause the tubular bubble to go off centre. Re-centre it by means of its adjusting screws, which will permit the raising or lowering of one end of the bubble.

The whole operation may be repeated if thought necessary.
  The above process has been dealt with in great detail, as collimation error is one of the main sources of error in the levelling process.
  The diagrams and much of the above detail can be dispensed with if the following is noted:

(1) ($\Delta H_{FALSE} - \Delta H_{TRUE}$) = the amount of collimation error.
(2) If $\Delta H_{FALSE} > \Delta H_{TRUE}$ then the line of sight is inclined *up* and vice versa.

An example will now be done to illustrate this approach.

Worked example

Example 2.1 Assume the same separation for *A*, *B* and *C*. With the instrument midway between *A* and *B* the readings are *A*(3.458), *B*(2.116). With the instrument at *C*, the readings are *A*(4.244), *B*(2.914).

  (i) From 'midway' readings, ΔH_{TRUE} = 1.342
  (ii) From readings at '*C*', ΔH_{FALSE} = 1.330
  _____

Amount of collimation error   = 0.012 m in 20 m
  _____

  (iii) $\Delta H_{FALSE} < \Delta H_{TRUE}$, therefore direction of line of sight is *down*
  (iv) With instrument at *C* the reading on *A*(4.244) must be raised by (0.012 × 30)/20 = 0.018 m to read 4.262 m

Some methods of adjustment advocate placing the instrument close to the staff at *B* rather than a distance away at *C*. This can result in error when using the reading on *B* and is not suitable for precise levels. The above method is satisfactory for all types of level.

For very precise levels, it may be necessary to account for the effect of curvature and refraction when carrying out the above test. Kukkamaki gives the following equation for it:

$$(c - r) = -1.68 \times 10^{-4} \times D^2$$

where D = distance in metres.

As a distance of 50 m would produce a correction to the staff readings of only −0.42 mm, it can be ignored for all but the most precise work.

2.6.2 Automatic level

There are two tests and adjustments necessary for an automatic level:

(1) To ensure that the line of collimation of the telescope is horizontal, within the limits of the bubble, when the circular bubble is central.
(2) The two-peg test for collimation error.

(1) Circular bubble

Although the circular bubble is relatively insensitive, it nevertheless plays an important part in the efficient functioning of the compensator:

(1) The compensator has a limited working range. If the circular bubble is out of adjustment, thereby resulting in excessive tilt of the line of collimation (and the vertical axis), the compensator may not function efficiently or, as it attempts to compensate, the large swing of the pendulum system may cause it to stick in the telescope tube.
(2) The compensator gives the most accurate results near the centre of its movement, so even if the bubble is in adjustment, it should be carefully and accurately centred.
(3) The plane of the pendulum swing of the freely suspended surfaces should be parallel to the line of sight, otherwise over- and undercompensation may occur. This would result if the circular bubble is in error transversely. Any residual error of adjustment can be eliminated by centring the bubble with the telescope pointing backwards, whilst at the next instrument set-up it is centred with the telescope pointing forward. This alternating process is continued throughout the levelling.
(4) Inclination of the telescope can cause an error in automatic levels which does not occur in tilting levels, known as 'height shift'. Due to the inclination of the telescope the centre of the object lens is displaced vertically above or below the centre of the cross-hair, resulting in very small reading errors, which cannot be tolerated in precise work.

From the above it can be seen that not only must the circular bubble be in adjustment but it should also be accurately centred when in use.

To adjust the bubble, bring it exactly to centre using the footscrews. Now rotate the bubble through 180° about the vertical axis. If the bubble moves off centre, bring it halfway back to centre with the footscrews and then exactly back to the centre using its adjusting screws.

(2) Two-peg test

This is carried out exactly as for the tilting level. However, the line of sight is raised or lowered to its correct reading by moving the cross-hair by means of its adjusting screws.

If the instrument is still unsatisfactory the fault may lie with the compensator, in which case it should be returned to the manufacturer.

2.7 PRINCIPLE OF LEVELLING

The instrument is set up and correctly levelled in order to make the line of sight through the telescope horizontal. If the telescope is turned through 360°, a horizontal plane of sight is swept out. Vertical measurements from this plane, using a graduated levelling staff, enable the relative elevations of ground points to be ascertained. Consider *Figure 2.18* with the instrument set up approximately midway between ground points *A* and *B*. If the reduced level (RL) of point *A* is known and equal to 100.000 m above OD (AOD), then the reading of 3.000 m on a vertically held staff at *A* gives the reduced level of the horizontal line of sight as 103.000 m AOD. This sight onto *A* is termed a backsight (BS) and the reduced level of the line of sight is called the height of the plane of collimation (HPC). Thus:

$$RL_A + BS = HPC$$

The reading of 1.000 m onto the staff at *B* is called a foresight (FS) and shows the ground point *B* to be 1.000 m below HPC; therefore its RL = (103.000 – 1.000) = 102.000 m AOD.

An alternative approach is to subtract the FS from the BS. If the result is positive then the difference is a *rise* from *A* to *B*, and if negative a *fall*, i.e.

$(3.000 – 1.000) = +2.000$ m rise from *A* to *B*;
therefore, $RL_B = 100.000 + 2.000 = 102.000$ m AOD

This then is the basic concept of levelling which is further developed in *Figure 2.19*.

It should be clearly noted that, in practice, the staff readings are taken to three places of decimals, that is to the nearest millimetre. However, in the following description only one place of decimals is used and the numbers kept very simple to prevent arithmetic interfering with an understanding of the concepts outlined.

The field data are entered into a field book which is pre-drawn into rows and columns as shown in *Figure* 2.20.

The field procedure for obtaining elevations at a series of ground points is as follows.

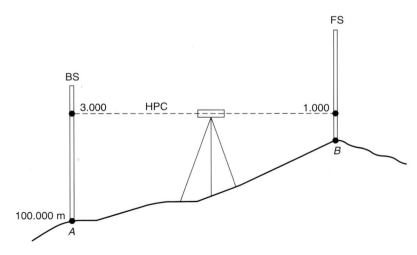

Fig. 2.18 *Basic principle of levelling*

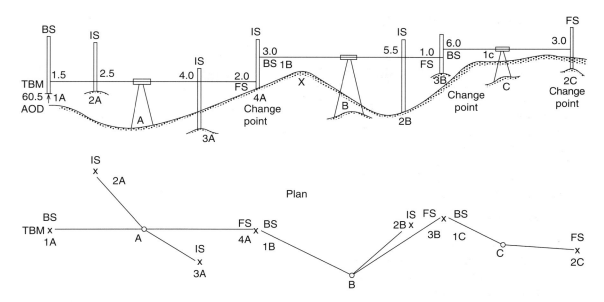

Fig. 2.19

Date 26/07/92   Levels   taken for *Road Alignment*
From *Chn. 2040 at Bridge N°6*   To *Chn 3040 at Bridge N°8*

Back sight	Inter-mediate	Fore sight	Rise	Fall	Reduced level	Distance	Remarks
2·856					35·688	0m	TBM on Bridge Abut. (N°6)
	1·432		1·424		37·112	20m	Chn. 2060m
		3·543		2·111	35·001	30m	C.P. at Chn. 2070m

Date 26/07/92 .... Levels   taken for *Road Alignment*
From *Chn 2040 at Bridge N°6*   To *Chn 3040 at Bridge N°8*

Back sight	Inter-mediate	Fore sight	Collimation or H.P.C.	Reduced level	Distance	Remarks
2·856			38·544	35·688	0m	TBM on Bridge Abut. (N°6)
	1·432			37·112	20m	Chn 2060m
		3·543		35·001	30m	C.P. at Chn 2070m

Fig. 2.20

The instrument is set up at *A* (as in *Figure 2.19*) from which point a horizontal line of sight is possible to the TBM at 1*A*. The first sight to be taken is to the staff held vertically on the TBM and this is called a backsight (BS), the value of which (1.5 m) would be entered in the appropriate column of a levelling book. Sights to points 2*A* and 3*A* where further levels relative to the TBM are required are called intermediate sights (IS) and are again entered in the appropriate column of the levelling book. The final sight from this instrument is set up at 4*A* and is called the foresight (FS). It can be seen from the figure that this is as far as one can go with this sight. If, for instance, the staff had been placed at *X*, it would not have been visible and would have had to be moved down the slope, towards the instrument at *A*, until it was visible. As foresight 4*A* is as far as one can see from *A*, it is also called the change point (CP), signifying a change of instrument position to *B*. To achieve continuity in the levelling the staff must remain at *exactly* the same point 4*A* although it must be turned to face the instrument at *B*. It now becomes the BS for the new instrument set-up and the whole procedure is repeated as before. Thus, one must remember that all levelling commences on a BS and finishes on a FS with as many IS in between as are required; and that CPs are always FS/BS. Also, it must be closed back into a known BM to ascertain the misclosure error.

2.7.1 Reduction of levels

From *Figure 2.19* realizing that the line of sight from the instrument at *A* is truly horizontal, it can be seen that the higher reading of 2.5 at point 2*A* indicates that the point is lower than the TBM by 1.0, giving 2*A* a level therefore of 59.5. This can be written as follows:

$1.5 - 2.5 = -1.0$, indicating a *fall* of 1.0 from 1*A* to 2*A*

Level of 2*A* = 60.5 − 1.0 = 59.5
Similarly between 2*A* and 3*A*, the higher reading on 3*A* shows it is 1.5 below 2*A*, thus:

$2.5 - 4.0 = -1.5$ (fall from 2*A* to 3*A*)

Level of 3*A* = level of 2*A* − 1.5 = 58.0

Finally the *lower* reading on 4*A* shows it to be *higher* than 3*A* by 2.0, thus:

$4.0 - 2.0 = + 2.0$, indicating a *rise* from 3*A* to 4*A*

Level of 4*A* = level of 3*A* + 2.0 = 60.0

Now, knowing the *reduced level* (RL) of 4*A*, i.e. 60.0, the process can be repeated for the new instrument position at *B*. This method of reduction is called the *rise-and-fall* (*R-and-F*) *method*.

2.7.2 Methods of booking

(1) Rise-and-fall

The following extract of booking is largely self-explanatory. Students should note:

(a) Each reading is booked on a separate line except for the BS and FS at change points. The BS is booked on the same line as the FS because it refers to the same point. As each line refers to a specific point it should be noted in the remarks column.
(b) Each reading is subtracted from the previous one, i.e. 2*A* from 1*A*, then 3*A* from 2*A*, 4*A* from 3*A* and stop; the procedure recommencing for the next instrument station, 2*B* from 1*B* and so on.

BS	IS	FS	Rise	Fall	RL	Distance	Remarks	
1.5					60.5	0	TBM (60.5)	1A
	2.5			1.0	59.5	30		2A
	4.0			1.5	58.0	50		3A
3.0		2.0	2.0		60.0	70	CP	4A (1B)
	5.5			2.5	57.5	95		2B
6.0		1.0	4.5		62.0	120	CP	3B (1C)
		3.0	3.0		65.0	160	TBM (65.1)	2C
10.5		6.0	9.5	5.0	65.0		Checks	
6.0				5.0	60.5		Misclosure	0.1
4.5			4.5		4.5		*Correct*	

(c) Three very important checks must be applied to the above reductions, namely:

The sum of BS – the sum of FS = sum of rises – sum of falls
= last reduced level – first reduced level

These checks are shown in the above table. It should be emphasized that they are nothing more than checks on the arithmetic of reducing the levelling results, they are in no way indicative of the accuracy of fieldwork.

(d) It follows from the above that the first two checks should be carried out and verified before working out the reduced levels (RL).

(e) Closing error = 0.1, and can be assessed only by connecting the levelling into a BM of known and proved value or connecting back into the starting BM.

(2) Height of collimation

This is the name given to an alternative method of booking. The reduced levels are found simply by subtracting the staff readings from the reduced level of the line of sight (plane of collimation). In *Figure 2.19*, for instance, the *height of the plane of collimation* (HPC) at *A* is obviously (60.5 + 1.5) = 62.0; now 2*A* is 2.5 below this plane, so its level must be (62.0 – 2.5) = 59.5; similarly for 3*A* and 4*A* to give 58.0 and 60.0 respectively. Now the procedure is repeated for *B*.

The tabulated form shows how simple this process is:

BS	IS	FS	HPC	RL	Remarks	
1.5			62.0	60.5	TBM (60.5)	1A
	2.5			59.5		2A
	4.0			58.0		3A
3.0		2.0	63.0	60.0	Change pt	4A (1B)
	5.5			57.5		2B
6.0		1.0	68.0	62.0	Change pt	3B (1C)
		3.0		65.0	TBM (65.1)	2C
10.5	12.0	6.0		65.0	Checks	
6.0				60.5	Misclosure	0.1
4.5				4.5	*Correct*	

Thus it can be seen that:

(a) BS is added to RL to give HPC, i.e., 1.5 + 60.5 = 62.0.
(b) Remaining staff readings are *subtracted* from HPC to give the RL.
(c) Procedure repeated for next instrument set-up at *B*, i.e., 3.0 + 60.0 = 63.0.
(d) Two checks same as R-and-F method, i.e:

   sum of BS − sum of FS = last RL − first RL.

(e) The above two checks are not complete; for instance, if when taking 2.5 from 62.0 to get RL of 59.5, one wrote it as 69.5, this error of 10 would remain undetected. Thus the *intermediate* sights are *not* checked by those procedures in (d) above and the following cumbersome check must be carried out:

   sum of all the RL except the first = (sum of each HPC multiplied by the number of IS or FS taken from it) − (sum of IS and FS).

   e.g. 362.0 = [(62.0 × 3.0) + (63.0 × 2.0) + (68.0 × 1.0)] − [12.0 + 6.0] = 362.0

2.7.3 Inverted sights

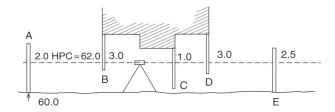

Fig. 2.21   *Inverted sights*

Figure 2.21 shows inverted sights at *B*, *C* and *D* to the underside of a structure. It is obvious from the drawing that the levels of these points are obtained by simply adding the staff readings to the HPC to give *B* = 65.0, *C* = 63.0 and *D* = 65.0; *E* is obtained in the usual way and equals 59.5. However, the problem of inverted sights is completely eliminated if one simply treats them as *negative* quantities and proceeds in the usual way:

BS	IS	FS	Rise	Fall	HPC	RL	Remarks	
2.0					62.0	60.0	TBM	A
	−3.0		5.0			65.0		B
	−1.0			2.0		63.0		C
	−3.0		2.0			65.0		D
		2.5		5.5		59.5	Misclosure	E (59.55)
2.0	−7.0	2.5	7.0	7.5		60.0	Checks	
		2.0		7.0		59.5	Misclosure	0.05
		0.5		0.5		0.5	*Correct*	

R-and-F method	HPC method
2.0–(–3.0) = +5.0 = Rise	62.0–(–3.0) = 65.0
–3.0–(–1.0) = –2.0 = Fall	62.0–(–1.0) = 63.0
–1.0–(–3.0) = +2.0 = Rise	62.0–(–3.0) = 65.0
–3.0– 2.5 = –5.5 = Fall	62.0–(+2.5) = 59.5

In the checks, inverted sights are treated as negative quantities; for example check for IS in HPC method gives

$$252.5 = (62.0 \times 4.0) - (-7.0 + 2.5)$$
$$= (248.0) - (-4.5) = 248.0 + 4.5 = 252.5$$

2.7.4 Comparison of methods

The rise-and-fall method of booking is recommended as it affords a complete arithmetical check on all the observations. Although the HPC method appears superior where there are a lot of intermediate sights, it must be remembered that there is no simple straightforward check on their reduction.

The HPC method is useful when setting out levels on site. For instance, assume that a construction level, for setting formwork, of 20 m AOD is required. A BS to an adjacent TBM results in an HPC of 20.834 m; a staff reading of 0.834 would then fix the bottom of the staff at the required level.

2.8 SOURCES OF ERROR

Any and all measurement processes will contain errors. In the case of levelling, these errors will be (1) instrumental, (2) observational and (3) natural.

2.8.1 Instrumental errors

(1) The main source of instrumental error is residual collimation error. As already indicated, keeping the horizontal lengths of the backsights and foresights at each instrument position equal will cancel this error. Where the observational distances are unequal, the error will be proportional to the difference in distances.

The easiest approach to equalizing the sight distances is to pace from backsight to instrument and then set up the foresight change point the same number of paces away from the instrument.
(2) Parallax error has already been described.
(3) Staff graduation errors may result from wear and tear or repairs and should be checked against a steel tape. Zero error of the staff, caused by excessive wear of the base, will cancel out on backsight and foresight differences. However, if two staffs are used, errors will result unless calibration corrections are applied.
(4) In the case of the tripod, loose wing nuts will cause twisting and movement of the tripod head. Overtight wing nuts make it difficult to open out the tripod correctly. Loose tripod shoes will also result in unstable set-ups.

2.8.2 *Observational errors*

(1) Since the basic concept of levelling involves vertical measurements relative to a horizontal plane, careful staff holding to ensure its verticality is fundamentally important.

Rocking the staff back and forth in the direction of the line of sight and accepting the minimum reading as the truly vertical one is frequently recommended. However, as shown in *Figure 2.22*, this concept is incorrect when using a flat-bottomed staff on flat ground, due to the fact that it is not being tilted about its face. Thus it is preferable to use a staff bubble, which should be frequently checked with the aid of a plumb-bob.

(2) Errors in reading the staff, particularly when using a tilting level which gives an inverted image. These errors may result from inexperience, poor observation conditions or overlong sights. Limit the length of sight to about 25–30 m, thereby ensuring clearly defined graduations.

(3) Ensure that the staff is correctly extended or assembled. In the case of extending staffs, listen for the click of the spring joint and check the face of the staff to ensure continuity of readings. This also applies to the jointed staff.

(4) Moving the staff off the CP position, particularly when turning it to face the new instrument position. Always use a well-defined and stable position for CPs. Levelling plates (*Figure 2.23*) should be used on soft ground

(5) Similarly with the tripod. To avoid tripod settlement, which may alter the height of collimation between sights or tilt the line of sight, set up on firm ground, with the tripod feet firmly thrust well into the ground. Even on pavements, locate the tripod shoes in existing cracks or joins. In precise levelling, the use of two staffs helps to reduce this effect.

Beginners should also refrain from touching or leaning on the tripod during observation.

(6) Booking errors can, of course, ruin good field work. Neat, clear, correct booking of field data is essential in any surveying operation. Typical booking errors in levelling are entering the values in the wrong columns or on the wrong lines, transposing figures such as 3.538 and 3.583 and making arithmetical errors in the reduction process. Very often, the use of pocket calculators simply enables the booker to make the errors quicker.

To avoid this error source, use neat, legible figures; read the booked value back to the observer and have him check the staff reading again; reduce the data as it is recorded.

(7) When using a tilting level remember to level the tubular bubble with the tilting screw prior to each new staff reading. With the automatic level, carefully centre the circular bubble and make sure the compensator is not sticking.

Residual compensator errors are counteracted by centring the circular bubble with the instrument pointing backwards at the first instrument set-up and forward at the next. This procedure is continued throughout the levelling.

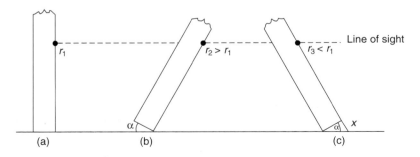

Fig. 2.22   *Showing staff readings* r_1, r_2, r_3

Fig. 2.23 *Levelling plate*

2.8.3 Natural errors

(1) Curvature and refraction have already been dealt with. Their effects are minimized by equal observation distances to backsight and foresight at each set-up and readings not less then 0.5 m above the ground.
(2) Wind can result in unsteady staff holding and instrument vibration. Precise levelling is impossible in strong winds. In tertiary levelling keep the staff to its shortest length and use a wind break to shelter the instrument.
(3) Heat shimmer can make staff reading difficult if not impossible and may result in delaying the work to an overcast day. In hot sunny climes, carry out the work early in the morning or evening.

Careful consideration of the above error sources, combined with regularly calibrated equipment, will ensure the best possible results but will never preclude random errors of observation.

2.9 CLOSURE TOLERANCES

It is important to realize that the amount of misclosure in levelling can only be assessed by:

(1) Connecting the levelling back to the BM from which it started, or
(2) Connecting into another BM of known and proved value.

When the misclosure is assessed, one must then decide if it is acceptable or not.

In many cases the engineer may make the decision based on his knowledge of the project and the tolerances required.

Alternatively the permissible criteria may be based on the distance levelled or the number of set-ups involved.

A common criterion used to assess the misclosure (*E*) is:

$$E = m(K)^{\frac{1}{2}}$$
(2.5)

where *K* = distance levelled in kilometres, *m* = a constant in millimetres, and *E* = the allowable misclosure in millimetres.

The value of *m* may vary from 2 mm for precise levelling to 12 mm or more for third-order engineering levelling.

In many cases in engineering, the distance involved is quite short but the number of set-ups quite high, in which case the following criterion may be used:

$$E = m(n)^{\frac{1}{2}}$$
(2.6)

where *n* = the number of set-ups, and *m* = a constant in millimetres.

As this criterion would tend to be used only for construction levelling, the value for *m* may be a matter of professional judgement. A value frequently used is ±5 mm.

2.10 ERROR DISTRIBUTION

In the case of a levelling circuit, a simple method of distribution is to allocate the error in proportion to the distance levelled. For instance, consider a levelling circuit commencing from a BM at *A*, to establish other BMs at *B*, *C*, *D* and *E* (*Figure 2.24*).

The observed value for the BM at *A*, is 20.018 m compared with its known value of 20.000 m, so the misclosure is 0.018 m. The distance levelled is 5.7 km. Considering the purpose of the work, the terrain and observational conditions, it is decided to adopt a value for *m* of 12 mm. Hence the acceptable misclosure is 12 (5.7)$^{1/2}$ = 29 mm, so the levelling is acceptable.

The difference in heights is corrected by (0.018/5.7) × distance involved. Therefore correction to *AB* = −0.005 m, to *BC* = −0.002 m, to *CD* = −0.003 m, to *DE* = −0.006 m and to *EA* = −0.002 m. The values of the BMs will then be *B* = 28.561 m, *C* = 35.003 m, *D* = 30.640 m, *E* = 22.829 m and *A* = 20.0000 m.

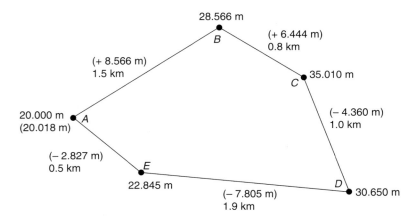

Fig. 2.24

In many instances, a closing loop with known distances is not the method used and each reduced level is adjusted in proportion to the cumulative number of set-ups to that point from the start. Consider the table below:

BS	IS	FS	Rise	Fall	R.L.	Adj.	Final R. L.	Remarks
1.361					20.842		20.842	TBM 'A'
	2.844			1.483	19.359	−0.002	19.357	
	2.018		0.826		20.185	−0.002	20.183	
0.855		3.015		0.997	19.188	−0.002	19.186	C.P.
	0.611			0.244	19.432	−0.004	19.428	
2.741		1.805		1.194	18.238	−0.004	18.234	C.P.
2.855		1.711	1.030		19.268	−0.006	19.262	C.P.
	1.362		1.493		20.761	−0.008	20.753	
	2.111			0.749	20.012	−0.008	20.004	
	0.856		1.255		21.267	−0.008	21.259	
		2.015		1.159	20.108	−0.008	20.100	TBM 'B' (20.100)
7.812		8.546	4.848	5.582	20.842			
		7.812		4.848	20.108			
		0.734		0.734	0.734			Arith. check

(1) There are four set-ups, and therefore $E = 5(4)^{\frac{1}{2}} = 0.010$ m. As the misclosure is only 0.008 m, the levelling is acceptable.
(2) The correction per set-up is $(0.008/4) = -0.002$ m and is cumulative as shown in the table.

2.11 LEVELLING APPLICATIONS

Of all the surveying operations used in construction, levelling is the most common. Practically every aspect of a construction project requires some application of the levelling process. The more general are as follows.

2.11.1 Sectional levelling

This type of levelling is used to produce ground profiles for use in the design of roads, railways and pipelines.

In the case of such projects, the route centre-line is set out using pegs at 10-m, 20-m or 30-m intervals. Levels are then taken at these peg positions and at critical points such as sudden changes in the ground profiles, road crossings, ditches, bridges, culverts, etc. The resultant plot of these elevations is called a longitudinal section. When plotting, the vertical scale is exaggerated compared with the horizontal, usually in the ratio of 10 : 1. The longitudinal section is then used in the vertical design process to produce formation levels for the proposed route design (*Figure 2.25*).

Whilst the above process produces information along a centre-line only, cross-sectional levelling extends that information at 90° to the centre-line for 20–30 m each side. At each centre-line peg the levels are taken to all points of interest on either side. Where the ground is featureless, levels at

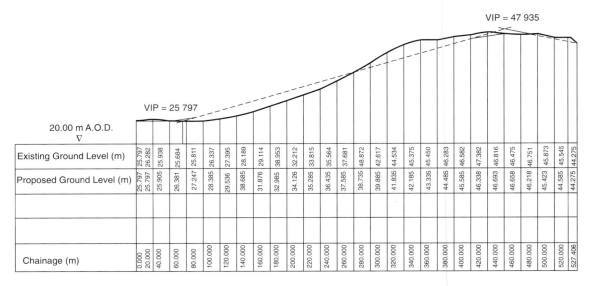

VIP = 47 935

VIP = 25 797

20.00 m A.O.D.

Existing Ground Level (m)	25.797 26.282 25.938 25.684 25.811 26.337 27.395 28.189 29.114 38.953 32.212 33.815 35.564 37.681 48.872 42.617 44.534 45.375 45.450 46.283 46.582 47.382 46.816 46.475 46.751 45.873 45.545 44.275	
Proposed Ground Level (m)	25.797 25.797 25.905 26.381 27.247 28.385 29.536 38.685 31.876 32.985 34.126 35.285 36.435 37.585 38.735 39.885 41.835 42.185 43.335 44.485 45.585 46.338 46.693 46.658 46.218 45.423 44.585 44.275	
Chainage (m)	0.000 20.000 40.000 60.000 80.000 100.000 120.000 140.000 160.000 180.000 200.000 220.000 240.000 260.000 280.000 300.000 320.000 340.000 360.000 380.000 400.000 420.000 440.000 460.000 480.000 500.000 520.000 527.406	

Fig. 2.25 *Longitudinal section of proposed route*

5-m intervals or less are taken. In this way a ground profile at right angles to the centre-line is obtained. When the design template showing the road details and side slopes is plotted at formation level, a cross-sectional area is produced, which can later be used to compute volumes of earthwork. When plotting cross-sections the vertical and horizontal scales are the same, to permit easy scaling of the area and side slopes (*Figure 2.26*).

From the above it can be seen that sectional levelling also requires the measurement of horizontal distance between the points whose elevations are obtained. As the process involves the observation of many points, it is imperative to connect into existing BMs at regular intervals. In most cases of route construction, one of the earliest tasks is to establish BMs at 100-m intervals throughout the area of interest.

Levelling which does not require the measurement of distance, such as establishing BMs at known positions, is sometimes called 'fly levelling'.

2.11.2 Contouring

A contour is a horizontal curve connecting points of equal elevation. They graphically represent, in a two-dimensional format on a plan or map, the shape or morphology of the terrain. The vertical distance between contour lines is called the contour interval. Depending on the accuracy required, they may be plotted at 0.1-m to 0.5-m intervals in flat terrain and 1-m to 10-m intervals in undulating terrain. The interval chosen depends on:

(1) The type of project involved; for instance, contouring an airstrip requires an extremely small contour interval.
(2) The type of terrain, flat or undulating
(3) The cost, for the smaller the interval the greater the amount of field data required, resulting in greater expense.

Contours are generally well understood so only a few of their most important properties will be outlined here.

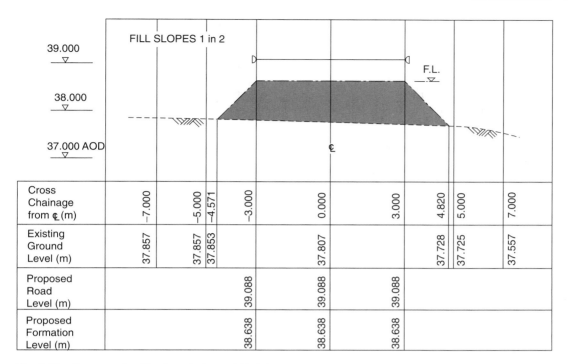

Cross Chainage from ¢ (m)	−7.000	−5.000	−4.571	−3.000	0.000	3.000	4.820	5.000	7.000
Existing Ground Level (m)	37.857	37.857	37.853		37.807		37.728	37.725	37.557
Proposed Road Level (m)					39.088	39.088	39.088		
Proposed Formation Level (m)					38.638	38.638	38.638		

Fig. 2.26 *Cross-section No. 3 at chainage 360.000 m*

(1) Contours are perpendicular to the direction of maximum slope.
(2) The horizontal separation between contour lines indicates the steepness of the ground. Close spacing defines steep slopes, wide spacing gentle slopes.
(3) Highly irregular contours define rugged, often mountainous terrain.
(4) Concentric closed contours represent hills or hollows, depending on the increase or decrease in elevation.
(5) The slope between contour lines is assumed to be regular.
(6) Contour lines crossing a stream form V's pointing upstream.
(7) The edge of a body of water forms a contour line.

Contours are used by engineers to:

(1) Construct longitudinal sections and cross-sections for initial investigation.
(2) Compute volumes.
(3) Construct route lines of constant gradient.
(4) Delineate the limits of constructed dams, road, railways, tunnels, etc.
(5) Delineate and measure drainage areas.

If the ground is reasonably flat, the optical level can be used for contouring using either the *direct* or *indirect* methods. In undulating areas it is more economical to use optical or electronic methods, as outlined later.

(1) Direct contouring

In this method the actual contour is pegged out on the ground and its planimetric position located.

A backsight is taken to an appropriate BM and the HPC of the instrument is obtained, say 34.800 m AOD. A staff reading of 0.800 m would then place the foot of the staff at the 34-m contour level. The staff is then moved throughout the terrain area, with its position pegged at every 0.800-m reading. In this way the 34-m contour is located. Similarly a staff reading of 1.800 m gives the 33-m contour and so on. The planimetric position of the contour needs to be located using an appropriate survey technique.

This method, although quite accurate, is tedious and uneconomical and could never be used over a large area. It is ideal, however, in certain construction projects which require excavation to a specific single contour line.

(2) Indirect contouring

This technique requires the establishment, over the site, of a grid of intersecting evenly spaced lines. The boundary of the grid is set out by theodolite and steel tape. The grid spacing will depend upon the rugosity of the ground and the purpose for which the data are required. All the points of intersection throughout the grid may be pegged or shown by means of paint from a spray canister. Alternatively ranging rods at the grid intervals around the periphery would permit the staff holder to align himself with appropriate pairs and thus fix the grid intersection point, for example, alignment with rods *B-B* and 2-2 fixes point *B*2 (*Figure 2.27*). When the RLs of all the intersection points are obtained, the contours are located by linear interpolation between the levels, on the assumption of a uniform ground slope between each value. The interpolation may be done arithmetically, using a pocket calculator, or graphically.

Consider grid points *B*2 and *B*3 with reduced levels of 30.20 m and 34.60 m respectively and a horizontal grid interval of 20 m (*Figure 2.28*).

Horizontal distance of the 31-m contour from $B2 = x_1$

where $(20/4.40) = 4.545 \text{ m} = K$

and $x_1 = K \times 0.80 \text{ m} = 3.64 \text{ m}$

Similarly for the 32-m contour

$x_2 = K \times 1.80 \text{ m} = 8.18 \text{ m}$

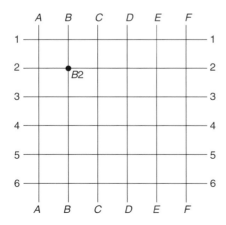

Fig. 2.27

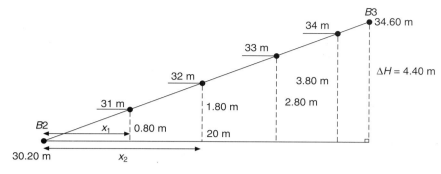

Fig. 2.28

and so on, where (20/4.40) is a constant K, multiplied each time by the difference in height from the reduced level of $B2$ to the required contour value. For the graphical interpolation, a sheet of transparent paper (*Figure 2.29*) with equally spaced horizontal lines is used. The paper is placed over the two points and rotated until $B2$ obtains a value of 30.20 m and $B3$ a value of 34.60 m. Any appropriate scale can be used for the line separation. As shown, the 31-, 32-, 33- and 34-m contour positions can now be pricked through onto the plan.

This procedure is carried out on other lines and the equal contour points joined up to form the contours required.

2.12 RECIPROCAL LEVELLING

When obtaining the relative levels of two points on opposite sides of a wide gap, it is impossible to keep the length of sights short and equal. The longer sight will be more diversely affected by collimation error and Earth curvature and refraction than the shorter one. In order to minimize these effects, the method of reciprocal levelling is used, as illustrated in *Figure 2.30*.

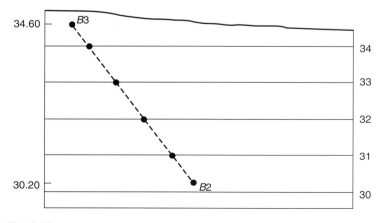

Fig. 2.29

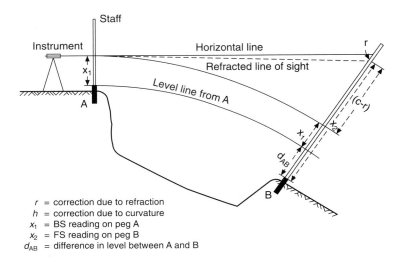

r = correction due to refraction
h = correction due to curvature
x_1 = BS reading on peg A
x_2 = FS reading on peg B
d_{AB} = difference in level between A and B

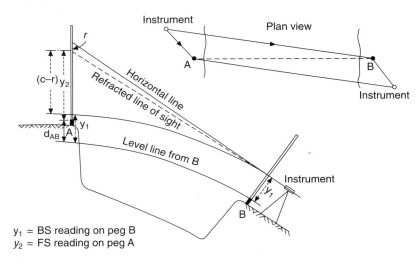

y_1 = BS reading on peg B
y_2 = FS reading on peg A

Fig. 2.30

With the instrument near *A*, backsighting onto *A* and foresighting onto *B*, the difference in level or elevation between *A* and *B* is:

$$\Delta H_{AB} = x_2 - x_1 - (c - r)$$

where: x_1 = BS on A

x_2 = FS on B

$(c - r)$ = the combined effect of curvature and refraction (with collimation error intrinsically built into r)

Similarly with the instrument moved to near *B*:

$$\Delta H_{AB} = y_1 - y_2 + (c - r)$$

where y_1 = BS on *B*

              y_2 = FS on *A*

then $2\Delta H_{AB} = (x_2 - x_1) + (y_1 - y_2)$

and $\Delta H_{AB} = \frac{1}{2}[(x_2 - x_1) + (y_1 - y_2)]$ (2.7)

This proves that the mean of the difference in level obtained with the instrument near *A* and then with the instrument near *B* is free from the errors due to curvature, refraction and collimation error. Normal errors of observation will still be present, however.

Equation (2.7) assumes the value of refraction equal in both cases. However, as refraction is a function of temperature and pressure, both may change during the time taken to transport the instrument from side *A* to side *B*, thus changing the value of refraction. To preclude this it is advisable to use two levels and take simultaneous reciprocal observations. However, this procedure creates the problem of each instrument having a different residual collimation error. They should therefore be interchanged and the whole procedure repeated. The mean of all the values obtained will then give the most probable value for the difference in level between *A* and *B*.

2.13 PRECISE LEVELLING

Precise levelling may be required in certain instances in construction such as in deformation monitoring, the provision of precise height control for large engineering projects such as long-span bridges, dams and hydroelectric schemes and in mining subsidence measurements.

2.13.1 Precise invar staff

The precise levelling staff has its graduations precisely marked (and checked by laser interferometry) on invar strips, which are attached to wooden or aluminium frames. The strip is rigidly fixed to the base of the staff and held in position by a spring-loaded tensioning device at the top. This arrangement provides support for the invar strip without restraining it in any way.

Each side of the supporting frame has its graduations offset to provide a check on the readings. Although the readings are offset, the difference in the BS and FS readings for each side should be equal to within ±1 mm.

For the most precise work, two staffs are used, in which case they should be carefully matched in every detail. A circular bubble built into the staff is essential to ensure verticality during observation. The staff should be supported by means of steadying poles or handles.

(1) The staff should have its circular bubble tested at frequent intervals using a plum-bob.
(2) Warping of the staff can be detected by stretching a fine wire from end to end.
(3) Graduation and zero error can be counteracted by regular calibration.
(4) For the highest accuracy the temperature of the strip should be measured by a field thermometer, in order to apply scale corrections.

2.13.2 Instruments

The instruments used should have precise levels of the highest accuracy as defined by the manufacturer. They should provide high-quality resolution with high magnification ($\times 40$) and extremely accurate horizontality of the line of sight. This latter facility would be provided by a highly sensitive tubular bubble with a large radius of curvature resulting in a greater horizontal bubble movement per angle of tilt. In the case of the automatic level a highly refined compensator would be necessary.

In either case a parallel plate micrometer, fitted in front of the object lens, would be used to obtain submillimetre resolution on the staff.

2.13.3 Parallel plate micrometer

For precise levelling, the estimation of 1 mm is not *sufficiently* accurate. A parallel plate glass micrometer in front of the object lens enables readings to be made direct to 0.1 mm, and estimated to 0.01 mm. The principle of the attachment is seen from *Figure 2.31*. Had the parallel plate been vertical the line of sight would have passed through without deviation and the reading would have been 1.026 m, the final figure being estimated. However, by manipulating the micrometer the parallel plate is tilted until the line of sight is displaced to the nearest indicated reading, which is 1.02 m. The amount of displacement *s* is measured on the micrometer and added to the exact reading to give 1.02647 m, only the last decimal place being estimated.

It can be seen from the figure that the plate could *equally* have moved in the opposite direction, displacing the line of sight up. To avoid the difficulty of whether to add or subtract *s*, the micrometer is always set to read zero before each sight. This will tilt the plate to its maximum position opposite to that shown in *Figure 2.31*, and so displace the line of sight upwards. This will not affect the levelling provided that it is done for every sight. In this position the micrometer screw will move only from zero to ten, and the line of sight is always displaced down so *s* is always added.

Parallel plate micrometers are also manufactured for use with 5-mm graduations.

2.13.4 Field procedure

In precise levelling of the highest accuracy, intermediate sights are avoided. The BS and FS observation distances are made equal in length to about 0.100 m and limited to 25 m. Very often

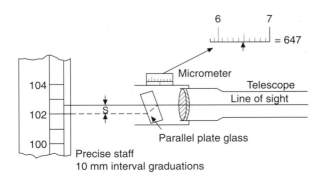

Fig. 2.31

these points are established in advance of the actual field work, with the instrument position clearly indicated. Using two double scale rods the sequence of observation would be:

(1) BS left-hand scale on staff *A*
(2) FS left-hand scale on staff *B*
(3) FS right-hand scale on staff *B*
(4) BS right-hand scale on staff *A*

Then $1 - 2 = \Delta H_1$ and $4 - 3 = \Delta H_2$; if these differences agree within the tolerances specified (± 1 mm), the mean is accepted.

Staff *A* is now leapfrogged to the next position and the above procedure repeated starting with staff *A* again (*Figure 2.32*)

Note also the procedure already outlined for levelling the circular bubble on automatic levels. This will happen as a matter of course if the telesocpe is aimed at staff *A* each time when centring the circular bubble.

The staff should never be sighted lower than 0.5 m above the ground for reasons already outlined.

The instrument should be shielded from the sun's heat to prevent differential expansion of its glass and metal parts.

Last, but by no means least, the levelling points, which will all be CPs and possible TBMs, must be constructed in such a way as to ensure their complete stability throughout the duration of their use. They should also be constructed so as to form rounded supports for the staff, thereby providing excellent CPs.

A useful adjunct to the above procedures is to use an electronic data logger, suitably programmed to compute the data as they are recorded, thereby providing useful checks at each instrument station.

A well-designed and rigorously observed levelling network, with interrelated and interdependent cross-checks to give extra degrees of freedom, would produce excellent results after a least squares estimation.

Typical tolerance limits for precise levelling vary from $\pm 2(K)^{\frac{1}{2}}$ mm to $\pm 4(K)^{\frac{1}{2}}$ mm, where K is the distance levelled in kilometres.

2.14 DIGITAL LEVELLING

As differential levelling is an extremely simple concept, much of the research and technological development has been in the measurement of distance and angle. Recently, however, the instrument manufacturer Wild has produced the world's first digital level, called the Wild NA 2000. This instrument uses electronic image processing to evaluate the staff reading. The observer is in effect replaced by a detector diode array, which derives a signal pattern from a bar-code-type levelling

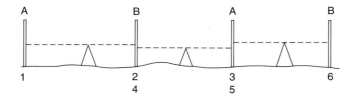

Fig. 2.32

staff. A correlation procedure within the instrument translates the pattern into the vertical staff reading and the horizontal distance of the instrument from the staff. Staff-reading errors by the observer are thus eliminated.

The basic field data are automatically stored by the instrument on its recording module, thus further eliminating booking errors (*Figure 2.33*).

2.14.1 Instrumentation

The design of both the staff and instrument are such that it can be used in the conventional way as well as digitally.

(1) The levelling staff

The staff is made from a glass-fibre-strengthened synthetic material, which has a coefficient of expansion of less than 10 ppm. It consists of three separate sections, each 1.35 m long, which slot together to give a maximum length of 4.05 m. On one side of the staff is a binary bar code for electronic measurement, and on the other side conventional graduations in metres. The black and white binary code comprises 2000 elements over the staff length with the basic element only 2 mm wide. As the correlation method is used to evaluate the image, the elements are arranged in a

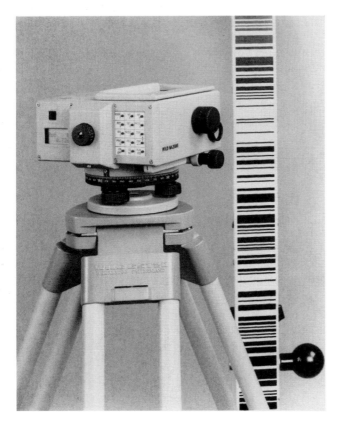

Fig. 2.33   *Wild NA 2000 digital level and staff*

pseudo-stochastic code. The code pattern is such that the correlation procedure can be used over a range from 1.8 m to 100 m.

The standard deviation of a single electronic staff reading is claimed to be 0.3 mm at a sighting distance of 50 m and 0.5 mm at 100 m.

The staff is fitted with a circular bubble and two holding knobs, as staff verticality is still very important. For more precise work a special lightweight tripod can be attached.

(2) The digital level (Figures 2.34 and 2.35)

The WILD NA 2000 digital level has the same optical and mechanical components as a normal automatic level. However, for the purpose of electronic staff reading a beam splitter is incorporated which transfers the bar code image to a detector diode array. The light, reflected from the white elements only of the bar code, is divided into infra-red and visible light components by the beam splitter. The visible light passes on to the observer, the infra-red to the diode array. The angular aperture of the instrument is 2°, resulting in 70 mm of the staff being imaged at a range of 1.80 m and 3.5 m at a range of 100 m. The bar code image received is converted into an analogous video signal, which is then compared with a stored reference code. The correlation procedure then obtains the height relationship by displacement of the codes, whilst the distance from instrument to staff is dependent on the image scale of the code.

The data processing is carried out on a single-chip microprocessor supported by a gate array. The evaluated data are then imaged on a two-line matrix display.

The measurement process is initiated by a very light touch on a key situated next to the focusing knob. A 15-position keypad on the eyepiece face of the instrument permits the entry of further numerical data and pre-programmed commands. The data can be stored in the WILD REC module. Alternatively, the instrument has a GS1 interface, which permits external control, data transfer and power supply.

2.14.2 Measuring procedure (Figure 2.36)

There are two external stages to the measuring procedure:

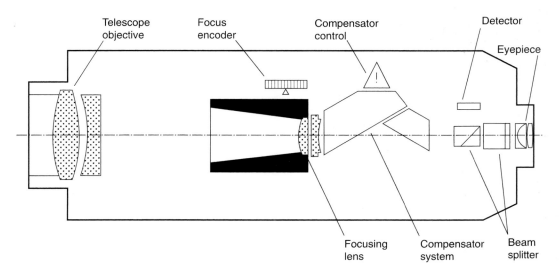

Fig. 2.34 *Optomechanical design of the WILD NA 2000*

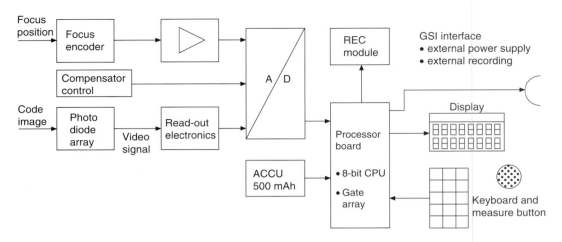

Fig. 2.35   *Electronics block diagram*

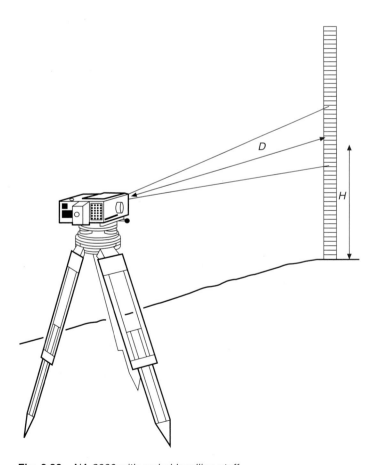

Fig. 2.36   *NA 2000 with coded levelling staff*

(1) Pointing and focusing on the staff.
(2) Triggering the digital measurement.

These are followed by two internal stages:

(1) Coarse correlation.
(2) Fine correlation.

The whole process takes about four seconds.

Triggering the measurement determines the focus position, from which the distance to the staff is measured, and initiates monitoring of the compensator.

The coarse correlation approximately determines the target height and the image scale. The process takes about one second. In one further second, the fine correlation with the aid of calibration constants produces the final staff reading and observation distance.

The results may be further processed in accordance with the accessed program and operating mode, displayed and recorded. The programs incorporated in the instrument are:

(1) MEASURE ONLY – Staff reading and horizontal distance.
(2) START LEVELLING – Commence line levelling.
(3) CONTINUE LEVELLING – Line levelling including intermediate sights. Automatic reduction of data. Setting out of levels.
(4) CHECK AND ADJUST – Facilitates the calibration of the instrument (two-peg test).
(5) ERASE DATA – Erases contents of REC module.
(6) INVERT – Permits use of inverted sights.
(7) SET – Enables the parameters of the instrument to be set and is similar to the initializing procedures used when setting up electronic theodolites.

2.14.3 Factors affecting the measuring procedure

Every operation in a measurement procedure is a possible error source and as such requires careful consideration in order to assess the effect on the final result.

(1) Pointing and focusing

It can be shown that at a range of 2 m only 0.3 mm of the staff width need be imaged and at 100 m only 14 mm. As the bar code is 50 mm wide, positioning the staff to face the instrument is not critical and results can be obtained even when the staff is at 45° to the line of sight.

The precision of the height measurement is independent of sharpness of image; however, a clear, sharply focused image reduces the time required for the measurement.

(2) Vibrations and heat shimmer

Vibration of the compensator caused by wind, traffic, etc. has a similar effect on the bar code image as that of heat shimmer. However, as digital levelling does not require a single reading, but instead is dependent on a section of the code, the effects of shimmer and vibration are not critical.

Similarly, scale errors on the staff are averaged.

(3) Illumination

As the method relies on reflected light from the white intervals of the bar code, illumination of the staff is important. During the day, this illumination will be affected by cloud, sun, twilight and the effects of shadows. All these variations are catered for by the instrument and are indicated by an increase in the measuring time as illumination decreases.

If used in artificial light, its spectral distribution must be comparable with daylight.

(4) Staff coverage

In some conditions part of the bar code section being interrogated by the instrument may be obscured. A minimum of 30 code elements are necessary to determine height and distance, requiring at least 70 mm of the staff section to be available. This means that for ranges greater than 5 m up to 30% of the staff section may be obscured. Below 5 m, all the section is required.

2.14.4 Operating features

(1) Resolution of the measuring system is 0.1 mm for height and 10 mm for distance.
(2) Range is from 1.8 m to 100 m.
(3) Standard deviation for 1-km double-run levelling at ranges below 50 m:

> Digital levelling ±1.5 mm.
> Optical levelling ±2.0 mm.

(4) Standard deviation of a single electronic staff reading at ranges of:

> 50 m = ±0.3 mm
> 100 m = ±0.5 mm

(5) Standard deviation of distance measurement:

> at   50 m = ±20 mm
> at 100 m = ±50 mm

(6) Duration of internal battery = 8 h.
(7) Weight including battery = 2.5 kg.

2.14.5 Advantages of digital levelling

(1) Fatigue-free observation, as visual staff reading by the observer is excluded.
(2) Easy to read, digital display of results, with the last digit selectable 1 mm or 0.1 mm.
(3) Measurement of consistent precision and reliability.
(4) Automatic data storage eliminates booking and its associated errors.
(5) Automatic reduction of data to produce ground levels, thereby eliminating arithmetical errors.
(6) User-friendly menus.
(7) Fast, economic surveys. As much as 50% saving in time.
(8) Increase in range up to 100 m.
(9) On-line link to computer, enables the computation and plotting of longitudinal sections and cross-sections in a very short time.
(10) Can be used in all the situations in which a conventional level is used.
(11) Can be used as a conventional level if necessary.

Worked examples

Example 2.2 The positions of the pegs which need to be set out for the construction of a sloping concrete slab are shown in the diagram. Because of site obstructions the tilting level which is used

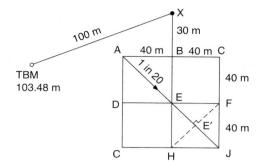

to set the pegs at their correct levels can only be set up at station *X* which is 100 m from the TBM. The reduced level of peg *A* is to be 100 m and the slab is to have a uniform diagonal slope from *A* towards *J* of 1 in 20 downwards.

To ensure accuracy in setting out the levels it was decided to adjust the instrument before using it, but it was found that the correct adjusting tools were missing from the instrument case. A test was therefore carried out to determine the magnitude of any collimation error that may have been present in the level, and this error was found to be 0.04 m per 100 m downwards.

Assuming that the backsight reading from station *X* to a staff held on the TBM was 1.46 m, determine to the nearest 0.01 m the staff readings which should be obtained on the pegs at *A*, *F* and *H*, in order that they may be set to correct levels.

Describe fully the procedure that should be adopted in the determination of the collimation error of the tilting level.
(ICE)

The simplest approach to this question is to work out the true readings at A, *F* and *H* and then adjust them for collimation error. Allowing for collimation error the true reading on TBM = 1.46 + 0.04 = 1.50 m.

HPC = 103.48 + 1.50 = 104.98 m

True reading on *A* to give a level of 100 m = 4.98 m
Distance *AX* = 50 m (ΔAXB = 3, 4, 5)
∴ Collimation error = 0.02 m per 50 m
Allowing for this error, actual reading at *A* = 4.98 – 0.02 = 4.96 m
Now referring to the diagram, line *HF* through *E'* will be a strike line
∴ *H* and *F* have the same level as *E'*

Distance $AE' = (60^2 + 60^2)^{\frac{1}{2}}$ = 84.85 m
Fall from *A* to *E'* = 84.85 ÷ 20 = 4.24 m
∴ Level at *E'* = level at *F* and *H* = 100 – 4.24 = 95.76 m
Thus true staff readings at *F* and *H* = 104.98 – 95.76 = 9.22 m

Distance $XF = (70^2 + 40^2)^{\frac{1}{2}}$ = 80.62 m
Collimation error ≈ 0.03 m
Actual reading at *F* = 9.22 – 0.03 = 9.19 m
Distance *XH* = 110 m, collimation error ≈ 0.04 m
Actual reading at *H* = 9.22 – 0.04 = 9.18 m

Example 2.3 The following readings were observed with a level: 1.143 (BM 112.28), 1.765, 2.566, 3.820 CP; 1.390, 2.262, 0.664, 0.433 CP; 3.722, 2.886, 1.618, 0.616 TBM.

(1) Reduce the levels by the R-and-F method.
(2) Calculate the level of the TBM if the line of collimation was tilted upwards at an angle of 6′ and each BS length was 100 m and FS length 30 m.
(3) Calculate the level of the TBM if in all cases the staff was held not upright but leaning backwards at 5° to the vertical.   (LU)

(1) The answer here relies on knowing once again that levelling always commences on a BS and ends on a FS, and that CPs are always FS/BS (see table on facing page).
(2) Due to collimation error

> the BS readings are too great by 100 tan 6′
> the FS readings are too great by   30 tan 6′
>   ─────────
> *net* error on BS is too great by   70 tan 6′

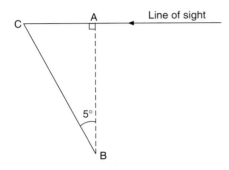

The student should note that the intermediate sights are unnecessary in calculating the value of the TBM; he can prove it for himself by simply covering up the IS column and calculating the value of TBM using BS and FS only.

There are three instrument set-ups, and therefore the *total net error* on BS = 3 × 70 tan 6′ = 0.366 m (too great).

> level of TBM = 113.666 − 0.366 = 113.300 m

(3) From the diagram it is seen that the true reading *AB* = actual reading *CB* × cos 5°. Thus each BS and FS needs to be corrected by multiplying it by cos 5°; however, this would be the same as multiplying the ΣBS and ΣFS by cos 5°, and as one subtracts BS from FS to get the difference, then

$$\textit{True difference in level} = \text{actual difference} \times \cos 5°$$

$$= 1.386 \cos 5° = 1.381 \text{ m}$$

$$\text{Level of TBM} = 112.28 + 1.381 = 113.661 \text{ m}$$

Example 2.4 One carriageway of a motorway running due N is 8 m wide between kerbs and the following surface levels were taken along a section of it, the chainage increasing from S to N. A concrete bridge 12 m in width and having a horizontal soffit, carries a minor road across the motorway from SW to NE, the centre-line of the minor road passing over that of the motorway carriageway at a chainage of 1550 m.

BS	IS	FS	Rise	Fall	RL	Remarks
1.143					112.280	BM
	1.765			0.622	111.658	
	2.566			0.801	110.857	
1.390		3.820		1.254	109.603	
	2.262			0.872	108.731	
	0.664		1.598		110.329	
3.722		0.433	0.231		110.560	
	2.886		0.836		111.396	
	1.618		1.268		112.664	
		0.616	1.002		113.666	TBM
6.255		4.869	4.935	3.549	113.666	
4.869			3.549		112.280	
1.386			1.386		1.386	Checks

Taking crown (i.e. centre-line) level of the motorway carriageway at 1550 m chainage to be 224.000 m:

(a) Reduce the above set of levels and apply the usual arithmetical checks.
(b) Assuming the motorway surface to consist of planes, determine the minimum vertical clearance between surface and the bridge soffit.   (LU)

The HPC method of booking is used because of the numerous intermediate sights.

BS	IS	FS	Chainage (m)	Location
1.591			1535	West channel
	1.490		1535	Crown
	1.582		1535	East channel
	−4.566			Bridge soffit*
	1.079		1550	West channel
	0.981		1550	Crown
	1.073		1550	East channel
2.256		0.844		CP
	1.981		1565	West channel
	1.884		1565	Crown
		1.975	1565	East channel

*Staff inverted

BS	IS	FS	HPC	RL	Remarks
1.591				223.390	1535 West channel
	1.490			223.491	1535 Crown
	1.582			223.399	1535 East channel
	−4.566			229.547	Bridge soffit
	1.079			223.902	1550 West channel
	0.981		224.981*	224.000	1550 Crown
	1.073			223.908	1550 East channel
2.256		0.844	226.393	224.137	CP
	1.981			224.412	1565 West channel
	1.884			224.509	1565 Crown
		1.975		224.418	1565 East channel
3.847	5.504	2.819		224.418	
2.819				223.390	
1.028				1.028	Checks

*Permissible to start here because this is the only known RL; also, in working back to 1535 m one still subtracts from HPC in the usual way.

Intermediate sight check

$$2245.723 = [(224.981 \times 7) + (226.393 \times 3) - (5.504 + 2.819)]$$
$$1574.867 + 679.179 - 8.323 = 2245.723$$

The student should now draw a sketch of the problem and add to it all the pertinent data as shown.

Examination of the sketch shows the road to be rising from S to N at a regular grade of 0.510 m in 15 m. This implies then, that the most northerly point (point *B* on east channel) should be the highest; however, as the crown of the road is higher than the channel, one should also check point

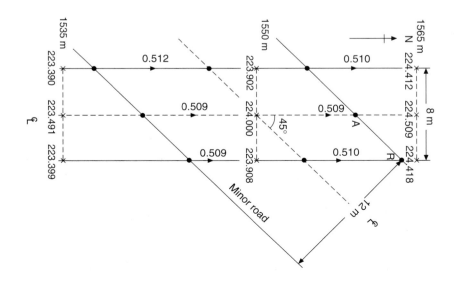

A on the crown; all other points can be ignored. Now, from the illustration the distance 1550 to *A* on the centre-line.

$$= 6 \times (2)^{\frac{1}{2}} = 8.5 \text{ m}$$

∴ Rise in level from 1550 to *A* = (0.509/15) × 8.5 = 0.288 m

∴ Level at *A* = 224.288 m giving a clearance of (229.547 − 224.288) = 5.259 m

   Distance 1550 to *B* along east channel = 8.5 + 4 = 12.5 m

∴ Rise in level from 1550 to *B* = (0.510/15) × 12.5 = 0.425 m

∴ Level at *B* = 223.908 + 0.425 = 224.333 m

∴ Clearance at *B* = 229.547 − 224.333 = 5.214 m

∴ Minimum clearance occurs at the most northerly point on the east channel, i.e. at *B*.

Example 2.5 In extending a triangulation survey of the mainland to a distant off-lying island, observations were made between two trig stations, one 3000 m and the other 1000 m above sea level. If the ray from one station to the other grazed the sea, what was the approximate distance between stations, (a) neglecting refraction, and (b) allowing for it? (*R* = 6400 km).   (ICE)
   Refer to equation (2.1).

(a) $D_1 = (2Rc_1)^{\frac{1}{2}} = (2 \times 6400 \times 1)^{\frac{1}{2}} = 113$ km

   $D_2 = (2Rc_2)^{\frac{1}{2}} = (2 \times 6400 \times 3)^{\frac{1}{2}} = 196$ km

   Total distance = 309 km

(b)   From p. 44 (2.6): $D_1 = (7/6 \times 2Rc_1)^{\frac{1}{2}}$, $D_2 = (7/6 \times 2Rc_2)^{\frac{1}{2}}$.

By comparison with the equation in (a) above, it can be seen that the effect of refraction is to increase distance by $(7/6)^{\frac{1}{2}}$:

∴   D = 309 $(7/6)^{\frac{1}{2}}$ = 334 km

Example 2.6 Obtain, from first principles, an expression giving the combined correction for the Earth's curvature and atmospheric refraction in levelling, assuming that the Earth is a sphere of 12 740 km diameter. Reciprocal levelling between two points *Y* and *Z* 730 m apart on opposite sides of a river gave the following results:

Instrument at	Height of instrument (m)	Staff at	Staff reading (m)
Y	1.463	Z	1.688
Z	1.436	Y	0.991

Determine the difference in level between Y and Z and the amount of any collimation error in the instrument.   (ICE)

(1) $(c - r) = \dfrac{6D^2}{14R} = 0.0673D^2$ m

(2) With instrument at Y, Z is lower by $(1.688 - 1.463) = 0.225$ m

 With instrument at Z, Z is lower by $(1.436 - 0.991) = 0.445$ m

 True height of Z below $Y = \dfrac{0.225 + 0.445}{2} = 0.335$ m

Instrument height at $Y = 1.463$ m; knowing now that Z is lower by 0.335 m, then a truly horizontal reading on Z should be $(1.463 + 0.335) = 1.798$ m; it was, however, 1.688 m, i.e. -0.11 m too low ($-$ indicates low). This error is due to curvature and refraction $(c - r)$ *and* collimation error of the instrument (e).

Thus: $(c - r) + e = -0.110$ m

Now $(c - r) = \dfrac{6D^2}{14R} = \dfrac{6 \times 730^2}{14 \times 6370 \times 1000} = 0.036$ m

$\therefore$ $e = -0.110 - 0.036 = -0.146$ m in 730 m

$\therefore$ Collimation error $e = 0.020$ m *down* in 110 m

Example 2.7 A and B are 2400 m apart. Observations with a level gave:

 A, height of instrument 1.372 m, reading at B 3.359 m
 B, height of instrument 1.402 m, reading at A 0.219 m

 Calculate the difference of level and the error of the instrument if refraction correction is one seventh that of curvature.   (LU)

 Instrument at A, B is lower by $(3.359 - 1.372) = 1.987$ m
 Instrument at B, B is lower by $(1.402 - 0.219) = \underline{1.183}$ m

 True height of B below $A = 0.5 \times 3.170$ m $= 1.585$ m

Combined error due to curvature and refraction

 $= 0.0673D^2$ m $= 0.0673 \times 2.4^2 = 0.388$ m

Now using same procedure as in *Example 2.6*:

 Instrument at $A = 1.372$, thus true reading at $B = (1.372 + 1.585)$
 $= 2.957$ m
 Actual reading at $B = \underline{3.359}$ m

 Actual reading at B too high by $+ 0.402$ m

Thus $(c - r) + e = +0.402$ m

 $e = +0.402 - 0.388 = +0.014$ m in 2400 m

 Collimation error $e = +0.001$ m *up* in 100 m

Exercises

(*2.1*) The following readings were taken with a level and a 4.25-m staff:

0.683, 1.109, 1.838, 3.398 [3.877 and 0.451] CP, 1.405, 1.896, 2.676 BM (102.120 AOD), 3.478 [4.039 and 1.835] CP, 0.649, 1.707, 3.722

Draw up a level book and reduce the levels by

(a) R-and-F,
(b) height of collimation.

What error would occur in the final level if the staff had been wrongly extended and a plain gap of 12 mm occurred at the 1.52-m section joint?   (LU)

Parts (a) and (b) are self checking. Error in final level = zero.
(Hint: all readings greater than 1.52 m will be too small by 12 mm. Error in final level will be calculated from BM only.)

(*2.2*) The following staff readings were observed (in the order given) when levelling up a hillside from a TBM 135.2 m AOD. Excepting the staff position immediately after the TBM, each staff position was higher than the preceding one.

1.408, 2.728, 1.856, 0.972, 3.789, 2.746, 1.597, 0.405, 3.280, 2.012, 0.625, 4.136, 2.664, 0.994, 3.901, 1.929, 3.478, 1.332

Enter the readings in level-book form by both the R-and-F and collimation systems (these may be combined into a single form to save copying).   (LU)

(*2.3*) The following staff readings in metres were obtained when levelling along the centre-line of a straight road *ABC*.

BS	IS	FS	Remarks
2.405			point *A* (RL = 250.05 m AOD)
1.954		1.128	CP
0.619		1.466	point *B*
	2.408		point *D*
	−1.515		point *E*
1.460		2.941	CP
		2.368	point *C*

D is the highest point on the road surface beneath a bridge crossing over the road at this point and the staff was held inverted on the underside of the bridge girder at *E*, immediately above *D*. Reduce the levels correctly by an approved method, applying the checks, and determine the headroom at *D*. If the road is to be regraded so that *AC* is a uniform gradient, what will be the new headroom at *D*? The distance *AD* = 240 m and *DC* = 60 m.   (LU)

(*Answer*: 3.923 m, 5.071 m)

(*2.4*) Distinguish, in construction and method of use, between dumpy and tilting levels. State in general terms the principle of an automatic level.

(*2.5*) The following levels were taken with a metric staff on a series of pegs at 100-m intervals along the line of a proposed trench.

BS	IS	FS	Remarks
2.10			TBM 28.75 m
	2.85		Peg *A*
1.80		3.51	Peg *B*
	1.58		Peg *C*
	2.24		Peg *D*
1.68		2.94	Peg *E*
	2.27		
	3.06		
		3.81	TBM 24.07 m

If the trench is to be excavated from peg *A* commencing at a formation level of 26.5 m and falling to peg *E* at a grade of 1 in 200, calculate the height of the sight rails in metres at *A*, *B*, *C*, *D* and *E*, if a 3-m boning rod is to be used.

Briefly discuss the techniques and advantages of using laser beams for the control of more precise work.   (KU)

(*Answer*: 1.50, 1.66, 0.94, 1.10, 1.30 m)

(*2.6*) (a) Determine from first principles the approximate distance at which correction for curvature and refraction in levelling amounts to 3 mm, assuming that the effect of refraction is one seventh that of the Earth's curvature and that the Earth is a sphere of 12 740 km diameter.
(b) Two survey stations *A* and *B* on opposite sides of a river are 780 m apart, and reciprocal levels have been taken between them with the following results:

Instrument at	Height of instrument (m)	Staff at	Staff reading (m)
A	1.472	*B*	1.835
B	1.496	*A*	1.213

Compute the ratio of refraction correction to curvature correction, and the difference in level between *A* and *B*:

(*Answer*: (a) 210 m (b) 0.14 to 1; *B* lower by 0.323 m).

2.15 TRIGONOMETRICAL LEVELLING

Trigonometrical levelling is used where difficult terrain, such as mountainous areas, precludes the use of conventional differential levelling.

   The modern approach is to measure the slope distance and vertical angle to the point in question. Slope distance is measured using electromagnetic distance measurers and the vertical (or zenith) angle using a theodolite.

   When these two instruments are integrated into a single instrument it is called a 'total station'. Total stations contain hard-wired algorithms which calculate and display the horizontal distance and vertical height, This latter facility has resulted in trigonometrical levelling being used for a wide variety of heighting procedures, including contouring. However, unless the observation distances are relatively short, the height values displayed by the total station are quite useless, if not highly dangerous, unless some attempt is made to apply corrections for curvature and refraction.

2.15.1 Short lines

From *Figure 2.37* it can be is seen that when measuring the vertical angle

$$\Delta h = S \sin \alpha \tag{2.8}$$

When using the zenith angle z

$$\Delta h = S \cos z \tag{2.9}$$

If the horizontal distance is used

$$\Delta h = D \tan \alpha = D \cot z \tag{2.10}$$

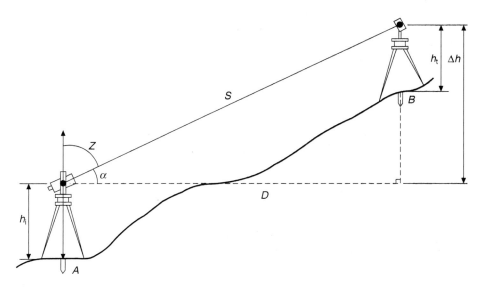

Fig. 2.37

The difference in elevation (ΔH) between ground points A and B is therefore

$$\Delta H = h_i + \Delta h - h_t$$
$$= \Delta h + h_i - h_t \qquad (2.11)$$

where h_i = vertical height of the measuring centre of the instrument above A

h_t = vertical height of the centre of the target above B

This is the basic concept of trigonometrical levelling. The vertical angles are positive for angles of elevation and negative for angles of depression. The zenith angles are always positive, but naturally when greater than 90° they will produce a negative result.

What constitutes a short line may be derived by considering the effect of curvature and refraction compared with the accuracy expected. The combined effect of curvature and refraction over 100 m = 0.7 mm, over 200 m = 3 mm, over 300 m = 6 mm, over 400 m = 11 mm and over 500 m = 17 mm.

If we apply the standard treatment for small errors to the basic equation we have

$$\Delta H = S \sin \alpha + h_i - h_t \qquad (2.12)$$

and then

$$\delta(\Delta H) = \sin \alpha \cdot \delta S + S \cos \alpha \, \delta\alpha + \delta h_i - \delta h_t \qquad (2.13)$$

and taking standard errors:

$$\sigma^2_{\Delta H} = (\sin \alpha \cdot \sigma_s)^2 + (S \cos \alpha \cdot \sigma_\alpha)^2 + \sigma_i^2 + \sigma_t^2$$

Consider a vertical angle of $\alpha = 5°$, with $\sigma_\alpha = \pm 5''$, $S = 300$ m with $\sigma_s = \pm 10$ mm and $\sigma_i = \sigma_t = \pm 2$ mm. Substituting in the above equation gives:

$$\sigma^2_{\Delta H} = 0.9^2 \text{ mm} + 7.2^2 \text{ mm} + 2^2 \text{ mm} + 2^2 \text{ mm}$$

$$= 7.8 \text{ mm}$$

This value balances out the effect of curvature and refraction over this distance and indicates that short sights should never be greater than 300 m. It also indicates that the accuracy of distance S is not critical. However, the accuracy of measuring the vertical angle is very critical and requires the use of a theodolite, with more than one measurement on each face.

2.15.2 Long lines

For long lines the effect of curvature (c) and refraction (r) must be considered. From *Figure 2.38*, it can be seen that the difference in elevation (ΔH) between A and B is

$$\Delta H = GB = GF + FE + EH - HD - DB$$
$$= h_i + c + \Delta h - r - h_t$$
$$= \Delta h + h_i - h_t + (c - r) \qquad (2.14)$$

Thus it can be seen that the only difference from the basic equation for short lines is the correction for curvature and refraction ($c - r$).

Although the line of sight is refracted to the target at D, the telescope is pointing to H, thereby

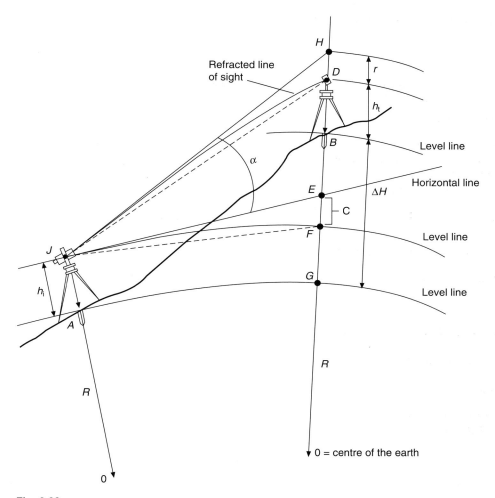

Fig. 2.38

measuring the angle α from the horizontal. It follows that $S \sin \alpha = \Delta h = EH$ and requires a correction for refraction equal to HD.

The correction for refraction is based on a quantity termed the 'coefficient of refraction' (K). Considering the atmosphere as comprising layers of air which decrease in density at higher elevations, the line of sight from the instrument will be refracted towards the denser layers. The line of sight therefore approximates to a circular arc of radius R_s roughly equal to $8R$, where R is the radius of the Earth. However, due to the uncertainty of refraction one cannot accept this relationship and the coefficient of refraction is defined as

$$K = R/R_s \tag{2.15}$$

An average value of $K = 0.15$ is frequently quoted but, as stated previously, this is most unreliable and is based on observations taken well above ground level. Recent investigation has shown that not only can K vary from -2.3 to $+3.5$ with values over ice as high as $+14.9$, but it also has a daily cycle. Near the ground, K is affected by the morphology of the ground, by the type of vegetation and by other assorted complex factors. Although much research has been devoted to modelling these effects, in order to arrive at an accurate value for K, the most practical method still appears to be by simultaneous reciprocal observations.

As already shown, curvature (c) can be approximately computed from $c = D^2/2R$, and as $D \approx S$ we can write

$$c = S^2/2R \tag{2.16}$$

Now considering *Figures 2.38* and *2.39*, the refracted ray *JD* has a radius R_s and a measured distance S and subtends angles δ at its centre, then

$$\delta = S/R_s$$
$$\delta/2 = S/2R_s$$

As the refraction $K = R/R_s$ we have

$$\delta/2 = SK/2R$$

Without loss of accuracy we can assume $JH = JD = S$ and treating the *HD* as the arc of a circle of radius S

$$HD = S \cdot \delta/2 = S^2 K/2R = r \tag{2.17}$$
$$(c - r) = S^2(1 - K)/2R \tag{2.18}$$

All the above equations express c and r in linear terms. To obtain the angles of curvature and refraction, *EJF* and *HJD* in *Figure 2.38*, reconsider *Figure 2.39*. Imagine *JH* is the horizontal line

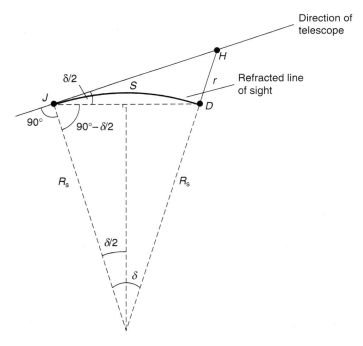

Fig. 2.39

JE in *Figure 2.38* and *JD* the level line *JF* of radius *R*. Then δ is the angle subtended at the centre of the Earth and the angle of curvature is half this value. To avoid confusion let $\delta = \theta$ and as already shown:

$$\theta/2 = S/2R = \hat{c} \tag{2.19}$$

where the arc distance at MSL approximates to *S*. Also, as shown:

$$\delta/2 = SK/2R = \hat{r} \tag{2.20}$$

Therefore in angular terms:

$$(\hat{c} - \hat{r}) = S(1 - K)/2R \text{ rads} \tag{2.21}$$

Note the difference between equations in linear terms and those in angular.

2.15.3 Reciprocal observations

Reciprocal observations are observations taken from *A* and *B*, the arithmetic mean result being accepted. If one assumes a symmetrical line of sight from each end and the observations are taken simultaneously, then the effect of curvature and refraction is cancelled out. For instance, for elevated sights, $(c - r)$ is added to a positive value to increase the height difference. For depressed or downhill sights, $(c - r)$ is added to a negative value and decreases the height difference. Thus the average of the two values is free from the effects of curvature and refraction. This statement is not entirely true as the assumption of symmetrical lines of sight from each end is dependent on uniform ground and atmospheric conditions at each end, at the instant of simultaneous observation.

In practice, sighting into each others' object lens forms an excellent target, with some form of intercommunication to ensure simultaneous observation.

The following numerical example is taken from an actual survey in which the elevation of *A* and *B* had been obtained by precise geodetic levelling and was checked by simultaneous reciprocal trigonometrical levelling.

Worked example

Example 2.8

> Zenith angle at $A = Z_A = 89° 59' 18.7''$ (VA $0° 00' 41.3''$)
> Zenith angle at $B = Z_B = 90° 02' 59.9''$ (VA $= -0° 02' 59.9''$)
> Height of instrument at $A = h_A = 1.290$ m
> Height of instrument at $B = h_B = 1.300$ m
> Slope distance corrected for meteorological conditions $= 4279.446$ m

As it is known that the observations are reciprocal, the values for curvature and refraction are ignored:

$$\Delta H_{AB} = S \cos Z_A + h_i - h_t$$
$$= 4279.446 \cos 89° 59' 18.7'' + 1.290 - 1.300 = 0.846 \text{ m}$$
$$\Delta H_{AB} = 4279.446 \cos 90° 02' 59.9'' + 1.300 - 1.290 = -3.722 \text{ m}$$

Mean value $\Delta H = 2.284$ m

This value compares favourably with 2.311 m obtained by precise levelling. However, the disparity between the two values 0.846 and -3.722 shows the danger inherent in single observations uncorrected for curvature and refraction. In this case the correction for curvature only is $+1.256$ m, which, when applied, brings the results to 2.102 m and -2.466 m, producing much closer agreement. To find K simply substitute the mean value $\Delta H = 2.284$ into the equation for a single observation.

From A to B:

$2.284 = 4279.446 \cos 89°59'\ 18.7'' + 1.290 - 1.300 + (c - r)$

where $(c - r) = S^2(1 - K)/2R$

and the local value of R for the area of observation $= 6364700$ m

$2.284 = 0.856 - 0.010 + S^2(1 - K)/2R$

$1.438 = 4279.446^2(1 - K)/2 \times 6364700$ m

$K = 0.0006$

From B to A:

$2.284 = -3.732 + 1.300 - 1.290 + S^2(1 - K)/2R$

$K = 0.0006$

Now this value for K could be used in single shots taken within the same area, to give improved results.

A variety of formulae are available for finding K direct. For example, using zenith angles:

$$K = 1 - \frac{Z_A + Z_B - 180°}{180°/\pi} \times \frac{R}{S} \qquad (2.22)$$

and using vertical angles:

$$K = (\theta + \alpha_0 + \beta_0)/\theta \qquad (2.23)$$

where $\theta =$ the angle subtended at the centre of the Earth by the arc distance $\approx S$ and is calculated using:

$$\theta'' = S\rho/R \quad \text{where } \rho = 206265$$

In the above formulae the values used for the angles must be those which would have been observed had $h_i = h_t$ and, in the case of vertical angles, entered with their appropriate sign. As shown in *Figure 2.40*, $\alpha_0 = \alpha - e$ and for an angle of depression it becomes $\beta_0 = \beta + e$.

By sine rule:

$$\sin e = \frac{h_{t-i} \sin (90° - \alpha)}{S}$$

$$e = \sin^{-1}\left(\frac{h_{t-i} \cos \alpha}{S} \right)$$

$$= \frac{h_{t-1}}{S} \cos \alpha + \frac{h^3_{t-i}}{6S^3} \cos^3 \alpha + \cdots$$

$$\therefore e = (h_{t-i} \cos \alpha)/S \qquad (2.24)$$

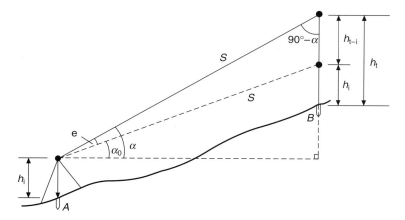

Fig. 2.40

For zenith angles:

$$e = (h_{t-i} \sin Z)/S \qquad (2.25)$$

2.15.4 Sources of error

Consider the formula for a single observation:

$$\Delta H = S \sin \alpha + h_i - h_t + S^2 (1 - K)/2R$$

The obvious sources of error lie in obtaining the slope distance S, the vertical angle α the heights of the instrument and target, the coefficient of refraction K and a value for the local radius of the Earth R. Differentiating gives:

$$\delta(\Delta H) = \delta S \sin \alpha + S \cos \alpha \cdot \delta\alpha + \delta h_i + \delta h_t + S^2 \, \delta K/2R + S^2(1 - K) \, \delta R/2R^2$$

and taking standard errors:

$$\sigma_{\Delta H}^2 = (\sigma_s \sin \alpha)^2 + (S \cos \alpha \cdot \sigma_\alpha)^2 + \sigma_i^2 + \sigma_t^2 + (S^2 \sigma_K/2R)^2 + (S^2 (1 + K) \, \sigma_R/2R^2)^2$$

Taking $S = 2000$ m ± 0.005 m, $\alpha = 8° \pm 07''$, $\sigma_i = \sigma_t = \pm 2$ mm, $K = 0.15 \pm 1$, $R = 6380$ km ± 10 km, we have

$$\sigma_{\Delta H}^2 = (0.7)^2 + (48.0)^2 + 2^2 + 2^2 + (156.7)^2 + (0.4)^2$$
$$\sigma_{\Delta H} = \pm 164 \text{ mm}$$

Once again it can be seen that the accuracy required to measure S is not a critical component.

However, the measurement of the vertical angle is and will increase with increase in distance. The error in the value of refraction is the most critical component and will increase rapidly as the square of the distance. Thus to achieve reasonable results over long sights, simultaneous reciprocal observations are essential.

2.15.5 *Contouring*

The ease with which electronic tacheometers or, as they are also called, total stations, produce horizontal distance, vertical height and horizontal direction makes them ideal instruments for rapid and accurate contouring in virtually any type of terrain. If used in conjunction with automatic data recorders, which in turn are interfaced with computers, the spatial data are transformed from direction, distance and elevation of a point, to its position and elevation in terms of three-dimensional coordinates. These points thus comprise a digital terrain or ground model (DTM/DGM) from which the contours are interpolated and automatically plotted.

The electronic tacheometer and a vertical rod that carries a single reflector are used to locate the ground points (*Figure 2.41*). A careful reconnaissance of the area is necessary, in order to plan the survey and define the necessary ground points that are required to represent the characteristic shape of the terrain. Break lines, the tops and bottoms of hills or depressions, the necessary features of water courses, etc., plus enough points to permit accurate interpolation of contour lines at the interval required, comprise the field data. As the observation distances are relatively short, curvature and refraction are ignored.

From *Figure 2.37*, it can be seen that if the reduced level of point A (RL_A) is known, then the reduced level of ground point B is:

$$RL_B = RL_A + h_i + \Delta h - h_t$$

When contouring, the height of the reflector is set to the same height as the instrument, i.e.

Fig. 2.41   *Contouring with a total station and detail pole*

$h_t = h_i$, and cancels out in the previous equation. Thus the height displayed by the instrument is the height of the ground point above A:

$$RL_B = RL_A + \Delta h$$

In this way the reduced levels of all the ground points are rapidly acquired and all that is needed are their positions. One method of carrying out the process is by radiation.

As shown in *Figure 2.42*, the instrument is set up on a control point A, whose reduced level is known, and sighted to a second control point (RO). The horizontal circle is zeroed. The instrument is then turned through a particular horizontal angle (θ) defining the direction of the first ray. Terrain points are then located by horizontal distance and height along this ray. This process is repeated along further rays until the area is covered. Unless a very experienced person is used to locate the ground points, there will obviously be a greater density of points near the instrument station. The method, however, is quite easy to organize in the field. The angle θ may vary from 20° to 60° depending on the terrain.

With the advent of computer plotting and contour interpolation, the locating of strings of linked terrain points is favoured by many ground-modelling systems. In this method the ground points are located in continuous strings throughout the area, approximately following the line of the contour. They would also follow the line of existing water courses, roads, hedges, kerbs, etc. (*Figure 2.43*). In this particular format the points are more easily processed by the computer.

Depending on the software package used, the string points may be transformed into a triangular or gridded structure. Heights can then be determined by linear interpolation and the terrain represented by simple planar triangular facets. Alternatively, high-order polynomials may be used to define three-dimensional surfaces fitted to the terrain points. From these data, contours are interpolated and a contour model of the terrain produced.

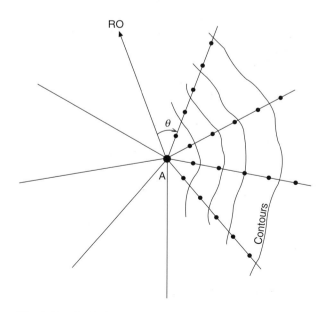

Fig. 2.42 *Radiation method*

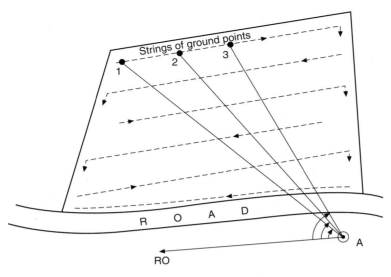

Fig. 2.43

Worked examples

Example 2.9 (a) Define the coefficient of refraction K, and show how its value may be obtained from simultaneous reciprocal trigonometric levelling observations.

  (b) Two triangulation stations A and B are 2856.85 m apart. Observations from A to B gave a mean vertical angle of $+01°35'38''$, the instrument height being 1.41 m and the target height 2.32 m. If the level of station A is 156.86 m OD and the value of K for the area is 0.16, calculate the reduced level of B (radius of Earth = 6372 km).   (KU)

(a) Refer to *Section 2.15.2*.
(b) This part will be answered using both the angular and the linear approaches.

Angular method
Difference in height of $AB = \Delta H = D \tan [\alpha + (\hat{c} - \hat{r})]$ where $\hat{c} = \theta/2$ and

$$\theta = \frac{D}{R} = \frac{2856.85}{6372\,000} = 0.000\,448 \text{ rad}$$

$$\therefore \ \hat{c} = 0.000\,224 \text{ rad}$$

$$\hat{r} = K(\theta/2) = 0.16 \times 0.000\,224 = 0.000\,036 \text{ rad}$$

$$\therefore \ (\hat{c} - \hat{r}) = 0.000\,188 \text{ rad} = 0°00'38.8''$$

$$\therefore \ \Delta H = 2856.85 \tan (01°35'38'' + 0°00'38.8'') = 80.03 \text{ m}$$

Refer to *Figure 2.37*.

$$\text{RL of } B = \text{RL of } A + h_i + \Delta H - h_t$$
$$= 156.86 + 1.41 + 80.03 - 2.32 = 235.98 \text{ m}$$

Linear method

$$\Delta H = D \tan \alpha + (c - r)$$

where $(c - r) = \left(\dfrac{D^2}{2R}\right)(1 - K) = \dfrac{2856.85^2}{2 \times 6\,372\,000} \times 0.84 = 0.54 \text{ m}$

$\qquad D \tan \alpha = 2856.85 \tan (01° \ 35' \ 39'') = 79.49 \text{ m}$

where $\therefore \ \Delta H = 79.49 + 0.54 = 80.03 \text{ m}$

Example 2.10 Two stations *A* and *B* are 1713 m apart. The following observations were recorded: height of instrument at *A* 1.392 m, and at *B* 1.464 m; height of signal at *A* 2.199 m, and at *B* 2 m. Elevation to signal at *B* 1°08′08″, depression angle to signal at *A* 1° 06′ 16″. If 1″ at the Earth's centre subtends 30.393 m at the Earth's surface, calculate the difference of level between *A* and *B* and the refraction correction.

(LU)

$$\Delta H = D \tan \left(\dfrac{\alpha + \beta}{2}\right) + \dfrac{(h_t' - h_i') - (h_t - h_i)}{2}$$

where h_i = height of instrument at *A*; h_t = height of target at *B*; h_i' = height of instrument at *B*; h_t' = height of target at *A*.

$$\therefore \ \Delta H = 1713 \tan \left(\dfrac{(1° \ 08' \ 08'') + (1° \ 06' \ 15'')}{2}\right) + \dfrac{(2.199 - 1.464) - (2.000 - 1.392)}{2}$$

$$= 33.490 + 0.064 = 33.55 \text{ m}$$

Using the alternative approach of reducing α and β to their values if $h_i = h_t$
Correction to angle of elevation

$$e'' \approx \dfrac{1.392 - 2.000}{1713.0} \times 206\,265 = -73.2''$$

$$\therefore \ \alpha = (1° \ 08' \ 08'') - (01' \ 13.2'') = 1° \ 06' \ 54.8''$$

Correction to angle of depression

$$e'' \approx \dfrac{(2.199 - 1.464)}{1713.0} \times 206\,265 = 88.5''$$

$$\therefore \ \beta = (1° \ 06' \ 15'') + (01' 28.5'') = 1° \ 07' \ 43.5''$$

$$\therefore \ \Delta H = 1713 \tan \left(\dfrac{1° \ 06' \ 54.8'' + 1° \ 07' \ 43.5''}{2}\right) = 33.55 \text{ m}$$

Refraction correction $\hat{r} = \frac{1}{2}(\theta + \alpha + \beta)$

where $\qquad\qquad \theta'' = 1713.0/30.393 = 56.4''$

$$= \tfrac{1}{2}[56.4'' + (1° 06' \ 54.8'') - (1° 07' \ 43.5'')] = 3.8''$$

and also $\qquad\qquad K = \dfrac{\hat{r}}{\theta/2} = \dfrac{3.8''}{28.2''} = 0.14$

Example 2.11 Two points *A* and *B* are 8 km apart and at levels of 102.50 m and 286.50 m OD, respectively. The height of the target at *A* is 1.50 m and at *B* 3.00 m, while the height of the instrument in both cases is 1.50 m on the Earth's surface subtends 1″ of arc at the Earth's centre and the effect of refraction is one seventh that of curvature, predict the observed angles from A to B and B to A.

Difference in level A and B = ΔH = 286.50 − 102.50 = 184.00 m

$\therefore$ by radians $\phi'' = \dfrac{184}{8000} \times 206\,265 = 4744'' = 1°19'04''$

Angle subtended at the centre of the Earth = $\theta'' = \dfrac{8000}{31} = 258''$

$\therefore$ Curvature correction $\qquad \hat{c} = \theta/2 = 129'' \qquad$ and $\qquad \hat{r} = \hat{c}/7 = 18''$

Now $\quad \Delta H = D \tan \phi$

where $\quad \phi = \alpha + (\hat{c} - \hat{r})$

$\qquad \therefore \alpha = \phi - (\hat{c} - \hat{r}) = 4744'' - (129'' - 18'') = 4633'' = 1°17'13''$

Similarly $\qquad \phi = \beta - (\hat{c} - \hat{r})$

$\qquad \therefore \beta = \phi + (\hat{c} - \hat{r}) = 4855'' = 1°20'55''$

The observed angle α must be corrected for variation in instrument and signal heights. Normally the correction is subtracted from the observed angle to give the truly reciprocal angle. In this example, α is the truly reciprocal angle, thus the correction must be added in this reverse situation

$e'' \approx [(h_t - h_i)/D] \times 206\,265 = [(3.00 - 1.50)/8000] \times 206\,265 = 39''$

$\therefore \alpha = 4633'' + 39'' = 4672'' = 1°17'52''$

Example 2.12 A gas drilling-rig is set up on the sea bed 48 km from each of two survey stations which are on the coast and several kilometres apart. In order that the exact position of the rig may be obtained, it is necessary to erect a beacon on the rig so that it may be clearly visible from theodolites situated at the survey stations, each at a height of 36 m above the high-water mark.

Neglecting the effects of refraction, and assuming that the minimum distance between the line of sight and calm water is to be 3 m at high water, calculate the least height of the beacon above the high-water mark, at the rig. Prove any equations used.

Calculate the angle of elevation that would be measured by the theodolite when sighted onto this beacon, taking refraction into account and assuming that the error due to refraction is one seventh of the error due to curvature of the Earth. Mean radius of Earth = 6273 km. (ICE)

From *Figure 2.44*

$$D_1 = (2c_1 R)^{\frac{1}{2}} \text{ (equation 2.1)}$$

$$\therefore D_1 = (2 \times 33 \times 6\,273\,000) = 20.35 \text{ km}$$

$$\therefore D_2 = 48 - D_1 = 27.65 \text{ km}$$

$$\therefore \text{ since } D_2 = (2c_2 R)^{\frac{1}{2}}$$

$$c_2 = 61 \text{ m, and to avoid grazing by 3 m, height of beacon} = 64 \text{ m}$$

Difference in height of beacon and theodolite = 64 − 36 = 28 m; observed vertical angle $\alpha = \phi - (\hat{c} - \hat{r})$ for angles of elevation, where

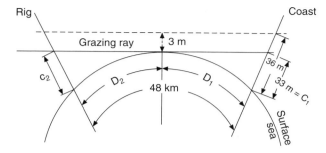

Fig. 2.44

$$\phi'' = \frac{28 \times 206\,265}{48\,000} = 120.3''$$

$$\hat{c} = \theta/2$$

where $\theta'' = \left(\dfrac{48}{6273}\right) \times 206\,265 = 1578.3''$

$\therefore \hat{c} = 789.2''$ and $\hat{r} = \hat{c}/7 = 112.7''$

$\therefore \alpha = 120.3'' - 789.2'' + 112.7'' = -556.2'' = -0°09' 16''$

The negative value indicates α to be an angle of depression, not elevation, as quoted in the question.

2.16 STADIA TACHEOMETRY

Stadia tacheometry can be used to locate the position of a point and its elevation, and as such it enables contouring to be carried out in a manner similar to that using total stations.

As the method fixes position it can also be used to locate detail and thereby produce a topographic plan. However, as it produces field data to a very low order of accuracy there are limitations on the scale of the plan.

It can be argued that the order of accuracy is so low that the method should be rendered obsolete. The reason it remains as a possible surveying technique is that it requires only the very basic instrumentation, namely a theodolite (accuracy to 1' suffices) and a levelling staff.

Using basic instrumentation a great deal of field data must be obtained per point, i.e. relative bearing, vertical angle, stadia readings and cross-hair reading. This in turn requires a great deal of data processing, which is only viable if suitable software is available. The use of a direct-reading tacheometer substantially reduces the amount of data processing necessary.

To summarize, the method has a very low order of accuracy; compared with modern data capture it is extremely slow and requires a great deal of processing, It should, therefore, only be used if there is no alternative.

2.16.1 Principles of stadia tacheometry

The principle of this form of tacheometry, in which the parallactic angle 2α remains fixed and the staff intercept S varies with distance D, is shown in *Figure 2.45*. The parallactic angle is defined

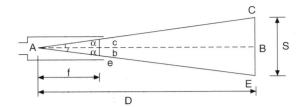

Fig. 2.45

by the position of the stadia hairs, c and e, each side of the main cross-hair b, then by similar triangles:

$$\frac{AB}{CE} = \frac{Ab}{ce}$$

put $ce = i$

then $D = (f/i) \, S = K_1 S$ $\qquad\qquad\qquad\qquad\qquad (2.26)$

In modern telescopes f and i are so arranged that $K_1 = 100$.

Equation (2.26) is basically correct for horizontal sights taken with any modern instrument. The telescope will now be examined in more detail. In *Figure 2.46 f* is the focal length of the object lens system, d is the distance from the object lens to the centre of the instrument, ce is the stadia interval. i, and D is the distance from the staff to the centre of the instrument; then by similar triangles:

$$\frac{Bp}{CE} = \frac{Op}{c'e'} \therefore Bp = S\left(\frac{f}{i}\right)$$

Now $D = Bp + (f + d) = S\,(f/i) + (f + d)$

The value $(f + d)$ is called the *additive constant*, K_2 and (f/i) is called the *multiplying constant*, K_1. Thus for horizontal sights:

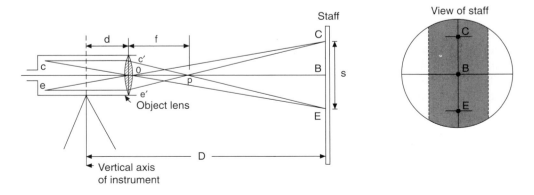

Fig. 2.46   *On right, view through telescope illustrating the stadia lines at* C *and* E

$$D = K_1 S + K_2 \qquad (2.27)$$

Tacheometry would have very little application if it was restricted to horizontal sights; thus the general formula will now be deduced. *Figure 2.47* illustrates an inclined sight.

By the sine rule in triangle *PCB*:

$$\frac{x_1}{\sin \alpha} = \frac{y \cot \alpha}{\sin [90° - (\theta + \alpha)]} = \frac{y \cot \alpha}{\cos (\theta + \alpha)}$$

Cross-multiply.

$$x_1 \cos (\theta + \alpha) = y \cot \alpha \sin \alpha = \frac{y \cos \alpha \sin \alpha}{\sin \alpha} = y \cos \alpha$$

from which $y = x_1 \cos \theta - x_1 \sin \theta \tan \alpha$ (a)

Similarly in triangle *PBE*

$$\frac{x_2}{\sin \alpha} = \frac{y \cot \alpha}{\sin [90° + (\theta - \alpha)]} = \frac{y \cot \alpha}{\cos (\theta - \alpha)}$$

$$x_2 \cos (\theta - \alpha) = y \cot \alpha \sin \alpha = y \cos \alpha$$

and $y = x_2 \cos \theta + x_2 \sin \theta \tan \alpha$ (b)

Adding (a) and (b)

$$2y = (x_1 + x_2) \cos \theta - (x_1 - x_2) \sin \theta \tan \alpha$$

i.e. $C'E' = S \cos \theta - (x_1 - x_2) \sin \theta \tan \alpha$ (c)

The maximum value for $\sin \theta$ would be 0.707 ($\theta = 45°$) and for $\tan \alpha$, 0.005 ($\alpha = 1/200$), whilst for the majority of work in practice $x_1 \approx x_2$. Thus, the second term may be neglected for all but the steepest sights.

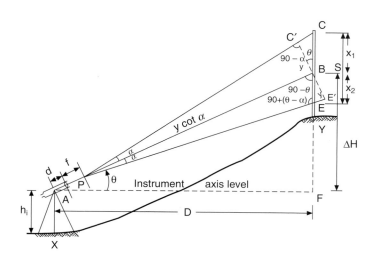

Fig. 2.47

Now, from *Figure 2.47*:

$$AB = K_1(C'E') + K_2 = K_1S \cos \theta + K_2$$

$$\therefore AF = D = AB \cos \theta = \underline{K_1S \cos^2 \theta + K_2 \cos \theta} \tag{d}$$

Similarly $FB = \Delta H = AB \sin \theta = \underline{K_1S \cos \theta \sin \theta + K_2 \sin \theta} \tag{e}$

Alternatively $\Delta H = D \tan \theta \tag{f}$

In 1823, an additional anallactic lens was built into the telescope, which reduced all observations to the centre of the instrument and thus eliminated the additive constant K_2. All modern internal focusing telescopes, although not strictly anallactic may be regarded as so. Equations (d) and (e) therefore reduce to

$$D = K_1S \cos^2 \theta \tag{g}$$

and $\Delta H = K_1S \cos \theta \sin \theta \tag{h}$

but $\cos \theta = \frac{1}{2} \sin 2\theta \tag{i}$

$$\therefore \Delta H = \frac{1}{2} K_1S \sin 2\theta$$

and where $K_1 = 100$

$$\underline{D = 100S \cos^2 \theta} \tag{2.28}$$

$$\underline{\Delta H = 50S \sin 2\theta} \tag{2.29}$$

With reference to *Figure 2.47* it can be seen that, given the reduced level of $X(\mathrm{RL}_x)$, then the level of Y is

$$\mathrm{RL}_x + h_i + \Delta H - BY \tag{2.30}$$

If the sight had been from Y to X then a simple sketch as in *Figure 2.48* will serve to show that

$$\mathrm{RL}_x = \mathrm{RL}_y + h_i - \Delta H - BX \tag{2.31}$$

where h_i = instrument height

    BY or BX = mid-staff reading

ΔH is positive when vertical angle is positive and *vice versa*.

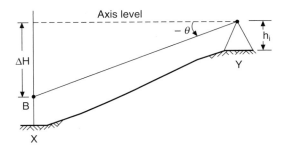

Fig. 2.48

Students should not attempt to commit equation (2.31) to memory, relying, if in doubt, on a quick sketch. Note that ΔH is always the vertical height from the centre of the transit axis to the mid-staff reading.

Thus, in general:

$$\text{RL}_x = (\text{RL}_y + h_i) \pm \Delta H - \text{mid-staff reading} \qquad (2.32)$$

where $(\text{RL}_y + h_i)$ = axis level

2.16.2 *Measurements of tacheometric constants*

The multiplying constant ($K_1 = 100$) and the additive constant ($K_2 = 0$) may vary due to ageing of the theodolite or temperature variations. It is therefore necessary to check them as follows.

Set up the instrument on fairly level ground giving horizontal sights to a series of pegs at known distances, D, from the instrument. Now, using the equation $D = K_1 S + K_2$ and substituting values for D and S, the equations may be solved:

(1) Simultaneously in pairs and the mean taken.
(2) As a whole by the method of least squares. For example:

Measured distance (m)	30	60	90	120	150	(*D*-values)
Staff intercept (m)	0.301	0.600	0.899	1.202	1.501	(*S*-values)

from which $K_1 = 100$ and $K_2 = 0$ by either of the above methods.

Errors in the region of 1/1000 can occur in the constants.

2.16.3 *Errors in staff holding*

The basic formula for horizontal distance (D) is $D = K_1 S \cdot \cos^2 \theta$. This is only so if we assume the vertical angle BAF (*Figure 2.47*) is equal to $C'BC$. If the staff is not held truly vertical, angle $C'BC \neq \theta$ and the formula is more correctly expessed as:

$$D = K_1 \cdot S \cdot \cos \theta_1 \cos \theta_2$$

where θ_1 = the vertical angle

θ_2 = angle $C'BC$.

Then, adopting the usual procedure for the treatment of small errors, the above expression is differentiated with respect to θ_2, giving

$$\delta D = -K_1 S \cos \theta_1 \sin \theta_2 \delta \theta_2$$

$$\therefore \quad \frac{\delta D}{D} = \left(\frac{-K_1 S \cos \theta_1 \sin \theta_2 \delta \theta_2}{K_1 S \cos \theta_1 \cos \theta_2} \right) - \tan \theta_2 \delta \theta_2 \qquad (2.33)$$

Using the above expression the following table may be drawn up assuming $\theta \approx \theta_1$ = angle of inclination = θ:

Table 2.1

θ	$\delta\theta_2 = 10'$	$\delta\theta_2 = 1°$	$\delta\theta_2 = 2°$
3°	1/6670	1/1090	1/550
5°	1/4000	1/650	1/330
10°	1/1960	1/325	1/160
15°	1/1280	1/215	1/110
20°	1/940	1/160	1/80
25°	1/740	1/120	1/60
30°	1/600	1/100	1/50

(a) Column 2 shows that if the staff is held reasonably plumb this source of error may be ignored.
(b) Column 3 shows that the accuracy falls off rapidly as the angle of inclination increases.
(c) Column 4 shows that where the staff is used carelessly, the accuracy is radically reduced, even on fairly level sights. It is obvious from this that all tacheometric staves should be fitted with a bubble and regularly checked.

2.16.4 Errors in horizontal distance

The errors in the resultant distance D will now be examined in more detail.

(1) Careless staff holding, which has already been discussed separately due to its importance.
(2) Error in reading the stadia intercept, which is immediately multiplied by 100 (K_1), thereby making it very significant. This source of error will decrease with increase in the length of sight, provided the reading error (δs) remains constant. Thus observation distances should be limited to a maximum of 100 m.
(3) Error in the determination of the instrument constants K_1 and K_2, resulting in an error in distance directly proportional to the error in the constant K_1 and directly as the error in K_2.
(4) Effect of differential refraction on the stadia intercept. This is minimized by keeping the lower reading 1–1.5 m above the ground.
(5) Random error in the measurement of the vertical angle. This has a negligible effect on the staff intercept and consequently on the horizontal distance.

In addition to the above sources of error, there are many others resulting from instrumental errors, failure to eliminate parallax, and natural errors due to high winds, heat shimmer, etc. The lack of statistical evidence makes it rather difficult to quote standards of accuracy; however, the usual treatment for small errors will give some basis for assessment.

Taking the equation for vertical staff tacheometry as in *Section 2.16.3* and differentiating with respect to each of the sources of error, in turn gives

$$D = K_1 S \cos \theta_1 \cos \theta_2$$

thus, $\delta D = S \cos \theta_1 \cos \theta_2 \, \delta K_1$

$$\therefore \frac{\delta D}{D} = \frac{S \cos \theta_1 \cos \theta_2 \, \delta K_1}{S \cos \theta_1 \cos \theta_2 \, K_1} = \frac{\delta K_1}{K_1} \tag{a}$$

Similarly, differentiating with respect to S, θ_1 and θ_2, in turn gives

$$\delta D/D = \delta S/S \tag{b}$$

$$\delta D/D = -\tan\theta_1\delta\theta_1 \tag{c}$$

$$\delta D/D = -\tan\theta_2\delta\theta_2 \tag{d}$$

From the theory of errors, the sum effect of the above errors will give a fractional or proportional standard error (PSE) of

$$\delta D/D = \pm\left[\left(\frac{\delta K_1}{K_1}\right)^2 + \left(\frac{\delta S}{S}\right)^2 + (\tan\theta_1\ \delta\theta_1)^2 + (\tan\theta_2\ \delta\theta_2)^2\right]^{\frac{1}{2}} \tag{2.34}$$

Assume now the following values: $D = 100$ m, $S = 1.008$ m. $\theta_1 = \theta_2 = 5°\cdot\delta S = \pm(2^2 + 2^2)^{\frac{1}{2}} = \pm3$ mm, $\delta K_1/K_1 = 1/1000$, $\delta\theta_1 = \pm10''$ (error in vertical angle); $\delta\theta_2 = \pm1°$ (error in staff holding).

N.B. $\delta\theta_1$ and $\delta\theta_2$ must always be expressed in radians (1 rad $= 206\,265''$)

$$\delta D/100 = \pm\left[(0.001)^2 + \left(\frac{0.003}{1.008}\right)^2 + (\tan 5°\cdot 10'')^2 + (\tan 5°\cdot 1°)^2\right]^{\frac{1}{2}}$$

$$= \pm[(100\times 10^{-8}) + (885\times 10^{-8}) + \text{zero} + (234\times 10^{-8})]^{\frac{1}{2}}$$

$$\therefore\ \delta D = \pm0.35\text{ m}\quad\text{and}\quad\delta D/D \approx 1\text{ in }300$$

It is obvious that the most serious sources of error result from careless staff holding and stadia intercept error, the error in vertical angles being negligible.

Thus in practice, the staff must be fitted with a circular bubble to ensure verticality and the sight distances limited to ensure accurate staff reading. Nevertheless, even with all precautions taken, the accuracy cannot be improved much beyond that indicated in the above analysis.

2.16.5 Errors in elevations

The main sources of error in elevation are (1) error in vertical angles and (2) additional errors arising from errors in the computed distance. *Figure 2.49* clearly shows that whilst the error resulting from (1) remains fairly constant, that resulting from (2) increases with increased elevation:

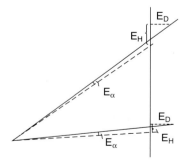

Fig. 2.49

$$\Delta H = D \tan \theta$$
$$\delta(\Delta H) = \delta D \tan \theta$$
$$\delta(\Delta H) = D \sec^2 \theta \, \delta\theta$$
$$\delta(\Delta H) = \pm [(\delta D \tan \theta)^2 + (D \sec^2 \theta \, \delta\theta)^2]^{\frac{1}{2}} \tag{2.35}$$
$$= \pm [(0.35 \tan 5°)^2 + (100 \sec^2 5° \times 10'' \sin 1'')^2]^{\frac{1}{2}}$$
$$= \pm 0.031 \text{ m}$$

This result indicates that elevations need be quoted only to the nearest 10 mm.

2.16.6 Application

Due to the low accuracy afforded by the method of stadia tacheometry, it is limited in its application to:

(1) Contouring
(2) Topographic detailing at small scales.

As the method gives distance and bearing to a point and its elevation, it can be used in exactly the same way as a total station, i.e. the radiation method or the method of strings (*Section 2.15.5*).

If topographic conditions permit, setting the mid-staff reading to the instrument height will result in their cancellation. The elevation of a ground point X will then simply equal the elevation of the instrument station plus or minus ΔH, as computed from equation (2.29).

Its application to detailing is dealt with in *Section 1.5.1*.

An example of the booking and reduction of the field data is shown in Table 2.2.

2.16.7 Direct-reading tacheometers

As already indicated, to obtain the distance to and reduced level of a single point it is necessary to use three equations, (2.28), (2.29) and (2.32). This excessive amount of data processing, can be virtually eliminated by the use of a direct-reading tacheometer.

An example of one such instrument which clearly illustrates the principles involved, is shown in *Figure 2.50*. The conventional stadia lines, used to obtain the staff intercept S, are replaced by

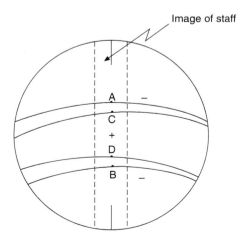

Fig. 2.50

Table 2.2

At station	A		Stn level (RL)	30.48 m OD		Survey	Canbury park
Grid ref	E 400, N 300		Ht of inst (h_i)	1.42 m		Surveyor	J. SMITH
Weather	Cloudy, cool		Axis level (RL + h_i)	31.90 m (Ax)		Date	12.12.93

Staff point	Angles observed					Staff readings	Staff intercept	Horizontal distance	Vertical height	Reduced level	Remarks
	Horizontal		Vertical circle		Vertical angle ±0			$K.S.\cos^2\theta$	$\dfrac{K}{2}\cdot S\cdot\sin 2\theta$		
	°	′	°	′	° ′	m	S	D	±H	Ax ± H − m	
RO	0	00									Station B
P1	48	12	95	20	−5 20	1.942 / m 1.404 / 0.866	1.076	106.67	−9.96	20.54	Edge of pond
P2	80	02	93	40	−3 40	0.998 / m 0.640 / 0.281	0.717	71.41	−4.58	26.68	Edge of pond
P3	107	56	83	20	+6 40	1.610 / m 1.216 / 0.822	0.788	77.74	+9.09	39.77	Edge of pond

curves. In *Figure 2.50*, the outer curves are of the function $\cos^2 \theta$, whilst the two inner are of the function $\sin \theta \cdot \cos \theta$. Thus the difference of the two outer curve staff readings A and B represents $S \cos^2 \theta$ and not just S. In this way the horizontal distance $D = (A - B)100$, which of course represents the basic formula. Similarly, $(C - D)100 = \Delta H$. The separation of the curves varies with variation in the vertical angle.

Different manufacturers' instruments have different methods of solving the problem. However, the objective remains the same, i.e. to eliminate computation. It should be noted that there is no improvement in accuracy.

Worked examples

Example 2.13 A theodolite has a tacheometric constant of 100 and an additive constant of zero. The centre reading on a vertical staff held on a point B was 2.292 m when sighted from A. If the vertical angle was $+25°$ and the horizontal distance AB 190.326 m, calculate the other staff readings and thus show that the two intercept intervals are not equal. Using these values calculate the level of B if A was 37.95 m and the height of the instrument 1.35 m.   (LU)

From basic equation
$$CD = 100S \cos^2 \theta$$
$$190.326 = 100S \cos^2 25°$$
$$\therefore S = 2.316 \text{ m}$$

From *Figure 2.51* $HJ = S \cos 25° = 2.1$ m

Inclined distance $CE = CD \sec 25° = 210$ m

$$\therefore 2\alpha = \frac{2.1}{210} \text{ rad} = 0°34'23''$$

$$\therefore \alpha = 0°17'11''$$

Now, by reference to *Figure 2.50*

$$DG = CD \tan (25° - \alpha) = 87.594$$
$$DE = CD \tan 25° \qquad = 88.749$$
$$DF = CD \tan (25° + \alpha) = 89.910$$

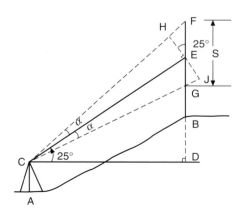

Fig. 2.51

It can be seen that the stadia intervals are

$$GE = S_1 = 1.155$$
$$EF = S_2 = 1.161$$
$$\left.\right\} = 2.316 \text{ (check)}$$

from which it is obvious that the

upper reading = (2.292 + 1.161) = 3.453

lower reading = (2.292 − 1.155) = 1.137

Vertical height $DE = \Delta H = CD \tan 25° = 88.749$ (as above)

∴ level of $B = 37.95 + 1.35 + 88.749 - 2.292 = 125.757$ m

Example 2.14 The following observations were taken with a tacheometer, having constants of 100 and zero, from a point *A* to *B* and *C*. The distance *BC* was measured as 157 m. Assuming the ground to be a plane within the triangle *ABC*, calculate the volume of filling required to make the area level with the highest point, assuming the sides to be supported by vertical concrete walls. Height of instrument was 1.4 m, the staff held vertically.   (LU)

At	To	Staff readings (m)	Vertical angle
A	B	1.48, 2.73, 3.98	+7°36′
	C	2.08, 2.82, 3.56	−5°24′

Horizontal distance $AB = 100 \times S \cos^2 \theta$

$$= 100 \times 2.50 \cos^2 7°36' = 246 \text{ m}$$

Vertical distance $AB = 246 \tan 7°36' = +32.8$ m

Similarly

Horizontal distance $AC = 148 \cos^2 5°24' = 147$ m

Vertical distance $AC = 147 \tan 5°24' = -13.9$ m

∴ Area of triangle $ABC = [S(S-a) \times (S-b) \times (S-c)]^{\frac{1}{2}}$

where $S = \frac{1}{2}(157 + 246 + 147) = 275$ m

∴ Area $= [275(275 - 157) \times (275 - 147) \times (275 - 246)]^{\frac{1}{2}}$

$$= 10975 \text{ m}^2$$

Assume level of $A = 100$ m

then level of $B = 100 + 1.4 + 32.8 - 2.73 = 131.47$ m

then level of $C = 100 + 1.4 - 13.9 - 2.82 = 84.68$ m

∴ Depth of fill at $A = 31.47$ m

Depth of fill at $C = 46.79$ m

Volume of fill = plan area × mean height

$$= 10\,975 \times \frac{1}{2}\,(31.47 + 46.79) = 286\,300 \text{ m}^3$$

Example 2.15 In order to find the radius of an existing road curve, three suitable points A, B and C were selected on its centre-line. The instrument was set at B and the following readings taken on A and C, the telescope being horizontal and the staff vertical.

Staff at	Horizontal bearing	Stadia readings (m)
A	0°00′	1.617 1.209 0.801
C	195°34′	2.412 1.926 1.440

If the instrument has a constant of 100 and 0, calculate the radius of the circular arc A, B, C. If the trunnion axis was 1.54 m above the road at B, find the gradients AB and BC.   (LU)

Note: As the theodolite is a clockwise-graduated instrument the angle ABC as shown in *Figure 2.52* equals 195°34′.

The angular relationships shown in the figure are from the geometry of angles at the centre being *twice* those at the circumference. It is therefore required to find angles BAC and BCA (α and β). From the formula for horizontal sights: $D = K_1 S + K_2$

$AB = 81.6$ m   and $BC = 97.2$ m

Assuming AB is 0°, then $BC = 15°34′$ for 97.2 m

$\therefore$ Coordinates of $BC = 97.2\frac{\sin}{\cos}15°34′ = +26.08(\Delta E), + 93.63(\Delta N)$

$\therefore$ Total coordinates of C relative to A = 26.08 E; (81.6 + 93.63) N = 175.23 N

Bearing $AC = \tan^{-1}\frac{26.08}{175.23} = 8°28′$

$\therefore \alpha = 8°28′$ and $\beta = (15°34′ - 8°28′) = 7°06′$

In triangle DCO, $R = 48.6/\sin 8°28′ = 330$ m

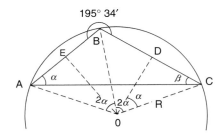

195° 34′

Fig. 2.52

Arc $AB = R \times 2\beta$ rad $= 330 \times 14°12'$ rad $= 81.78$ m

Arc $BC = R \times 2\alpha$ rad $= 330 \times 16°56'$ rad $= 97.53$ m

Grade $AB = (1.54 - 1.209)$ in $81.78 = 1$ in 250 falling from A to B

Grade $BC = (1.926 - 1.54)$ in $97.53 = 1$ in 250 falling from B to C

At alternative method of finding α and β would have been to use the equation:

$$\tan \frac{A - C}{2} = \frac{a - c}{a + c} \tan \frac{A + C}{2}$$

However, using coordinates involves less computation and precludes the memorizing of the equation in this case. This is particularly so in the next question where the above equation plus the sine rule would be necessary to find *CD*.

Example 2.16 The following readings were taken by a theodolite from station B on to stations A, C and D.

Sight	Horizontal angle	Vertical angle	Stadia readings (m)		
			Top	Centre	Bottom
A	301°10′				
C	152°36′	−5°00′	1.044	2.283	3.522
D	205°06′	+2°30′	0.645	2.376	4.110

The line *BA* in *Figure 2.53* has a bearing of $28°46'$ and the instrument constants are 100 and 0. Find the slope and bearing of line *CD*.   (LU)

Distance $BC = 100S \cos^2 \theta = 247.8 \cos^2 5° = 246$ m

Height $BC = 246 \tan 5° = -21.51$ m

Distance $BD = 346.5 \cos^2 2°30' = 345.9$ m

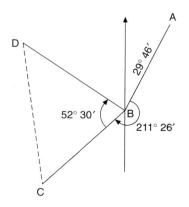

Fig. 2.53

Height $BD = 345.9 \tan 2°30' = 15.1$ m
Bearing $BC = (28°46' + 211°26') = 240°12'$
Bearing $BD = (240°12' + 52°30') = 292°42'$

$\therefore$ Coordinates of $BC = 246\genfrac{}{}{0pt}{}{\sin}{\cos} 240°12' = -213.5\ (\Delta E);\ -122.2\ (\Delta N)$

$\therefore$ Coordinates of $BD = 345.9\genfrac{}{}{0pt}{}{\sin}{\cos} 292°42' = -319.2\ (\Delta E);\ +133.5\ (\Delta N)$

$\therefore$ Coordinates of C relative to $D = -105.7\ (\Delta E);\ +255.7\ (\Delta N)$

$\therefore$ Bearing $CD = \tan^{-1}\dfrac{-105.7}{+255.7} = 327°32'$

Length $CD = 255.7/\cos 22°28' = 276.75$ m
Difference in level between C and $D = -(21.51 + 2.283) - (15.1 - 2.376) = 36.52$ m

$\therefore$ Grade $CD = 36.52$ in $276.75 = 1$ in 7.6 rising

Exercises

(2.7) In order to survey an existing road, three points A, B, and C were selected on its centre-line. The instrument was set at A and the following observations were taken.

Staff	Horizontal angle	Vertical angle	Stadia readings (m)
B	0°00'	−1°11' 20"	1.695 1.230 0.765
C	6°29'	−1°04' 20"	2.340 1.500 0.660

If the staff was vertical and the instrument constants 100 and 0, calculate the radius of the curve ABC. If the instrument was 1.353 m above A, find the falls A to B and B to C. (LU)

(Answer: $R = 337.8$ m, $A - B = 1.806$ m, $B - C = 1.482$ m)

(2.8) Readings were taken on a vertical staff held at points A, B and C with a tacheometer whose constants were 100 and 0. If the horizontal distances from instrument to staff were respectively 45.9, 63.6 and 89.4 m, and the vertical angles likewise $+5°$, $+6°$ and $-5°$, calculate the staff intercepts. If the mid-hair reading was 2.100 m in each case, what was the difference in level between A, B and C?

(Answer: $S_A = 0.462$, $S_B = 0.642$, $S_C = 0.900$, B is 2.670 m above A, C is 11.835 m below A)

(2.9) A theodolite has a multiplying constant of 100 and an additive constant of zero. When set 1.35 m above station B, the following readings were obtained.

Station	Sight	Horizontal circle	Vertical circle	Stadia readings (m)
B	A	28° 21' 00"		
B	C	82° 03' 00"	20°30'	1.140 2.292 3.420

The coordinates of A are E 163.86, N 0.0, and those of B, E 163.86, N 118.41. Find the coordinates of C and its height above datum if the level of B is 27.3 m AOD. (LU)

(Answer: E 2.64 N 0.0, 101.15 m AOD)

3
Distance

Distance is one of the fundamental measurements in surveying. Although frequently measured as a spatial distance (sloping distance) in three-dimensional space, inevitably it is the horizontal equivalent which is required.

Distance is required in many instances, e.g. to give scale to a network of control points, to fix the position of topographic detail by offsets or polar coordinates, to set out the position of a point in construction work, etc.

The basic methods of measuring distance are, at the present time, by *taping* or by *electro-magnetic* (or *electro-optical*) *distance measurement*, generally designated as EDM. For very rough reconnaissance surveys or approximate estimates, pacing may be suitable, whilst, in the absence of any alternative, optical methods may be used.

It may be that in the near future, EDM will be rendered obsolete for distances over 5 km by the use of GPS satellite methods. These methods can obtain vectors between two points accurate to 0.1 ppm.

3.1 TAPES

Tapes come in a variety of lengths and materials. For engineering work the lengths are generally 10 m, 30 m, 50 m and 100 m.

For general use, where precision is not a prime consideration, linen or glass fibre tapes may be used. The linen tapes are made from high-class linen, combined with metal fibres to increase their strength. They are usually encased in plastic boxes with recessed handles. These tapes are graduated in 5-mm intervals only.

More precise versions of the above tapes are made of steel and graduated in millimetres.

For high-accuracy work, steel bands mounted in an open frame are used. They are standardized so that they measure their nominal length when the temperature is 20°C and the applied tension between 50 N to 80 N. This information is clearly printed on the zero end of the tape. *Figure 3.1* shows a sample of the equipment.

For the very highest calibre of work, invar tapes made from 35% nickel and 65% steel are available. The singular advantage of such tapes is that they have a negligible coefficient of expansion compared with steel, and hence temperature variations are not critical. Their disadvantages are that the metal is soft and weak, whilst the price is more than ten times that of steel tapes. An alternative tape, called a Lovar tape, is roughly, midway between steel and invar.

Much ancillary equipment is necessary in the actual taping process, e.g.

(1) Ranging rods, made of wood or steel, 2 m long and 25 mm in diameter, painted alternately red and white, with a pointed metal shoe to allow it to be thrust into the ground. They are generally used to align a straight line between two points.

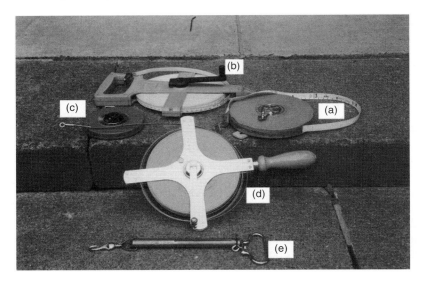

Fig. 3.1   *(a) Linen tape, (b) fibreglass, (c) steel, (d) steel band, (e) spring balance*

(2) Chaining arrows made from No. 12 steel wire are also used to mark the tape lengths (*Figure 3.2*).

(3) Spring balances generally used with roller-grips or tapeclamps to firmly grip the tape when the standard tension is applied. As it is quite difficult to maintain the exact tension required with a spring balance, it may be replaced by a tension handle, which ensures the application of correct tension.

(4) Field thermometers are also necessary to record the tape temperature at the time of measurement, thereby permitting the computation of tape corrections when the temperature varies from

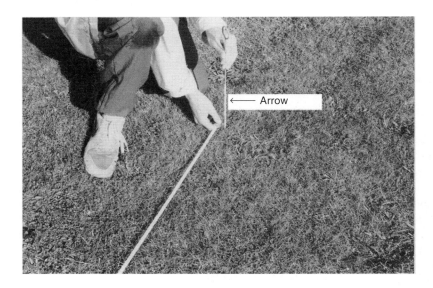

Fig. 3.2   *The use of a chaining arrow to mark the position of the end of the tape*

standard. These thermometers are metal cased and can be clipped onto the tape if necessary, or simply laid on the ground alongside the tape.
(5) Hand levels may be used to establish the tape horizontal. This is basically a hand-held tube incorporating a spirit bubble to ensure a horizontal line of sight. Alternatively, an Abney level may be used to measure the slope of the ground.
(6) Plumb-bobs may be necessary if stepped taping is used.
(7) Measuring plates are necessary in rough ground, to afford a mark against which the tape may be read. *Figure 3.3* shows the tensioned tape being read against the edge of such a plate. The corners of the triangular plate are turned down to form grips, when it is pressed into the earth and thereby prevent its movement.

In addition to the above, light oil and cleaning rags should always be available to clean and oil the tape after use.

3.2 FIELD WORK

3.2.1 Measuring along the ground (Figures 3.3 and 3.4)

The most accurate way to measure distance with a steel band is to measure the distance between pre-set measuring marks, rather than attempt to mark the end of each tape length. The procedure is as follows:

(1) The survey points to be measured should be set flush with the ground surface. Ranging rods are then set behind each peg, in the line of measurement.
(2) Using a linen tape, arrows are aligned between the two points at intervals less than a tape length. Measuring plates are then set firmly in the ground at these points, with their measuring edge normal to the direction of taping.
(3) The steel band is then carefully laid out, in a straight line between the survey point and the first plate. One end of the tape is firmly anchored, whilst tension is slowly applied at the other end.

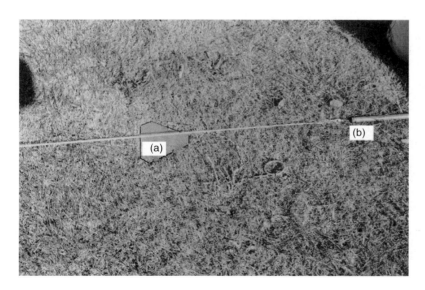

Fig. 3.3 *(a) Measuring plate, (b) spring balance tensioning the tape*

Fig. 3.4 *Plan view*

At the exact instant of standard tension, both ends of the tape are read simultaneously against the survey station point and the measuring plate edge respectively, on command from the person applying the tension. The tension is eased and the whole process repeated at least four times or until a good set of results is obtained.

(4) When reading the tape, the metres, decimetres and centimetres should be noted as the tension is being applied; thus on the command 'to read', only the millimetres are required.

(5) The readings are noted by the booker and quickly subtracted from each other to give the length of the measured bay.

(6) In addition to 'rear' and 'fore' readings, the tape temperature is recorded, the value of the applied tension, which may in some instances be greater than standard, and the slope or difference in level of the tape ends.

(7) This method requires five operatives:
one to anchor the tape end
one to apply tension
two observers to read the tape and one booker

(8) The process is repeated for each bay of the line being measured, care being taken not to move the first measuring plate, which is the start of the second bay, and so on.

(9) The data may be booked as follows:

Bay	Rear	Fore	Difference	Temp.	Tension	Slope	Remarks
A–1	0.244	29.368	29.124	08°C	70 N	5°30′	Standard Values
	0.271	29.393	29.122				20°C, 70 N
	0.265	29.389	29.124				
	0.259	29.382	29.123				Range 2 mm
	Mean =		29.123				
1–2							2nd bay

The mean result is then corrected for:

(1) Tape standardization.
(2) Slope.
(3) Temperature.
(4) Tension (if necessary).

The final total distance may then be reduced to its equivalent MSL or mean site level.

3.2.2 Measuring in catenary

Although the measurement of base lines in catenary is virtually obsolete, it is still the most accurate method of obtaining relatively short distances over rough terrain. The only difference from the procedures outlined above is that the tape is raised off the ground between two measuring marks and so the tape sags in catenary.

Figure 3.5 shows the basic set-up, with tension applied by levering back on a ranging rod held through the handle of the tape (*Figure 3.6*).

Figure 3.7 shows a typical measuring head with magnifier attached. In addition to the corrections already outlined, a further correction for sag in the tape is necessary.

For extra precision the measuring heads may be aligned in a straight line by theodolite, the difference in height of the heads being obtained by levelling to a staff held directly on the heads.

3.2.3 *Step measurement*

The process of step measurement has already been outlined in *Chapter 1*. This method of measurement over sloping ground should be avoided if high accuracy is required. The main source of error lies in attempting to accurately locate the suspended end of the tape, as shown in *Figure 3.8*.

The steps should be kept short enough to minimize sag in the tape, and thus the sum of the steps equals the horizontal distance required.

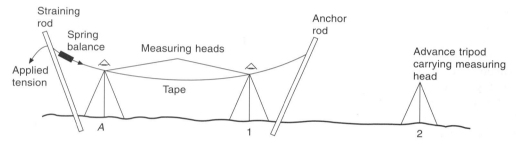

Fig. 3.5

Fig. 3.6 *Tension being applied with the aid of a ranging rod to a steel band suspended in catenary*

Fig. 3.7   *Measuring head*

Fig. 3.8   *Step measurement*

3.3 DISTANCE ADJUSTMENT

To eliminate or minimize the systematic errors of taping, it is necessary to adjust each measured bay to its final horizontal equivalent as follows.

3.3.1 Standardization

During a period of use, a tape will gradually alter in length for a variety of reasons. The amount of change can be found by having the tape standardized at either the National Physical Laboratory (NPL) (invar) or the Department of Trade and Industry (DTI) (steel), or by comparing it with a reference tape kept purely for this purpose. The tape may then be specified as being 30.003 m at 20°C and 70 N tension or as 30 m exactly, at a temperature other than standard.

Worked examples

Example 3.1 A distance of 220.450 m was measured with a steel band of nominal length 30 m. On standardization the tape was found to be 30.003 m. Calculate the correct measured distance, assuming the error is evenly distributed throughout the tape.

Error per 30 m = 3 mm

$\therefore$ Correction for total length $= \left(\dfrac{220.450}{30}\right) \times 3 \text{ mm} = 22 \text{ mm}$

$\therefore$ Correct length is 220.450 + 0.022 = 220.472 m

Student notes

(1) *Figure 3.9* shows that when the tape is too long, the distance measured appears too short, and the correction is therefore positive. The reverse is the case when the tape is too short.
(2) When setting out a distance with a tape the rules in (1) are reversed.
(3) It is better to compute *Example 3.1* on the basis of the correction (as shown), rather than the total corrected length. In this way fewer significant figures are used.

Example 3.2 A 30-m band standardized at 20°C was found to be 30.003 m. At what temperature is the tape exactly 30 m? Coefficient of expansion of steel = 0.000 011/°C.

Expansion per 30 m per °C = 0.000 011 × 30 = 0.000 33 m

Expansion per 30 m per 9°C = 0.000 33 × 9 = 0.003 m

$\therefore$ Tape is 30 m at 20°C − 9°C = 11°C

Alternatively, using equation (3.1) where $\Delta t = (t_s - t_a)$, then

$$t_a = \frac{C_t}{KL} + t_s = -\left(\frac{0.003}{0.000\ 011 \times 30}\right) + 20°C = 11°C$$

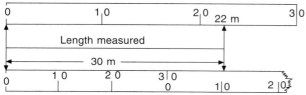

Fig. 3.9

where t_a = actual temperature and t_s = standard temperature.

This then becomes the standard temperature for future temperature corrections.

3.3.2 Temperature

Tapes are usually standardized at 20°C. Any variation above or below this value will cause the tape to expand or contract, giving rise to systematic errors. The difficulty of obtaining the true temperature of the tape resulted in the use of invar tapes. Invar is a nickel-steel alloy with a very low coefficient of expansion.

Coefficint of expansion of steel	$K = 11.2 \times 10^{-6}$ per °C	
Coefficient of expansion of invar	$K = 0.5 \times 10^{-6}$ per °C	
Temperature correction	$C_t = KL\Delta t$	(3.1)

where L = measured length (m) and Δt = difference between the standard and field temperatures (°C) = $(t_s - t_a)$

The sign of the correction is in accordance with the rule specified in (1) of the student notes mentioned earlier.

3.3.3 Tension

Generally the tape is used under standard tension, in which case there is no correction. It may, however, be necessary in certain instances to apply a tension greater than standard. From Hooke's law:

stress = strain × a constant

This constant is the same for a given material and is called the *modulus of elasticity* (E). Since strain is a non-dimensional quantity, E has the same dimensions as stress, i.e. N/mm²:

$$\therefore E = \frac{\text{Direct stress}}{\text{Corresponding strain}} = \frac{\Delta T}{A} \div \frac{C_T}{L}$$

$$\therefore C_T = L \times \frac{\Delta T}{AE} \tag{3.2}$$

ΔT is normally the total stress acting on the cross-section, but as the tape would be standardized under tension, ΔT in this case is the amount of stress *greater* than standard. Therefore ΔT is the difference between field and standard tension. This value may be measured in the field in kilograms and should be converted to newtons (N) for compatibility with the other units used in the formula, i.e. 1 kgf = 9.806 65 N.

E is modulus of elasticity in N/mm²; A is cross-sectional area of the tape in mm²; L is measured length in m; and C_T is the extension and thus correction to the tape length in m. As the tape is stretched under the extra tension, the correction is positive.

3.3.4 Sag

When a tape is suspended between two measuring heads, A and B, both at the same level, the shape it takes up is a catenary (*Figure 3.10*). If C is the lowest point on the curve, then on length CB there are three forces acting, namely the tension T at B, T_0 at C and the weight of portion CB, where w

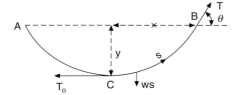

Fig. 3.10

is the weight per unit length and s is the arc length CB. Thus CB must be in equilibrium under the action of these three forces. Hence

Resolving vertically $\qquad T \sin \theta = ws$

Resolving horizontally $\qquad T \cos \theta = T_0$

$$\therefore \tan \theta = \frac{ws}{T_0}$$

For a small increment of the tape

$$\frac{dx}{ds} = \cos \theta = (1 + \tan^2\theta)^{-\frac{1}{2}} = \left(1 + \frac{w^2s^2}{T_0^2}\right)^{-\frac{1}{2}} = \left(1 - \frac{w^2s^2}{2T_0^2} \cdots\right)$$

$$\therefore x = \int \left(1 - \frac{w^2s^2}{2T_0^2}\right) ds$$

$$= s - \frac{w^2s^3}{6T_0^2} + K$$

When $x = 0$, $s = 0$, $\quad \therefore K = 0 \quad \therefore x = s - \frac{w^2s^3}{6T_0^2}$

The sag correction for the whole span $\quad ACB = C_s = 2(s - x) = 2\left(\frac{w^2s^3}{6T_0^2}\right)$

but $s = L/2$ $\qquad \therefore C_s = \frac{w^2L^3}{24T_0^2} = \frac{w^2L^3}{24T^2}$ for small values of θ $\qquad$ (3.3)

i.e. $T \cos \theta \approx T \approx T_0$

where $\quad w =$ weight per unit length (N/m)
$\qquad T =$ tension applied (N)
$\qquad L =$ recorded length (m)
$\qquad C_s =$ correction (m)

As $w = W/L$, where W is the *total* weight of the tape, then by substitution in equation (3.3):

$$C_s = \frac{W^2L}{24T^2} \qquad\qquad (3.4)$$

Although this equation is correct, the sag correction is proportional to the cube of the length.

Equations (3.3) and (3.4) apply only to tapes standardized on the flat and are always negative. When a tape is standardized in catenary, i.e. it records the horizontal distance when hanging in sag, no correction is necessary provided the applied tension, say T_A, equals the standard tension T_s. Should the tension T_A exceed the standard, then a sag correction is necessary for the excess tension $(T_A - T_s)$ and

$$C_s = \frac{w^2 L^3}{24} \left(\frac{1}{T_A^2} - \frac{1}{T_S^2} \right) \tag{3.5}$$

In this case the correction will be positive, in accordance with the basic rule. The sag y of the tape may also be found as follows:

$$\frac{dy}{ds} = \sin \theta \approx \tan \theta = \frac{ws}{T_0}, \text{ when } \theta \text{ is small}$$

$$\therefore y = \int \frac{ws}{T_0} \, ds = \frac{ws^2}{2T_0}$$

If y is the maximum sag at the centre of the tape, then

$$s = \frac{L}{2} \quad \text{and} \quad y = \frac{wL^2}{8T} \tag{3.6}$$

Equation (3.6) enables w to be found from field measurement of sag, i.e.

$$w = \frac{8Ty}{L^2} \tag{3.7}$$

which on substitution in equation (3.4) gives

$$C_S = -\frac{8y^2}{3L} \tag{3.8}$$

Equation (3.8) gives the sag correction by measuring sag y and is independent of w and T.

3.3.5 Slope

If the difference in height of the two measuring heads is h, the slope distance L and the horizontal equivalent D, then by Pythagoras

$$D = (L^2 - h^2)^{\frac{1}{2}} \tag{3.9}$$

Prior to the use of pocket calculators the following alternative approach was generally used, due to the tedium of obtaining square roots

$$\therefore D = (L^2 - h^2)^{\frac{1}{2}} = L \left(1 - \frac{h^2}{L^2} \right)^{\frac{1}{2}} = L \left(1 - \frac{h^2}{2L^2} - \frac{h^4}{8L^4} \right)$$

$$\therefore \text{Slope correction} \quad C_h = D - L = - \left(\frac{h^2}{2L} + \frac{h^4}{8L^3} \right) \tag{3.10}$$

The use of Pythagoras is advocated due to the small error that can arise when using only two terms of the above expansion on long lines measured by EDM.

On the relatively short lines involved in taping, the first term $-h^2/2L$ will generally suffice. Alternatively if the vertical angle of the slope of the ground is measured then:

$$D = L \cos \theta \tag{3.11}$$

and the correction $C_\theta = L - D$.

$$C_\theta = L (1 - \cos \theta) \tag{3.12}$$

3.3.6 Altitude

If the surveys are to be connected to the national mapping system of a country, the distances will need to be reduced to the common datum of that system, namely MSL. Alternatively, if the engineering scheme is of a local nature, distances may be reduced to the mean level of the area. This has the advantage that setting-out distances on the ground are, without sensible error, equal to distances computed from coordinates in the mean datum plane.

Consider *Figure 3.11* in which a distance L is measured in a plane situated at a height H above MSL.

By similar triangles $M = \dfrac{R}{R + H} \times L$

$\therefore$ Correction $C_M = L - M = L - \dfrac{RL}{R + H} = L \left(1 - \dfrac{R}{R + H} \right) = \dfrac{LH}{R + H}$

As H is negligible compared with R in the denominator

$$C_M = \frac{LH}{R} \tag{3.13}$$

The correction is negative for surface work but may be positive for tunnelling or mining work below MSL.

3.4 ERRORS IN TAPING

Methods of measuring with a tape have been dealt with, although it must be said that training in the methods is best learnt in the field. The quality of the end results, however, can only be appreciated by an understanding of the errors involved. Of all the methods of measuring, taping is probably the least automated and therefore most susceptible to personal and natural errors. The majority of errors affecting taping are systematic, not random, and their effect will therefore increase with the number of bays measured.

The errors basically arise due to defects in the equipment used; natural errors due to weather conditions and human errors resulting in tape-reading errors etc. They will now be dealt with individually.

3.4.1 Standardization

Taping cannot be more accurate than the accuracy to which the tape is standardized. It should therefore be routine practice to have one tape standardized by the appropriate authority.

This is done on payment of a small fee; the tape is returned with a certificate of standardization quoting the 'true' length of the tape and standard conditions of temperature and tension. This tape is then kept purely as a standard with which to compare working tapes.

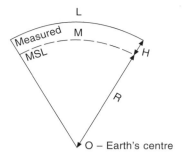

Fig. 3.11

Alternatively a base line may be established on site and its length obtained by repeated measurements using, say, an invar tape hired purely for that purpose. The calibration base should be then checked at regular intervals to confirm its stability.

3.4.2 Temperature

Neglecting temperature effects could constitute a main source of error in measurement with a steel tape. For example, measuring in winter conditions in the UK, with temperatures at 0°C, would cause a 50-m tape, standard at 20°C, to contract by

$$11.2 \times 10^{-6} \times 50 \times 20 = 11.2 \text{ mm per 50 m}$$

Thus even for ordinary precision measurement, this cannot be ignored.

Even if the tape temperature is measured there may be an index error in the thermometer used, part of the tape may be in shade and part in the sun, or the thermometer may record ground or air temperature which may not be the tape temperature. Although the use of an invar tape would resolve the problem, this is rarely, if ever, a solution applied on site. This is due to the high cost of such tapes and their fragility. The effect of an error in temperature measurement can be assessed by differentiating the basic equation, i.e.

$$\delta C_t = KL \ \delta(\Delta t)$$

If $L = 50$ m and the error in temperature is ± 2°C then $\delta C_t = \pm 1.1$ mm. However, if this error remained constant it would be proportional to the number of tape lengths. Every effort should therefore be made to obtain an accurate value for tape temperature using calibrated thermometers.

3.4.3 Tension

If the tension on the tape is greater or less than standard, the tape will stretch or shorten accordingly. Tension applied without the aid of a spring balance or tension handle may vary from length to length, resulting in random error. Tensioning equipment containing error would produce a systematic error proportional to the number of tape lengths. The effect of this error is greater on a light tape having a small cross-sectional area than on a heavy tape.

Consider a 50-m tape with a cross-sectional area of 4 mm^2, a standard tension of 50 N and a value for the modulus of elasticity of $E = 210$ kN/mm^2. Under a pull of 90 N the tape would stretch by

$$C_T = \frac{50\,000 \times 40}{4 \times 210 \times 10^3} = 2.4 \text{ mm}$$

As this value is proportional to the number of tape lengths it is very necessary to cater for it in precision measurement, using calibrated tensioning equipment.

3.4.4 Sag

The correction for sag is equal to the difference in length between the arc and its subtended chord and is always negative. As the sag correction is a function of the weight of the tape, it will be greater for heavy tapes than light ones. Correct tension is also very important.

Consider a 50-m heavy tape of $W = 1.7$ kg with a standard tension of 80 N:

$$C_S = \frac{(1.7 \times 9.81)^2 \times 50}{24 \times 80^2} = 0.090 \text{ m}$$

and indicates the large corrections necessary.

If the above tape was supported throughout its length to form three equal spans, the correction per span reduces to 0.003 m. This important result shows that the sag correction could be virtually eliminated by the choice of appropriate support.

The effect of an error in tensioning can be found by differentiating with respect to T:

$$C_s = W^2 L / 24 T^2$$
$$\delta C_s = -2 W^2 L \, \delta T / 24 T^3$$

In the above case, if the error in tensioning was ± 5 N, then the error in the correction for sag would be ± 0.01 m. This result indicates the importance of calibrating the tensioning equipment.

The effect of error in the weight (W) of the tape can be found using

$$\delta C_s = 2 W L \, \delta W / 24 T^2$$

and shows that an error of ± 0.1 kg in W produces an error of ± 0.011 m in the sag correction. The compounded effect of both these errors would be ± 0.016 m and cannot be ignored.

3.4.5 Slope

Correction for slope is always important.

Consider a 50-m tape measuring on a slope with a difference in height of 5 m. The correction for slope is

$$C_h = -h^2/2L = -25/100 = -0.250 \text{ m}$$

and would constitute a major source of error if ignored. The second-order error resulting from non-use of the second term $h^4/8L^3$ is less than 1 mm.

Error in the measurement of the difference in height (h) can be assessed using

$$\delta C_h = h \, \delta h / L$$

Assuming an error of ± 0.005 m, it would produce an error of ± 0.0005 m (δC_h). Thus error in obtaining the difference in height is negligible and as it is proportional to h, would get smaller on less steep slopes.

Considering slope reduction using the vertical angle θ, we have

$$\delta C_\theta = L \sin \theta \, \delta\theta$$
$$\text{and } \delta \theta'' = \delta C_\theta \times 206\,265/L \sin \theta$$

If $L = 50$ m is required to an accuracy of ± 5 mm on a slope of $5°$ then

$$\delta\,\theta'' = 0.005 \times 206265/50 \sin 5° = 237'' \approx 04'$$

This level of accuracy could easily be achieved using an Abney level to measure slope. As the slopes get less steep the accuracy required is further reduced; however, for the much greater distances obtained using EDM, the measurement of vertical angles is much more critical. Indeed, if the accuracy required above is increased to that of precision measurement, say ± 1 mm, the angle accuracy required rises to $\pm 47''$ and would require the use of a theodolite.

3.4.6 Misalignment

If the tape is not in a straight line between the two points whose distance apart is being measured, then the error in the horizontal plane will be equivalent to that of slope in the vertical plane. Taking the amount by which the end of the tape is off line equal to e, then the resultant error is $e^2/2L$.

A 50-m tape, off line by 0.500 m (an excessive amount), would be in error by 2.5 mm. The error is systematic and will obviously result in a recorded distance longer than the actual. If we consider a more realistic error in misalignment of, say, 0.05 m, the resulting error is 0.025 mm and completely negligible. Thus for the majority of taping, alignment by eye is quite adequate.

3.4.7 Plumbing

If stepped measurement is used, locating the end of the tape by plumb-bob will inevitably result in error. Plumbing at its best will probably produce a random error of about ± 3 mm. In difficult, windy conditions it would be impossible to use, unless sheltered from the wind by some kind of makeshift wind break, combined with careful steadying of the bob itself.

3.4.8 Incorrect reading of the tape

Reading errors are random and quite prevalent amongst beginners. Care and practice are needed to obviate them.

3.4.9 Booking error

Booking error can be reduced by adopting the process of the booker reading the measurement back to the observer for checking purposes. However, when measuring to millimetres with tensioning equipment, the reading has usually altered by the time it comes to check it. Repeated measurements will generally reveal booking errors, and thus distances should always be measured more than once.

3.5 ACCURACIES

If a great deal of taping measurement is taking place, then it is advisable to construct graphs of all the corrections for slope, temperature, tension and sag, for a variety of different conditions. In this way the values can be rapidly obtained and more easily considered. Such an approach rapidly produces in the measurer a 'feel' for the effect of errors in taping.

As random errors increase as the square root of the distance, systematic errors are proportional to distance and reading errors are independent of distance, it is not easy to produce a precise assessment of taping accuracies under variable conditions. Considering taping with a standardized tape corrected only for slope, one could expect an accuracy in the region of 1 in 5000 to 1 in 10 000. With extra care and correcting for all error sources the accuracy would rise to the region of 1 in 30 000. Precise measurement in catenary is capable of very high accuracies of 1 in 100 000 and above. However, the type of catenary measurement carried out on general site work would probably achieve about 1 in 50 000.

The worked examples should now be carefully studied as they illustrate the methods of applying these corrections both for measurement and setting out.

As previously mentioned, the signs of the corrections for measurement are reversed when setting out. As shown, measuring with a tape which is too long produces a smaller distance which requires positive correction. However, a tape that is too long will set out a distance that is too long and will require a negative correction. This can be expressed as follows:

Horizontal distance (D) = Measured distance (M) + The algebraic sum of the corrections (C)

i.e. $D = M + C$

When setting out, the horizontal distance to be set out is known and the engineer needs to know its equivalent measured distance on sloping ground to the accuracy required. Therefore $M = D - C$, which has the effect of automatically reversing the signs of the correction. Therefore, compute the corrections in the normal way as if for a measured distance and then substitute the algebraic sum in the above equation.

Worked examples

Example 3.3 A base line was measured in catenary in four lengths giving 30.126, 29.973, 30.066 and 22.536 m. The differences of level were respectively 0.45, 0.60, 0.30 and 0.45 m. The temperature during observation was 10°C and the tension applied 15 kgf. The tape was standardized as 30 m, at 20°C, on the flat with a tension of 5 kg. The coefficient of expansion was 0.000 011 per °C, the weight of the tape 1 kg, the cross-sectional area 3 mm², $E = 210 \times 10^3$ N/mm² (210 kN/mm²), gravitational acceleration g = 9.806 65 m/s².

(a) Quote each equation used and calculate the length of the base.
(b) What tension should have been applied to eliminate the sag correction?   (LU)

(a) As the field tension and temperature are constant throughout, the first three corrections may be applied to the base as a whole, i.e. L = 112.701 m, with negligible error.

	+	−
Tension $C_T = \dfrac{L\Delta_T}{AE} = \dfrac{112.701 \times (10 \times 9.806\,65)}{3 \times 210 \times 10^3} =$	+0.0176	
Temperature $C_t = LK\Delta t = 112.701 \times 0.000\,011 \times 10 =$		−0.0124
Sag $C_s = \dfrac{LW^2}{24T^2} = \dfrac{112.701 \times 1^2}{24 \times 15^2}$		−0.0210
Slope $C_h = \dfrac{h^2}{2L} = \dfrac{1}{2 \times 30}(0.45^2 + 0.60^2 + 0.30^2) + \dfrac{0.45^2}{2 \times 22.536} =$		−0.0154
	+0.0176	−0.0488

Horizontal length of base (D) = measured length (M) + sum of corrections (C)

$$= 112.701 \text{ m} + (-0.031)$$

$$= 112.670 \text{ m}$$

N.B. In the slope correction the first three bays have been rounded off to 30 m, the resultant second order error being negligible.

Consider the situation where 112.670 m is the horizontal distance to be set out on site. The equivalent measured distance would be:

$$M = D - C$$

$$= 112.670 - (-0.031) = 112.701 \text{ m}$$

(b)   To find the applied tension necessary to eliminate the sag correction, equate the two equations:

$$\frac{\Delta T}{AE} = \frac{W^2}{24T_A^2}$$

where ΔT is the difference between the applied and standard tensions, i.e. $(T_A - T_S)$.

$$\therefore \frac{(T_A - T_S)}{AE} = \frac{W^2}{24T_A^2}$$

$$\therefore T_A^3 - T_A^2 T_S - \frac{AEW^2}{24} = 0$$

Substituting for T_S, W, A and E, making sure to convert T_S and W to newtons

gives $$T_A^3 - 49T_A^2 - 2\,524\,653 = 0$$

Let $$T_A = (T + x)$$

then $$(T + x)^3 - 49(T + x)^2 - 2\,524\,653 = 0$$

$$T^3 \left(1 + \frac{x}{T}\right)^3 - 49T^2 \left(1 + \frac{x}{T}\right)^2 - 2\,524\,653 = 0$$

Expanding the brackets binomially gives

$$T^3 \left(1 + \frac{3x}{T}\right) - 49T^2 \left(1 + \frac{2x}{T}\right) - 2\,524\,653 = 0$$

$$\therefore T^3 + 3T^2 x - 49T^2 - 98Tx - 2\,524\,653 = 0$$

$$\therefore x = \frac{2\,524\,653 - T^3 + 49T^2}{3T^2 - 98T}$$

assuming $T = 15$ kgf $= 147$ N, then $x = 75$ N

$\therefore$ at the first approximation $T_A = (T + x) = 222$ N

Example 3.4 A base line was measured in catenary as shown below, with a tape of nominal length 30 m. The tape measured 30.015 m when standardized in catenary at 20°C and 5 kgf tension. If the mean reduced level of the base was 30.50 m OD, calculate its true length at mean sea level.

Given: weight per unit length of tape = 0.03 kg/m (w); density of steel = 7690 kg/m^3 (ρ); coefficient of expansion = 11×10^{-6} per °C; $E = 210 \times 10^3$ N/mm^2; gravitational acceleration $g = 9.806\,65$ m/s^2; radius of the Earth = 6.4×10^6 m (R).   (KU)

Bay	Measured length (m)	Temperature (°C)	Applied Tension (kgf)	Difference in level (m)
1	30.050	21.6	5	0.750
2	30.064	21.6	5	0.345
3	30.095	24.0	5	1.420
4	30.047	24.0	5	0.400
5	30.041	24.0	7	–

Standardization:

		+	−

Error/30 m = 0.015 m

Total length of base = 150.97 m

$$\therefore \text{Correction} = \frac{150.297}{30} \times 0.015 = \qquad +0.0752$$

Temperature:

Bays 1 and 2 $C_t = 60 \times 11 \times 10^{-6} \times 1.6 = 0.0010 \text{ m}$

Bays 3, 4 and 5 $C_t = 90 \times 11 \times 10^{-6} \times 4 = 0.0040 \text{ m}$ $\Big\}$ +0.0050

(Second-order error negligible in rounding off bays to 30 m.)

Tension:

Bay 5 only $C_T = \dfrac{L\Delta T}{AE}$, changing ΔT to newtons

where cross-sectional area $A = \dfrac{w}{\rho}$

$$\therefore A = \frac{0.03}{7690} \times 10^6 = 4 \text{ mm}^2$$

$$\therefore C_T = \frac{30 \times 2 \times 9.81}{4 \times 210 \times 10^3} = \qquad +0.0007$$

Slope:

$$C_h = \frac{h^2}{2L} - \frac{1}{2 \times 30}(0.750^2 + 0.345^2 + 1.420^2 + 0.400^2) = \qquad -0.0476$$

The second-order error in rounding off to 30 m is negligible in this case also. However, care should be taken when many bays are involved, as their accumulative effect may be significant.

Sag:

Bay 5 only $C_s = \dfrac{L^3 w^2}{24}\left(\dfrac{1}{T_s^2} - \dfrac{1}{T_A^2}\right)$

$$= \frac{30^3 \times 0.03^2}{24}\left(\frac{1}{5^2} - \frac{1}{7^2}\right) = \qquad +0.0006$$

Altitude:

$$C_M = \frac{LH}{R} = \frac{150 \times 30.5}{6.4 \times 10^6} = \qquad -0.0007$$

+0.0815 | −0.0483

Therefore   Total correction = +0.0332 m
Hence   Corrected length = 150.297 + 0.0332 = 150.3302 m

Example 3.5 (a) A standard base was established by accurately measuring with a steel tape the distance between fixed marks on a level bed. The mean distance recorded was 24.984 m at a temperature of 18°C and an applied tension of 155 N. The tape used had recently been standardized in catenary and was 30 m in length at 20°C and 100 N tension. Calculate the true length between the fixed marks given: total weight of the tape = 0.90 kg; coefficient of expansion of steel = 11×10^{-6} per °C; cross-sectional area = 2 mm^2; $E = 210 \times 10^3$ N/mm^2; gravitational acceleration = 9.807 m/s^2.

   (b)   At a later date the tape was used to measure a 30-m bay in catenary. The difference in level of the measuring heads was 1 m, with an error of 3 mm. Tests carried out on the spring balance indicated that the applied tension of 100 N had an error of 2 N. Ignoring all other sources of error, what is the probable error in the measured bay?   (KU)

   (a) If the tape was standardized in catenary, then when laid on the flat it would be too long by an amount equal to the sag correction. This amount, in effect, then becomes the standardization correction:

$$\text{Error per 30 m} = \frac{LW^2}{24T_s^2} = \frac{30 \times (0.90 \times 9.807)^2}{24 \times 100^2} = 0.0097 \text{ m}$$

$$\therefore \text{ Correction} = \frac{0.0097 \times 24.984}{30} = 0.0081 \text{ m}$$

$$\text{Tension} = \frac{24.984 \times 55}{2 \times 210 \times 10^3} = 0.0033 \text{ m}$$

$$\text{Temperature} = 24.984 \times 11 \times 10^{-6} \times 2 = -0.0006 \text{ m}$$

$\therefore$ Total correction = 0.0108 m

$\therefore$ Corrected length = 24.984 + 0.011 = 24.995 m

(b) *Effect of levelling error:* $C_h = \dfrac{h^2}{2L}$

$$\therefore \ \delta C_h = \frac{h \times \delta h}{L} = \frac{1 \times 0.003}{30} = 0.0001 \text{ m}$$

Effect of tensioning error: Sag $C_s = \dfrac{LW^2}{24T^2}$

$$\therefore \ \delta C_s = -\frac{LW^2}{12T^3} \ \delta T$$

$$\therefore \ \delta C_s = \frac{30 \times (0.9 \times 9.807)^2 \times 2}{12 \times 100^3} = 0.0004 \text{ m}$$

Tension $C_T = \dfrac{L\Delta T}{AE}$

$$\therefore \ \delta C_T = \frac{L \times \delta(\Delta T)}{A \times E} = \frac{30 \times 2}{2 \times 210 \times 10^3} = 0.0001 \text{ m}$$

$\therefore$ Total error = 0.0006 m

Example 3.6 A 30-m invar reference tape was standardized on the flat and found to be 30.0501 m at 20°C and 88 N tension. It was used to measure the first bay of a base line in catenary, the mean recorded length being 30.4500 m.

Using a field tape, the mean length of the same bay was found to be 30.4588 m. The applied tension was 88 N at a constant temperature of 15°C in both cases.

The remaining bays were now measured in catenary, using the field tape only. The mean length of the second bay was 30.5500 m at 13°C and 100 N tension. Calculate its reduced length given: cross-sectional area = 2 mm^2; coefficient of expansion of invar = 6×10^{-7} per °C; mass of tape per unit length = 0.02 kg/m; difference in height of the measuring heads = 0.5 m; mean altitude of the base = 250 m OD; radius of the Earth = 6.4×10^6 m; gravitational acceleration = 9.807 m/s^2; Young's modulus of elasticity = 210 kN/mm^2. (KU)

To find the corrected length of the first bay using the reference tape:

Standardization:

	+	−
Error per 30 m = 0.0501 m		
$\therefore$ Correction for 30.4500 m =	+0.0508	
Temperature = $30 \times 6 \times 10^{-7} \times 5$ =		−0.0001
Sag $= \dfrac{30^3 \times (0.02 \times 9.807)^2}{24 \times 88^2}$ =		−0.0056
	+0.0508	−0.0057

Therefore Total correction = +0.0451 m
Hence   Corrected length = 30.4500 + 0.0451 = 30.4951 m

(using reference tape). Field tape corrected for sag measures 30.4588 − 0.0056 = 30.4532 m.

Thus the field tape is measuring too short by 0.0419 m (30.4951 − 30.4532) and is therefore too long by this amount. Therefore field tape is 30.0419 m at 15°C and 88 N.

To find length of second bay:

Standardization:

	+	−
Error per 30 m = 0.0419		
$\therefore$ Correction $= \dfrac{30.5500}{30} \times 0.0419$ =	+0.0427	
Temperature = $30 \times 6 \times 10^{-7} \times 2$ =		−0.000 04
Tension $= \dfrac{30 \times 12}{2 \times 210 \times 10^3}$ =	+0.0009	
Sag $= \dfrac{30^3 \times (0.02 \times 9.807)^2}{24 \times 100^2}$ =		−0.0043
Slope $= \dfrac{0.500^2}{2 \times 30.5500}$ =		−0.0041
Altitude $= \dfrac{30.5500 \times 250}{6.4 \times 10^6}$ =		−0.0093
	+0.0436	−0.0177

Therefore Total correction = +0.0259 m
Hence   Corrected length of second bay = 30.5500 + 0.0259 = 30.5759 m

N.B. Rounding off the measured length to 30 m is permissible only when the resulting error has a negligible effect on the final distance.

Example 3.7 A copper transmission line of 12 mm diameter is stretched between two points 300 m apart, at the same level with a tension of 5 kN, when the temperature is 32°C. It is necessary to define its limiting positions when the temperature varies. Making use of the corrections for sag, temperature and elasticity normally applied to base-line measurements by a tape in catenary, find the tension at a temperature of − 12°C and the sag in the two cases.

Young's modulus for copper is 70 kN/mm^2, its density 9000 kg/m^3 and its coefficient of linear expansion 17.0 × 10^{-6}/°C. (LU)

In order first of all to find the amount of sag in the above two cases, one must find (a) the weight per unit length and (b) the sag length of the wire.

(a) w = area × density = $\pi r^2 \rho$

$\qquad$ = 3.142 × 0.006^2 × 9000 = 1.02 kg/m

(b) at 32°C, the sag length of wire = $L_H + \left(\dfrac{L^3 w^2}{24 T^2} \right)$

where L is itself the sag length. Thus the first approximation for L of 300 m must be used.

$\therefore$ Sag length = $300 + \left(\dfrac{300^3 \times (1.02 \times 9.807)^2}{24 \times 5000^2} \right)$ = 304.5 m

Second approximation = $300 + \left(\dfrac{304.5^3 \times (1.02 \times 9.807)^2}{24 \times 5000^2} \right)$

$\qquad\qquad\qquad\qquad$ = 304.71 m = L_1

$\therefore$ Sag = $y_1 = \dfrac{wL_1^2}{8T} = \dfrac{(1.02 \times 9.807) \times 304.71^2}{8 \times 5000}$ = 23.22 m

At − 12°C there will be a reduction in L_1 of

$\qquad (L_1 K \Delta t)$ = 304.71 × 17.0 × 10^{-6} × 44 = 0.23 m

$\qquad \therefore L_2$ = 304.71 − 0.23 = 304.48 m

From equation (2.7) $y_1 \propto L_1^2$ $\qquad \therefore y_2 = y_1 \left(\dfrac{L_2}{L_1} \right)^2$ = 23.22 $\dfrac{(304.48)^2}{(304.71)^2}$ = 23.18 m

Similarly, $y_1 \propto 1/T_1$ $\qquad \therefore T_2 = T_1 \left(\dfrac{y_1}{y_2} \right)$ = 5000 $\left(\dfrac{23.22}{23.18} \right)$ = 5009 N or 5.009 kN

Exercises

(*3.1*) A tape of nominal length 30 m was standardized on the flat at the NPL, and found to be

30.0520 m at 20°C and 44 N of tension. It was then used to measure a reference bay in catenary and gave a mean distance of 30.5500 m at 15°C and 88 N tension. As the weight of the tape was unknown, the sag at the mid-point of the tape was measured and found to be 0.170 m.

Given: cross-sectional area of tape = 2 mm²; Young's modulus of elasticity = 200 × 10³ N/mm²; coefficient of expansion = 11.25 × 10⁻⁶ per °C; and difference in height of measuring heads = 0.320 m. Find the horizontal length of the bay. If the error in the measurement of sag was ±0.001 m, what is the resultant error in the sag correction? What does this resultant error indicate about the accuracy to which the sag at the mid-point of the tape was measured?   (KU)

(*Answer*: 30.5995 m and ±0.000 03 m)

(*3.2*) The three bays of a base line were measured by a steel tape in catenary as 30.084, 29.973 and 25.233 m, under respective pulls of 7, 7 and 5 kg, temperatures of 12°, 13° and 17°C and differences of level of supports of 0.3, 0.7 and 0.7 m. If the tape was standardized on the flat at a temperature of 15°C under a pull of 4.5 kg, what are the lengths of the bays? 30 m of tape is exactly 1 kg with steel at 8300 kg/m³, with a coefficient of expansion of 0.000 011 per °C and $E = 210 \times 10^3$ N/mm².   (LU)

(*Answer*: 30.057 m, 29.940 m and 25.194 m)

(*3.3*) The details given below refer to the measurement of the first 30-m bay of a base line. Determine the correct length of the bay reduced to mean sea level.

With the tape hanging in a catenary at a tension of 10 kg and at a mean temperature of 13°C, the recorded length was 30.0247 m. The difference in height between the ends was 0.456 m and the site was 500 m above MSL. The tape had previously been standardized in catenary at a tension of 7 kg and a temperature of 16°C, and the distance between zeros was 30.0126 m.

$R = 6.4 \times 10^6$ m; weight of tape per m = 0.02 kg; sectional area of tape = 3.6 mm²; $E = 210 \times 10^3$ N/mm²; temperature coefficient of expansion of tape = 0.000 011 per °C.   (ICE)

(*Answer*: 30.0364 m)

(*3.4*) The following data refer to a section of base line measured by a tape hung in catenary.

Bay	Observed length (m)	Mean temperature (°C)	Reduced levels of index marks (m)	
1	30.034	25.2	293.235	293.610
2	30.109	25.4	293.610	294.030
3	30.198	25.1	294.030	294.498
4	30.075	25.0	294.498	294.000
5	30.121	24.8	294.000	293.355

Length of tape between 0 and 30 m graduations when horizontal at 20°C and under 5 kg tension is 29.9988 m; cross-sectional area of tape = 2.68 mm²; tension used in the field = 10 kg; temperature coefficient of expansion of tape = 11.16 × 10⁻⁶ per °C; elastic modulus for material of tape = 20.4 × 10⁴ N/mm²; weight of tape per metre length = 0.02 kg; mean radius of the Earth = 6.4 × 10⁶ m. Calculate the corrected length of this section of the line.   (LU)

(*Answer*: 150.507 m)

3.6 ELECTROMAGNETIC DISTANCE MEASUREMENT (EDM)

The advent of EDM equipment has completely revolutionized all surveying procedures, resulting in a change of emphasis and techniques. Taping distance, with all its associated problems, has been rendered obsolete for all base-line measurement. Distance can now be measured easily, quickly and with great accuracy, regardless of terrain conditions. Modern EDM equipment contains hard-wired algorithms for reducing the slope distance to its horizontal and vertical equivalent. For most engineering surveys, 'total stations' combined with electronic data loggers are now virtually standard equipment on site. Basic theodolites can be transformed into total stations by add-on, top-mounted EDM modules (*Figure 3.12(a) (b)*). A standard measurement of distance takes between 1.5 and 3 s. Automatic repeated measurements can be used to improve reliability in difficult atmospheric conditions. Tracking modes, for the setting out of distance, repeat the measurement every 0.3 s. The development of EDM has produced fundamental changes in surveying procedures, e.g.

(1) Traversing on a grandiose scale, with much greater control of swing errors, is now a standard procedure.
(2) The inclusion of many more measured distances into triangulation, rendering classical triangulation obsolete. This results in much greater control of scale error.
(3) Setting-out and photogrammetric control, over large areas, by polar coordinates from a single base line.
(4) Offshore position fixing by such techniques as the Tellurometer Hydrodist System.
(5) Deformation monitoring to sub-millimetre accuracies using high-precision EDM, such as the Com-Rad Geomensor CR234. This instrument has a range of 10 km and an accuracy of ±0.1 mm ±0.1 mm/km of the distance measured.

The latest developments in EDM equipment provide plug-in recording modules (*Figure 3.13(a) (b)*), capable of recording many thousand blocks of data for direct transfer to the computer. There is practically no surveying operation which does not utilize the speed, economy, accuracy and reliability of modern EDM equipment.

3.6.1 Classification of instruments

EDM instruments may be classified according to the type and wavelength of the electromagnetic energy generated or according to their operational range. Very often one is a function of the other.
  Considering the energy generated, the classification is as follows:

(1) Infra-red radiation (IR) classifies those instruments most commonly used in engineering. The IR has wavelengths of 0.8–0.9 μm transmitted by gallium arsenide (GaAs) luminescent diodes, at a high, constant frequency. The accuracies required in distance measurement are such that the measuring wave connot be used directly due to its poor propagation characteristics. The measuring wave is therefore superimposed on the high-frequency waves generated, called carrier waves. The superimposition is achieved by amplitude (*Figure 3.14*), frequency (*Figure 3.15*) or impulse modulation (*Figure 3.16*). In the case of IR instruments, amplitude modulation is used. Thus the carrier wave develops the necessary measuring characteristics whilst maintaining the high-frequency propagation characteristics that can be measured with the requisite accuracy.
  In addition to IR, visible light, with extremely small wavelengths, can also be used as a carrier. Many of the instruments using visual light waves have a greater range and a much greater accuracy than that required for more general surveying work. Typical of such instruments are the KernMekometer ME 5000, accurate to ±0.2 mm ±0.2 mm/km, with a range of 8 km, and the Com-Rad Geomensor CR234.

(a)

(b)

Fig. 3.12 *(a) Wild DI 1600 E.D.M (b) Wild DI 1600 top-mounted on the telescope of the Wild T 1000 Electronic theodolite*

(a)

(b)

Fig. 3.13   *(a) Wild GRM 10 Rec. Module (b) TC 1600K Total Station with GRM 10 Rec. Module fitted*

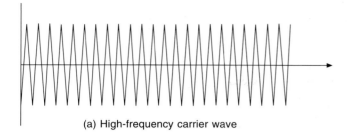

(a) High-frequency carrier wave

(b) Lower-frequency measuring wave

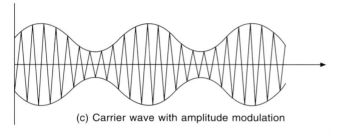

(c) Carrier wave with amplitude modulation

Fig. 3.14 (*a*) *Carrier wave modulation*

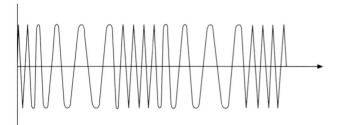

Fig. 3.15 *Frequency modulation of the carrier wave*

(2) Microwave instruments use radio wavelengths as carriers and therefore require two instruments, one at each end of the line to be measured, that are capable of receiving and transmitting the signals. The microwave carrier is always frequency modulated (*Figure 3.15*) for measuring purposes and has wavelengths generally in the order of 10 cm and 3 cm. As these instruments do not rely on light being returned to the master instrument by a reflector, they can be used day or night in most weather conditions. These instruments are capable of long ranges up to 25 km and beyond, with typical accuracies of ±10 mm ±5 mm/km.

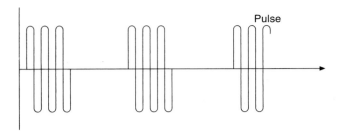

Fig. 3.16 *Impulse modulation of the carrier wave*

(3) The final classification of equipment refers to those instruments which use very long radio waves with wavelengths of 150 m to 2 km. They are primarily used for position-fixing systems in hydrographic and oceanographic surveying. Typical examples are the Pulse 8 system and the Syledis system, for offshore position fixing.

The above classification shows that it would also be possible to classify by range, such as:

(a) Short-range, electro-optical instruments using amplitude-modulated infra-red or visible light with ranges up to 5 km.
(b) Medium-range microwave equipment, frequency modulated to give ranges up to 25 km.
(c) Long-range radio wave equipment with ranges up to 100 km.

3.7 MEASURING PRINCIPLES

Although there is a wide variety of EDM instruments available, there are basically only two methods of measurement employed, namely the *pulse method* and the more popular *phase difference method*.

3.7.1 Pulse method (Figure 3.17)

A short, intensive pulse of radiation is transmitted to a reflector target, which immediately transmits it back, along a parallel path, to the receiver. The measured distance is computed from the velocity of the signal multiplied by the time it took to complete its journey, i.e.

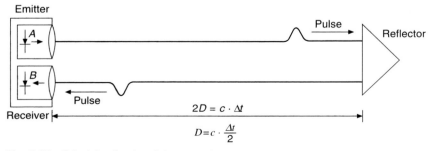

Fig. 3.17 *Principle of pulse distance meter*

$$2D = c \cdot \Delta t$$
$$D = c \cdot \Delta t/2 \tag{3.14}$$

If the time of departure of the pulse from gate A is t_A and the time of its reception at gate B is t_B, then $(t_B - t_A) = \Delta t$.

c = the velocity of light in the medium through which it travelled

D = the distance between instrument and target

It can be seen from equation (3.14) that the distance is dependent on the velocity of light in the medium and the accuracy of its transit time. Taking an approximate value of 300 000 km/s for the speed of light, 10^{-10} s would be equivalent to 15 mm of distance.

The distance that can be measured is largely a function of the power of the pulse. Powerful laser systems can obtain tremendous distances when used with retrodirective prisms and even medium distances when the pulse is 'bounced' off natural or man-made features.

A typical distance meter for general surveying is the Wild DI3000 (*Figure 3.18*), which has a range of 14 km and an accuracy of ± 3 mm ± 1 mm/km when used with prisms. Distances of 250 m can be measured to natural features. This latter facility can be extremely useful when surveying the face of quarries or the facades of buildings.

Fig. 3.18 *Wild Distomat DI 3000/T 1600*

3.7.2 Phase difference method

The majority of EDM instruments, whether infra-red, light or microwave, use this form of measurement. Basically the instrument measures the amount ($\delta\lambda$) by which the reflected signal is out of phase with the emitted signal. *Figure 3.19(a)* shows the signals in phase whilst (*b*) shows the amount ($\delta\lambda$) by which they are out of phase. The double distance is equal to the number (*M*) of full wavelengths (λ) plus the fraction of a wavelength ($\delta\lambda$). The phase difference can be measured by analog or digital methods. *Figure 3.20* illustrates the digital phase measurement of $\delta\lambda$.

Basically, all the equipment used works on the principle of 'distance "equals" velocity × time'. However, as time is required to such very high accuracies, recourse is made to the measurement of phase difference.

As shown in *Figure 3.20*, as the emitted and reflected signals are in continuous motion, the only constant is the phase difference $\delta\lambda$.

Figure 3.21 shows the path of the emitted radiation from instrument to reflector and back to instrument, and hence it represents twice the required distance from instrument to reflector. Any periodic phenomenon which oscillates regularly between maximum and minimum values may be analysed as a simple harmonic motion. Thus, if *P* moves in a circle with a constant angular velocity *w*, the radius vector *A* makes a phase angle ϕ with the *x*-axis. A graph of values, computed from

$$y = A \sin (wt) \tag{3.15}$$
$$= A \sin \phi \tag{3.16}$$

for various values of ϕ produces the sine wave illustrated and shows

A = amplitude or maximum strength of the signal

w = angular velocity

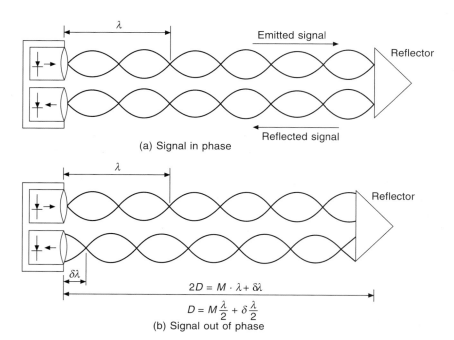

(a) Signal in phase

$$2D = M \cdot \lambda + \delta\lambda$$
$$D = M\frac{\lambda}{2} + \delta\frac{\lambda}{2}$$
(b) Signal out of phase

Fig. 3.19 *Principle of phase difference method*

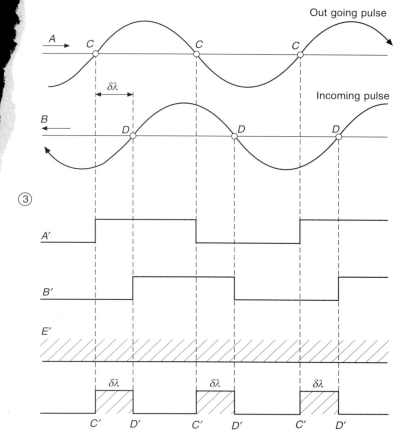

Fig. 3.20 *Digital phase-measurement*

t = time

ϕ = phase angle

The value of B is plotted when $\phi = \pi/2 = 90°$, C when $\phi = \pi = 180°$, D when $\phi = 1.5\pi = 270°$ and $A' = 2\pi = 360°$. Thus 2π represents a complete wavelength (λ) and $\phi/2\pi$ a fraction of a wavelength ($\delta\lambda$). The time taken for A to make one complete revolution or cycle is the period of the oscillation and is represented by t s. Hence the phase angle is a function of time. The number of revolutions per second at which the radius vector rotates is called the frequency f and is measured in hertz, where one hertz is one cycle per second.

With reference to *Figure 3.21*, it can be seen that the double path measurement (2D) from instrument to reflector and back to instrument is equal to

$$2D = M\lambda + \delta\lambda \qquad\qquad (3.17)$$

where

M = the integer number of wavelengths in the medium

$\delta\lambda$ = the fraction of a wavelength = $\dfrac{\phi}{2\pi}\lambda$

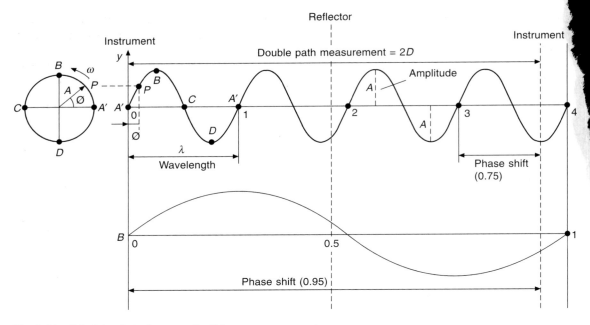

Fig. 3.21   *Principle of electromagnetic distance measurement*

As the phase difference method measures only the fraction of wavelength that is out of phase, a second wavelength of different frequency is used to obtain a value for M.

Consider *Figure 3.21*. The double distance is 3.75λ; however, the instrument will only record the phase difference of 0.75. A second frequency is now generated with a wavelength four times greater, producing a phase difference of 0.95. In terms of the basic measuring unit, this is equal to $0.95 \times 4 = 3.80$, and hence the value for M is 3. The smaller wavelength provides a more accurate assessment of the fractional portion, and hence the double path measurement is $3 + 0.75 = 3.75\lambda$. Knowing the value of λ in units of length would thus produce the distance. This then is the basic principle of the phase difference method, further illustrated below:

	f	λ	$\lambda/2$	$\delta\lambda$
Fine reading	15 MHz	20 m	10 m	6.325
1st rough reading	1.5 MHz	200 m	100 m	76.33
2nd rough reading	150 MHz	2000 m	1000 m	876.3
		Measured distance		876.325

The first number of each rough reading is added to the initial fine reading to give the total distance. Thus the single distance from instrument to reflector is

$$D = M(\lambda/2) + \frac{\phi}{2\pi}(\lambda/2) \qquad (3.18)$$

where it can be seen that $\lambda/2$ is the main unit length of the instrument. The value chosen for the main unit length has a fundamental effect on the precision of the instrument due to the limited resolution of phase measurement. The majority of EDM instruments use $\lambda/2$ equal 10 m. With a phase resolution of 3×10^{-4} errors of 3 mm would result. The Tellurometer MA100 uses $\lambda/2 = 2$ m, resulting in 0.6 mm error, whilst the Kern Mekometer uses $\lambda/2 = 0.3$ m and a resulting error of 0.1 mm.

Implicit in the above equation is the assumption that λ is known and constant. However, in most EDM equipment, this is not so only the frequency f is known and is related to λ as follows:

$$\lambda = c/f \tag{3.19}$$

where

c = the velocity of electromagnetic waves (light) in a medium

This velocity c can only be calculated if the refractive index n of the medium is known, and also the velocity of light c_0 in a vacuum:

$$n = c_0/c \tag{3.20}$$

where n is always greater than unity

At the XVIth General Assembly of the International Union for Geodesy and Geophysics in 1975 the following value for c_0 was recommended:

$$c_0 = 299\,792\,458 \pm 1.2 \text{ m/s} \tag{3.21}$$

From the standard deviation quoted, it can be seen that this value is accurate to 0.004 mm/km and can therefore be regarded as error free compared with the most accurate EDM measurement.

The value of n can be computed from basic formulae available. However, *Figure 3.14* shows the carrier wave contained within the modulation envelope. The carrier travels at what is termed the phase velocity, whilst the group of frequencies travel at the slower group velocity. The measurement procedure is concerned with the modulation and so it is the group refractive index n_g with which we are concerned.

From equation (3.20) $c = c_0/n$ and from equation (3.19) $\lambda = c_0/nf$

Replace n with n_g and substitute in equation (3.18):

$$D = M \frac{(c_0)}{(2n_g f)} + \frac{\phi}{2\pi} \frac{(c_0)}{(2n_g f)} \tag{3.22}$$

Two further considerations are necessary before the final formula can be stated:

(1) The physical centre of the instrument which is plumbed over the survey station does not coincide with the position within the instrument to which the measurements are made. This results in an instrument constant K_1.
(2) Similarly with the reflector. The light wave passing through the atmosphere, whose refractive index is approximately unity, enters the glass prism of refractive index about 1.6 and is accordingly slowed down. This change in velocity combined with the light path through the reflector results in a correction to the measured distance called the reflector constant K_2.

Both these constants are combined and catered for in the instrument; then

$$D = M \frac{(c_0)}{(2n_g f)} + \frac{\phi}{2\pi} \frac{(c_0)}{(2n_g f)} + (K_1 + K_2) \tag{3.23}$$

is the fundamental equation for distances measured with EDM equipment.

An examination of the equation shows the errors sources to be:

(1) In the measurement of phase ϕ.
(2) In the measurement of group refractive index n_g.
(3) In the stability of the frequency f.
(4) In the instrument/reflector constants K_1 and K_2.

3.8 METEOROLOGICAL CORRECTIONS

Using EDM equipment, the measurement of distance is obtained by measuring the time of propagation of electromagnetic waves through the atmosphere. Whilst the velocity of these waves in a vacuum (c_0) is known, its value will be reduced according to the atmospheric conditions through which the waves travel at the time of measurement. As shown in equation (3.20), the refractive index (n) of the prevailing atmosphere is necessary in order to apply a velocity correction to the measured distance. Thus if D' is the measured distance, the corrected distance D is obtained from

$$D = D'/n \tag{3.24}$$

The value of n is affected by the temperature, pressure and water vapour content of the atmosphere as well as by the wavelength λ of the transmitted electromagnetic waves. It follows from this that measurements of these atmospheric conditions are required at the time of measurement.

As already shown, steel tapes are standardized under certain conditions of temperature and tension. In a similar way, EDM equipment is standardized under certain conditions of temperature and pressure. For instance, all the Geodimeters (AgaGeotronics) are standardized at 20°C and 1013.25 mbar of pressure, whilst Kern equipment is standardized at 12°C and 1013.25 mbar. Under these respective atmospheric conditions, the measured distances would not require a velocity correction. However, measuring with a Geodimeter under typical winter conditions of 0°C and 1035 mbar would require a correction of −25 mm/km or −2.5 mm per 100 m. As this error is systematic it would accumulate over a traverse containing many stations. It follows that even on low-order surveys the measurement of temperature and pressure is important.

The refractive index is related to wavelength via the Cauchy equation:

$$n = A + \frac{B}{\lambda^2} + \frac{C}{\lambda^4} \tag{3.25}$$

where A, B and C are constants relative to specific atmospheric conditions. To afford a correction in parts per million (ppm) the refractive number or refractivity (N) is used:

$$N = (n - 1) \times 10^6 \tag{3.26}$$

If $n = 1.000\,300$, then $N = 300$.

A value for n for standard air (0°C temperature, 1013.25 mbar pressure and dry air with 0.03% CO_2) is given by Barrel and Sears as

$$(n - 1) \times 10^6 = \left(287.604 + \frac{1.6288}{\lambda^2} + \frac{0.0136}{\lambda^4} \right) \tag{3.27}$$

However, as stated in *Section 3.7.2*, it is the refractive index of the modulated beam, not the carrier, that is required; hence the use of group refractive index where

$$N_g = A + \frac{3B}{\lambda^2} + \frac{5C}{\lambda^4} \tag{3.28}$$

and therefore, for group velocity in standard air with λ in μm:

$$N_g = (n_g - 1) \times 10^6 = 287.604 + \frac{3 \times 1.6288}{\lambda^2} + \frac{5 \times 0.0136}{\lambda^4} \tag{3.29}$$

The above formula is accurate to ± 0.1 ppm at wavelengths between 560 and 900 nm. An alternative formula derived by B. Edlen is

$$N_g = (n_g - 1) \times 10^6 = 287.569 + \frac{3 \times 1.6206}{\lambda^2} + \frac{5 \times 0.0139}{\lambda^4} \tag{3.30}$$

It follows that different instruments using different wavelengths will have different values for refractivity; for example:

$\lambda = 0.910\ \mu$m gives $N_g = 293.6$
$\lambda = 0.820\ \mu$m gives $N_g = 295.0$

To accomodate the actual atmospheric conditions under which the distances are measured, equation (3.29) was modified by Barrel and Sears:

$$N'_g = \frac{N_g \times Q \times P}{T} - \frac{V \times e}{T} \tag{3.31}$$

where T = absolute temperature in Kelvins (K) = 273.15 + t,
t = dry bulb temperature in °C
P = atmospheric pressure
e = partial water vapour pressure
with P and e in mbar, $Q = 0.2696$ and $V = 11.27$
with P and e in mmHg, $Q = 0.3594$ and $V = 15.02$

The value for e can be calculated from

$$e = e_s - 0.000\ 662 \times P \times (t - t_w) \tag{3.32}$$

where t = dry bulb temperature
t_w = wet bulb temperature
e_s = saturation water vapour pressure

The value for e_s can be calculated from

$$\log e_s = \frac{7.5 t_w}{t_w + 237.3} + 0.7857 \tag{3.33}$$

Using equation (3.33), the following table can be produced:

t_w (°C)	0	10	20	30	40
e_s	6.2	12.3	23.4	42.4	73.7

e_s is therefore quite significant at high temperatures.

Working back through the previous equations it can be shown that at 100% humidity and 30°C temperature a correction of approximately 2 ppm is necessary for the distance measured. In practice, humidity is generally ignored in the velocity corrections for instruments using light waves as it is insignificant compared with other error sources. However, for long lines being measured to very high accuracies in hot, humid conditions, it may be necessary to apply corrections for humidity.

The velocity correction is normally applied by means of a nomogram supplied with the equipment or by entry of prevailing temperature and pressure into the instrument. In the case of infra-red/light waves, the humidity term is ignored as it is only relevant in conditions of high humidity and temperature. Under normal conditions the error would be in the region of 0.7 ppm. The more important aspect concerning the use of nomograms is that they are not interchangeable between instruments. As already shown, instruments having different wavelengths will have different values for N, which in turn would be used in the formula to produce a nomogram specific to that instrument. Also, different instruments are standardized to different atmospheric conditions, which again renders nomograms specific to the instrument. As an example of the error which could arise, Wild use $N = 282$, whilst Tellurometer use $N = 274$. At 20°C and 1012.5 mbar, the Wild nomogram gives a correction of +8 mm/km, and the Tellurometer a correction of zero.

For maximum accuracy, it may be necessary to compute the velocity correction from first principles, and if producing computer software to reduce the data, this would certainly be the best approach. When using this approach, the ppm velocity correction dial must be set to zero when, measuring, or the standard values for t and P entered into the instrument. The following example will now be computed in detail in order to illustrate the process involved.

Worked example

Example 3.8 An EDM instrument has a carrier wave of 0.91 μm and is standardized at 20°C and 1013.25 mbar. A distance of 1885.864 m was measured with the mean values $P = 1030$ mbar, dry bulb temperature $t = 30$°C, wet bulb temperature $t_w = 25$°C. Calculate the velocity correction.

Step 1. Compute the value for partial water vapour pressure e:

$$\log e_s = \frac{7.5t_w}{t_w + 237.3} + 0.7857$$

$$= \frac{7.5 \times 25}{25 + 237.3} + 0.7857$$

$$e_s = 31.66 \text{ mbar}$$

Using equation (3.32):

$$e = e_s - 0.000\ 662 \times P \times (t - t_w)$$
$$= 31.66 - 0.000\ 662 \times 1030 \times (30 - 25) = 28.25 \text{ mbar}$$

Step 2. Compute refractivity (N_g) for standard atmosphere using equation (3.29):

$$N_g = 287.604 + \frac{3 \times 1.6288}{\lambda^2} + \frac{5 \times 0.0136}{\lambda^4}$$

$$= 287.604 + \frac{4.8864}{0.91^2} + \frac{0.0680}{0.91^4}$$

$$= 293.604$$

Step 3. Compute refractivity for the standard conditions of the instrument, i.e. 20°C and 1013.25 mbar, using equation (3.31):

$$N_g' = \frac{N_g \times Q \times P}{T} - \frac{V \times e}{T}$$

which for P and e in mbar becomes

$$N_g' = \frac{N_g\,(0.2696P)}{273.15 + t} - \frac{11.27e}{273.15 + t}$$

$$= \frac{79.156P}{273.15 + t} - \frac{11.27e}{273.15 + t}$$

This in effect is the equation for computing what is called the reference or nominal refractivity for the instrument:

$$N_g' = \frac{79.156 \times 1013.25}{273.15 + 20} - \frac{11.27 \times 28.25}{273.15 + 20}$$

$$= 273.60 - 1.09$$

$$= 272.51$$

Step 4. The reference refractivity now becomes the base from which the velocity correction is obtained. Now compute refractivity (N_g'') under the prevailing atmospheric conditions at the time of measurement:

$$N_g'' = \frac{79.156 \times 1030}{273.15 + 30} - \frac{11.27 \times 28.25}{273.15 + 30} = 267.90$$

$$\therefore \text{ Velolcity correction in ppm} = 272.51 - 267.90 = 4.6 \text{ ppm}$$

$$\text{Correction in mm} = 1885.864 \times 4.6 \times 10^{-6} = 8.7 \text{ mm}$$

$$\text{Corrected distance} = 1885.873 \text{ m}$$

Using the nomogram the correction in ppm = 4.2.

  Using equation (3.20):

(1) Velocity of light waves under standard conditions:

$$c_s = c_0/n_g = 299\ 792\ 458/1.000\ 2725 = 299\ 710\ 770 \text{ m/s}$$

(2) Velocity under prevailing conditions:

$$C_p = 299\ 792\ 458/1.000\ 2679 = 299\ 712\ 150 \text{ m/s}$$

As distance = (velocity × time), the increased velocity of the measuring waves at the time of measuring would produce a positive velocity correction, as shown.

  Now considering equation (3.31) and differentiating with respect to t, P and e, we have

$$\delta N_g' = \frac{0.2696\,N_g\,P\,\delta t}{(273.15 + t)^2} + \frac{0.2696\,N_g\,\delta P}{(273.15 + t)} - \frac{11.27\,\delta e}{(273.15 + t)}$$

At $t = 15°C$, $P = 1013$ mbar, $e = 10$ mbar and $N_g = 294$, an error of 1 ppm in N_g' and therefore in distance will occur for an error in temperature of $\pm 1°C$, in pressure of ± 3 mbar and in e of ± 39 mbar.

From this it can be seen that the measurement of humidity can ordinarily be ignored for instruments using light. Temperature and pressure should be measured using carefully calibrated, good quality equipment, but not necessarily very expensive, highly accurate instruments. The measurements are usually taken at each end of the line being measured, on the assumption that the mean value would be equal to those values measured at the mid-point of the line. However, tests on a 3-km test line showed the above assumption to be in error by 2°C and −3 mbar. The following measuring procedure is therefore recommended:

(1) Temperature and pressure should be measured at each end of the line.
(2) The above measurements should be measured well above the ground (3 m if possible) to avoid ground radiation effects and properly reflect mid-line conditions.
(3) The measurements should be synchronized with the EDM measurements.
(4) If possible, mid-line observations should be included.
(5) Ground-grazing lines should be avoided.

In the case of microwave instruments, humidity becomes a serious consideration and is catered for in the following formula given by Essen and Froome:

$$N_{\mathrm{m}} = (n_{\mathrm{m}} - 1) \times 10^6 = \frac{77.64(P - e)}{T} + \frac{64.68}{T}\left(1 + \frac{5748}{T}\right)e \qquad (3.34)$$

where n_{m} = group refractive index for microwaves and the units are °C and mbar.

The formula is suitable for values of λ ranging from 0.03 m to 1.00 m, and is accurate to 0.1 ppm under normal conditions. In the direct computation of the velocity correction it simply replaces the equivalent equation for light waves.

3.8.1 *The second velocity correction*

As shown in *Section 2.15.2*, the path of light waves through the atmosphere is on a curve roughly eight times the radius of the ellipsoid. In the case of microwaves the path is roughly 4R. This is catered for by a second velocity correction equal to

$$-(K - K^2)\frac{D^3}{12R^2} \qquad (3.35)$$

where
K = coefficient of refraction (as derived in equation (2.15))
D = distance displayed by the instrument
R = mean radius of the ellipsoid in the direction (α) of the line measured

R may be calculated from

$$R = \rho v/(v\cos^2 \alpha + \rho \sin^2 \alpha) \qquad (3.36)$$

where ρ = radius of curvature of the ellipsoid in the meridian plane
 v = radius of curvature of the ellipsoid in the prime vertical plane
 α = the azimuth of the measured line

In the case of light waves with an average value $K = 0.15$, the correction would equal 0.25 mm for a distance of 10 km and could safely be ignored. However, for the longer distances measured with microwaves, with average K-values of 0.25, a distance of 30 km would require a correction of −10 mm. Also it should be remembered that K can vary enormously from its average value.

3.9 GEOMETRICAL REDUCTIONS

The measured distance, after the velocity corrections have been applied, is the spatial distance from instrument to target. This distance will most certainly have to be reduced to the horizontal and then to its equivalent on the ellipsoid of reference.

From *Figure* 3.22, D_1 represents the measured distance (after the velocity corrections have been applied) which is reduced to its chord equivalent D_2; this is in turn reduced to D_3; the chord equivalent of the ellipsoidal distance (at MSL) D_4. Strictly speaking the ellipsoid of reference may be different to the geoid at MSL. A geoid – ellipsoid separation of 6 m would affect the distance by 1 mm/km. In the UK, for instance, maximum separation is in the region of 4.2 m, producing a scale error of 0.7 mm/km. Where such information is unavailable, the geoid and ellipsoid are assumed coincident.

(1) Reduction from D_1 to D_2

$$D_2 = D_1 - C_1$$

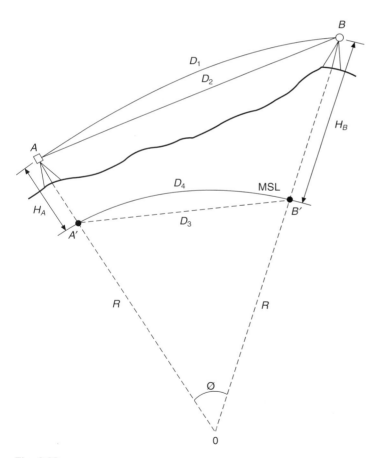

Fig. 3.22

where $C_1 = K^2 D_1^3 / 24 R^2$ (3.37)

and K = coefficient of refraction

R, the radius of the ellipsoid in the direction α from A to B, may be calculated from equation (3.36) or with sufficient accuracy for lines less than 10 km from

$$R = (\rho v)^{\frac{1}{2}}$$ (3.38)

For the majority of lines in engineering surveying this first correction may be ignored.

The value for K is best obtained by simultaneous reciprocal observations, although it can be obtained from

$$K = (n_1 - n_2)R/|H_1 - H_2|$$ (3.39)

where n_1 and n_2 are the refractive indices at each end of the line and H_1 and H_2 the respective heights above MSL.

(2) Reduction from D_2 to D_3

In triangle *ABO* using the cosine rule:

$$\cos \theta = (R + H_B)^2 + (R + H_A)^2 - D_2^2 / 2(R + H_B)(R + H_A)$$

but $\cos \theta = 1 - 2\sin^2(\theta/2)$ and $\sin \theta/2 = D_3/2R$

$$\therefore \cos \theta = 1 - D_3^2 / 2R^2$$

$$1 - D_3^2 / 2R^2 = (R + H_B)^2 + (R + H_A)^2 - D_2^2 / 2(R + H_B)(R + H_A)$$

$$D_3 = R \left\{ \frac{[D_2 - (H_B - H_A)][D_2 + (H_B - H_A)]}{(R + H_A)(R + H_B)} \right\}^{\frac{1}{2}}$$

$$= \left[\frac{D_2^2 - (H_B - H_A)^2}{(1 + H_A/R)(1 + H_B/R)} \right]^{\frac{1}{2}}$$ (3.40)

where H_A = height of the instrument centre above the ellipsoid (or MSL)
$\quad\quad H_B$ = height of the target centre above the ellipsoid (or MSL)

The above rigorous approach may be relaxed for the relatively short lines measured in the majority of engineering surveys.

Pythagorus may be used as follows:

$$D_M = [D_2^2 - (H_B - H_A)^2]^{\frac{1}{2}}$$ (3.41)

where D_M = the horizontal distance at the mean height H_M where $H_M = (H_B + H_A)/2$.

D_M would then be reduced to MSL using the altitude correction:

$$C_A = (D_M H_M)/R$$ (3.42)

then $D_3 = D_4 = D_M - C_A$

(3) Reduction of D_3 to D_4

Figure 3.22 shows

$$\theta/2 = D_4/2R = \sin^{-1}(D_3/2R)$$

where

$$\sin^{-1}(D_3/2R) = D_3/2R + D_3^3/8R \times 3! + 9D_3^5/32R^5 \times 5!$$

$$D_4/2R = D_3/2R + D_3^3/48R^3$$

$$D_4 = D_3 + (D_3^3/24R^2) \tag{3.43}$$

For lines less than 10 km, the correction is 1 mm and is generally ignored.

It can be seen from the above that for the majority of lines encountered in engineering (< 10 km), the procedure is simply:

(1) Reduce the measured distance D_2 to the mean altitude using equation (3.41) $= D_M$.
(2) Reduce the horizontal distance D_M to D_4 at MSL using equation (3.42).

(4) Reduction by vertical angle

Total stations have the facility to reduce the measured slope distance to the horizontal using the vertical angle, i.e.

$$D = S \cos \alpha \tag{3.44}$$

where α = the vertical angle
S = the slope distance
D = the horizontal distance

However, the use of this simple relationship will be limited to short distances if the effect of refraction is ignored.

Whilst S may be measured to an accuracy of, say, ± 5 mm, reduction to the horizontal should not result in further degradation of this accuracy. Assume then that the accuracy of reduction must be ± 1 mm.

Then $\quad \delta D = -S \sin \alpha \, \delta\alpha$

and $\quad \delta\alpha'' = \delta D \times 206\,265/S \sin \alpha \tag{3.45}$

where $\quad \delta D$ = the accuracy of reduction

$\delta\alpha''$ = the accuracy of the vertical angle

Consider $S = 1000$ m, measured on a slope of $\alpha = 5°$; if the accuracy of reduction to the horizontal is to be practically error free, let $\delta D = \pm 0.001$ m and then

$$\delta\alpha'' = 0.001 \times 206\,265/1000 \sin 5° = \pm 2.4''$$

This implies that to achieve error-free reduction the vertical angle must be measured to an accuracy of $\pm 2.4''$. If the accuracy of the double face mean of a vertical angle is, say, $\pm 4''$, then a further such determination is required to reduce it to $4''/(2)^{\frac{1}{2}} = \pm 2.8''$. However, the effect of refraction assuming an average value of $K = 0.15$ is $2.4''$ over 1000 m. Hence the limit has been reached where refraction can be ignored. For $\alpha = 10°$, $\delta\alpha = \pm 1.2''$, and hence refraction would need to be considered over this relatively steep sight. At $\alpha = 2°$, $\delta \alpha'' = \pm 6''$ and the effect of refraction is negligible. It is necessary to carry out this simple appraisal depending on the distances and the slopes involved, in order to assess the effect of ignoring refraction.

In some cases, standard corrections for refraction (and curvature) are built into the instrument which may or may not help the reduction accuracy. The value of refraction can be so variable that it cannot be catered for unless it is cancelled by using the mean of simultaneous reciprocal vertical angles. If the distances involved are long, then the consideration is shown in *Figure 3.23*. If α and β are the reciprocal angles, corrected for any difference of height in instrument and target (refer page 93), then

$$\alpha_0 = (\alpha - \beta)/2 \qquad (3.46)$$

where β is negative for angle of depression
and $S \cos \alpha_0 = AC$ $\qquad (3.47)$
However, the distance required is AB' where

$$AB' = AC - B'C \qquad (3.48)$$

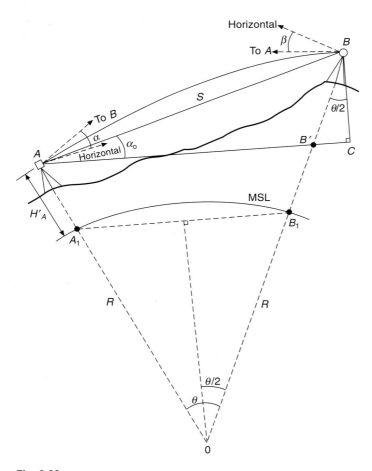

Fig. 3.23

but $BC = S \sin \alpha_0$
and $B'C = BC \tan \theta/2$
where

$$\theta'' = \frac{AC \times 206\,265}{R + H'_A} = \frac{S \times 206\,265}{R + H'_A} \tag{3.49}$$

This will give the horizontal distance (AB') at the height of the instrument axis above MSL, i.e. H'_A.

AB' can then be reduced to A_1B_1 at MSL using the altitude correction (equation (3.42)). Alternatively, the above procedure can be rearranged to give

$$AB' = S \cos (\alpha_0 + \theta/2) \sec \theta/2 \tag{3.50}$$

Both procedures will now be demonstrated by an example.

Worked example

Example 3.9 Consider $S = 5643.856$ m and reciprocal vertical angles of $\alpha = 5°00'22''$, $\beta = 5°08'35''$. If station A is 42.95 m AOD and the instrument height is 1.54 m, compute the horizontal distance (AB') at instrument height above MSL (*Figure 3.23*) ($R = 6380$ km).

$\alpha_0 = [5°00'\,22'' - (-5°08'\,35'')]/2 = 5°04'\,29''$

$AC = 5643.856 \cos 5°04'\,29'' = 6521.732$ m

$BC = 5643.856 \sin 5°04'\,29'' = 499.227$ m

$\theta' = (5621.732 \times 206\,265)/(6\,380\,000 + 44.49) = 03'\,02''$

$B'C = 499.227 \tan 01'\,31'' = 0.220$ m

$AB' = 5621.732 - 0.220 = 5621.512$ m

Alternatively from equation (3.50):

$AB' = S \cos (\alpha_0 + \theta/2) \sec \theta/2$

$\quad = 5643.856 \cos (-5°04'\,29'' + 01'\,31'') \sec 01'\,31''$

$\quad = 5621.512$ m

AB' is now reduced to A_1B_1 using the correction (equation (3.42)) where the height above MSL includes the instrument height, i.e. $42.95 + 1.54 = 44.49$ m.

If a value for the angle of refraction ($\hat{r}$) is known and only one vertical angle α is observed then

$$AB' = S \cos (\alpha' + \theta) \sec \theta/2 \tag{3.51}$$

where $\alpha' = \alpha - \hat{r}$

Adapting the above example and assuming an average value for K of 0.15:

$r'' = SK\rho/2R$

$\quad = \dfrac{5643.856 \times 0.15 \times 206\,265}{2 \times 6\,380\,000} = 14''$

$\alpha' = 5°\,00'\,22'' - 14'' = 5°\,00'\,08''$

$AB' = 5643.856 \cos (5°\,00'\,08'' + 3'\,02'') \sec 01'\,31''$

$\quad = 5621.924$ m

The difference of 0.412 m once again illustrates the dangers of assuming a value for K.
   If zenith angles are used then:

$$AB' = S\left[\frac{\sin Z_A - \theta(2 - K)/2}{\cos \theta/2}\right] \qquad (3.52)$$

where K = the coefficient of refraction

For most practical purposes equation (3.52) is reduced to

$$AB' = S \sin Z_A - \frac{S^2(2 - K) \sin 2Z_A}{4R} \qquad (3.53)$$

3.10 ERRORS AND CALIBRATION

Although modern EDM equipment is exceptionally well constructed, the effects of age and general
wear and tear may alter its performance. It is essential therefore that all field equipment should be
regularly calibrated. In the light of legislation on quality assurance, calibration to ensure accuracy
of performance to the standards demanded is virtually mandatory. From the point of view of
calibration, the errors have been classified under three main headings.

3.10.1 *Zero error (independent of distance)*

Zero error results from changes in the instrument/reflector constant due to ageing of the instrument
or as a result of repairs. The built-in correction for instrument/reflector constants is usually correct
to 1 or 2 mm but may change between reflectors and so should be assessed for a particular
instrument/reflector combination. A variety of other matters may affect the value of the constant
and these matters may vary from instrument to instrument. Some instruments have constants which
are signal strength dependent, while others are voltage dependent. The signal strength may be
affected by the accuracy of the pointing or by prevailing atmospheric conditions. It is very important,
therefore, that periodical testing is carried out.
   A simple procedure can be adopted to obtain the zero error for a specific instrument/reflector
combination. Consider three points A, B, C set out in a straight line such that $AB = 10$ m, $BC =$
20 m and $AC = 30$ m. Assume a zero error of +0.3 m exists in the instrument; the measured lengths
will then be 10.3, 20.3 and 30.3. Now:

$$AB + BC = AC$$
$$10.3 + 20.3 = 30.3$$
$$30.6 - 30.3 = +0.3$$

and the error is obtained. Now as

   Correction = –Error

every measured distance will have a correction of –0.3 m.

$$\text{Zero error} = k_o = l_{AB} + l_{BC} - l_{AC} \qquad (3.54)$$

from which it can be seen that the base-line lengths do not need to be known prior to measurement.
If there are more than two bays in the base line of total length L, then

$$k_o = L - \Sigma l_i/n - 1 \qquad (3.55)$$

where l_i is the measured length of each of the n sections.

Alternatively the initial approach may be adopted with as many combinations as possible. For example, if the base line comprises three bays AB, BC and CD, we have

$$AB + BC - AC = k_o$$
$$AC + CD - AD = k_o$$
$$AB + BD - AD = k_o$$
$$BC + CD - BD = k_o$$

with the arithmetic mean of all four values being accepted.

The most accurate approach is a least squares solution of the observation equations. Let the above bays be a, b and c, with measured lengths l and residual errors of measurement r.

Observation equations:

$$a + k_o = l_{AB} + r_1$$
$$a + b + k_o = l_{AC} + r_2$$
$$a + b + c + k_o = l_{AD} + r_3$$
$$b + k_o = l_{BC} + r_4$$
$$b + c + k_o = l_{BD} + r_5$$
$$c + k_o = l_{CD} + r_6$$

Normal equations:

$$3a + 2b + c + 3k_o = l_{AB} + l_{AC} + l_{AD}$$
$$2a + 4b + 2c + 4k_o = l_{AC} + l_{AD} + l_{BC} + l_{BD}$$
$$a + 2b + 3c + 3k_o = l_{AB} + l_{BD} + l_{CD}$$
$$3a + 4b + 3c + 6k_o = l_{AB} + l_{BC} + l_{CD} + l_{AC} + l_{AD} + l_{BD}$$

Solution of the normal equations gives the values for bays a, b and c plus the zero error k_o. Substituting these values back into the observation equations gives the residual values r, which can be plotted to give an indication of cyclic error.

The distances measured should, of course, be corrected for slope and velocity, before they are used to find k_o. If possible, the bays should be in multiples of $\lambda/2$ if the effect of cyclic errors is to be cancelled.

3.10.2 Cyclic error (varies with distance)

As already shown, the measurement of the phase difference between the transmitted and received waves enables the fractional part of the wavelength to be determined. Thus, errors in the measurement of phase difference will produce errors in the measured distance. Phase errors are cyclic and not proportional to the distance measured and may be non-instrumental and/or instrumental.

The non-instrumental cause of phase error is spurious signals from reflective objects illuminated by the beam. Normally the signal returned by the reflector will be sufficiently strong to ensure complete dominance over spurious reflections. However, care should be exercised when using vehicle reflectors or Scotchlite for short-range work.

The main cause of phase error is instrumental and derives from two possible sources. In the first instance, if the phase detector were to deviate from linearity around a particular phase value, the resulting error would repeat each time the distance resulted in that phase. Excluding gross malfunctioning, the phase readout is reliably accurate, so maximum errors from this source should not exceed 2 or 3 mm. The more significant source of phase error arises from electrical cross-talk, or spurious coupling, between the transmit and receive channels. This produces an error which varies sinusoidally with distance and is inversely proportional to the signal strength.

Cyclic errors in phase measurement can be determined by observing to a series of positions distributed over a half wavelength. A bar or rail accurately divided into 10-cm intervals over a distance of 10 m would cover the requirements of most short-range instruments. Details of such an arrangement are given in Hodges (1968). A micrometer on the bar capable of very accurate displacements of the reflector of +0.1 mm over 20 cm would enable any part of the error curve to be more closely examined.

The error curve plotted as a function of the distance should be done for strong and weakest signal conditions and may then be used to apply corrections to the measured distance. For the majority of short-range instruments the maximum error will not exceed ±5 mm.

Most short-range EDM instruments have values for $\lambda/2$ equal to 10 m. A simple arrangement for the detection of cyclic error which has proved satisfactory is to lay a steel band under standard tension on a horizontal surface. The reflector is placed at the start of 10-m section and the distance from instrument to reflector obtained. The reflector is displaced precisely 100 mm and the distance is re-measured. The difference between the first and second measurement should be 100 mm; if not, the error is plotted at the 0.100 m value of the graph. The procedure is repeated every 100 mm throughout the 10-m section and an error curve produced. If, in the field, a distance of 836.545 m is measured, the 'cyclic error' correction is abstracted from point 6.545 m on the error curve.

3.10.3 *Scale error (proportional to distance)*

Scale errors in EDM instruments are largely due to the fact that the oscillator is temperature dependent. The quartz crystal oscillator ensures the frequency (f) remains stable to within ±5 ppm over an operational temperature range of $-20°C$ to $50°C$. The modulation frequency can, however, vary from its nominal value due to incorrect factory setting, ageing of the crystal and lack of temperature stabilization. Most modern short-range instruments have temperature-compensated crystal oscillators which have been shown to perform well. However, warm-up effects have been shown to vary from 1 to 5 ppm during the first hour of operation.

Diode errors also cause scale error, as they could result in the emitted wavelength being different from its nominal value.

The magnitude of the resultant errors may be obtained by field or laboratory methods.

The laboratory method involves comparing the actual modulation frequency of the instrument with a reference frequency. The reference frequency may be obtained from off-air radio transmissions such as Droitwich MSF in the UK or from a crystal-generated laboratory standard. The correction for frequency is equal to

$$\left(\frac{\text{Nominal frequency} - \text{Actual frequency}}{\text{Nominal frequency}} \right) \text{ppm}$$

A simple field test is to measure a base line whose length is known to an accuracy greater than the measurements under test. The base line should be equal to an integral number of modulation half wavelengths. The base line AB should be measured from a point C in line with AB; then $CB - CA = AB$. This differential form of measurement will eliminate any zero error, whilst the use of an integral number of half wavelengths will minimize the effect of cyclic error. The ratio of the measured length to the known length will provide the scale error.

3.10.4 Multi-pillar base lines

The establishment of multi-pillar base lines for EDM calibration requires careful thought, time and money. Not only must a suitable area be found to permit a base line of, in some cases, over 1 km to be established, but suitable ground conditions must also be present. If possible the bedrock should be near the surface to permit the construction of the measurement pillars on a sound solid foundation. The ground surface should be reasonably horizontal, free from growing trees and vegetation and easily accessible. The construction of the pillars themselves should be carefully considered to provide maximum stability in all conditions of wetting and drying, heat and cold, sun and cloud, etc. The pillar-centring system for instruments and reflectors should be carefully thought out to avoid any hint of centring error. When all these possible error sources have been carefully considered, the pillar separations must then be devised.

The total length of the base line is obviously the first decision, followed by the unit length of the equipment to be calibrated. The interpillar distances should be spread over the measuring range of equipment, with their fractional elements evenly distributed over the half wavelength of the basic measuring wave.

Finally, the method of obtaining interpillar distances to the accuracy required has to be considered. The accuracy of the distance measurement must obviously be greater than the equipment it is intended to calibrate. For general equipment with accuracies in the range of 3–5 mm, the base line could be measured with equipment of superior accuracy such as those already mentioned.

For even greater accuracy, laser interferometry accurate to 0.1 ppm may be necessary.

When such a base line is established, a system of regular and periodic checking must be instituted and maintained to monitor short- and long-term movement of the pillars. Appropriate computer software must also be written to produce zero, cyclic and scale errors per instrument from the input of the measured field data.

Several such base lines have been established throughout the UK, the most recent one (1991) by Thames Water at Ashford in Middlesex, in conjunction with the National Physical Laboratory at Teddington, Middlesex. This is an eight-pillar base line, with a total length of 818.93 m and interpillar distances affording a good spread over a 10-m period, as shown below:

	2	3	4	5	6	7	8
1	260.85	301.92	384.10	416.73	480.33	491.88	818..93
2		41.07	123.25	155.88	219.48	231.03	558.08
3			82.18	114.81	178.41	189.96	517.01
4				32.63	96.23	107.78	434.83
5					63.60	75.15	402.20
6						11.55	338.60
7							327.05

With soil conditions comprising about 5 m of gravel over London clay, the pillars were constructed by inserting 8×0.410 m steel pipe into a 9-m borehole and filling with reinforced concrete to within 0.6 m of the pillar top. Each pillar top contains two electrolevels and a platinum resistance thermometer to monitor thermal movement. The pillars are surrounded by 3×0.510 m PVC pipe, to reduce such movement to a minimum. The pillar tops are all at the same level, with Kern baseplates attached. Measurement of the distances has been carried out using a Kern ME5000 Mekometer and checked by a Terrameter. The Mekometer has in turn been calibrated by laser interferometry. The above brief description serves to illustrate the care and planning needed to produce a base line for commercial calibration of the majority of EDM equipment.

3.11 OTHER ERROR SOURCES

3.11.1 Reduction from slope to horizontal

The reduction process using vertical angles has already been dealt with in *Section 3.10.4*. On steep slopes the accuracy of angle measurement may be impossible to achieve, particularly when refraction effects are considered. An alternative procedure is, of course, to obtain the difference in height (h) of the two measuring sources and correct for slope using Pythagorus. If the correction is C_h, then the first term of a binomial expansion of Pythagorus gives

$$C_L = h^2/2S$$

where S = the slope length measured

Then $\delta C_h = h \, \delta h/S$ (3.56)

and for $S = 1000$ m, $h = 100$ m and the accuracy of reduction $\delta C_h = \pm 0.001$ m, substituting in equation (3.56) gives $\delta h = \pm 0.010$ m. This implies that the difference in level should be obtained to an accuracy of ± 0.010 m, which is within the accuracy criteria of tertiary levelling. For $h = 10$ m, $\delta h = \pm 0.100$ m, and for $h = 1$ m, $\delta h = \pm 1$ m.

Analysis of this sort will enable the observer to decide on the method of reduction, i.e. vertical angles or differential levelling, in order to achieve the required accuracy.

3.11.2 Reduction to the plane of projection

Many engineering networks are connected to the national grid system of their country; a process which involves reducing the horizontal lengths of the network to mean sea level (MSL) and then to the projection using local scale factors (LSF).

Reduction to MSL is carried out using

$$C_M = \frac{LH}{R}$$ (3.57)

where C_M = the altitude correction, H = the mean height of the line above MSL or the height of the measuring station above MSL and R = mean radius of the Earth (6.38×10^6 m).

Differentiating equation (3.57) gives $\delta C_M = L\delta H/R$

and for $L = 1000$ m, $\delta C_M = \pm 1$ mm, then $\delta H = \pm 6.38$ m. As Ordnance Survey (OS) tertiary bench marks are guaranteed to ± 10 mm, and the levelling process is of more than comparable accuracy, the errors from this source may be ignored.

Reduction of the horizontal distance to MSL theoretically produces the chord distance, not the arc or spheroidal distance. However, the chord/arc correction is negligible at distances of up to 10 km and will not therefore be considered further.

To convert the ellipsoidal distance to grid distance it is necessary to calculate the LSF and multiply the distance by it. The LSF changes from point to point. Considering the OS national grid (NG) system of the UK, it changes from one side of a 10-km square to the other by about 6 parts in 100 000 (Ordnance Survey, 1950). Thus the value for the middle of the square would be in error by approximately 1 in 30 000.

For details of scale factors, their derivation and application, refer to Chapter 5.

The following approximate formula for scale factors will now be used for error analysis, of the UK system.

$$F = F_0[1 + (E_m^2/2R^2)] \tag{3.58}$$

where E_m = the NG easting of the mid-point of the line = 4 000 000 m
F_0 = the scale factor at the central meridian = 0.999 601 27
R = the mean radius of the Earth (6.38×10^6 m)

Then the scale factor correction is $C = LF_0\{1 + (E_m^2/2R^2)\} - L$

and

$$\delta C/\delta E_m = LF_0(E_m/R^2) \tag{3.59}$$

Then for $L = 1000$ m, $\delta C = \pm 1$ mm and $E_m = 120$ km, $\delta E_m = \pm 333$ m; thus the accuracy of assessing one's position on the NG is not critical. Now, differentiating with respect to R

$$\delta C/\delta R = LF_0 E_m^2/R^3 \tag{3.60}$$

and for the same parameters as above, $\delta R = \pm 18$ km. The value for $R = 6.38 \times 10^6$ m is a mean value for the whole Earth and is accurate to about 10 km between latitudes 30° and 60°, while below 30° a more representative value is 6362 km.

It can be seen therefore that reduction to MSL and thereafter to NG will have a negligible effect on the accuracy of the reduced horizontal distance.

3.11.3 Eccentricity errors

These errors may arise from the manner in which the EDM equipment is mounted on a theodolite and the type of prism used.

(1) Consider telescope-mounted EDM instruments used with a tilting reflector which is offset the same distance, h, above the target as the centre of the EDM equipment is above the line of sight of the telescope (*Figure 3.24*).

In this case the measured distance S is equal to the distance from the centre of the theodolite to the target and the eccentricity e is self-cancelling at instrument and reflector. Hence D and ΔH are obtained in the usual way without further correction.

(2) Consider now a telescope-mounted EDM unit with a non-tilting reflector, as in *Figure 3.25*.

The measured slope distance S will be greater than S' by length $AB = h \tan \alpha$. If α is negative, S will be less than S' by $h \tan \alpha$.

Thus if S is used in the reduction to the horizontal D will be too long by $AF = h \sin \alpha$ when α is positive, and too small when α is negative.

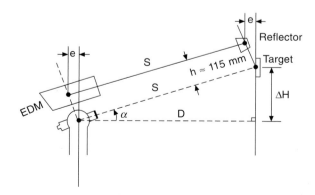

Fig. 3.24

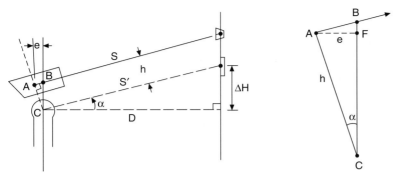

Fig 3.25

If we assume an approximate value of $h = 115$ mm then the error in D when $\alpha = 5°$ is 10 mm, at 10° it is 20 mm and so on to 30° when it is 58 mm. The errors in ΔH for the above vertical angles are 1 mm, 14 mm and 33 mm, respectively.

(3) Instruments mounted on a yoke on the theodolite are generally used with non-tilting reflectors and offset target (*Figure 3.26*). As shown, there is no eccentricity error as the measuring centre of the EDM unit coincides with the axis of tilt.

  If used with a tiltable reflector there will be an eccentricity error $e = h \tan \alpha$ on the slope distance, as in the previous example. However, as in this case the prism is tilting, the slope distance will be *too small* when α is positive and vice versa.

(4) If a yoke-mounted EDM unit is used with a reflector, the centre of which is also the target (*Figure 3.27*), then eccentricity error results because the measured angle of elevation α is not that of the measured distance S.

In triangle *ABC* $h/\sin \theta = S/\sin(90° - \alpha)$

$$\therefore \sin \theta = h \cos \alpha /S$$

Thus, having obtained a value for θ, the horizontal distance D is obtained from

$$D = S \cos(\alpha - \theta)$$

when α is positive.

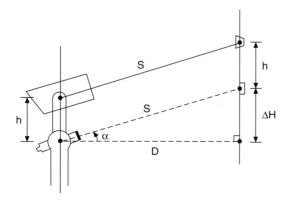

Fig. 3.26

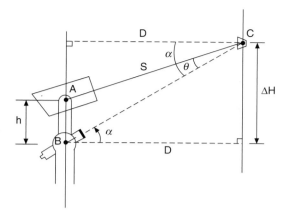

Fig. 3.27

For an angle of depression, i.e. when α is negative

$$D = S \cos(\alpha + \theta)$$

(5) When the EDM unit is co-axial with the telescope line of sight and observations are direct to the centre of the reflector, there are no eccentricity corrections.

3.12 INSTRUMENT SPECIFICATIONS

The measuring accuracy of all EDM equipment is specified in manufacturers' literature as

$$\pm(a \text{ mm} + b \text{ mm/km})$$

with a typical example being

$$\pm(3 \text{ mm} + 3 \text{ mm/km of the distance measured})$$

Thus for a distance of 2 km, the accuracy is

$$\pm(3^2 + 6^2)^{\frac{1}{2}} = \pm 7 \text{ mm}$$

In the above specification, a is a result of errors in phase measurement (θ) and zero error (z), i.e.

$$a^2 = \sigma_\theta^2 + \sigma_z^2$$

In the case of b, the resultant error sources are error in the modulation frequency f and in the group refractive index n_g, i.e.

$$b^2 = (\sigma_f / f)^2 + (\sigma_{n_g} / n_g)^2$$

The reason why the specification is expressed in two parts is that θ and z are independent of distance, whilst f and n_g are a function of distance.

For short distances, frequently encountered in engineering, part a is more significant and would require greater consideration.

3.13 DEVELOPMENTS IN EDM

Improvements in technology and the discovery of the microchip have transformed EDM instruments from large, cumbersome units, which measured distance only, to smaller units which could be mounted on a theodolite, to the present state of the art represented by the total station.

The average total station is a fully integrated instrument that captures all the spatial data necessary for a three-dimensional position fix. The angles and distances are displayed on a digital readout and can be recorded at the press of a button. The more advanced instruments clearly indicate the developments that have taken place, and are still taking place. For example:

- *Dual axis compensators* built in to the vertical axis of the instrument which constantly monitor the inclination of the vertical axis in two directions. These tilt sensors have a range of 3'. The horizontal and vertical angles are automatically corrected, thus permitting single-face observations without loss of accuracy.
- *Graphic electronic levelling display*, illustrating the levelling situation parallel to a pair of footscrews and at right angles, enables rapid, precise levelling without rotation of the alidade. The problems caused by direct sunlight on plate bubbles are also eradicated.
- On-board *PCMCIA memory cards* using SRAM or FLASH technology are available in various capacities for the logging of observations. Capacities up to 8.0 Mb capable of storing 250 000 surveyed points are available. The card memory unit can be connected to any external computer or to a special card reader for data transfer. Alternatively, the observations can be downloaded directly into intelligent electronic data loggers. Both systems can be used in reverse to load information into the instruments.

  Some instruments and/or data loggers can be interfaced directly with a computer for immediate processing and plotting of data.

- *Friction clutch and endless drive* eliminates the need for horizontal and vertical circle clamps plus the problem of running out of thread on slow motion screws.
- *Laser plummet*, incorporated into the vertical axis, replaces the optical plummet. A clearly visible laser dot is projected on to the ground permitting quick and convenient centring of the instrument.
- *Extensive keyboards* (*Figure 3.31*) with multi-line LCD displays of alphanumeric and graphic data control every function of the instrument. Built in software with menu and edit facilities, they automatically reduce angular and linear observations to three-dimensional coordinates of the vector observed. This facility can be reversed for setting-out purposes. Detachable control units are available on particular instruments (*Figure 3.28*).
- *Guide light* fitted to the telescope of the instrument enables the target operator to maintain alignment when setting-out points. This light changes colour when the operator moves off-line. With the instrument in the tracking mode, taking measurements every 0.3 s, the guide light speeds up the setting-out process.
- *Automatic target recognition* (ATR) is incorporated in most robotic instruments and is more accurate and consistent than human sighting. The telescope is pointed in the general direction of the target, and the ATR module completes the fine pointing with excellent precision and minimum measuring time as there is no need to focus. It can also be used on a moving reflector. After initial measurement, the reflector is tracked automatically (*Figure 3.29*). A single key touch records all data without interrupting the tracking process. To ensure that the prism is always pointed to the instrument, *360° prisms* are available from certain manufacturers. ATR recognizes targets up to 1000 m away and maintains lock on prisms moving at a speed of 5 m s^{-1}. A further advantage of ATR is that it can operate in darkness.

Fig. 3.28 *Detachable control units (Courtesy of Spectra-Precision)*

- In order to utilize ATR, the instrument must be fitted with *servo motors* to drive both the horizontal and vertical movements of the instrument. It also permits the instrument to automatically turn to a specific bearing (direction) when setting out, calculated from the up-loaded design co-ordinates of the point.

- *Reflectorless measurement* is also available on many instruments, typically using two different coaxial red laser systems. One laser is invisible and is used to measure long distances (6 km to a single reflector), the other is visible, does not require a reflector, and has a limited range of about 200 m. A single key stroke allows one to alternate between the visible or invisible laser. With the latest Geodimeter 600s DR 200+, distances of almost 500 m have been recorded.

   Possible uses for this technique include surveying the facades of buildings, tunnel profiling, cooling tower profiling, bridge components, dam faces – indeed any situation which is difficult or impossible to access directly. The extremely narrow laser used clearly defines the target point (*Figure 3.30*).

- *Built-in programmes* are available with most total stations. Examples of which are:

   traverse coordinate computation with Bowditch adjustment;
   missing line measurement in the horizontal and vertical planes to any two points sighted from the instrument;
   remote object elevation determines the heights of inaccessible points;
   offset measurement gives the distance and bearing to an inaccessible point close to the reflector by obtaining the vector to the reflector and relative direction to the inaccessible point;
   resection to a minimum of two known points determines the position of the total station observing those points;
   building facade survey allows for the co-ordination of points on the face of a building or structure using angles only;
   three-dimensional coordinate values of points observed;

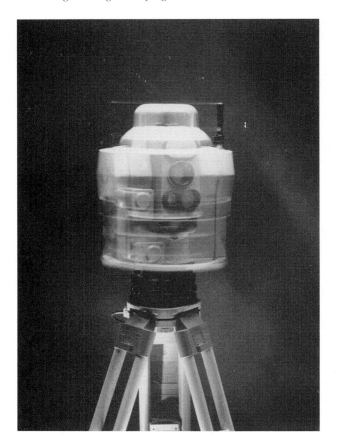

Fig. 3.29   *Geodimeter, rotating automatically in its search for the reflector pole*

Fig. 3.30   *Reflectorless surveying (Courtesy of Leica)*

Setting-out data to points whose coordinates have been uploaded into the total station; Coding of the topographic detail with automatic point number incrementation.

The above has detailed many of the developments in total station design and construction which have led to the development of fully automated one-man systems, frequently referred to as robotic surveying systems. Robotic surveying produces high productivity since the fully automatic instrument can be left unmanned and all operations controlled from the target point that is being measured or set-out. The equipment used at the target point would consist of an extendible reflector pole with a circular bubble. It would carry a 360° prism and a control unit incorporating a battery and radio modem. Two types are illustrated (in *Figures 3.31* and *3.32*) and are for the remote control of the

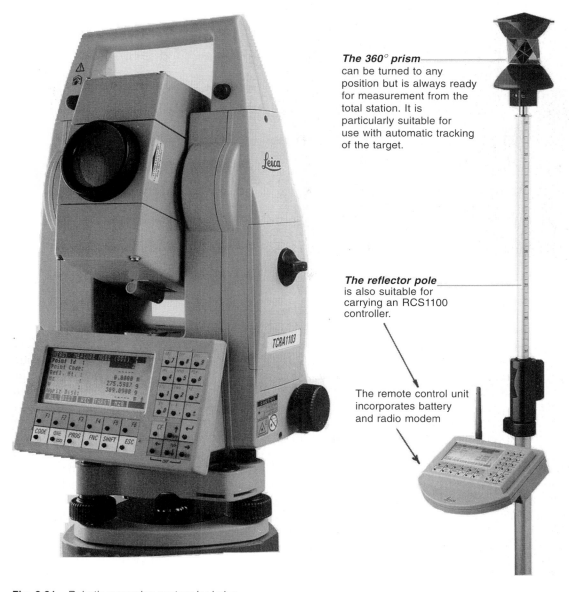

The 360° prism can be turned to any position but is always ready for measurement from the total station. It is particularly suitable for use with automatic tracking of the target.

The reflector pole is also suitable for carrying an RCS1100 controller.

The remote control unit incorporates battery and radio modem

Fig. 3.31   *Robotic surveying system by Leica*

Fig. 3.32   *Robotic surveying system by Spectra-Precision*

total station and the storage of data. They have an alphanumeric LCD display and are graphics capable. Import and export of data is via the radio modem. Storage capacity would be equal to at least 10 000 surveyed points, plus customized software and the usual facilities to view, edit, code, set-out, etc. The entire measurement procedure is controlled from the reflector pole with facilities for keying the start/stop operation, aiming, changing modes, data registration, calculations and data input

The control unit on the Geodimeter system (*Figure 3.32*), which is detachable between the total station and reflector pole, is called Geodat Win as it runs under Windows 95. Standard features include file handling, desktop icons, pull-down menus, etc. Its graphics capability allows the survey to be drawn in real time, cross-sections displayed, and points to be set-out shown on the screen. The software, Geodat Win Base, contains several different modules including Database, Calculations Engine, Coordinate Transformation Engine, Surveying Applications, CAD Applications and Instrument Control Manager. In addition to a library of applications programs, Geodat Win Base supports all major coordinate systems, datums and ellipsoids and can supply satellite information and sky plots. Thus, it can be used with GPS antennae.

The total station unit in robotic surveying must have servo motors, ATR, dual axis compensators and a guide light. In the case of the Leica system, a radio modem fits to the tripod and permits cable-less exchange of data between the controllers. The robotic system can be used for both surveying and setting-out, is a one-man operation, and is reported to increase productivity by as much as 200%.

3.13.1 Machine guidance (Figure 3.33)

Robotic surveying has resulted in the development of several customized systems, not the least of which are those for the control of construction plant on site. These systems, produced by the major companies Leica and Spectra Precision, are capable of controlling slip-form pavers, rollers, motorgraders and even road headers in tunnelling. In each case the method is fundamentally the same.

(a)

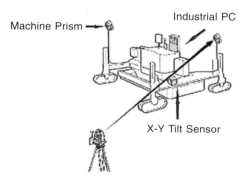

(b)

Fig. 3.33 *Machine guidance by robotic EDM*

The machine is fitted with a customized 360° prism strategically positioned on the machine. The total station is placed some distance outside the working area and continuously monitors the three-dimensional position of the prism. This data is transmitted via the radio link to an industrial PC on board the construction machine. The PC compares the construction project data with the machine's current position and automatically and continuously sends the appropriate control commands to the machine controller to give the necessary construction position required. All the information is clearly displayed on a large screen in alphanumeric and graphics format. Such information comprises actual and required grading profiles; compression factors for each surface area being rolled and the exact location of the roller; tunnel profiles showing the actual position of the cutter head relative to the required position. In slip forming, for example, complex profiles, radii and routes are quickly completed to accuracies of 2 mm and 5 mm in vertical and lateral positions respectively. Not only do these systems provide extraordinary precision, they also afford greater safety, speedier construction and higher quality.

Normally all these operations are controlled using stringlines, profile boards, batter boards, etc. As these would no longer be required, their installation and maintenance costs are eliminated, they do not interfere with the machines and construction site logistics, and so errors due to displacement of 'wood' and 'string' are precluded.

3.14 OPTICAL DISTANCE MEASUREMENT (ODM)

ODM, in all its forms, has been rendered obsolete by EDM. The limitations of stadia tacheometry have been dealt with in *Chapter 2*. Its application to contouring (*Chapter 2*) and topographic detailing (*Chapter 1*) has been dealt with, but included the proviso that it should only be used if there is no alternative.

A more serious contender in the measurement of distance by optical method was the subtense bar. This instrument is also obsolete (used only in obtaining scale in electronic coordinate determination sytems (ECS)) and is mentioned here purely because it provides an interesting measuring concept.

3.14.1 Subtense tacheometry

This method uses a horizontal subtense bar with targets at each end precisely 2 m apart. Although the bar is of steel construction, the targets are connected to an invar wire in such a way as to compensate for temperature changes. The bar can be set up horizontally on a normal theodolite tribrach and set at 90° to the line of sight by means of a small sighting device at its centre (*Figure 3.34*).

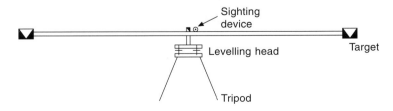

Fig. 3.34

3.14.1.1 Principle of operation

The principle is illustrated by *Figure 3.35*. Regardless of the elevation, the angle θ subtended by the bar is measured in the horizontal plane by the theodolite. The horizontal distance *TB* is then given by

$$D = b/2 \cot \theta/2 \tag{3.61}$$
$$= \cot \theta/2, \text{ when } b = 2 \text{ m} \tag{3.62}$$

The vertical distance is given by

$$H = D \tan \alpha \tag{3.63}$$

and the level of *B* relative to *T*, would therefore be

level of B = level of $T + h_1 + H - h_2$

showing that in the computation of levels, one would require the instrument heights.

3.14.1.2 Errors

The three sources of error in the distance *D* are:

(1) Variation in the length of the subtense bar.
(2) Error in setting the bar at 90° to the line of sight, and horizontally.
(3) Error in the measurement of the subtense angle.

To simplify the differentiation of each variable the basic formula is reduced to a form as follows:

$$D = b/2 \cot \theta/2, \text{ but since } \theta/2 \text{ is very small}$$
$$\tan \theta/2 \approx \theta \text{ rad, thus } \cot \theta/2 = 2/\theta$$
$$\therefore D = (b/2)(2/\theta) = b/\theta$$

It can be shown that the error in this approximation is roughly 1 in $3D^2$ and should therefore never be used for the reduction of sights; for example when $D = 40$ m, b/θ is accurate to only 1 in 4800.

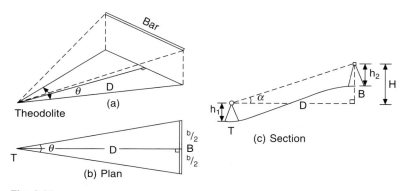

Fig. 3.35

(1) Error in bar length

$$D = b/\theta$$

$$\therefore \delta D = \frac{\delta b}{\theta} \quad \text{and} \quad \frac{\delta D}{D} = \frac{\delta b}{D}\frac{\theta}{b}$$

Thus $\dfrac{\delta D}{D} = \dfrac{\delta b}{b}$ \hfill (3.64)

Manufacturers of the various subtense bars claim a value of 1/100 000 for $\delta b/b$ due to a 20°C change in temperature. This source of error may therefore be ignored.

(2) Error in bar setting

Failure to align the bar at 90° to the line of sight results in the length AC being reduced to $A'C' \approx b \cos \phi$ (*Figure 3.36*). Misalignment in the vertical plane, however, shows $A'C' = b \cos \phi$. Thus, the error in the bar length $= b - b \cos \phi$ in both cases

i.e. $\quad \delta b = b (1 - \cos \phi)$, then from equation (3.64) above

$$\delta D/D = \delta b/b = (1 - \cos \phi)$$

but $\quad \cos \phi = 1 - \phi^2/2! + \phi^4/4! \ldots$

$$\therefore \delta D/D = \phi^2/2 \hfill (3.65)$$

If $\delta D/D$ is not to exceed 1/20 000 then

$$\phi = \left(\frac{2}{20\ 000}\right)^{\frac{1}{2}} = 1/100 \text{ rad} \approx 0°\ 34'$$

Alignment to this accuracy is easily obtained by using the standard sighting devices. This source of error may therefore be ignored.

(3) Error in the measurement of the subtense angle

$$D = \frac{b}{\theta}$$

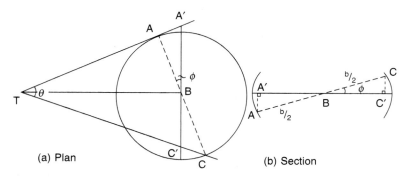

(a) Plan   (b) Section

Fig. 3.36

$$\therefore \delta D = \frac{-b}{\theta^2}\,\delta\theta = \frac{-b}{\theta}\,\frac{\delta\theta}{\theta} = -D\,\frac{\delta\theta}{\theta}$$

$$\therefore \frac{\delta D}{D} = \frac{\delta\theta}{\theta} \tag{3.66}$$

Using the above relationship *Table 3.1* may be deduced, assuming a 2-m bar and an error of ±1″ in the measurement of θ, illustrating that the accuracy falls off rapidly with increase in distance. By further manipulation of the above equation it can be shown that the error in D varies as the square of the distance:

$$\delta D = -(b/\theta^2)\delta\theta \qquad \text{but } \theta^2 = b^2/D^2$$

$$\therefore \delta D = (D^2/b)\delta\theta \tag{3.67}$$

Thus, an error of ±1″ produces four times the error at 80 m than it does at 40 m. This can be further clarified from *Table 3.1*, where 40/10 000 = 40 mm, 80/5000 = 16 mm.

To achieve a PSE of 1/10 000 the distance must be limited to 40 m and an accuracy of ±1″ attained in the measurement of the angle. This is possible only with a 1″-reading theodolite.

Research has proved that the subtense angle should be measured at least eight times to achieve the necessary accuracy. As a 1″ instrument would be used, there is no need to change face between observations to eliminate instrumental errors, as each end of the bar is at the same elevation. However, to eliminate graduation errors they should be observed on different parts of the horizontal circle.

Worked examples

Example 3.10 The majority of short-range EDM equipment measures the difference in phase of the transmitted and reflected light waves on two different frequencies, in order to obtain distance.

The frequencies generally used are 15 MHz and 150 kHz. Taking the velocity of light as 299 793 km/s and a measure distance of 346.73 m, show the computational processes necessary to obtain this distance, clearly illustrating the phase difference technique.

Travel distance = 2 × 346.73 = 693.46 m
Travel time of a single pulse = $t = D/V$ = 693.46/299 793 000

−2.313 μs = 2313 ns

Standard frequency = f = 15 MHz = 15 × 10^6 cycles/s
Time duration of a single pulse = 1/15 × 10^6 s = 66.$\bar{6}$ ns = t_p.
∴ No. of pulses in the measured distance = t/t_p = 2313/66.$\bar{6}$ = 34.673

i.e. $2D = M\lambda + \delta\lambda = 34\lambda + 0.673\lambda$

However, only the phase difference $\delta\lambda$ is known and not the value of M; hence the use of a second frequency.

A single pulse (λ) takes 66.6 ns, which at 15 MHz = 20 m, and $\lambda/2$ = 10 m.

Table 3.1

D(m)	20	40	60	80	100
$\delta D/D$	1 in 20 626	1 in 10 313	1 in 6875	1 in 5106	1 in 4125

$$\therefore D = M(\lambda/2) + 0.673\ (\lambda/2) = 6.73\ \text{m}$$

Now using $f = 150$ kHz $= 150 \times 10^3$ cycles/s, the time duration of a single pulse $= 1/150 \times 10^3 = 6.667\ \mu\text{s}$.

> At 150 kHz, $6.6674\ \mu\text{s} = 1000$ m
>
> No. of pulses $= t/t_\text{p} = 2.313/6.667 = 0.347$
>
> $\therefore D = 0.347 \times 1000\ \text{m} = 347\ \text{m}$

> Fine measurement using 15 MHz $= 6.73$ m
>
> Coarse measurement using 15 kHz $= 347$ m
>
> Measured distance $= 346.73$ m

Example 3.11 (a)   Using EDM, top-mounted on a theodolite, a distance of 1000 m is measured on an angle of inclination of $09°00'\ 00''$. Compute the horizontal distance.

Now, taking $R = 6.37 \times 10^6$ m and the coefficient of refraction $K = 1.10$, correct the vertical angle for refraction effects, and recompute the horizontal distance.

(b)   If the EDM equipment used above was accurate to $\pm(3$ mm $+ 5$ ppm), calculate the required accuracy of the vertical angle, and thereby indicate whether or not it is necessary to correct it for refraction.

(c)   Calculate the equivalent error allowable in levelling the two ends of the above measured line.   (KU)

(a) Horizontal distance $= D = S \cos \alpha = 1000 \cos 9° = 987.688$ m

> $\hat{r}'' = SK\rho/2R = 987.69 \times 1.10 \times 206\ 265/2 \times 6\ 370\ 000 = 17.6''$
>
> Corrected angle $= 8°59'\ 42''$
>
> $\therefore D = 1000 \cos 8°59'\ 42'' = 987.702$ m
>
> Difference $= 14$ mm

(b) Distance is accurate to $\pm(3^2 + 5^2)^{\frac{1}{2}} = \pm5.8$ mm

> $D = S \cos \alpha$
>
> $\delta D = -S \sin \alpha\ \delta\alpha$
>
> $\delta\alpha'' = \delta D\rho/S \sin \alpha = 0.0058 \times 206\ 265/10\ 000 \sin 9° = \pm7.7''$

Vertical angle needs to be accurate to $\pm7.7''$, so refraction must be catered for.

(c) To reduce S to D, apply the correction $-h^2/2S = C_h$

> $\delta C_h = h\ \delta h/S$
>
> and $\delta h = \delta C_h S/h$
>
> where $h = S \sin \alpha = 1000 \sin 9° = 156.43$ m
>
> $\delta h = 0.0058 \times 1000/156.43 = \pm0.037$ m

Example 3.12
(a) Modern total stations supply horizontal distance (D) and vertical height (ΔH) at the press of a button.

What corrections must be applied to the initial field data of slope distance and vertical angle to obtain the best possible values for *D* and Δ*H*?

(b) When using EDM equipment of a particular make, why is it inadvisable to use reflectors and nomograms from other makes of instrument?

(c) To obtain the zero error of a particular EDM instrument, a base line *AD* is split into three sections *AB*, *BC* and *CD* and measured in the following combinations:

$AB = 20.512, AC = 63.192, AD = 153.303$

$BC = 42.690, BD = 132.803, CD = 90.1201$

Using all possible combinations, compute the zero error: (KU)

(a) Velocity correction using temperature and pressure measurements at the time of measurement. Vertical angle corrected for refraction to give *D* and for Earth curvature to give Δ*H*.

(b) Instrument has a built-in correction for the reflector constant and may have been calibrated for that particular instrument/reflector combination. Instrument standardization may be different from that used to produce a nomogram for other instruments.

(c) $20.512 + 42.690 - 63.192 = +0.010$

    $63.192 + 90.120 - 153.303 = +0.009$

    $20.512 + 132.803 - 153.303 = +0.012$

    $42.690 + 90.120 - 132.803 = +0.007$

    Mean = +0.009 m

    Correction = −0.009 m

Example 3.13 Manufacturers specify the accuracy of EDM equipment as

$\pm(a + bD)$ mm

where *b* is in ppm of the distance measured, *D*.

Describe in detail the various errors defined by the variables *a* and *b*. Discuss the relative importance of *a* and *b* with regard to the majority of measurements taken in engineering surveys.

What calibration procedures are required to minimize the effect of the above errors in EDM measurement. (KU)

For answer, refer to appropriate sections of the text.

REFERENCES

Hodges, D.J. (1968) 'Errors in Model 6 Geodimeter Measurements and a Method for Increased Accuracy', *The Mining Engineer,* December.

Ordnance Survey (1950) *Constants, Formulae and Methods Used in Transverse Mercator Projection*, HMSO.

4

Angles

As shown in Chapter 1, horizontal and vertical angles are fundamental measurements in surveying.

The vertical angle, as already illustrated, is used in obtaining the elevation of points (trig levelling) and in the reduction of slant distance to the horizontal.

The horizontal angle is used primarily to obtain relative direction to a survey control point, or to topographic detail points, or to points to be set out.

The instrument used in the measurement of angles is called a theodolite, the horizontal and vertical circles of which can be likened to circular protractors set in horizontal and vertical planes. It follows that, although the points observed are at different elevations, it is always the horizontal angle and not the space angle which is measured. For example, observations to points A and C from B (*Figure 4.1*) will give the horizontal angle $ABC = \theta$. The vertical angle of elevation to A is α and its zenith angle is Z_A.

4.1 THE THEODOLITE

There are basically two types of theodolite, the optical microptic type or the electronic digital type, both of which are capable of resolving angles to $1'$, $20''$, $1''$ or $0.1''$ of arc, depending upon the accuracy requirements of the work in hand. The finesse of selecting an instrument specific to the survey tolerances is usually overridden by the commercial aspects of the company and a $1''$ instrument may be used for all work. When one considers that $1''$ of arc subtends 1 mm in 200 m, it is sufficiently accurate for practically all work carried out in engineering.

Figure 4.2 shows a typical theodolite, whilst *Figure 4.3* shows the main components of the new obsolete vernier-type theodolite. This exploded diagram enables the relationships of the various parts to be more clearly understood along with the relationships of the main axes. In a correctly adjusted instrument these axes should all be normal to each other, with their point of intersection being the point about which the angles are measured. Neither figure illustrates the complexity of a modern theodolite or the very high calibre of the process of its production. This can be clearly seen from *Figure 4.4*.

The basic features of a typical theodolite are, with reference to *Figure 4.3*, as follows:

(1) The trivet stage forming the base of the instrument connects it to the tripod head.
(2) The tribrach supports the rest of the instrument and with reference to the plate bubble can be levelled using the footscrews which act against the fixed trivet stage.
(3) The lower plate carries the horizontal circle which is made of glass, with graduations from 0° to 360° photographically etched around the perimeter. This process enables lines of only 0.004 mm thickness to be sharply defined on a small-diameter circle (100 mm), thereby resulting in very compact instruments.

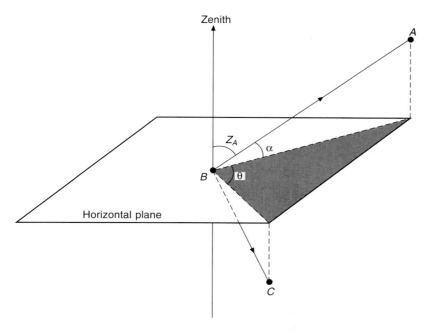

Fig. 4.1

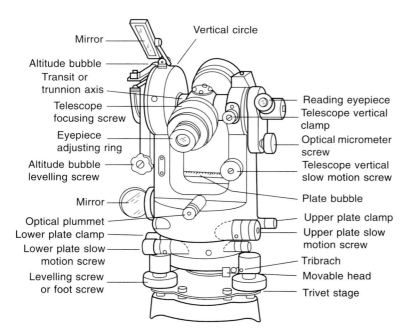

Fig. 4.2

(4) The upper plate carries the horizontal circle index and fits concentric with the lower plate.
(5) The plate bubble is attached to the upper plate and when centred, using the footscrews, establishes the instrument axis vertical. Some modern digital or electronic theodolites have replaced the spirit bubble with an electronic bubble.

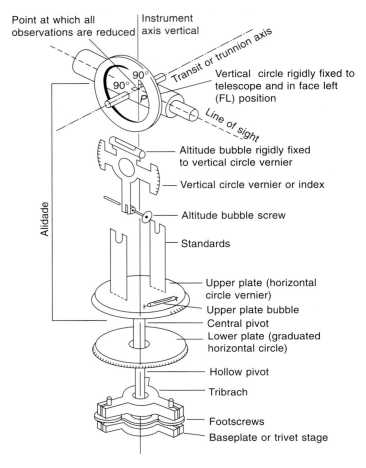

Fig. 4.3   *Simplified vernier theodolite*

(6) The upper plate also carries the standards which support the telescope by means of its transit axis. The standards are tall enough to allow the telescope to be fully rotated about its transit axis.

(7) The vertical circle similar in construction to the horizontal circle is fixed to the telescope axis and rotates with rotation of the telescope.

(8) The vertical circle index, against which the vertical angles are measured, is set normal to gravity by means of (a) an altitude bubble attached to it, or (b) an automatic compensator. The latter method is now universally employed in modern theodolites.

(9) The lower plate clamp (*Figure 4.2*) enables the horizontal circle to be clamped into a fixed position. The lower plate slow motion screw permits slow movement of the theodolite around its vertical axis, when the lower plate clamp is clamped. Most modern theodolites have replaced the lower plate clamp and slow motion screw with a horizontal circle-setting screw. This single screw rotates the horizontal circle to any reading required.

(10) Similarly, the upper plate clamp and slow motion screw have the same effect on the horizontal circle index.

(11) The telescope clamp and slow motion screw fix and allow fine movement of the telescope in the vertical plane.

(12) The altitude bubble screw centres the altitude bubble, which, as it is attached to the vertical circle index, establishes it horizontal prior to reading the vertical circle. As stated in (8), this is now done by means of an automatic compensator.

(13) The optical plummet, built into either the base of the instrument or the tribrach (*Figure 4.13*), enables the instrument to be centred precisely over the survey point. The line of sight through the plummet is coincident with the vertical axis of the instrument.

(14) The telescopes are similar to those of the optical level but usually shorter in length. They also possess rifle sights or collimators for initial pointing.

4.1.1 Reading systems

The theodolite circles are generally read by means of a small auxiliary reading telescope at the side of the main telescope (*Figure 4.2*). The small circular mirrors, as shown in *Figure 4.4*, reflect light into the complex system of lenses and prisms used to read the circles.

There are basically three reading systems in use at the present time.

(a) Optical scale reading.
(b) Optical micrometer reading.
(c) Electronic digital display.

(1) The optical scale reading system is generally used on theodolites with a resolution of 20″ or

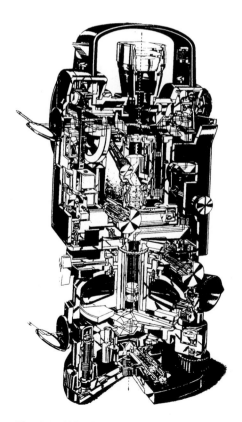

Fig. 4.4 *Wild theodolite by Leica*

less. Both horizontal and vertical scales are simultaneously displayed and are read directly with the aid of the auxiliary telescope.

The telescope used to give the direct reading may be a 'line microscope' or a 'scale microscope'.

The line microscope uses a fine line etched onto the graticule as an index against which to read the circle.

The scale microscope has a scale in its image plane, whose length corresponds to the line separation of the graduated circle. *Figure 4.5* illustrates this type of reading system and shows the scale from 0′ to 60′ equal in scale of one degree on the circle. This type of instrument is frequently referred to as a direct-reading theodolite and, at best, can be read, by estimation, to 20″.

(2) The optical micrometer system generally uses a line microscope, combined with an optical micrometer using exactly the same principle as the parallel plate micrometer on a precise level.

Figure 4.6 illustrates the principle involved. If the observer's line of sight passes at 90° through the parallel plate glass, the circle reading would be $23°20' + S$, with the value of S unknown. The parallel plate is rotated using the optical micrometer screw (*Figure 4.2*) until the line of sight is at an exact reading of $23°20'$ on the circle. This is as a result of the line of sight being refracted towards the normal and emerging on a parallel path. The distance S through which the observer's line of sight was displaced is recorded on the micrometer scale as $11'40''$.

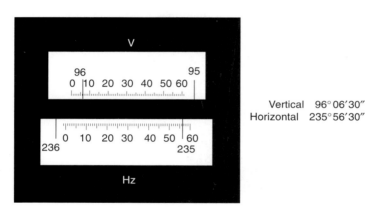

Vertical   96° 06′30″
Horizontal   235° 56′30″

Fig. 4.5   *Wild T16 direct reading theodolite*

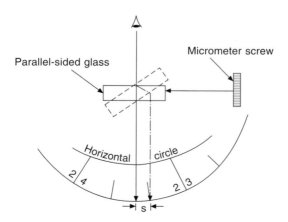

Fig. 4.6

The shift of the image is proportional to the angle of tilt of the parallel plate and is read on the micrometer scale. Before the scale can be read, the micrometer must be set to give an exact reading (23° 20′), as shown on *Figure 4.7*, and the micrometer scale reading (11′ 40″) added on. Thus the total reading is 23° 31′ 40″. In this instance the optical micrometer reads only one side of the horizontal cricle, which is common to 20″ instruments.

On more precise theodolites, reading to 1″ of arc and above, a coincidence microscope is used. This enables diametrically opposite sides of the circle to be combined and a single mean reading taken. This mean reading is therefore free from circle eccentricity error.

Figure 4.8 shows the diametrically opposite scales brought into coincidence by means of the optical micrometer screw. The number of divisions on the main scale between 94° and 95° is three; therefore each division represents 20′. The indicator mark can only take up one of two positions, either mid-division or on a full division. In this case it is mid-division and represents

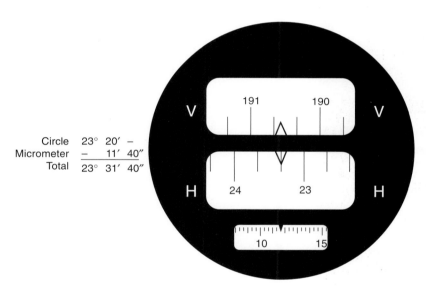

Circle	23° 20′	−
Micrometer	−	11′ 40″
Total	23° 31′ 40″	

Fig. 4.7 *The reading system of a Watts Microptic No. 1 theodolite*

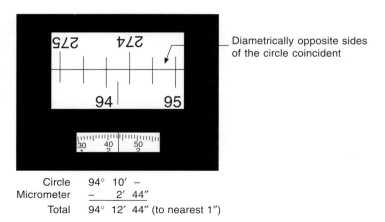

Circle	94° 10′	−
Micrometer	−	2′ 44″
Total	94° 12′ 44″ (to nearest 1″)	

Fig. 4.8 *Wild T2 (old pattern) theodolite reading system*

a reading of 94°10′; the micrometer scale reads 2′ 44″ to the nearest second, giving a total reading of 94° 12′ 44″. An improved version of this instrument is shown in *Figure 4.9*.

The above process is achieved using two parallel plates rotating in opposite directions, until the diametrically opposite sides of the circle coincide.

(3) There are basically two systems used in the electro-optical scanning process, either the incremental method or the code method (*Figure 4.10*).

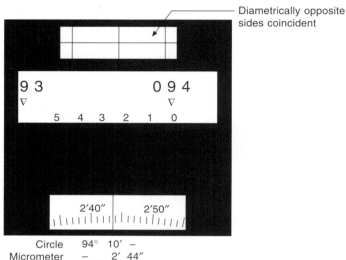

Circle	94°	10′	–
Micrometer	–	2′	44″
Total	94°	12′	44″ (to nearest 1″)

Fig. 4.9 *Wild T2 (new pattern)*

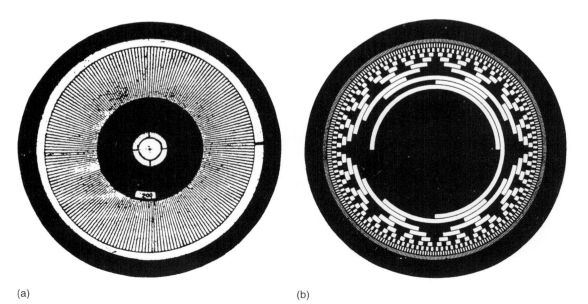

(a)   (b)

Fig. 4.10 *(a) Incremental disk. (b) Binary coded disk*

The basic concept of the incremental method can be illustrated by considering a glass circle of 70–100 mm diameter, graduated into a series of radial lines. The width of these photographically etched lines is equal to their spacing. The circle is illuminated by a light diode; a photodiode, equal in width to a graduation, forms an index mark. As the alidade of the instrument rotates, the glass circle moves in relation to the diode. The light intensity signal radiated approximates to a sine curve. The diode converts this to an electrical signal correspondingly modulated to a square wave signal (*Figure 4.11*). The number of signal periods is counted by means of the leading and trailing edges of the square wave signal and illustrated digitally in degress, minutes and seconds on the LCD. This simplified arrangement would produce a relatively coarse least count resolution, requiring further refinement.

For example, in the case of the Kern E2 electronic theodolite, the glass circle contains 20 000 radial marks, each 5.5 μm thick, with equal width spacing. A section of the circle comprising 200 marks is superimposed on the diametrically opposite section, forming a moiré pattern. A full period (light–dark variation) corresponds to an angular value of approximately 1 min of arc, with a physical length of 2 mm. A magnification of this period by two, provides a length of 4 mm over which the brightness pattern can be electronically scanned. Thus the coarse measurement can be obtained from 40 000 periods per full circle, equivalent to 30″ per period.

The fine reading to 0.3″ is obtained by monitoring the brightness distribution of the moiré pattern using the four diodes shown (*Figure 4.12*). The fine measurement obtains the scanning position location with respect to the leading edge of the square wave form within the last moiré pattern. It is analogous to measuring the fraction of a wavelength using the phase angle in EDM measurement.

The code methods use coded graduated circles (*Figure 4.10(b)*). Luminescent diodes above the glass circle and photodiodes below, one per track, detect the light pattern emitted, depending on whether a transparent track (signal 1) or an opaque track (signal 0) is opposite the diode at that instant. The signal is transferred to the computer for processing into a digital display. If there are n tracks, the full circle is divided into 2″equal sectors. Thus a 16-track disk has an angular resolution of 2^{16}, which is 65 532 parts of a full circle and is equivalent to a 20″ resolution.

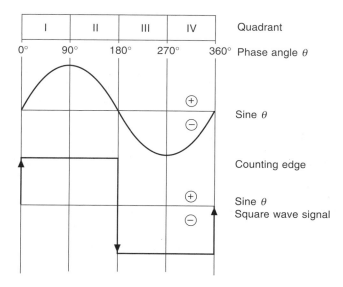

Fig. 4.11

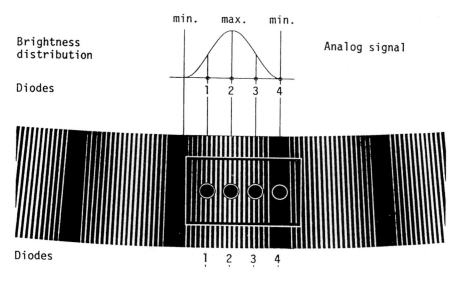

Fig. 4.12

The advantage of the electronic systems over the glass arc scales is that they produce a digital output free from misreading errors and in a form suitable for automatic data recording and processing. *Figure 4.13* illustrates the glass arc and electronic theodolites.

4.2 INSTRUMENTAL ERRORS

In order to achieve reliable measurement of the horizontal and vertical angles, one must use an instrument that has been properly adjusted and adopt the correct field procedure.

In a properly adjusted instrument, the following geometrical relationships should be maintained (*Figure 4.3*):

(1) The plane of the horizontal circle should be normal to the vertical axis of rotation.
(2) The plane of the vertical circle should be normal to the horizontal transit axis.
(3) The vertical axis of rotation should pass through the point from which the graduations of the horizontal circle radiate.
(4) The transit axis of rotation should pass through the point from which the graduations of the vertical circle radiate.
(5) The principal tangent to the plate bubble should be normal to the main axis of rotation.
(6) The line of sight should be normal to the transit axis.
(7) The transit axis should be normal to the main axis of rotation.
(8) When the telescope is horizontal, the vertical circle indices should be horizontal and reading zero, and the principal tangent of the altitude bubble should, at the same instance, be horizontal.
(9) The main axis of rotation should meet the transit axis at the same point as the line of sight meets this axis.
(10) The line of sight should maintain the same position with change of focus (an important fact when coplaning wires).

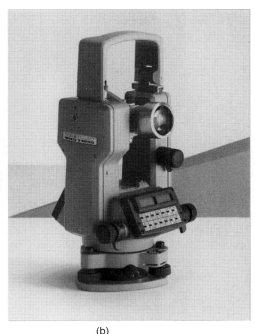

(a)   (b)

Fig. 4.13   *(a) Wild TI glass arc theodolite with optical plummet in the alidade. (b) Wild T1600 electronic theodolite with optical plummet in the tribrach*

Items (1), (2), (3) and (4) above are virtually achieved by the instrument manufacturer and no provision is made for their adjustment. Similarly, (9) and (10) are dealt with, as accurately as possible, in the manufacturing process and in any event are minimized by double face observations. Items (5), (6), (7) and (8) can, of course, be achieved by the usual adjustment procedures carried out by the operator.

The procedure referred to above as 'double face observation' is fundamental to the accurate measurement of angles. An examination of *Figure 4.2* shows that an observer looking through the eyepiece of the telescope would have the vertical circle on the left-hand side of his face; this would be termed a 'face left' (FL) observation. If the telescope is now transmitted through 180° about its transit axis and then the instrument rotated through 180° about its vertical axis, the vertical circle would be on the right-hand side of the observer's face as he looked through the telescope eyepiece. This is called a 'face right' (FR) observation. The mean result of a FL and FR observation, called a double face observation, is free from the majority of instrumental errors present in the theodolite.

The main instrumental errors will now be dealt with in more detail and will serve to emphasize the necessity for double face observation.

4.2.1 Eccentricity of centres

This error is due to the centre of the central pivot carrying the alidade (upper part of the instrument) not coinciding with the centre of the hollow pivot carrying the graduated circle (*Figures 4.3* and *4.14*)

The effect of this error on readings is periodic. If B is the centre of the graduated circle and A is the centre about which the alidade revolves, then distance AB is interpreted as an arc ab in seconds on the graduated circle and is called the *error of eccentricity*. If a vernier is at D, on the line of the

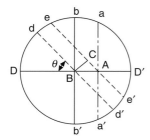

Fig. 4.14

two centres, it reads the same as it would if there were no error. If, at *b*, it is in error by *ba* = *E*, the maximum error. In an intermediate position *d*, the error will be *de* = *BC* = *AB* sin θ = *E* sin, θ, θ being the horizontal angle of rotation.

The horizontal circle is graduated clockwise, so the vernier supposedly at *b* will be at *a*, giving a reading too great by +*E*. The opposite vernier supposedly at *b'* will be at *a'*, thereby reading too small by – *E*. Similarly for the intermediate positions at *d* and *d'*, the errors will be + *E* sin θ and – *E* sin θ. *Thus the mean of the two verniers 180° apart, will be free of error.*

Modern glass-arc instruments in the 20″ class can be read on one side of the graduated circle only, thus producing an error which varies sinusoidally with the angle of rotation. Readings on both faces of the instrument would establish verniers 180° apart. Thus the mean of readings on both faces of the instrument will be free of error. With 1″ theodolites the readings 180° apart on the circle are automatically averaged and so are free of this error.

Manufacturers claim that this source of error does not arise in the construction of modern glass-arc instruments.

4.2.2 Collimation in azimuth

Collimation in azimuth error refers to the error which occurs in the observed angle due to the line of sight, or more correctly, the line of collimation, not being at 90° to the transit axis (*Figure 4.3*). If the line of sight in *Figure 4.15* is at right angles to the transit axis it will sweep out the vertical plane *VOA* when the telescope is depressed through the vertical angle α.

If the line of sight is not at right angles but in error by an amount *e*, the vertical plane swept out will be *VOB*. Thus the pointing is in error by – ϕ (negative because the horizontal circle is graduated clockwise).

$$\tan \phi = \frac{AB}{VA} = \frac{OA \tan e}{VA} \qquad \text{but} \quad \frac{OA}{VA} = \sec \alpha$$

$$\therefore \tan \phi = \sec \alpha \tan e$$

as ϕ and *e* are very small, the above may be written

$$\phi = e \sec \alpha \qquad\qquad (4.1)$$

On changing face *VOB* will fall to the other side of *A* and give an equal error of opposite sign, i.e. +ϕ. *Thus the mean of readings on both faces of the instrument will be free of this error.*

ϕ is the error of one sighting to a target of elevation α. An angle, however, is the difference between two sightings; therefore the error in an angle between two objects of elevation, α_1 and α_2, will be $e(\sec \alpha_1 - \sec \alpha_2)$ and will obviously be zero if $\alpha_1 = \alpha_2$, or if measured in the horizontal plane, ($\alpha = 0$).

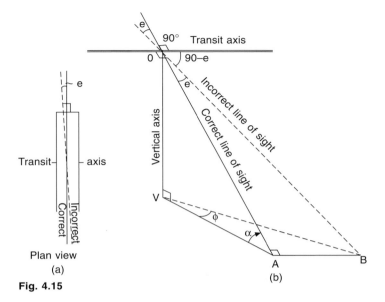

Fig. 4.15

On the opposite face the error in the angle simply changes sign to $-e(\sec \alpha_1 - \sec \alpha_2)$, indicating that the *mean of the two angles taken on each face will be free of error regardless of elevation.*

Vertical angles: it can be illustrated that the error in the measurement of the vertical angles is $\sin \alpha = \sin \alpha_1 \cos e$ where α is the measured altitude and α_1 the true altitude. However, as e is very small, $\cos e \approx 1$, hence $\alpha_1 \approx \alpha$, proving that the effect of this error on vertical angles is negligible.

4.2.3 Transit axis error

Error will occur in the measurement of the horizontal angle if the transit axis is not at 90° to the instrument axis (*Figure 4.3*). At the time of measurement the instrument axis should be vertical. If the transit axis is set correctly at right angles to the vertical axis, then when the telescope is depressed it will sweep out the truly vertical plane *VOA* (*Figure 4.16*). Assuming the transit axis is inclined to the horizontal by e, it will sweep out the plane *COB* which is inclined to the vertical by e. This will create an error $-\phi$ in the horizontal reading of the theodolite (negative as the horizontal circle is graduated clockwise).

If α is the angle of inclination then

$$\sin \phi = \frac{AB}{VB} = \frac{VC}{VB} = \frac{OV}{VB} \tan e = \tan \alpha \tan e \qquad (4.2)$$

Now, as ϕ and e are small, $\phi = e \tan \alpha$

From *Figure 4.16* it can be seen that the correction ϕ to the reading at B, to give the correct reading at A, is positive because of the clockwise graduations of the horizontal circle. Thus, when looking through the telescope towards the object, if the left-hand end of the transit axis is high, then the correction to the reading is positive, and *vice versa.*

On changing face, *COB* will fall to the other side of A and give an equal error of opposite sign. *Thus, the mean of the readings on both faces of the instrument will be free of error.*

As previously, the error in the measurement of an angle between two objects of elevations α_1 and α_2 will be

$$e(\tan \alpha_1 - \tan \alpha_2)$$

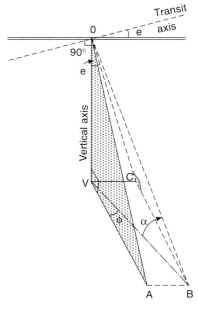

Fig. 4.16

which on changing face becomes $-e(\tan \alpha_1 - \tan \alpha_2)$ indicating *that the mean of two angles, taken one on each face, will be free of error regardless of elevation*. Also, if $\alpha_1 = \alpha_2$, or the angle is measured in the horizontal plane ($\alpha = 0$), it will be free of error. Note if α_1 is positive and α_2, negative, then the correction is $e[\tan \alpha_1 - (-\tan \alpha_2)] = e(\tan \alpha_1 + \tan \alpha_2)$.

Vertical angles: errors in the measurement of vertical angles can be shown to be $\sin \alpha = \sin \alpha_1 \sec e$. As e is very small $\sec e \approx 1$, thus $\alpha_1 = \alpha$, proving that the effect of this error on vertical angles is negligible.

4.2.4 *Effect of non-verticality of the instrument axis*

If the plate levels of the theodolite are not in adjustment, then the instrument axis will be inclined to the vertical, and hence measured azimuth angles will not be truly horizontal. Assuming the transit axis is in adjustment, i.e. perpendicular to the vertical axis, then error in the vertical axis of e will cause the transit axis to be inclined to the horizontal by e, producing an error in pointing of $\phi = e \tan \alpha$ as in the previous case. Here, however, the *error is not eliminated by double-face observations* (*Figure 4.17*), but varies with different pointings of the telescope. For example, *Figure 4.18(a)* shows the instrument axis truly vertical and the transit axis truly horizontal. Imagine now that the instrument axis is inclined through e in a plane at 90° to the plane of the paper (*Figure 4.18(b)*). There is no error in the transit axis. If the alidade is now rotated clockwise through 90° into the plane of the paper, it will be as in *Figure 4.18(c)*, and when viewed in the direction of the arrow, will appear as in *Figure 4.18(d)* with the transit axis inclined to the horizontal by the same amount as the vertical axis, e. Thus, the error in the transit axis varies from zero to maximum through 90°. At 180° it will be zero again, and at 270° back to maximum in exactly the same position.

   If the horizontal angle between the plane of the transit axis and the plane of dislevelment of the

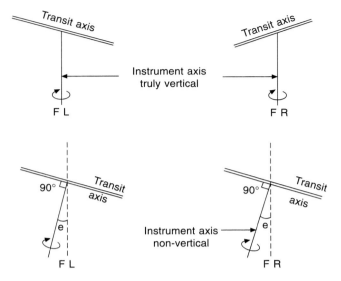

Fig. 4.17

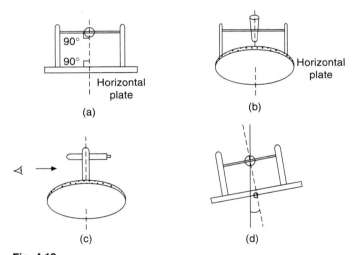

Fig. 4.18

vertical axis is δ, then the transit axis will be inclined to the horizontal by $e \cos \delta$. For example, in *Figure 4.18(b)*, $\delta = 90°$, and therefore as $\cos 90° = 0$, the inclination of the transit axis is zero, as shown.

For an angle between two targets at elevations α_1 and α_2, in directions δ_1 and δ_2, the correction will be $e(\cos \delta_1 \tan \alpha_1 - \cos \delta_2 \tan \alpha_2)$. When $\delta_1 = \delta_2$, the correction is a maximum when α_1 and α_2 have opposite signs. When $\delta_1 = -\delta_2$, that is in opposite directions, the correction is maximum when α_1 and α_1 have the same sign.

If the instrument axis is inclined to the vertical by an amount e and the transit axis further inclined to the horizontal by an amount i, both in the same plane, then the maximum dislevelment of the transit axis on one face will be $(e + i)$, and $(e - i)$ on the reverse face (*Figure 4.19*) Thus, the correction to a pointing on one face will be $(e + i) \tan \alpha$ and on the other $(e - i) \tan \alpha$, resulting in a correction of $e \tan \alpha$ to the mean of both face readings.

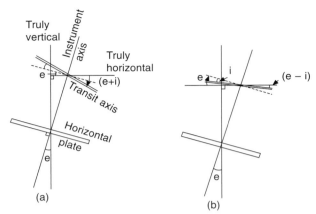

Fig. 4.19   *(a) Face left, and (b) face right*

As shown, the resultant error increases as the angle of elevation α increases and is not eliminated by double face observations. As steep sights frequently occur in mining and civil engineering surveys, it is very important to recognize this source of error and adopt the correct procedures.

Thus, as already illustrated, the correction for a specific direction δ due to non-verticality (e) of the instrument axis is $e \cos \delta \tan \alpha$. The value of $e \cos \delta = E$ can be obtained from

$$E'' = S'' \frac{(L - R)}{2} \tag{4.3}$$

where S'' = the sensitivity of the plate bubble in seconds of arc per bubble division

L and R = the left- and right-hand readings of the ends of the plate bubble when viewed from the eyepiece end of the telescope

Then the correction to each horizontal circle reading is $C'' = E'' \tan \alpha$, and is plus when $L > R$ and *vice versa*.

For high-accuracy survey work, the accuracy of the correction C will depend upon how accurately E can be assessed. This, in turn, will depend on the sensitivity of the plate bubble and how accurately the ends can be read. For very high accuracy involving extremely steep sights, an Electrolevel attached to the theodolite will measure axis tilt direct. This instrument has a sensitivity of 1 scale division equal to $1''$ of tilt and can be read to 0.25 div. The average plate bubble has a sensitivity of $20''$ per div.

Assuming that one can read each end of the plate bubble to an accuracy of ± 0.5 mm, then for a bubble sensitivity of $20''$ per (2 mm) div, on a vertical angle of $45°$, the error in levelling the instrument (i.e. in the vertical axis) would be $\pm 0.35 \times 20'' \tan 45° = \pm 7''$. It has been shown that the accuracy of reading a bubble through a split-image coincidence system is ten times greater. Thus, if the altitude bubble, usually viewed through a coincidence system, was used to level the theodolite, error in the axis tilt would be reduced to $\pm 0.7''$.

Modern theodolites are rapidly replacing the altitude bubble with automatic vertical circle indexing with stabilization accuracies of $\pm 0.3''$, and which may therefore be used for high-accuracy levelling of the instrument as follows:

(1) Accurately level the instrument using its plate bubble in the normal way.
(2) Clamp the telescope in any position, place the plane of the vertical circle parallel to two footscrews and note the vertical circle reading.
(3) With telescope remaining clamped, rotate the alidade through $180°$ and note vertical circle reading.

(4) Using the two footscrews of (2) above, set the vertical circle to the mean of the two readings obtained in (2) and (3).

(5) Rotate through 90° and by using only the remaining footscrew obtain the same mean vertical circle reading.

The instrument is now precisely levelled to minimize axis tilt and virtually eliminate this source of error on steep sights.

Vertical angles are not affected significantly by non-verticality of the instrument axis as their horizontal axis of reference is established independently of the plate bubble.

4.2.5 Circle graduation errors

In the construction of the horizontal and vertical circles of the theodolite, the graduation lines on a 100-mm-diameter circle have to be set with an accuracy of 0.4 μm. In spite of the sophisticated manufacturing processes available, both regular and irregular errors of graduation occur.

It is possible to produce error curves for each instrument. However, such curves showed maximum errors in the region of only ±0.3″. In practice, therefore, such errors are dealt with by observing the same angle on different parts of the circle, distributed symmetrically around the circumference. If the angle is to be observed 2, 4 or n times, where a double face measurement is regarded as a single observation, then the alidade is rotated through 180°/n prior to each new measurement.

4.2.6 Optical micrometer errors

When the optical micrometer is rotated from zero to its maximum position, then the displacement of the circle should equal the least count of the main scale reading. However, due to circle graduation errror, plus optical and mechanical defects, this may not be so. Investigation of the resultant errors revealed a cyclic variation, the effects of which can be minimized by using different micrometer settings.

4.2.7 Vertical circle index errror

In the measurement of a vertical angle it is important to note that the vertical circle is attached to and rotates with the telescope. The vertical circle reading is relevant to a fixed vertical circle index which is rendered horizontal by means of its attached altitude bubble (*Figure 4.3*) or by automatic vertical circle indexing.

Vertical circle index error occurs when the index is not horizontal. *Figure 4.20* shows the index inclined at e to the horizontal. The measured vertical angle on FL is M, which requires a correction of +e, while on FR the required correction is −e. The index error is thus eliminated by taking the mean of the FL and FR readings.

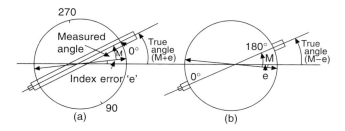

Fig. 4.20 *(a) Face left, and (b) face right*

4.3 INSTRUMENT ADJUSTMENT

In order to maintain the primary axes of the theodolite in their correct geometrical relationship (*Figure 4.3*), the instrument should be regularly tested and adjusted. Although the majority of the resultant errors are minimized by double face procedures, this does not apply to plate bubble error. Also, many operations in engineering surveying are carried out on a single face of the instrument, and hence regular calibration is important.

4.3.1 Tests and adjustments

(1) Plate level test

The instrument axis must be truly vertical when the plate bubble is centralized. The vertical axis of the instrument is perpendicular to the horizontal plate which carries the plate bubble. Thus to ensure that the vertical axis of the instrument is truly vertical, as defined by the bubble, it is necessary to align the bubble axis parallel to the horizontal plate.

Test: Assume the bubble is not parallel to the horizontal plate but is in error by angle *e*. It is set parallel to a pair of footscrews, levelled approximately, then turned through 90° and levelled again using the third footscrew only. It is now returned to its former position, accurately levelled using the pair of footscrews, and will appear as in *Figure 4.21(a)*. The instrument is now turned through 180° and will appear as in *Figure 4.21(b)*, i.e. the bubble will move off centre by an amount representing twice the error in the instrument (2*e*).

Adjustment: The bubble is brought half-way back to the centre using the pair of footscrews. This will cause the instrument axis to move through *e*, thereby making it truly vertical and, in the event of there being no adjusting tools available, the instrument may be used at this stage. The bubble will still be off centre by an amount proportional to *e*, and should now be centralized by raising or lowering one end of the bubble using its capstan adjusting screws.

(2) Collimation in azimuth

The purpose of this test is to ensure that the line of sight is perpendicular to the transit axis.

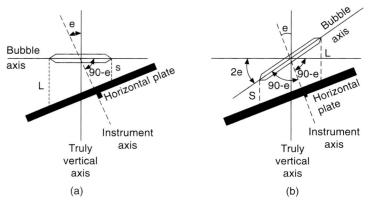

(a)   (b)

Fig. 4.21   *(a) When levelled over two footscrews. (b) When turned through 180°*

Test: The instrument is set up, and levelled, and the telescope directed to bisect a fine mark at *A*, situated at instrument height about 50 m away (*Figure 4.22*). If the line of sight is perpendicular to the transit axis, then when the telescope is rotated vertically through 180°, it will intersect at A_1. However, assume that the line of sight makes an angle of (90° − *e*) with the transit axis, as shown dotted in the face left (FL) and face right (FR) positions. Then in the FL position the instrument would establish a fine mark at A_L. Change face, re-bisect point *A*, transit the telescope and establish a fine mark at A_R. From the sketch it is obvious that distance A_LA_R represents four times the error in the instrument (4*e*).

Adjustment: The cross-hairs are now moved in azimuth using their horizontal capstan adjusting screws, from A_R to a point mid-way between A_R and A_1; this is one-quarter of the distance A_LA_R.

This movement of the reticule carrying the cross-hair may cause the position of the vertical hair to be disturbed in relation to the transit axis; i.e. it should be perpendicular to the transit axis. It can be tested by traversing the telescope vertically over a fine dot. If the vertical cross-hair moves off the dot then it is not at right angles to the transit axis and is corrected with the adjusting screws.

This test is frequently referred to as one which ensures the verticality of the vertical hair, which will be true only if the transit axis is truly horizontal. However, it can be carried out when the theodolite is not levelled, and it is for this reason that a dot should be used and not a plumb line as is sometimes advocated.

(3) Spire test (transit axis test)

This test ensures that the transit axis is perpendicular to the vertical axis of the instrument.

Test: The instrument is set up and carefully levelled approximately 50 m from a well-defined point of high elevation, preferably greater than 30° (*Figure 4.23*). The well-defined point *A* is bisected and the telescope then lowered to its horizontal position and a further point made. If the transit axis is in adjustment the point will appear at A_1 directly below *A*. If, however, it is in error by the amount *e* (transit axis shown dotted in FL and FR positions), the mark will be made at A_L. The instrument is now changed to FR, point *A* bisected again and the telescope lowered to the horizontal, to fix point A_R. The distance A_LA_R is twice the error in the instrument (2*e*).

Adjustment: Length A_LA_R is bisected and a fine mark made at A_1. The instrument is now moved in azimuth, using a plate slow-motion screw until A_1 is bisected. The student should note that no adjustment of any kind has yet been made to the instrument. Thus, when the telescope is raised back to *A* it will be in error by the horizontal distance $A_LA_R/2$. By moving one end of the transit axis using

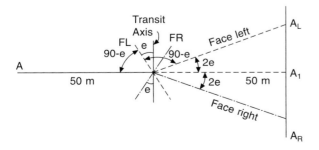

Fig. 4.22 *Collimation in azimuth*

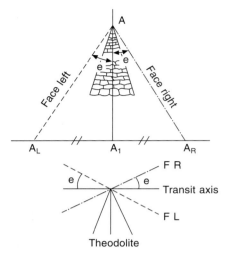

Fig. 4.23   *Spire test (transit axis test)*

the adjusting screws, the line of sight is made to bisect A. This can only be made to bisect A when the line of sight is elevated. Movement of the transit axis when the telescope is in the horizontal plane $A_L A_R$, will not move the line of sight to A_1, hence the need to incline steeply the line of sight.

It should be noted that in modern instruments this adjustment cannot be carried out, i.e. there is no facility for moving the transit axis. Manufacturers claim that this error does not occur in modern equipment.

(4) Vertical circle index test

This is to ensure that when the telescope is horizontal and the altitude bubble central, the vertical circle reads zero (or its equivalent).

Test: Centralize the altitude bubble using the clip screw (altitude bubble screw) and, by rotating the telescope, set the vertical circle to read zero (or its equivalent for a horizontal sight).

Note the reading on a vertical staff held about 50 m away. Change face and repeat the whole procedure. If error is present, a different reading on each face is obtained, namely A_L and A_R in *Figure 4.24*.

Adjustment: Set the telescope to read the mean of the above two readings, thus making it truly horizontal. The vertical circle will then no longer read zero, and must be brought back to zero

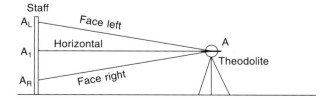

Fig. 4.24   *Vertical circle index test*

without affecting the horizontal level of the telescope. This is done by moving the verniers to read zero using the clip screw.

Movement of the clip screw will cause the altitude bubble to move off centre. It is recentralized by means of its capstan adjusting screws.

(5) Optical plummet

The line of sight through the optical plummet must coincide with the vertical instrument axis of the theodolite.

If the optical plummet is fitted in the alidade of the theodolite (*Figure 4.14(a)*), rotate the instrument through 360° in 90° intervals and make four marks on the ground. If the plummet is out of adjustment, the four marks will form a square, intersecting diagonals of which will give the correct point. Adjust the plummet cross-hairs to bisect this point

If the plummet is in the tribrach it cannot be rotated. It is set on its side, on a table with the plummet viewing a nearby wall and a mark aligned on the wall. The instrument is then turned through 180° and the procedure repeated. If the plummet is out of adjustment a second mark will be aligned. The plummet is adjusted to intersect the point midway between the two marks.

4.3.2 Alternative approach

(1) Plate level test

The procedure for this is as already described.

(2) Collimation in azimuth

With the telescope horizontal and the instrument carefully levelled, sight a fine mark and note the reading. Change face and repeat the procedure. If the instrument is in adjustment, the two readings should differ by exactly 180°. If not, the instrument is set to the corrected reading as shown below using the upper plate slow-motion screw; the line of sight is brought back on to the fine mark by adjusting the cross-hairs.

e.g. FL reading $\quad$ 01°30′ 20″
$\quad$ FR reading $\quad$ 181°31′ 40″

Difference $= 2e = 01′\ 20″$
∴ $\qquad e = \pm40″$
Corrected reading $= 181°31′\ 00″$ or $01°31′\ 00″$

(3) Spire test

With the instrument carefully levelled, sight a fine point of high elevation and note the horizontal circle reading. Change face and repeat. If error is present, set the horizontal circle to the corrected reading, as above. Adjust the line of sight back on to the mark by raising or lowering the transit axis. (Not all modern instruments are capable of this adjustment.)

(4) Vertical circle index test

Assume the instrument reads 0° on the vertical circle when the telescope is horizontal and in FL position. Carefully level the instrument, horizontalize the altitude bubble and sight a fine point of high elevation. Change face and repeat. The two vertical circle readings should *sum* to 180°, any difference being twice the index error.

e.g. FL reading (*Figure 4.25(a)*)   09°58′ 00″
FR reading (*Figure 4.25(b)*) 170°00′ 20″

$$\text{Sum} = 179°58′\ 20″$$
$$\text{Correct sum} = 180°00′\ 00″$$

$$2e = \quad -01′\ 40″$$
$$e = \qquad -50″$$

Thus with the target still bisected, the vernier is set to read 170° 00′ 20″ + 50″ = 170°01′10″ by means of the clip or altitude bubble screw. The altitude bubble is then centralized using its capstan adjusting screws. If the vertical circle reads 90° and 270° instead of 0° and 180°, the readings sum to 360°.

These alternative procedures have the great advantage of using the theodolite's own scales rather than external scales, and can therefore be carried out by one person.

4.4 FIELD PROCEDURE

The methods of setting up the theodolite and observing angles will now be dealt with. It should be emphasized, however, that these instructions are no substitute for practical experience.

4.4.1 Setting up using a plumb-bob

Figure 4.26 shows a theodolite set up with the plumb-bob suspended over the survey station. The procedure is as follows:

(1) Extend the tripod legs to the height required to provide comfortable viewing through the theodolite. It is important to leave at least 100 mm of leg extension to facilitate positioning of the plumb-bob.
(2) Attach the plumb-bob to the tripod head, so that it is hanging freely from the centre of the head.
(3) Stand the tripod approximately over the survey station, keeping the head reasonably horizontal.
(4) Tighten the wing units at the top of the tripod legs and move the whole tripod until the plumb-bob is over the station.
(5) Now tread the tripod feet firmly into the ground.

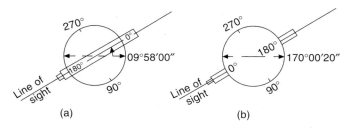

Fig. 4.25   (a) *Face left, and* (b) *face right*

Fig. 4.26

(6) Unclamp a tripod leg and slide it in or out until the plumb-bob is exactly over the station. If this cannot be achieved in one movement, then use the slide extension to bring the plumb-bob in line with the survey point and another tripod leg. Using this latter leg, slide in or out to bring the plumb-bob onto the survey point.
(7) Remove the theodolite from the box and, holding it by its standard, attach it to the tripod head.
(8) The instrument axis is now set truly vertical using the plate bubble as follows:

   (a) Set the plate bubble parallel to two footscrews *A* and *B* as shown in (*Figure 4.27(a)*) and centre it by equal amounts of simultaneous contra-rotation of both screws. (The bubble follows the direction of the left thumb.)
   (b) Rotate alidade through 90°(*Figure 4.27(b)*) and centre the bubble using footscrew *C* only.
   (c) Repeat (a) and (b) until bubble remains central in both positions. If there is no bubble error this procedure will suffice. If there is slight bubble error present, proceed as follows.
   (d) From the initial position at *B* (*Figure 4.27(a)*), rotate the alidade through 180°; if the bubble moves off centre bring if half-way back using the footscrews *A* and *B*.
   (e) Rotate through a further 90°, placing the bubble 180° different to its position in *Figure 4.27(b)*. If the bubble moves off centre, bring it half-way back with footscrew *C* only.
   (f) Although the bubble is off centre, the instrument axis will be truly vertical and will remain so as long as the bubble remains the same amount off centre (*Section 4.3.2*).

(9) Check the plumb-bob; if it is off the survey point, slacken off the whole theodolite and shift it laterally across the tripod head until the plumb-bob is exactly over the survey point.
(10) Repeat (8) and (9) until the instrument is centred and levelled.

4.4.2 Setting up using the optical plumb-bob

It is rare, if ever, that a theodolite is centred over the survey station using only a plumb-bob. All modern instruments have an optical plummet built into the alidade section of the instrument (*Figure 4.2* and *4.13(a)*), or into the tribrach section (*Figures 4.4* and *4.13(b)*). Proceed as follows:

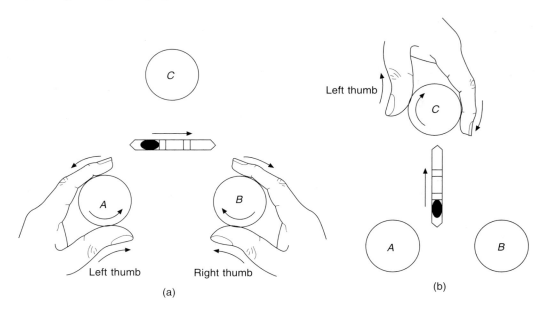

Fig. 4.27   *Footscrews*

(1) Establish the tripod roughly over the survey point using a plumb-bob as in (1) to (5) of *Section 4.4.1*.
(2) Attach the tribrach only or theodolite, depending on the situation of the optical plummet, to the tripod.
(3) Using the footscrews to incline the line of sight through the plummet, centre the plummet exactly on the survey point.
(4) Using the leg extension, slide the legs in or out until the circular bubble of the tribrach/theodolite is exactly centre. Even though the tripod movement may be excessive, the plummet will still be on the survey point. Thus the instrument is now approximately centred and levelled.
(5) Precisely level the instrument using the plate bubble, as shown.
(6) Unclamp and move the whole instrument laterally over the tripod until the plummet cross-hair is exactly on the survey point.
(7) Repeat (5) and (6) until the instrument is exactly centred and levelled.

4.4.3 Setting up using a centring rod

Kern tripods are fitted with centring rods, which makes the setting-up process extremely quick and easy.

(1) Set the tripod roughly over the survey point.
(2) Unclamp the telescopic joint at the head of the rod, extend the rod and place its centre point on the survey point. Tread the feet of the tripod firmly into the ground.
(3) Adjust the leg extensions up or down until the circular bubble on the rod is exactly central. Re-tighten the telescopic joint clamp.
(4) Attach the theodolite to the tripod head and level it in the usual way. No further centring is necessary.

4.4.4 Centring errors

Provided there is no wind, centring with a plumb-bob is accurate to $\pm 3-5$ mm. In windy conditions it is impossible to use unless protected in some way.

The optical plummet is accurate to $\pm 1-0.5$ mm, provided the instrument axis is truly vertical and is not affected by adverse weather conditions.

The centring rod is extremely quick and easy to use and provided it is in adjustment will give centring accuracies of about ± 1 mm.

Forced centring or constrained centring systems are used to control the propagation of centring error in precise traversing. Such systems give accuracies in the region of $\pm 0.1-0.3$ mm. They will be dealt with in Chapter 6.

The effect of centring errors on the measured horizontal angle (θ) is shown in *Figure 4.28*.

Due to a miscentring error, the theodolite is established at B', not the actual station B, and the angle θ is observed, not θ'. The maximum angular error (e_θ) occurs when the centring error BB' lies on the bisector of the measured angle and can be shown to be equal to:

$$(\theta - \theta') = e_\theta = \pm e_c L_{Ac}/L_{AB}L_{Bc}(2)^{\frac{1}{2}} \tag{4.4}$$

where e_c = the centring error
L_{AB}, L_{Bc}, L_{Ac} = horizontal lengths AB, BC and AC

The effect of target-centring errors on the horizontal angle at B can be obtained as follows. If one assumes $AB = AB'$, then $B'AB = e_c/L_{AB} = e_{tA}$, and similarly $BCD' = e_c/L_{Bc} = e_{tc}$. Error in the angle would therefore be equal to the sum of these two errors:

$$e_{\theta t} = [(e_c/L_{AB})^2 + (e_c/L_{Bc})^2]^{\frac{1}{2}} \tag{4.5}$$

It can be seen from equations (4.4) and (4.5) that as the lengths L decrease, the error in the measured angles will increase. Consider the following examples.

Worked examples

Example 4.1 In the horizontal angle *ABC*, *AB* is 700 m and *BC* is 1000 m. If the error in centring the targets at *A* and *C* is ± 5 mm in both cases, what will be the resultant error in the measured angle?

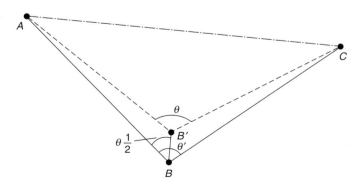

Fig. 4.28

$$e''_{tA} = \pm e_c/L_{AB} = \pm(0.005/700) \; 206\;265 = \pm1.5''$$

$$e''_{tc} = \pm e_c/L_{Bc} = \pm(0.005/1000) \; 206\;265 = \pm1.0''$$

$$e''_{\theta t} = \pm(1.0^2 + 1.5^2)^{\frac{1}{2}} = \pm1.8''$$

Example 4.2 Consider the same question as in *Example 4.1* with $AB = 70$ m, $BC = 100$ m.

$$e''_{tA} = \pm(0.005/70) \; 206\;265 = \pm14.7''$$

$$e''_{tc} = \pm(0.005/100) \; 206\;265 = \pm10.3''$$

$$e''_{\theta t} = \pm(14.7^2 + 10.3^2)^{1/2} = \pm18''$$

It can be seen that decreasing the lengths by a factor of 10 increases the angular error by the same factor.

Example 4.3 Assuming angle *ABC* is 90° and the centring error of the theodolite is ±3 mm, if the remaining data are as in *Example 4.2*, what is the maximum error in the observed angle due to centring errors only?

$$L_{Ac} = (L^2_{AB} + L^2_{Bc})^{\frac{1}{2}} = 122 \text{ m}$$

Then from equation (4.4):

$$e''_\theta = \pm[(0.003 \times 122)/70 \times 100 \times 2^{\frac{1}{2}}] \; 206\;265 = \pm7.6''$$

From example (4.2), error due to target miscentring = ±18″.

$$\text{Total error} = \pm(7.6^2 + 18.0^2)^{\frac{1}{2}} = \pm 19.5''$$

4.5 MEASURING ANGLES

Although the theodolite or total station is a very complex instrument the measurement of horizontal and vertical angles is a simple concept. The horizontal and vertical circles of the instrument should be regarded as circular protractors graduated from 0° to 360° in a clockwise manner. Then a simple horizontal angle measurement between three survey points *A*, *B* and *C* would be as shown in *Figure 4.29*.

(1) Instrument is set up and centred and levelled on survey point *B*. Parallax is dealt with.
(2) Commencing on, say, 'face left', the target set at survey point *A* is carefully bisected and the horizontal scale reading noted = 25°.
(3) The upper plate clamp is released and survey point *C* is bisected. The horizontal scale reading is noted = 145°.
(4) The horizontal angle is then the difference of the two directions, i.e. (FS − BS) = (145° − 25°) = 120°.
(5) Change face and observe survey point *C* on 'face right', and note the reading = 325°.
(6) Release upper plate and swing to point *A*, and note the reading = 205°.
(7) The readings or directions must be subtracted in the same order as in (5), i.e. *C* − *A*.

Thus (325° − 205°) = 120°

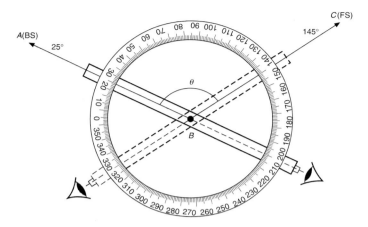

Fig. 4.29 *Measuring a horizontal angle*

(8) Note how changing face changes the readings by 180°, thus affording a check on the observations. The mean of the two values would be accepted if they are in acceptable agreement.

Had the BS to *A* read 350° and the FS to *C* 110°, it can be seen that 10° has been swept out from 350° to 360° and then from 360° or 0° to 110°, would sweep out a further 110°. The total angle is therefore 10° + 110° = 120° or (FS − BS) = [(110° + 360°) − 350°] = 120°.

A further examination of the protractor shows that (BS − FS) = [(25° + 360°) − 145°] = 240°, producing the external angle. It is thus the manner in which the data are reduced that determines whether or not it is the internal or external angle which is obtained.

A method of booking the data for an angle measured in this manner is shown in *Table 4.1*. This approach constitutes the standard method of measuring single angles in traversing, for instance.

4.5.1 Measurement by directions

The method of directions is generally used when observing a set of angles as in *Figure 4.30*. The

Table 4.1

Sight to	Face	Reading			Angle		
		°	′	″	°	′	″
A	*L*	020	46	28	80	12	06
C	*L*	100	58	34			
C	*R*	280	58	32	80	12	08
A	*R*	200	46	24			
A	*R*	292	10	21	80	12	07
C	*R*	012	22	28			
C	*L*	192	22	23	80	12	04
A	*L*	112	10	19			
					Mean = 80	12	06

Note the built-in checks supplied by changing face, i.e. the reading should change by 180°. Note that to obtain the clockwise angle one always deducts BS (*A*) reading from the FS (*C*) reading, regardless of the order in which they are observed.

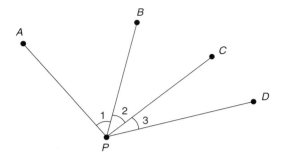

Fig. 4.30

angles are observed, commencing from *A* and noting all the readings, as the instrument moves from point to point in a clockwise manner. On completion at *D*, face is changed and the observations repeated moving from *D* in an anticlockwise manner. Finally the mean directions are reduced relative to the starting direction for *PA* by applying the 'orientation correction'. For example, if the mean horizontal circle reading for *PA* is 48° 54′ 36″ and the known bearing for *PA* is 40° 50′ 32″, then the orientation correction applied to all the mean bearings is obviously −8° 04′ 04″.

The observations as above, carried out on both faces of the instrument, constitute a full set. If measuring *n* sets the reading is altered by 180°/*n* each time.

4.5.2 Measurement by repetition

This method of measurement requires the use of a lower plate clamp and slow motion arrangement. As stated previously, modern theodolites have, in the majority of cases, replaced this arrangement with a horizontal circle-setting screw, thereby rendering this method obsolete to a large extent.

Consider angle *ABC* (*Figure 4.29*); the procedure would be as follows:

(1) Observe *A* on, say, FL and record the reading R_1.
(2) Release upper plate and observe *C* (record the reading purely as a check).
(3) Release lower plate and move clockwise back to *A*.
(4) Repeat (2) and (3) *n* times, and record the *n*th reading R_n.
(5) Change face and repeat, commencing from *C* and moving anticlockwise.

The angle deduced from the FL readings only would be $(R_n − R_1)/n$, and similarly for the FR observations, the mean value comprises the value of one set.

It is advisable to commence with the instrument reading approximately zero, but not set specifically to zero, as this would introduce a setting error. If the instrument commenced with an initial reading to *A* of, say, 350° and after four repetitions read 70°, the angle is obviously

$$[(70° + 360°) − 350°]/4 = 20°$$

which can be expressed in the general form as

$$(R_n + N(360°) − R_1)/n \qquad (4.6)$$

where R_n = the final reading
  R_1 = the initial reading
  N = the number of times 360° is passed
  n = the number of repetitions

4.5.3 *Comparison of the methods*

The basic errors in the observation of an angle are 'pointing' error and 'reading' error, both of which are affected by a variety of factors.

Observing an angle by the method of direction involves two pointings and two readings per single face. Thus we have

$$\sigma_\theta^2 = 2\sigma_p^2 + 2\sigma_R^2 \tag{4.7}$$

where σ_p^2 = the pointing variance

σ_R^2 = the reading variance

σ_θ^2 = the resulting angle variance

For the mean of n such angles:

$$\sigma_{M\theta}^2 = \frac{2\sigma_p^2}{n} + \frac{2\sigma_R^2}{n} \tag{4.8}$$

Considering repetition measurement, if θ is measured n times ($n/2$ on each face) we have $2n$ pointings but only 2 readings. The variance of the sum of the observations is

$$\sigma_\theta^2 = 2n\sigma_p^2 + 2\sigma_R^2 \tag{4.9}$$

and the variance of the mean of n repetitions is

$$\sigma_{M\theta}^2 = \frac{2\sigma_p^2}{n} + \frac{2\sigma_R^2}{n^2} \tag{4.10}$$

Comparison of equations (4.8) and (4.10) shows the method of repetition to be more accurate as the reading error is reduced by n^2 compared with n. This is particularly significant for instruments which have a large least count. Their performance would be significantly improved by the method of repetition. However, as previously stated, most instruments are direction theodolites, as they lack the lower plate clamping facility.

4.5.4 *Further considerations in angular measurement*

Considering *Figure 4.30*, the angles may be measured by 'closing the horizon'. This involves observing the points in order from A to D and continuing clockwise back to A, thereby completing the full circle. The difference between the sum of all the angles and 360° is distributed evenly amongst all the angles to bring their sum to 360°. Repeat anticlockwise on the opposite face.

A method favoured in the measurement of precise networks is to measure all the combinations of angles. In the case above it would involve measuring *APB*, *APC*, *APD*, *BPC*, *BPD* and *CPD*. The angles could then be resolved by forming condition equations in a least squares solution (see Chapter 7).

4.5.5 *Vertical angles*

In the measurement of horizontal angles the concept is of a measuring index moving around a protractor. In the case of a vertical angle, the situation is reversed and the protractor moves relative to a fixed horizontal index.

Figure 4.31(a) shows the telescope horizontal and reading 90°; changing face would result in a reading of 270°. In *Figure 4.31(b)*, the vertical circle index remains horizontal whilst the protractor rotates with the telescope, as the top of the spire is observed. The vertical circle reading of 65° is the zenith angle, equivalent to a vertical angle of $(90° - 65°) = +25° = \alpha$. This illustrates the basic concept of vertical angle measurement.

4.6 SOURCES OF ERROR

Error in the measurement of angle results from instrument, personal or natural sources.

The instrumental errors have been dealt with and, as indicated, can be minimized by taking several measurements of the angle on each face of the theodolite. Regular calibration of the equipment is also of prime importance. The remaining sources will now be dealt with.

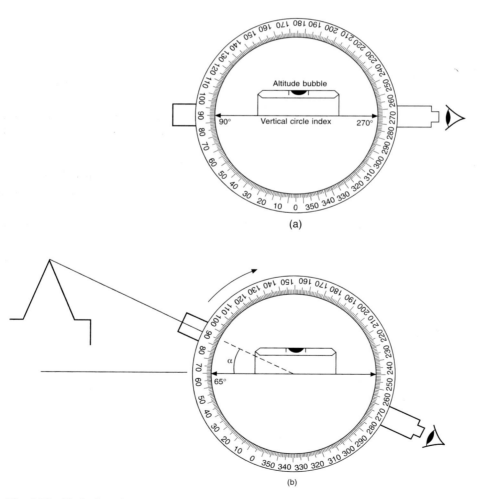

Fig. 4.31 *Vertical angles*

4.6.1 *Personal error*

(1) Careless centring of the instrument over the survey point. Always ensure that the optical plummet or centring rod is in adjustment. Similarly for the targets.
(2) Light clamping of the horizontal and vertical movement. Hard clamping can affect the pointing and is unnecessary.
(3) The final movement of the slow motion screws should be clockwise, thus producing a positive movement against the spring. An anticlockwise movement which releases the spring may cause backlash.
(4) Failure to eliminate parallax and poor focusing on the target can affect accurate pointing. Keep the observed target near the centre of the field of view.
(5) Incorrect levelling of the altitude bubble will produce vertical angle error.
(6) The plate bubble must also be carefully levelled and regularly checked throughout the measuring process.
(7) Make quick, decisive observations. Too much care can be counterproductive.
(8) All movement of the theodolite should be done gently whilst movement around the tripod should be reduced to a minimum.

4.6.2 *Natural errors*

(1) Wind vibration may require some form of wind shield to be erected to protect the instrument. Dual axis tilt sensors in modern total stations have greatly minimized this effect.
(2) Vertical and lateral refraction of the line of sight is always a problem. The effect on the vertical angle has already been discussed in Chapter 2. Lateral refraction, particularly in tunnels, can cause excessive error in the horizontal angle. A practical solution in tunnels is to use zig-zag traverses with frequent gyro-theodolite azimuths included.
(3) Temperature differentials can cause unequal expansion of the various parts of the instrument. Plate bubbles will move off centre towards the hottest part of the bubble tube. Heat shimmer may make accurate pointing impossible. Sheltering the instrument by means of a large survey umbrella will greatly help in this situation.
(4) Avoid tripod settlement by careful selection of ground conditions. If necessary use pegs to pile the ground on which the tripod feet are set.

All the above procedures should be included in a pre-set survey routine, which should be strictly adhered to. Inexperienced observers should guard against such common mistakes as:

(1) Turning the wrong screw.
(2) Sighting the wrong target.
(3) Using the stadia instead of the cross-hair.
(4) Forgetting to set the micrometer.
(5) Misreading the circles.
(6) Transposing figures when booking the data.

5

Position

5.1 INTRODUCTION

Engineering surveying is concerned essentially with fixing the position of a point in two or three dimensions.

For example, in the production of a plan or map, one is concerned in the first instance with the accurate location of the relative position of survey points forming a framework, from which the position of topographic detail is fixed. Such a framework of points is referred to as a control network.

The same network used to locate topographic detail may also be used to set out points, defining the position, size and shape of the designed elements of the construction project.

Precise control networks are also used in the monitoring of deformation movements on all types of structures.

In all these situations the engineer is concerned with relative position, to varying degress of accuracy and over areas of varying extent. In order to define position to the high accuracies required in engineering surveying, a suitable homogeneous coordinate system and reference datum must be adopted.

Consideration of *Figure 5.1* illustrates that if the area under consideration is of limited extent, the orthogonal projection of *AB* onto a plane surface may result in negligible distortion. Plane surveying techniques could be used to capture field data and plane trigonometry used to compute position. This is the case in the majority of engineering surveys. However, if the area extended from *C* to *D*, the effect of the Earth's curvature is such as to produce unacceptable distortion if treated as a flat surface. It can also be clearly seen that the use of a plane surface as a reference datum for the elevations of points is totally unacceptable.

If *Figure 5.2* is now considered, it can be seen that projecting *CD* onto a surface (*cd*) that was the same shape and parallel to *CD* would be more acceptable. Further, if that surface was brought closer to *CD*, say *c'd'*, the distortion would be even less. This then is the problem of the geodetic surveyor: that of defining a mathematical surface that approximates to the shape of the area under consideration and then fitting and orientating it to the Earth's surface. Such a surface is referred to in surveying as a 'reference ellipsoid'.

5.2 REFERENCE ELLIPSOID

To arrive at the concept of a reference ellipsoid, the various surfaces involved must be reviewed.

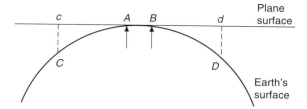

Fig. 5.1

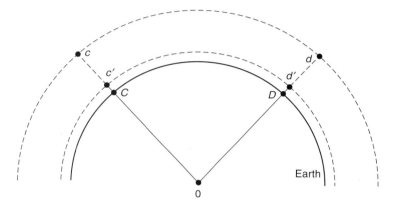

Fig. 5.2

5.2.1 Earth's surface

The Earth's physical surface is a reality upon which the surveying observations are made and points located. However, due to its variable topographic surface and overall shape, it connot be defined mathematically and so position cannot be computed on its surface. It is for this reason that in surveys of limited extent, the Earth is treated as flat and plane trigonometry used to define position.

5.2.2 The geoid

Having rejected the physical surface of the Earth as a computational surface, one is instinctively drawn to a consideration of a mean sea level surface. This is not surprising, as 70% of the Earth's surface is ocean.

If these oceans were imagined to flow in interconnecting channels throughout the land masses, then, ignoring the effects of friction, tides, wind stress, etc., an equipotential surface, approximately at MSL would be formed. Such a surface is called the 'geoid', a physical reality, the shape of which can be measured. Although the gravity potential is everywhere the same and the surface is smoother than the physical surface of the Earth, it still contains many irregularities which render it unsuitable for the mathematical location of planimetric position. These irregularities are thought to be due to the mass anomalies throughout the Earth. Measurements have shown the global shape of the geoid to be as in *Figure 5.3*. The resultant pear shape or lumpy potato is due to the displacement of the poles by about 20 m.

In spite of this, the geoid remains important to the surveyor as it is the surface to which all terrestrial measurements are related.

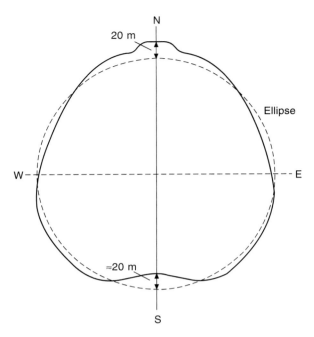

Fig. 5.3 *The geoid (solid line)*

As the direction of the gravity vector (termed the 'vertical') is everywhere normal to the geoid, it defines the direction of the surveyor's plumb-bob line. Thus any instrument which is horizontalized by means of a spirit bubble will be referenced to the local equipotential surface. Elevations in Great Britain, as described in Chapter 2, are related to the equipotential surface passing through MSL, as defined at Newlyn, Cornwall. Such elevations or heights are called orthometric heights (H) and are the linear distances measured along the gravity vector from a point to the equipotential surface used as a reference datum. As such, the geoid is the equipotential surface that best fits MSL and the heights in question, referred to as heights above or below MSL. It can be seen from this that orthometric heights are datum dependent. Therefore, elevations related to the Newlyn datum cannot be related to elevations that are relative to other datums established for use in other countries. A global MSL varies from the geoid by as much as 3 m in places, and hence it is not possible to have all countries on the same datum.

5.2.3 The ellipsoid

The ellipsoid of rotation is the closest mathematically definable shape to the figure of the Earth. It is represented by an ellipse rotated about its minor axis and is defined by its semi-major axis a (*Figure 5.4*) or the flattening f. Although the ellipsoid is a concept and not a physical reality, it represents a smooth surface for which formulae can be developed to compute ellipsoidal distance, azimuth and ellipsoidal coordinates. Due to the variable shape of the geoid, it is not possible to have a global ellipsoid of reference for use by all countries. The best-fitting global geocentric ellipsoid is the Geodetic Reference System 1980 (GRS80), which has the following dimensions:

semi-major axis   6 378 137.0 m
semi-minor axis   6 356 752.314 m

the difference being approximately 21 km.

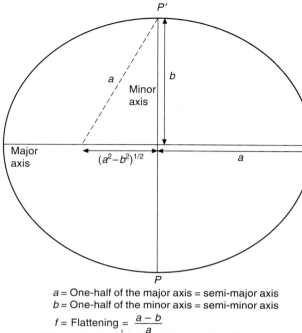

a = One-half of the major axis = semi-major axis
b = One-half of the minor axis = semi-minor axis

f = Flattening = $\dfrac{a-b}{a}$

PP' = Axis of revolution of the Earth's ellipsoid

Fig. 5.4 *Elements of an ellipse*

The most precise global geoid is the Earth Gravitational Model 1996 (EGM96). However, it still remains a complex, undulating figure which varies from the GRS80 ellipsoid by more than 100 m in places. In the UK the geoid–ellipsoid separation is as much as 57 m in the region of the Hebrides. As a 6-m vertical separation between geoid and ellipsoid would result in a scale error of 1 ppm, different countries have adopted local ellipsoids that give the best fit in their particular situation. A small sample of ellipsoids used by different countries is shown below:

Ellipsoid	*a* metres	*1/f*	Where used
Airy (1830)	6 377 563	299.3	Great Britain
Everest (1830)	6 377 276	300.8	India, Pakistan
Bessel (1841)	6 377 397	299.2	East Indies, Japan
Clarke (1866)	6 378 206	295.0	North and Central America
Australian National (1965)	6 378 160	298.2	Australia
South American (1969)	6 378 160	298.2	South America

When $f = 0$, the figure described is a circle, and the flattening of this circle is described by $f = (a - b)/a$. A further parameter used in the definition of an ellipsoid is e, referred to as the first eccentricity of the ellipse, and is equal to $(a^2 - b^2)^{\frac{1}{2}}/a$.

Figure 5.5 illustrates the relationship of all three surfaces. It can be seen that if the geoid and ellipsoid were parallel at A, then the deviation of the vertical would be zero in the plane shown. If

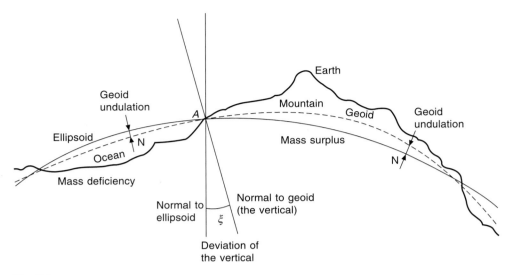

Fig. 5.5

the value for geoid–ellipsoid separation (N) was zero, then not only would the surfaces be parallel, they would fit each other exactly. As the ellipsoid is a smooth surface and the geoid is not, perfect fit can never be achieved. However, the values for deviation of the vertical and geoid–ellipsoid separation can be used as indicators of the closeness of fit.

5.3 COORDINATE SYSTEMS

5.3.1 Astronomical coordinates

As shown in *Figure 5.6*, astronomical latitude ϕ_A defines the latitude of the vertical (gravity vector) through the point in question (*P*) to the plane of the equator, whilst the astronomical longitude λ_A is the angle in the plane of the equator between the zero meridian plane (Greenwich) and the meridian plane through *P*, both of which contain the spin axis.

The common concept of a line through the North and South Poles comprising the spin axis of the Earth is not acceptable as part of a coordinate system, as it is constantly moving with respect to the solid body of the Earth. This results in the North Pole changing position by as much as 5–10 m per year due to the polar motion of the Earth's spin axis. It is thus necessary to define a mean spin axis which does not change position. Such an axis has been defined (and internationally agreed) by the International Earth Rotation Service (IERS) based in Paris, and has an IERS Reference Pole (IRP). Similarly, the Greenwich Meridian adopted is not the one passing through the centre of the observatory telescope. It is one defined as the mean value of the longitudes of a large number of participating observatories throughout the world and is called the IERS Reference Meridian (IRM).

The instantaneous position of the Earth with respect to this axis is constantly monitored by the IERS and published for the benefit of those who need it.

Astronomical latitude and longitude do not define position on the Earth's surface but rather the direction and inclination of the vertical through the point in question. Due to the undulation of the equipotential surface it is possible to have verticals through different points which are parallel and therefore have the same coordinates. An astronomical coordinate system is therefore unsatisfactory for precise positioning.

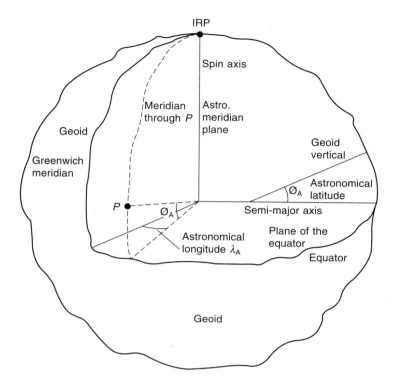

Fig. 5.6 *Astronomical coordinates*

5.3.2 Geodetic coordinates

Considering a point P at height h, measured along the normal through P, above the ellipsoid, the ellipsoidal latitude and longitude will be ϕ_G and λ_G, as shown in *Figure 5.7*. Thus the ellipsoidal latitude is the angle describing the inclination of the normal to the ellipsoidal equatorial plane. The ellipsoidal longitude is the angle in the equatorial plane between the IRM and the geodetic meridian plane through the point in question P. The height h of P above the ellipsoid is called the ellipsoidal height. As the ellipsoid has a conceptually smooth surface no two points will have the same coordinates, as in the previous system. Also, the ellipsoidal coordinates can be used to compute azimuth and ellipsoidal distance. These are the coordinates used in classical geodesy to describe position on an ellipsoid of reference.

5.3.3 Cartesian coordinates

As shown in *Figure 5.8*, if the IERS spin axis is regarded as the Z-axis, the X-axis is in the direction of the zero meridian (IRM) and the Y-axis is perpendicular to both, a conventional three-dimensional coordinate system is formed. If we regard the origin of the cartesian system and the ellipsoidal coordinate system as coincident at the mass centre of the Earth then transformation between the two systems may be carried out as follows:

(1) Ellipsoidal to cartesian

$$X = (v + h) \cos \phi_G \cos \lambda_G \tag{5.1}$$

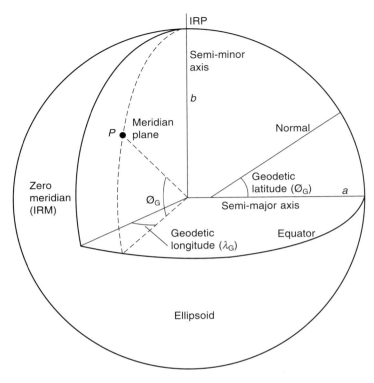

Fig. 5.7 *Geodetic coordinates*

$$Y = (v + h) \cos \phi_G \sin \lambda_G \tag{5.2}$$
$$Z = [(1 - e^2)v + h] \sin \phi_G \tag{5.3}$$

(2) Cartesian to ellipsoidal

$$\tan \lambda_G = Y/X \tag{5.4}$$

$$\tan \phi_G = (Z + e^2 v \sin \phi_G)/(X^2 + Y^2)^{\frac{1}{2}} \tag{5.5}$$
$$h = X \sec \phi_G \sec \lambda_G - v \tag{5.6}$$
$$= Y \sec \phi_G \operatorname{cosec} \lambda_G - v \tag{5.7}$$

where

$$v = a/(1 - e^2 \sin^2 \phi_G)^{\frac{1}{2}}$$

$$e = (a^2 - b^2)^{\frac{1}{2}}/a$$
a = semi-major axis
b = semi-minor axis
h = ellipsoidal height

The transformation in equation (5.5) is complicated by the fact that v is dependent on ϕ_G and so an iterative procedure is necessary.

This procedure converges rapidly if an initial value for ϕ_G is obtained from

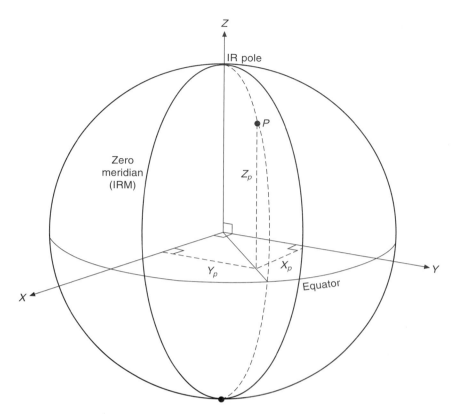

Fig. 5.8 *Geocentric cartesian coordinates*

$$\phi_G = \sin^{-1}(Z/a) \tag{5.8}$$

Alternatively, ϕ_G can be found direct from

$$\tan \phi_G = \frac{Z + \varepsilon b \sin^3 \theta}{(X^2 + Y^2)^{\frac{1}{2}} - e^2 a \cos^3 \theta} \tag{5.9}$$

where $\varepsilon = (a^2/b^2) - 1$

$\tan \theta = a \cdot Z/b \,(X^2 + Y^2)^{\frac{1}{2}}$

Cartesian coordinates are used in satellite position fixing. Where the various systems have parallel axes but different origins, translation from one to the other will be related by simple translation parameters in X, Y and Z, i.e. ΔX, ΔY and ΔZ.

Whilst the cartesian coordinate system provides a simple, well-defined method of defining position, it is not always convenient in terms of heights. As the Z ordinate is vertical from the horizontal equatorial plane and ellipsoidal heights (h) are in a direction normal to the surface of the reference ellipsoid, an increase in h will not produce an equal increase in Z (except at the poles). Indeed, Z ordinates can appear to result in water flowing uphill in some cases.

The increasing use of satellites makes a study of cartesian coordinates and their transformation to ellipsoidal coordinates important.

5.3.4 Plane rectangular coordinates

The geodetic surveys required to establish the ellipsoidal or cartesian coordinates of points over a large area require very high precision, not only in the capture of the field data but also in their processing. The mathematical models involved must of necessity be complete and hence are quite involved. To avoid this the area of interest on the ellipsoid of reference, if of limited extent, may be regarded as a plane surface or curvature catered for by the mathematical projection of ellipsoidal position onto a plane surface. These coordinates in the UK are termed eastings (E) and northings (N) and are obtained from

$$E = f_E\ (\phi_G, \lambda_G)\ \text{(ellipsoid parameters)}$$

$$N = f_N\ (\phi_G, \lambda_G)\ \text{(ellipsoid parameters)}$$

The result is the definition of position by plane coordinates (E, N) which can be utilized using plane trigonometry. These positions will contain some distortion compared with their position on the ellipsoid, which is an inevitable result of projecting a curved surface onto a plane. However, the overriding advantage is that only small adjustments need to be made to the observed field data to produce the plane coordinates.

Figure 5.9 illustrates the concept involved and shows the plane tangential to the ellipsoid at the local origin 0. Generally, the projection, used to transform observations in a plane reference system is an orthomorphic projection, which will ensure that at any point in the projection the scale is the same in all directions. The result of this is that, for small areas, shape and direction are preserved. Thus when connecting engineering surveys into such a reference system, the observed distance,

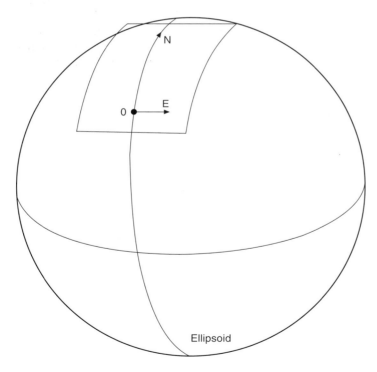

Fig. 5.9 *Plane rectangular coordinates*

when reduced to its horizontal equivalent at MSL, simply requires multiplication by a local scale factor, and observed horizontal angles generally require no further correction.

5.3.5 Height

In outlining the coordinate systems in general use, the elevation or height of a point has been defined as 'orthometric', 'ellipsoidal' or by the Z ordinate. With the increasing use of satellites in engineering surveys, it is important to understand the different categories.

Orthometric height (H) is the one most used in engineering surveys and has been defined in *Section 5.2.2*; in general terms, it is referred to as height above MSL.

Ellipsoidal height has been defined in *Section 5.3.2* and is rarely used in engineering surveys for most practical purposes. However, satellite systems define position and height in X, Y and Z coordinates, which for use in local systems are first transformed to ϕ_G, λ_G and h using the equations of *Section 5.3.3*. The value of h is the ellipsoidal height, which, as it is not related to gravity, is of no practical use, particularly when dealing with the direction of water flow. It is therefore necessary to transform h to H, the relationship of which is shown in *Figure 5.10*:

$$h = N + H \cos \xi \tag{5.10}$$

However, as ξ is always less then $60''$, it can be ignored:

$$\therefore h = N + H \tag{5.11}$$

with an error of less than 0.4 mm at the worst.

The term N is referred to as the 'geoid–ellipsoid separation' or 'geoid height' and to transform ellipsoidal heights to orthometric, must be known to a high degree of accuracy for the particular reference system in use. In global terms N is known (relevant to the WGS84 ellipsoid) to an accuracy of 2–6 m. However, for use in local engineering projects N would need to be known to an accuracy greater than h, in order to provide precise orthometric heights from satellite data. To this end, many national mapping organizations, such as the Ordnance Survey in Great Britain, have

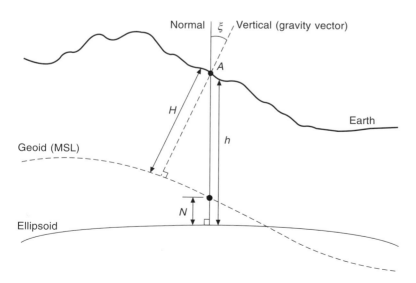

Fig. 5.10

carried out extensive work to produce an accurate model of the geoid and its relationship to the local ellipsoid.

5.4 LOCAL SYSTEMS

The many systems established by various countries throughout the world for positioning on and mapping of the Earth's surface are astrogeodetic systems. Such systems have endeavoured to use reference ellipsoids which most closely fit the geoid of that area and are defined by the following eight parameters:

(1) The size and shape of the ellipsoid, defined by the semi-major axis a, and one other chosen from the semi-minor axis b or the flattening f or the eccentricity e (2 parameters).
(2) The minor axis of the ellipsoid is orientated parallel to the mean spin axis of the Earth as defined by IERS (2 parameters).
(3) The centre of the ellipsoid is implicitly defined with respect to the mass centre of the Earth, by choice of geodetic latitude, longitude and ellipsoidal height at the origin of the system (3 parameters).
(4) The zero meridian or X-axis of the system is chosen to be parallel to the mean (Greenwich) meridian, as defined by the IERS (1 parameter).

It follows that all properly defined geodetic systems will have their axes parallel and can be related to each other by simple translations in X, Y and Z.

The goodness of fit can be indicated by an examination of values for the deviation of the vertical (ξ) and geoid–ellipsoid separation (N), as indicated in *Figure 5.5*. Considering a meridianal section through the geoid–ellipsoid (*Figure 5.11*), it is obvious that the north–south component of the deviation of the vertical is a function of the ellipsoidal latitude (ϕ_G) and astronomical latitude (ϕ_A), i.e.

$$\xi = \phi_A - \phi_G \tag{5.12}$$

The deviation of the vertical in the east-west direction (prime vertical) is

$$\eta = (\lambda_A - \lambda_G) \cos \phi \tag{5.13}$$

where ϕ is ϕ_A or ϕ_G, the difference being negligible. It can be shown that the deviation in any azimuth α is given by

$$\psi = -(\xi \cos \alpha + \eta \sin \alpha) \tag{5.14}$$

whilst at 90° to α the deviation is

$$\zeta = (\xi \sin \alpha - \eta \cos \alpha) \tag{5.15}$$

Thus, in very general terms, the process is briefly as follows. A network of points is established throughout the country to a high degree of observational accuracy. One point in the network is defined as the origin, and its astronomical coordinates, height above the geoid and azimuth to a second point are obtained. The ellipsoidal coordinates of the origin can now be defined as

$$\phi_G = \phi_A - \xi \tag{5.16}$$

$$\lambda_G = \lambda_A - \eta \sec \phi \tag{5.17}$$

$$h = H + N \tag{5.18}$$

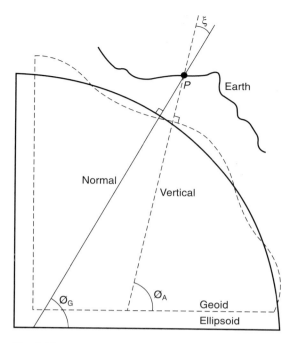

Fig. 5.11

However, at this stage of the proceedings there are no values available for ξ, η and N, so they are assumed equal to zero and an established ellipsoid is used as a reference datum, i.e.

$$\phi_G = \phi_A$$
$$\lambda_G = \lambda_A$$
$$h = H$$

plus a and f, comprising five parameters.

As field observations are automatically referenced to gravity (geoid), then directions and angles will be measured about the vertical, with distance observed at ground level. In order to compute ellipsoidal coordinates, the directions and angles must be reduced to their equivalent about the normal on the ellipsoid and distance reduced to the ellipsoid. It follows that as ξ, η and N will be unknown at this stage, an iterative process is involved, commencing with observations reduced to the geoid.

The corrections involved are briefly outlined as follows:

(1) Deviation of the vertical

As the horizontal axis of the theodolite is perpendicular to the vertical (geoid) and not the normal (ellipsoid), a correction similar to plate bubble error is applied to the directions, i.e.

$$- \zeta \tan \beta \tag{5.19}$$

where ζ is as in equation (5.15)
and β is the vertical angle.

It may be ignored for small values of β.

(2) Skew normal (Figure 5.12)

This correction results from the fact that the normals passing through A and B are not coplanar, due to the ellipsoidal shape involved. Thus plane $AA'B$ containing the observed direction AB does not coincide with the plane $AA'B'$. The observed direction AB must be reduced to the direction $A'B'$ on the ellipsoid by

$$C'' = 0.00011 \cdot H_B \sin 2\alpha \cdot \cos^2 \phi_m \qquad (5.20)$$

where H_B = the ellipsoid height of B in metres
  α = the azimuth of the observed line
  ϕ_m = the mean latitude

This correction may only be necessary in very mountainous areas producing large values for H_B.

(3) Normal–Geodesic

From *Figure 5.13* it can be seen that the plane containing the normal at A and point B cuts the surface along the normal section AB. Similarly, the normal section BA contains the normal at B and point A. In other words, the line of sight from A to B on the ellipsoid would be the normal section AB; from B to A it would be the normal section BA. However, the shortest distance between A and B is the line of double curvature called the geodesic. The geodesic divides the angle between the two normal sections in the ratio of $2:1$, as shown. If α_{AB} is the azimuth of the normal section AB and α'_{AB} the azimuth of the geodesic from A to B then

$$(\alpha'_{AB} - \alpha_{AB})'' = -[0.028(L/100)^2 \sin 2\alpha_{AB} \cos^2 \phi_A] \qquad (5.21)$$

where L = length in km along the normal section AB

  ϕ_A = latitude of A

The difference in length of the geodesic and normal section is

$$\delta L \text{ mm} = 7.7 \times 10^{-17} \times L^2 \sin^2 2\alpha \cos \phi \qquad (5.22)$$

For all but the very longest of lines (>1000 km), the above correction is negligible.

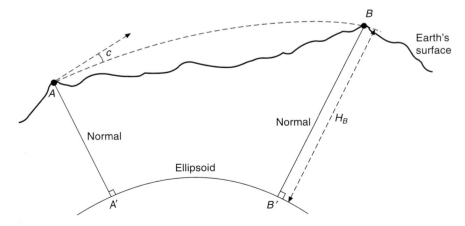

Fig. 5.12   *Skew normals*

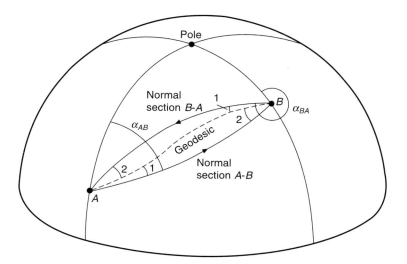

Fig. 5.13 *Normal sections and geodesic (exaggerated)*

(4) Reduction of distance to the ellipsoid

The measured distance is reduced to its horizontal equivalent by applying all the corrections appropriate to the method of measurement.

It is then reduced to MSL (geoid), A_1B_1 in *Figure 5.14*:

$$A_1B_1 = L_1 = L - LH/(R + H) \tag{5.23}$$

where $L = AB$, the mean horizontal distance at ground level
  H = mean height above MSL
  R = mean local radius of the Earth

To reduce it to the ellipsoid, however, requires a knowledge of the geoid–ellipsoid separation (N); then

$$A_2B_2 = L_2 = L_1 - L_1N/(R_\alpha + N) \tag{5.24}$$

where L_1 = the above geoidal distance
  R_α = the average radius of curvature of the ellipsoid in the direction α of the line

where:

$$R_\alpha = \rho v/\rho \sin^2 \alpha + v\cos^2 \alpha \tag{5.25}$$

$$\rho = a(1 - e^2)/(1 - e^2 \sin^2 \phi)^{\frac{3}{2}} = \text{meridional radius of curvature} \tag{5.26}$$

$$v = a/(1 - e^2 \sin^2 \phi)^{\frac{1}{2}} = \text{prime vertical radius of curvature at } 90° \text{ to } \rho. \tag{5.27}$$

As already stated, N values may not be available and hence the geoidal distance may have to be accepted. It should be remembered that $N = 6$ m will produce a scale error of 1 ppm if ignored. In the UK, the maximum value for N is about 4.5 m, resulting in a scale error of only 0.7 ppm, and may therefore be ignored for scale purposes. Obviously it cannot be ignored in heighting.

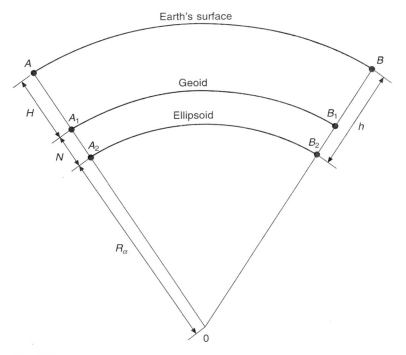

Fig. 5.14

(5) Laplace correction

The observed astronomical azimuths (α_A) must be reduced to their equivalent on the ellipsoid (α_G), using the important Laplace equation:

$$\alpha_G = \alpha_A - (\lambda_G - \lambda_A) \sin \phi \tag{5.28}$$

Astronomical azimuths can be observed to a very high accuracy ($<1''$) and can be used to control the propagation of error through the control network, where the geodetic azimuths may have been reduced in accuracy by 10-fold.

(6) Geoid–ellipsoid separation

In the past, geoidal profiles were obtained at a large number of points at intervals of 20–30 km using astrogeodetic levelling.

From *Figure 5.15* it can be seen that, provided a value for N is known (usually $N = 0$) at the origin A, then the value for N at B is $N_A + \Delta N_1$, and so on, where

$$\Delta N \approx L \tan \psi \tag{5.29}$$

and ψ is given in equation (5.14).

The main error source in this instance is in assuming a regular deviation of the geoid from the ellipsoid (*Figure 5.15*). This, of course, may not be so, particularly in mountainous areas where the shape of the geoid may be quite complicated. To minimize the error, the spacing between stations should be kept to a minimum and the mean value of ψ at each end of the line used, i.e.

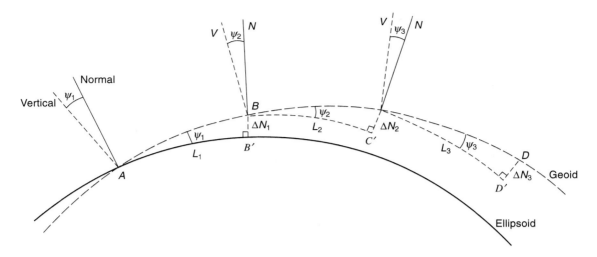

Fig. 5.15

$$N_B - N_A = \Delta N = \tfrac{1}{2}(\psi''_A + \psi''_B) L_{AB} \sin 1'' \tag{5.30}$$

In addition to the above process, gravimetric and satellite data are now used to obtain detailed information about the local geoid.

When the field data are finally reduced (by an iterative process) to the ellipsoid, the ellipsoidal coordinates can be computed. The formulae involved in this process will depend on the distances involved and the accuracy requirements. For relatively short lines (<100 km) the mid-latitude formula is the simplest to use and will serve to illustrate the procedures involved.

5.5 COMPUTATION ON THE ELLIPSOID

In the first instance, geodetic (ellipsoidal) azimuths are computed through the network from one Laplace azimuth to the next, using the Laplace azimuths to control and adjust the observed directions.

Before proceeding with the computation of ellipsoidal coordinates, it is necessary to consider certain aspects of direction. In plane surveying, for instance, the direction of *BA* differs from that of *AB* by exactly 180°. However, as shown in *Figure 5.16*,

$$\text{Azimuth } BA = \alpha_{AB} + 180° + \Delta\alpha = \alpha_{BA} \tag{5.31}$$

where $\Delta\alpha$ is the additional correction due to the convergence of the meridians *AP* and *BP*.

Using the corrected ellipsoidal azimuths and distances, the coordinates are now calculated relative to a selected point of origin.

The basic problems are known as the 'direct' and 'reverse' problems and are analogous to computing the 'polar' and 'join' in plane surveying.

The mid-latitude formulae are generally expressed as

$$\Delta\phi'' = \frac{L\cos\alpha_m}{\rho_m \sin 1''}\left(1 + \frac{\Delta\lambda^2}{12} + \frac{\Delta\lambda^2 \sin^2 \phi_m}{24}\right) \tag{5.32}$$

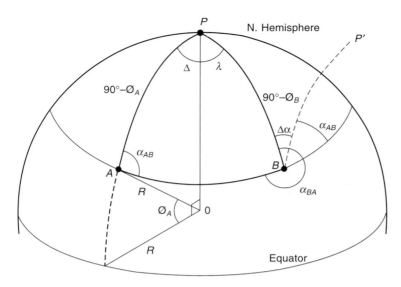

Fig. 5.16

$$\Delta\lambda'' = \frac{L \sin \alpha_m \cdot \sec \phi_m}{\nu \sin 1''} \left(1 + \frac{\Delta\lambda^2 \sin^2 \phi_m}{24} - \frac{\Delta\phi^2}{24} \right) \tag{5.33}$$

$$\Delta\alpha'' = \Delta\lambda'' \sin \phi_m \left(1 + \frac{\Delta\lambda^2 \sin^2 \phi_m}{24} + \frac{\Delta\lambda^2 \cos^2 \phi_m}{12} + \frac{\Delta\phi^2}{12} \right) \tag{5.34}$$

where $\Delta\phi = \phi_A - \phi_B$

$\Delta\lambda = \lambda_A - \lambda_B$

$\Delta\alpha = \alpha_{BA} - \alpha_{AB} \pm 180°$

$\alpha_m = \alpha_{AB} + \dfrac{\Delta\alpha}{2}$

$\phi_m = (\phi_A + \phi_B)/2$

The above formula is accurate to 1 ppm for lines up to 100 km in length.

(1) The direct problem

Given:

(a) Ellipsoidal coordinates of $A = \phi_A, \lambda_A$
(b) Ellipsoidal distance $AB = L_{AB}$
(c) Ellipsoidal azimuth $AB = \alpha_{AB}$
(d) Ellipsoidal parameters $= a, e^2$

determine:

(a) Ellipsoidal coordinates of $B = \phi_B, \lambda_B$
(b) Ellipsoidal azimuth $BA = \alpha_{BA}$

As the mean values of ρ, ν and ϕ are required, the process must be an iterative one and will be outlined using the first term only of the mid-latitude formula:

(a) Determine ρ_A and v_A from equations (5.26) and (5.27) using ϕ_A
(b) Determine $\Delta\phi'' = L_{AB} \cos \alpha_{AB}/\rho_A \sin 1''$
(c) Determine the first value for $\phi_m = \phi_A + (\Delta\phi/2)$
(d) Determine improved values for ρ and v using ϕ_m and iterate until negligible change in $\Delta\phi$
(e) Determine $\Delta\lambda'' = L_{AB} \sin \alpha_{AB} \sec \phi_m/v_m \sin 1''$
(f) Determine $\Delta\alpha'' = \Delta\lambda'' \sin \phi_m$ and so deduce

$$\alpha_m = \alpha_{AB} + (\Delta\alpha/2)$$

(g) Iterate the whole procedure until the differences between successive values of $\Delta\phi$, $\Delta\lambda$ and $\Delta\alpha$ are insignificant. Three iterations normally suffice.
(h) Using the final accepted values, we have:

$$\phi_B = \phi_A + \Delta\phi$$
$$\lambda_B = \lambda_B + \Delta\lambda$$
$$\alpha_{BA} = \alpha_{AB} + \Delta\alpha \pm 180°$$

(2) The reverse problem

Given:

(a) Ellipsoidal coordinates of $A = \phi_A, \lambda_A$
(b) Ellipsoidal coordinates of $B = \phi_B, \lambda_B$
(c) Ellipsoidal parameters $= a, e^2$

determine:

(a) Ellipsoidal azimuths $= \alpha_{AB}$ and α_{BA}
(b) Ellipsoidal distance $= L_{AB}$

This procedure does not require iteration:

(a) Determine $\phi_m = (\phi_A + \phi_B)/2$
(b) Determine $v_m = a/(1 - e^2 \sin^2 \phi_m)^{\frac{1}{2}}$ and
$$\rho_m = a(1 - e^2)/(1 - e^2 \sin^2 \phi_m)^{\frac{3}{2}}$$
(c) Determine $\alpha_m = v_m \cdot \Delta\lambda \cos \phi_m /(\rho_m \cdot \Delta\phi)$
(d) Determine L_{AB} from $\Delta\phi'' = L_{AB} \cos \alpha_m/\rho_m \sin 1''$ which can be checked using $\Delta\lambda''$
(e) Determine $\Delta\alpha'' = \Delta\lambda'' \sin \phi_m$, then
$$\alpha_{AB} = \alpha_m - (\Delta\alpha/2) \text{ and } \alpha_{BA} = \alpha_{AB} + \Delta\alpha \pm 180°$$

Whilst the mid-latitude formula serves to illustrate the procedures involved, computers now permit the use of the more accurate equations. Such formulae may be obtained direct from the national mapping agency for the area concerned.

On completion of all the computation throughout the network, values of ξ, η and N can be obtained at selected stations. The best-fitting ellipsoid is the one for which values of ΣN^2 or $\Sigma(\xi^2 - \eta^2)$ are a minimum. If the fit is not satisfactory, then the values of ξ, η and N as chosen at the origin could be altered or a different ellipsoid selected. Although it would be no problem to change the ellipsoid due to the present use of computers, in the past it was usual to change the values of ξ, η and N to improve the fit.

Although the above is a brief description of the classical geodetic approach, the majority of ellipsoids in use throughout the world were chosen on the basis of their availability at the time. For instance, the Airy ellipsoid adopted by Great Britain was chosen in honour of Professor Airy who was Astronomer Royal at the time and had just announced the parameters of his ellipsoid. In fact, recent tests have shown that the fit is quite good, with maximum values of N equal to 4.5 m and maximum values for deviation of the vertical equal to $10''$.

5.6 DATUM TRANSFORMATIONS

Coordinate transformations are quite common in surveying. They range from simple translations between coordinates and setting-out grids on a construction site, to transformation between global systems.

The increasing use of satellites will require transformation from a XYZ system on a global ellipsoid to a local system on a local ellipsoid.

Whilst the mathematical procedures are well defined in all manner of transformations, problems can arise due to varying scale throughout the network used to establish position. Thus in a local system, there may be a variety of parameters, established empirically, to be used in different areas of the system.

5.6.1 Basic concept

From *Figure 5.17* it can be seen that the basic parameters in a conventional transformation between similar XYZ systems would be:

(1) Translation of the origin 0, which would involve shifts in X, Y and Z i.e. ΔX, ΔY, ΔZ.
(2) Rotation about the three axes, θx, θy and θz, in order to render the axes of the systems involved parallel. θx and θy would change the polar axes, and θz the zero meridian.
(3) One scale parameter $(1 + S)$ to equalize the scales of the different coordinate systems.

In addition to the above, the size (a) of the ellipsoid and its shape (f) may also need to be included. However, not all the parameters are generally used in practice. The most common transformation is the translation in X, Y and Z only (three parameters). Also common is the four-parameter (ΔX, ΔY, ΔZ + scale) and the five-parameter (ΔX, ΔY, ΔZ + scale + θz). A full transformation would entail seven parameters.

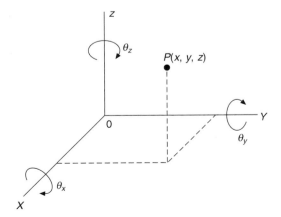

Fig. 5.17

A simple illustration of the process can be made by considering the transformation of the coordinates of P (X', Y', Z') to (X, Y, Z) due to rotation θx about axis OX (*Figure 5.18*):

$$X = \qquad\qquad = X'$$

$$Y = Or - qr = Y' \cos\theta - Z' \sin\theta \qquad\qquad (5.35)$$

$$Z = mr + Pn = Y' \sin\theta + Z' \cos\theta \qquad\qquad (5.36)$$

In matrix form:

$$\begin{bmatrix} X \\ Y \\ Z \end{bmatrix} = \begin{bmatrix} 1 & 0 & 0 \\ 0 & \cos\theta & -\sin\theta \\ 0 & \sin\theta & \cos\theta \end{bmatrix} \begin{bmatrix} X' \\ Y' \\ Z' \end{bmatrix} \qquad\qquad (5.37)$$

$$x = R_\theta \cdot x'$$

where R_θ = rotational matrix for angle θ
x' = the vector of original coordinates

Similar rotation matrices can be produced for rotations about axes OY (α) and OZ (β), giving

$$x = R_\theta R_\alpha R_\beta x' \qquad\qquad (5.38)$$

If a scale change and translation of the origin to X_o, Y_o, Z_o is made, the coordinates of P would be

$$\begin{bmatrix} X \\ Y \\ Z \end{bmatrix} = \begin{bmatrix} X_o \\ Y_o \\ Z_o \end{bmatrix} + (1 + S) \begin{bmatrix} a_{11} & a_{12} & a_{13} \\ a_{21} & a_{22} & a_{23} \\ a_{31} & a_{32} & a_{33} \end{bmatrix} \begin{bmatrix} X' \\ Y' \\ Z' \end{bmatrix} \qquad\qquad (5.39)$$

The a coefficients of the rotation matrix would involve the sines and cosines of the angles of rotation, obtained from the matrix multiplication of R_θ, R_α and R_β.

For the small angles of rotation the sines of the angles may be taken as their radian measure ($\sin\theta = \theta$) and the cosines made equal to unity, with sufficient accuracy. The above equation is

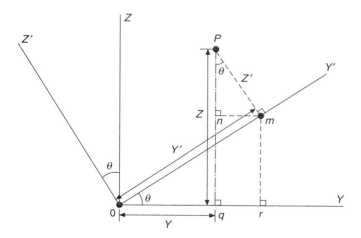

Fig. 5.18

referred to in surveying as the Helmert transformation and describes the full transformation between the two geodetic datums.

Whilst the X, Y, Z coordinates of three points would be sufficient to determine the seven parameters, in practice as many points as possible are used in a least squares solution. Ellipsoidal coordinates (ϕ, λ, h) would need to be transformed to X, Y and Z for use in the transformations.

As a translation of the origin of the reference system is the most common, a Molodenskii transform permits the transformation of ellipsoidal coordinates from one system to another in a single operation i.e.

$$\phi = \phi' + \Delta\phi''$$
$$\lambda = \lambda' + \Delta\lambda''$$
$$h = h' + \Delta h$$

where $\Delta\phi'' = (-\Delta X \sin \phi' \cos \lambda' - \Delta Y \sin \phi' \sin \lambda' + \Delta Z \cos \phi'$

$$+ (a' \Delta f + f' \Delta a) \sin 2\phi')/(\rho \sin 1'') \tag{5.40}$$

$$\Delta\lambda'' = (-\Delta X \sin \lambda' + \Delta Y \cos \lambda')/(v \sin 1'') \tag{5.41}$$

$$\Delta h = (\Delta X \cos \phi' \cos \lambda' + \Delta Y \cos \phi' \cdot \sin \lambda' + \Delta Z \sin \phi'$$

$$+ (a' \Delta f + f' \Delta a) \sin^2 \phi' - \Delta a) \tag{5.42}$$

In the above formulae:

ϕ', λ', h' = ellipsoidal coordinates in the first system
ϕ, λ, h = ellipsoidal coordinates in the required system
a', f' = ellipsoidal parameters of the first system
$\Delta a, \Delta f$ = difference between the parameters in each system
$\Delta X, \Delta Y, \Delta Z$ = origin translation values
v = radius of curvature in the prime vertical (equation (5.27))
ρ = radius of curvature in the meridian (equation (5.26))

It must be emphasized once again that whilst the mathematics of transformation are rigorously defined, the practical problems of varying scale etc. must always be considered.

5.7 ORTHOMORPHIC PROJECTION

The ellipsoidal surface, representing a portion of the Earth's surface, may be represented on a plane using a specific form of projection, i.e.

$$E = f_E (\phi, \lambda) \tag{5.43}$$
$$N = f_N (\phi, \lambda) \tag{5.44}$$

where E and N on the plane of the projection represent ϕ, λ on the reference ellipsoid.

Representation of a curved surface on a plane must result in some form of distortion, and therefore the properties required of the projection must be carefully considered. In surveying, the properties desired are usually:

(1) A line on the projection must contain the same intermediate points as that on the ellipsoid.
(2) The angle between the tangents to any two lines on the ellipsoid should have a corresponding

angle on the projection. This property is termed orthomorphism and results in small areas retaining their shape.

Using the appropriate projection mathematics the geodesic *AB* in *Figure 5.19* is projected to the curved dotted line *ab*; point *C* on the geodesic will appear at *c* on the projection. The meridian *AP* is represented by the dotted line 'geodetic north', and then:

(1) The angle γ between grid and geodetic north is called the 'grid convergence' resulting from the convergence of meridians.
(2) The angle α is the azimuth of *AB* measured clockwise from north.
(3) The angle *t* is the grid bearing of the chord *ab*.
(4) The angle *T* is the angle between grid north and the projected geodesic. From (3) and (4) we have the $(t - T)$ correction.
(5) The line scale factor (F) is the ratio between the length (S) of the geodesic *AB* as calculated from ellipsoidal coordinates and its grid distance (G) calculated from the plane rectangular coordinates, i.e.

$$F = G/S \tag{5.45}$$

Similarly the point scale factor can be obtained from the ratio between a small element of the geodesic and a corresponding element of the grid distance.
(6) It should be noted that the project geodesic is always *concave to the central meridian*.

It can be seen from the above that:

(1) The geodetic azimuth can be transformed to grid bearing by the application of 'grid convergence' and the '$t - T$' correction.
(2) The ellipsoidal distance can be transformed to grid distance by multiplying it by the scale factor.

The plane coordinates may now be computed, using this adjusted data, by the application of plane trigonometry. Thus apart from the cartographic aspects of producing a map or plan, the engineering

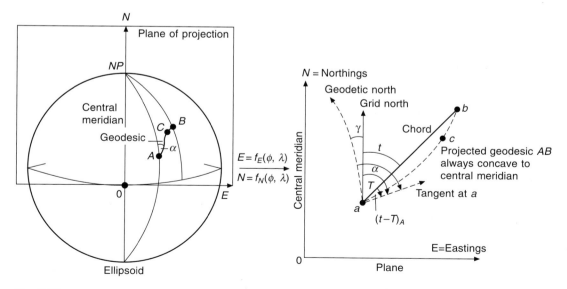

Fig. 5.19

surveyor now has an extremely simple mathematical process for transforming field data to grid data and vice versa.

The orthomorphic projection that is now used virtually throughout the world is the transverse Mercator projection, which is ideal for countries having their greatest extent in a north–south direction. This can be envisaged as a cylinder surrounding the ellipsoid (*Figure 5.20*) onto which the ellipsoid positions are projected. The cylinder is in contact with the ellipsoid along a meridian of longitude and the lines of latitude and longitude are projected onto the cylinder from a point source at the centre of the ellipsoid. Orthomorphism is achieved by stretching the scale along the meridians to keep pace with the increasing scale along the parallels. By opening up the cylinder and spreading it out flat, the lines of latitude and longitude form a graticule of complex curves intersecting at right angles, the central meridian being straight.

It is obvious from *Figure 5.20* that the ratio of distance on the ellipsoid to that on the projection would only be correct along the central meridian, where the cylinder and ellipsoid are in contact, and thus the scale factor would be unity ($F = 1$).

The following illustrates a sample of the basic transverse Mercator projection formulae:

$$(1) \quad E = F_0 \left[v \cdot \Delta\lambda \cos\phi + v \frac{\Delta\lambda^3}{6} \cos^3\phi (\psi - t^2) + v \frac{\Delta\lambda^5}{120} \cos^5\phi (4\psi^3(1 - 6t^2) \right.$$

$$\left. + \psi^2 (1 + 8t^2) - \psi(2t^2) + t^4) + v \cdot \frac{\Delta\lambda^7}{5040} \cos^7\phi (61 - 479t^2 + 179t^4 - t^6) \right] \qquad (5.46)$$

$$(2) \quad N = F_0 \left[M + v \sin\phi \frac{\Delta\lambda^2}{2} \cos\phi + v \sin\phi \frac{\Delta\lambda^4}{24} \cos^3\phi(4\psi^2 + \psi - t^2) \right.$$

$$+ v \sin\phi \frac{\Delta\lambda^6}{720} \cos^5\phi (8\psi^4 (11 - 24t^2) - 28\psi^3 (1 - 6t^2) + \psi^2 (1 - 32t^2)$$

$$\left. - \psi(2t^2 + t^4) + v \sin\phi \frac{\Delta\lambda^8}{40\,320} \cos^7\phi (1385 - 3111t^2 + 543t^4 - t^6) \right] \qquad (5.47)$$

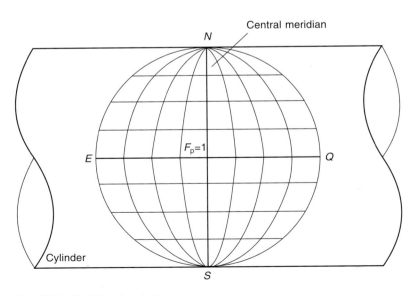

Fig. 5.20   *Cylindrical projection*

where:

F_o = scale factor on the central meridian

$\Delta\lambda$ = longitude measured from the central meridian

$t = \tan \phi$

$\psi = v/\rho$

M = meridian distance from the latitude of the origin, and is obtained from:

$$M = a(A_o\phi - A_2 \sin 2\phi + A_4 \sin 4\phi - A_6 \sin 6\phi) \tag{5.48}$$

and $A_o = 1 - e^2/4 - 3e^4/64 - 5e^6/256$

$A_2 = (3/8)(e^2 + e^4/4 + 15e^6/128)$

$A_4 = (15/256)(e^4 + 3e^6/4)$

$A_6 = 35e^6/3072$

(3) Grid convergence $= \gamma = -\Delta\lambda \sin \phi - \dfrac{\Delta\lambda^3}{3} \sin \phi \cos^2 \phi (2\psi^2 - \psi)$ \hfill (5.49)

which is sufficiently accurate for most applications in engineering surveying.

(4) The point scale factor can be computed from:

$$F = F_o[1 + (E^2/2R_m^2) + (E^4/24R_m^4)] \tag{5.50a}$$

The scale factor for the line AB can be computed from:

$$F = F_o[1 + (E_A^2 + E_A E_B + E_B^2)]/6R_m^2 \tag{5.50b}$$

where $R_m^2 = \rho v F_o^2$

E = distance of the point from the central meridian in terms of the difference in Eastings.

In the majority of cases in practice, it is sufficient to take the distance to the easting of the mid-point of the line (E_m), and then:

$$F = F_o(1 + E_m^2/2R^2) \tag{5.51}$$

and $R^2 = \rho v$

(5) The 'arc-to-chord' correction or the $(t - T)$ correction, as it is more commonly called, from A to B, in seconds of arc, is

$$(t - T)_A'' = -(N_B - N_A)(2E_A - E_B)/6R^2 \sin 1'' \tag{5.52}$$

with sufficient accuracy for most purposes.

5.8 ORDNANCE SURVEY NATIONAL GRID

The Ordnance Survey (OS) is the national mapping agency for Great Britain; its maps are based on a transverse Mercator projection of Airy's ellipsoid called the OSGB (36) datum.

The central meridian selected is 2° W, with the point of origin at 49°N on this meridian. The scale factor varies as the square of the distance from the central meridian, and therefore in order to reduce scale error at the extreme east and west edges of the country the scale factor on the central meridian was reduced by a factor of 2499/2500. The effect of this is to reduce the scale factor on the central meridian to 0.99960127 and conceptually reduce the radius of the enclosing cylinder as shown in *Figure 5.21*.

The projection cylinder cuts the ellipsoid at two sub-parallels, 180 km each side of the central meridian, where the scale factor will be unity. Inside these two parallels the scale is too small by 0.04%, and outside of them too large by 0.04%.

The central meridian (2°W) which constitutes the *N*-axis (*Y*-axis) was assigned a large easting value of E 400 000 m. The *E*-axis (*X*-axis) was assigned a value of N 100 000 m relative to the 49°N parallel of latitude. Thus a rectangular grid is superimposed on the developed cylinder and is called the OS National Grid (NG) (*Figure 5.22*). The assigned values result in a 'false origin' and positive values only throughout, what is now, a plane rectangular coordinate system. Such a grid thereby establishes the direction of grid north, which differs from geodetic north by γ, a variable amount called the grid convergence. At the central meridian grid north and geodetic north are the same direction.

5.8.1 Scale factors

The concept of scale factors has been fully dealt with and it only remains to deal with their application. It should be clearly understood that scale factors transform distance on the ellipsoid to distance on the plane of projection. From *Figure 5.23*, it can be seen that a horizontal distance at ground level *AB* must first be reduced to its equivalent at MSL (geoid) A_1B_1, using the altitude correction, thence to the ellipsoid $A_1'B_1'$ using the geoid–ellipsoid value (*N*) and then multiplied by the scale factor to produce the projection distance A_2B_2.

Whilst this is theoretically the correct approach, lack of knowledge of *N* may result in this step being ignored. In Great Britain, the maximum value is 4.5 m, resulting in a scale error of only 0.7 ppm if ignored. Thus the practical approach is to reduce to MSL and then to the projection plane, i.e. from *D* to *S* to *G*, as in *Figure 5.24*.

The basic equation for scale factor is given in equation 5.50, where the size of the ellipsoid and the value of the scale factor on the central meridian (F_o) are considered. Specific to the OSGB (36) system, the following formula may be developed, which is sufficiently accurate for most purposes.

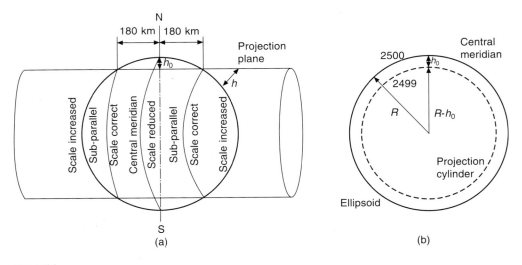

Fig. 5.21

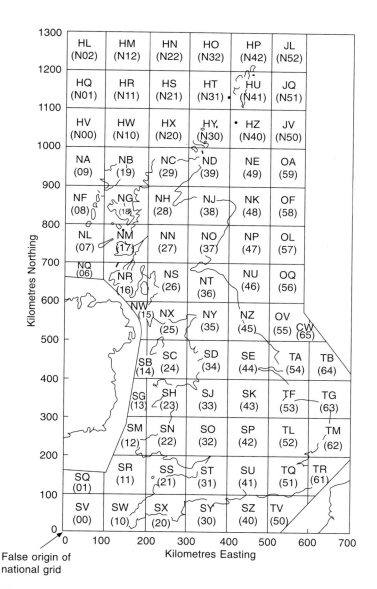

Fig. 5.22 *National reference system of Great Britain showing 100-km squares, the figures used to designate them in the former system, and the letters which have replaced the figures. (Courtesy Ordnance Survey, Crown Copyright Reserved)*

Scale error (SE) is the difference between the scale factor at any point (F) and that at the central meridian (F_o) and varies as the square of the distance from the central meridian, i.e.

$$SE = K(\Delta E)^2$$

where ΔE is the difference in easting between the central meridian and the point in question:

$$F = F_o + SE = 0.99960127 + K(\Delta E)^2$$

Consider a point 180 km east or west of the central meridian where $F = 1$:

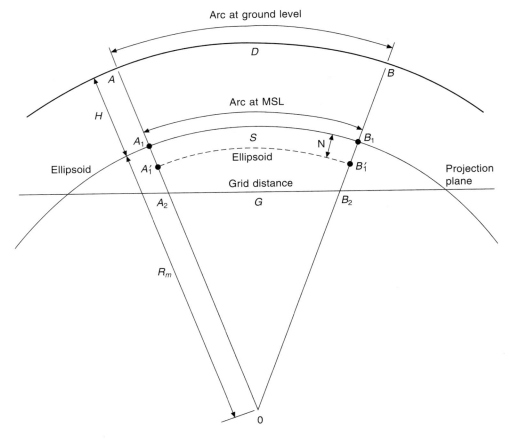

Fig. 5.23

$$1 = 0.99960127 + K(180 \times 10^3)^2$$

$$K = 1.228 \times 10^{-14}$$

and $\qquad F = F_0 + (1.228 \times 10^{-14} \times \Delta E^2)$ $\qquad\qquad\qquad$ (5.53)

where $\quad F_0 = 0.99960127$

$\qquad\quad \Delta E = E - 400\,000$

Thus the value of F for a point whose NG coordinates are E 638824 N 309912 is:

$$F = 0.99\,960\,127 + (1.228 \times 10^{-14} \times 238\,824^2) = 1.0003016$$

As already intimated in equation (5.50), the treatment for highly accurate work is to compute F for each end of the line and in the middle, and then obtain the mean value from Simpson's rule. However, for all practical purposes, it is sufficient to compute F at the mid-point of a line. On the OS system the scale factor varies, at the most, by only 60 ppm in 10 km, and hence a single value for F at the centre of the site can be regarded as constant throughout the area. On long motorway or route projects, however, one would need to use a different scale factor for every 5 to 10 km section.

Whilst the above formula may be sufficiently accurate for most purposes, it may not be adequate

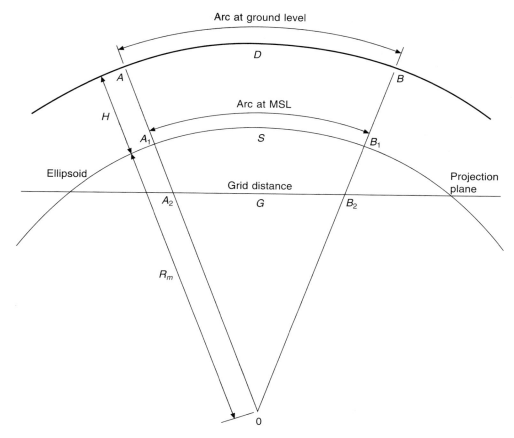

Fig. 5.24

for distance measured to extremely fine sub-millimetric accuracies by such instruments as the Kern Mekometer and the Com-Rad Geomensor. In this case recourse should be made to the complete formula available from the OS. The computer has made the use of the formula more applicable than projection tables.

The following examples will serve to illustrate the classical application of scale factors.

Worked examples

Example 5.1 Grid to ground distance
Any distance calculated from NG coordinates will be grid distance. If this distance is to be set out on the ground it must:

(a) Be *divided* by the LSF to give the ellipsoidal distance at MSL, i.e. $S = G/F$.
(b) Have the altitude correction applied to give the horizontal ground distance.

Consider two points, *A* and *B*, whose coordinates are

A: E 638 824.076	N 307 911.843	
B: E 644 601.011	N 313 000.421	

$\therefore \Delta E = 5\,776.935 \qquad \therefore \Delta N = 5\,088.578$

Grid distance $= (\Delta E^2 + \Delta N^2)^{\frac{1}{2}} = 7698.481$ m $= G$

Mid-easting of $AB = $ E 641 712 m

$\therefore F = 1.000\ 3188$ (from equation (5.53))

$\therefore$ Ellipsoidal distance at MSL $= S = G/F = 7696.027$ m

Now assuming AB at a mean height (H) of 250 m above MSL, the altitude correction C_m is

$$C_m = \frac{SH}{R} = \frac{7696 \times 250}{6\ 384\ 100} = +0.301 \text{ m}$$

$\therefore$ Horizontal distance at ground level $= 7696.328$ m

This situation could arise where the survey and design coordinates of a project are on the OSNG. Distances calculated from the grid coordinates would need to be transformed to their equivalent on the ground for setting-out purposes.

Example 5.2 Ground to grid distance
When connecting surveys to the national grid, horizontal distances measured on the ground must be:

(a) Reduced to their equivalent on the ellipsoid at MSL.
(b) *Multiplied* by the LSF to produce the equivalent grid distance, i.e. $G = S \times F$.

Consider now the previous problem worked in reverse

Horizontal ground distance $= 7696.328$ m
Altitude correction C_m $=\quad -0.301$ m
_____

$\therefore$ Ellipsoidal distance S at MSL $= 7696.027$ m
$F = 1.000\ 3188$
_____

$\therefore$ Grid distance $G = S \times F = 7698.481$ m

This situation could arise in the case of a link traverse connected into the OSNG system. The length of each leg of the traverse would need to be reduced from its horizontal distance at ground level to its equivalent distance on the NG.

There is no application of grid convergence as the traverse commences from a grid bearing and connects into another grid bearing. The application of the $(t - T)$ correction to the angles would generally be negligible, being a maximum of 7″ for a 10-km line and much less than the observational errors of the traverse angles. It would only be necessary to consider both the above effects if the angular error was being controlled by taking gyro-theodolite observations on intermediate lines in the traverse.

The two applications illustrated in the examples of first reducing to MSL and then to the plane of the projection (NG) can be combined to give:

$$F_a = F(1 - H/R) \tag{5.54}$$

where H is the ground height relative to MSL and is positive when above and negative when below MSL.

Then from *Example 5.1*:

$$F_a = 1.0003188\ (1 - 250/6\ 384\ 100) = 1.0002797$$

F_a is then the scale factor adjusted for altitude and can be used directly to transform from ground to grid and vice versa.

From *Example 5.2*:

$$7696.328 \times 1.0002797 = 7698.481 \text{ m}$$

5.8.2 Grid convergence

All grid north lines on the NG are parallel to the central meridian (E400000 m), whilst the meridians converge to the pole. The difference between these directions is termed the grid convergence γ.

An approximate formula may be derived from the first term of equation (5.49).

$$\gamma = \Delta\lambda \sin\phi$$

but $\Delta\lambda = \Delta E/R \cos\phi_m$

$$\gamma'' = \frac{\Delta E \tan\phi_m \times 206\,265}{R} \tag{5.55}$$

where ΔE = distance from the central meridian

R = mean radius of Airy's ellipsoid = $(\rho v)^{\frac{1}{2}}$

ϕ_m = mean latitude of the line

The approximate method of computing γ is acceptable only for lines close to the central meridian, where values correct to a few seconds may be obtained. As the distance from the central meridian increases, so too does the error in the approximate formula and the more rigorous methods are necessary.

If the NG coordinates of a point are E 626 238 and N 302 646 and the latitude, scaled from an OS map, is N 52° 34′, then taking $R = 6\,380\,847$ m gives

$$\gamma'' = \frac{226\,238 \tan 52°34'}{6\,380\,847} \times 206\,265 = 9554'' = 2°39'14''$$

5.8.3 (t − T) correction

As already shown, the $(t - T)$ correction results from the fact that the geodesic appears as a curved line on the projection and differs from the direction of the chord, as computed from plane trigonometry, by a small correction. *Figure 5.25* illustrates the angle θ as 'observed' and the angle β as computed from the grid coordinates; then:

$$\beta = \theta - (t - T)_{BA} - (t - T)_{BC}$$

An approximate formula for $(t - T)$ specific to the NG is as follows:

$$(t - T)''_A = (2\Delta E_A + \Delta E_B)(N_A - N_B)K \tag{5.56}$$

where ΔE = NG easting −400 000, expressed in km

N = NG northing expressed in km

A = station at which the correction is required

B = station observed

$K = 845 \times 10^{-6}$

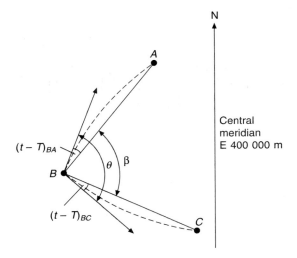

Fig. 5.25

A maximum value for a 10-km line would be about 7″.

The signs of the corrections for $(t - T)$ and grid convergence are best obtained from a diagram similar to that of *Figure 5.26*, where for line *AB*:

$$\phi = \text{grid bearing } AB$$

$$\theta = \text{azimuth } AB$$

then $\quad \theta = \phi - \gamma - (t - T)_A$, or

$$\phi = \theta + \gamma + (t - T)_A$$

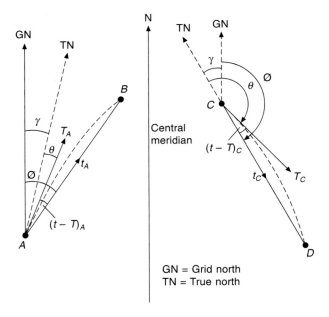

Fig. 5.26

For line *CD*:

$$\theta = \phi + \gamma - (t - T)_C, \text{ or}$$

$$\phi = \theta - \gamma + (t - T)_C$$

A careful study of the *Worked examples* will further illustrate the application of the projection corrections.

5.9 PRACTICAL APPLICATIONS

All surveys connected to the NG should have their measured distances reduced to the horizontal, and then to MSL; and should then be multiplied by the local scale factor to reduce them to grid distance.

Consider *Figure 5.27* in which stations *A*, *B* and *C* are connected into the NG via a link traverse from OSNG stations *W*, *X* and *Y*, *Z*:

(1) The measured distance D_1 to D_4 would be treated as above.
(2) The observed angles should in theory be corrected by appropriate $(t - T)$ corrections as in *Figure 5.25*. These would generally be negligible but could be quickly checked using

$$(t - T)'' = (\Delta N_{AB} \cdot E/2R^2)\ 206\,265 \tag{5.57}$$

where E = easting of the mid-point of the line
R = an approximate value for the radius of the ellipsoid for the area

(3) There is no correction for grid convergence as the survey has commenced from a grid bearing and has connected into another.
(4) Grid convergence and $(t - T)$ would need to be applied to the bearing of, say, line *BC* if its bearing had been found using a gyro-theodolite and was therefore relative to true north (TN) (see *Figure 5.26*). This procedure is sometimes adopted on long traverses to control the propagation of angular error.

When the control survey and design coordinates are on the NG, the setting out by bearing and distance will require the grid distance, as computed from the design coordinates, to be corrected to its equivalent distance on the ground. Thus grid distance must be changed to its MSL value and then divided by the local scale factor to give horizontal ground distance.

The setting-out angle, as computed from the design (grid) coordinates, will require no correction.

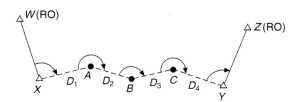

Fig. 5.27 *Link traverse*

5.10 THE UNIVERSAL TRANSVERSE MERCATOR PROJECTION (UTM)

UTM is a world-wide system of transverse Mercator projections based on the International Earth Ellipsoid 1924. It comprises about 60 zones, each 6° in longitude wide, with central meridian at 3°, 9°, etc. from zero meridian. The zones are numbered from 1 to 60, starting with 180° to 174° W as zone 1 and proceeding eastwards to zone 60. In latitude, the UTM system extends from 80° N to 80° S, with the polar caps covered by a polar stereographic projection.

The scale factor at each central meridian is 0.9996 to counteract the enlargement ratio at the edges of the strips. The false origin of northings is zero at the equator for the northern hemisphere and 10^6 m south of the equator for the southern hemisphere. The false origin for eastings is 5×10^5 m west of the zone central meridian.

Projection tables are available for the system and all NATO maps are based on it. However, as there is no continuity across the zones, one cannot compute between points in different zones.

5.11 PLANE RECTANGULAR COORDINATES

A plane rectangular coordinates system is as defined in *Figure 5.28*.

It is split into four quadrants with the typical mathematical convention of the axis to the north and east being positive and to the south and west, negative.

In pure mathematics, the axis is defined as x and y, with angles measured anticlockwise from the

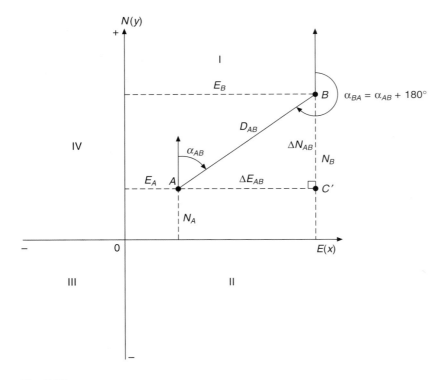

Fig. 5.28

x-axis. In surveying, the *x*-axis is referred to as the east-axis (*E*) and the *y*-axis as the north-axis (*N*), with angles (α) measured clockwise from the *N*-axis.

From *Figure 5.28*, it can be seen that to obtain the coordinates of point *B*, we require the coordinates of point *A* and the coordinates of the line *AB*, i.e.

$$E_B = E_A + \Delta E_{AB} \text{ and}$$
$$N_B = N_A + \Delta N_{AB}$$

It can further be seen that to obtain the coordinates of the line *AB* we require its horizontal distance and direction.

The system used to define a direction is called the *whole circle bearing system* (WCB). A WCB is the direction measured clockwise from 0° full circle to 360°. It is therefore always positive and never greater than 360°.

Figure 5.29 shows the WCB of the lines as follows:

WCB *OA* = 40°
WCB *OB* = 120°
WCB *OC* = 195°
WCB *OD* = 330°

As shown in *Figure 5.29*, the reverse or back bearing is 180° different to the forward bearing, thus:

WCB *AO* = 40° + 180° = 220°
WCB *BO* = 120° + 180° = 300°
WCB *CO* = 195° − 180° = 15°
WCB *DO* = 330° − 180° = 150°

Thus if WCB < 180° it is easier to add 180° to get the reverse bearing, and if > 180° subtract, as shown.

The above statement should not be confused with a similar rule for finding WCBs from the observed angles. For instance (*Figure 5.30*), if the WCB of *AB* is 0° and the observed angle *ABC* is 140°, then the relative WCB of *BC* is 320°, i.e.

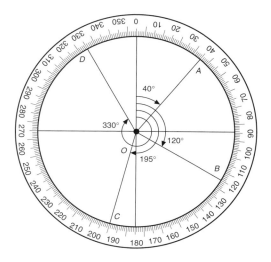

Fig. 5.29

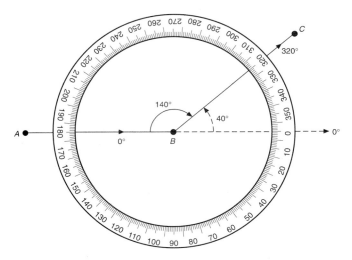

Fig. 5.30

WCB of AB =   0°
Angle ABC = 140°
————
Sum = 140°
+ 180°
————
WCB of BC = 320°
————

Similarly (*Figure 5.31*), if WCB of AB is 0° and the observed angle ABC is 220°, then the relative WCB of BC is 40°, i.e.

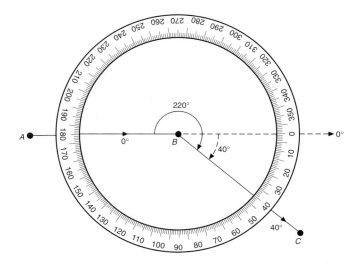

Fig. 5.31

$$
\begin{aligned}
\text{WCB of } AB = \quad & 0° \\
\text{Angle } ABC = \quad & 220° \\
\hline
\text{Sum} = \quad & 220° \\
& -180° \\
\hline
\text{WCB of } BC = \quad & 40°
\end{aligned}
$$

Occasionally, when subtracting 180°, the resulting WCB is still greater than 360°, in this case, one would need to subtract a further 360°. However, this problem is eliminated if the following rule is used.

Add the angle to the previous WCB:

If the *sum* < 180°, then *add* 180°
If the *sum* > 180°, then *subtract* 180°
If the *sum* > 540°, then *subtract* 540°

The application of this rule to traverse networks is shown in Chapter 6.

It should be noted that if both bearings are pointing out from *B*, then

WCB *BC* = WCB *BA* + angle *ABC*

as shown in *Figure 5.32*, i.e.

WCB *BC* = WCB *BA* (30°) + angle *ABC* (110°) = 140°

Having now obtained the WCB of a line and its horizontal distance (polar coordinates), it is possible to transform them to ΔE and ΔN, the rectangular coordinates. From *Figure 5.28*, it can clearly be seen that from right-angled triangle *ABC*:

$$\Delta E = D \sin \alpha \qquad\qquad\qquad (5.58a)$$
$$\Delta N = D \cos \alpha \qquad\qquad\qquad (5.58b)$$

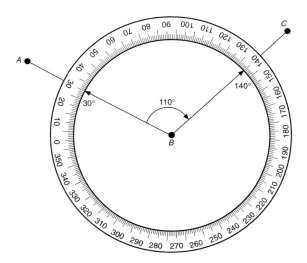

Fig. 5.32

where D = horizontal length of the line
$\quad\quad\alpha$ = WCB of the line
$\quad\quad\Delta E$ = difference in eastings of the line
$\quad\quad\Delta N$ = difference in northings of the line

It is important to appreciate that ΔE, ΔN define a *line*, whilst E, N define a *point*.
From the above basic equations can be derived the following:

$$\alpha = \tan^{-1}(\Delta E/\Delta N) = \cot^{-1}(\Delta N/\Delta E) \tag{5.59}$$

$$D = (\Delta E^2 + \Delta N^2)^{\frac{1}{2}} \tag{5.60}$$

$$D = \Delta E/\sin\alpha = \Delta N/\cos\alpha \tag{5.61}$$

In equation (5.59) it should be noted that the trigonometrical functions of tan and cot can become very unrealiable on pocket calculators as α approaches 0° (180°) and 90° (270°) respectively. To obviate this problem:

Use tan when $|\Delta N| > |\Delta E|$, and
use cot when $|\Delta N| < |\Delta E|$

Exactly the same situation prevails for sin and cos in equation (5.61); thus:

Use cos when $|\Delta N| > |\Delta E|$, and
use sin when $|\Delta N| < |\Delta E|$

The two most fundamental calculations in surveying are computing the 'polar' and the 'join'.

5.11.1 Computing the polar

Computing the polar for a line involves calculating ΔE, ΔN given the horizontal distance (D) and WCB (α) of the line.

Worked example

Example 5.3 Given the coordinates of A and the distance and bearing of AB, calculate the coordinates of point B.

$\quad\quad E_A$ = 48 964.38 m, N_A = 69 866.75 m, WCB AB = 299° 58′ 46″
$\quad\quad$Horizontal distance = 1325.64 m

From the WCB of AB, the line is obviously in the fourth quadrant and signs of ΔE, ΔN are (−, +) respectively. The pocket calculator will automatically provide the correct signs.

$$\Delta E_{AB} = D\sin\alpha = 1325.64 \sin 299° 58′ 46″ = -1148.28 \text{ m}$$

$$\Delta N_{AB} = D\cos\alpha = 1325.64 \cos 299° 58′ 46″ = +662.41 \text{ m}$$

$$\therefore E_B = E_A + \Delta E_{AB} = 48\,964.38 - 1148.28 = 47\,816.10 \text{ m}$$

$$N_B = N_A + \Delta N_{AB} = 69\,866.75 + 662.41 = 70\,529.16 \text{ m}$$

This computation is best carried out using the P (Polar) to R (Rectangular) keys of the pocket calculator. However, as these keys work on a pure math basis and not a surveying basis, one must know the order in which the data are input and the order in which the data are output.

The following methods apply to the majority of pocket calculators. However, as new types are being developed all the time, then individuals may have to adapt to their own specific make.

Using P and R keys;

(1) Enter *horizontal distance (D)*; press $\boxed{\text{P} \rightarrow \text{R}}$ or $(x \leftrightarrow y)$

(2) Enter WCB (α); press $\boxed{=}$ or (R)

(3) Value displayed is $\pm \Delta N$

(4) Press $\boxed{x \leftrightarrow y}$ to get $\pm \Delta E$

Operations in brackets are for an alternative type of calculator.

5.11.2 Computing the join

This involves computing the horizontal distance (D) and WCB (α) from the difference in coordinates $(\Delta E, \Delta N)$ of a line.

Worked examples

Example 5.4 Given the following coordinates for two points A and B, compute the length and bearing of AB.

$E_A = 48\ 964.38$ m	$N_A = 69\,866.75$ m
$E_B = 48\ 988.66$ m	$N_B = 62\,583.18$ m
$\Delta E_{AB} = 24.28$ m	$\Delta N_{AB} = -7\,283.57$ m

Note:

(1) A rough plot of the E, N of each point will show B to be south-east of A, and line AB is therefore in the second quadrant.

(2) If the direction is from A to B then:

$$\Delta E_{AB} = E_B - E_A$$
$$\Delta N_{AB} = N_B - N_A$$

If the required direction is B to A then:

$$\Delta E_{BA} = E_A - E_B$$
$$\Delta N_{BA} = N_A - N_B$$

(3) As $\Delta N > \Delta E$ use tan:

$$\alpha_{AB} = \tan^{-1}(\Delta E/\Delta N) = \tan^{-1}(24.28/-7283.57)$$
$$= -0°\ 11'\ 27''$$

It is obvious that as α_{AB} is in the second quadrant and must therefore have a WCB between 90° and 180°, and as we cannot have a negative WCB, $-0°\ 11'\ 27''$ is unacceptable. Depending on the signs of the coordinates entered into the pocket calculator, it will supply the angles as shown in *Figure 5.33*.

If in quadrant I + α_1 = WCB

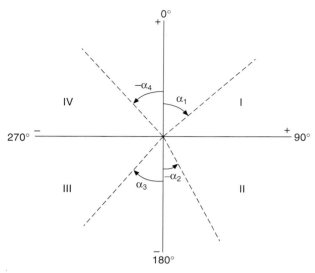

Fig. 5.33

If in quadrant II − α_2, then $(-\alpha_2 + 180) = $ WCB
If in quadrant III + α_3, then $(\alpha_3 + 180) = $ WCB
If in quadrant IV − α_4, then $(-\alpha_4 + 360) = $ WCB
∴ WCB $AB = -0° \ 11' \ 27'' + 180° = 179° \ 48' \ 33''$

Horizontal distance $AB = D_{AB} = (\Delta E^2 + \Delta N^2)^{\frac{1}{2}}$

$(24.28^2 + 7283.57^2)^{\frac{1}{2}} = 7283.61$ m

Also as $\Delta N > \Delta E$ use $D_{AB} = \Delta N/\cos \alpha = 7283.57/\cos 179° \ 48' \ 33'' = 7283.61$ m

Note what happens with some pocket calculators when $\Delta E/\sin \alpha$ is used:

$D_{AB} = \Delta E/\sin \alpha = 24.28/\sin 179° \ 48' \ 33'' = 7 \ 289.84$ m

This enormous error of more than 6 m proves that when computing distance it is advisable to use the Pythagoras equation $D = (\Delta E^2 + \Delta N^2)^{\frac{1}{2}}$, at all times. Of the remaining two equations, the appropriate one may be used as a check.

Using R and P keys:

(1) Enter ±ΔN; press $\boxed{\text{R} \rightarrow \text{P}}$ or (x ↔ y)

(2) Enter ±ΔE; press $\boxed{=}$ or (P)

(3) Value displayed is *horizontal distance (D)*

(4) Press $\boxed{\text{x} \leftrightarrow \text{y}}$ to obtain WCB in degrees and decimals

(5) If value in '4' is negative, *add* 360°

(6) Change to d.m.s. (°, ′, ″)

When using the P and R keys, the angular values displayed in the four quadrants are as in *Figure 5.34*; thus only a single 'IF' statement is necessary as in (5) above.

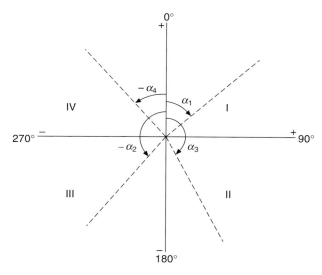

Fig. 5.34

Worked examples

Example 5.5 The national grid coordinates of two points, A and B, are A: E_A 238 824.076, N_A 307 911.843; and B: E_B 244 601.011, N_B 313 000.421

Calculate (1) The grid bearing and length of $\overrightarrow{AB}$.

(2) The azimuth of $\overrightarrow{BA}$ and $\overrightarrow{AB}$

(3) The ground length AB.

Given: (a) Mean latitude of the line = N 54° 00′.
(b) Mean altitude of the line = 250 m AOD.
(c) Local radius of the Earth = 6 384 100 m. (KU)

(1)

E_A = 238 824.076	N_A = 307 911.843
E_B = 244 601.011	N_B = 313 000.421
ΔE = 5776.935	ΔN = 5088.578

Grid distance $= (\Delta E^2 + \Delta N^2)^{\frac{1}{2}} = 7698.481$ m

Grid bearing $\overrightarrow{AB} = \tan^{-1} \dfrac{\Delta E}{\Delta N} = 48° \; 37′ \; 30″$

(2) In order to calculate the azimuth, i.e. the direction relative to true north, one must compute (a) the grid convergence at A and $B(\gamma)$ and (b) the $(t - T)$ correction at A and B (*Figure 5.35*).

(a) Grid convergence at $A = \gamma_A = \dfrac{\Delta E_A \tan \phi_m}{R}$

where ΔE_A = Distance from the central meridian

$= 400\,000 - E_A = 161\,175.924$ m

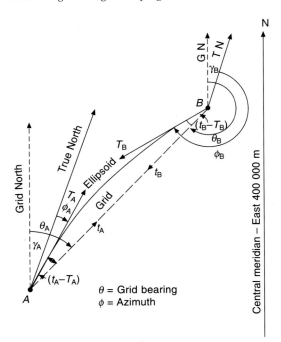

Fig. 5.35

$$\therefore \gamma_A'' = \frac{161\,176 \tan 54°}{6\,384\,100} \times 206\,265 = 7167'' = 1°\,59'\,27''$$

Similarly $\quad \gamma_B'' = \dfrac{155\,399 \tan 54°}{6\,384\,100} \times 206\,265 = 6911'' = 1°\,55'\,11''$

(b) $(t_A - T_A)'' = (2\Delta E_A + \Delta E_B)(N_A - N_B)K$
$$= 477.751 \times -5.089 \times 845 \times 10^{-6} = -2.05''$$

N.B. The eastings and northings are in km.

$(t_B - T_B)'' = (2\Delta E_B + \Delta E_A)(N_B - N_A)K$
$$= 471.974 \times 5.089 \times 845 \times 10^{-6} = +2.03''$$

Although the signs of the $(t - T)$ correction are obtained from the equation the student is advised always to draw a sketch of the situation.

Referring to *Figure 5.35*

Azimuth $\overrightarrow{AB} = \phi_A = \theta_A - \gamma_A - (t_A - T_A)$
$$= 48°\,37'\,30'' - 1°\,59'\,27'' - 02'' = 46°\,38'\,01''$$

Azimuth $\overrightarrow{BA} = \phi_B = \theta_B - \gamma_B - (t_B - T_B)$
$$= (48°\,37'\,30'' + 180°) - 1°\,55'\,11'' + 2''$$
$$= 226°\,42'\,21''$$

(3) To obtain ground length from grid length one must obtain the LSF adjusted for altitude.

$$\text{Mid-easting of } AB = 241712.544 \text{ m} = E$$
$$\text{LSF} = 0.999\,601 + [1.228 \times 10^{-14} \times (E - 400\,000)^2] = F$$
$$\therefore F = 0.999\,908$$

The altitude is 250 m OD, i.e. $H = +250$. LSF F_a adjusted for altitude is

$$F_a = F\left(1 - \frac{H}{R}\right) = 0.999\,908\left(1 - \frac{250}{6\,384\,100}\right) = 0.999\,869$$

$$\therefore \text{ Ground length } AB = \text{grid length}/F_a$$
$$\therefore AB = 7698.481/0.999\,869 = 7699.483 \text{ km}$$

Example 5.6 As part of the surveys required for the extension of a large underground transport system, a baseline was established in an existing tunnel and connected to the national grid via a wire correlation in the shaft and precise traversing therefrom.

Thereafter, the azimuth of the base was checked by gyro-theodolite using the reversal point method of observation as follows:

Reversal points	Horizontal circle readings			Remarks
	°	′	″	
r_1	330	20	40	Left reversal
r_2	338	42	50	Right reversal
r_3	330	27	18	Left reversal
r_4	338	22	20	Right reversal

$$\text{Horizontal circle reading of the baseline} = 28° \, 32' \, 46''$$
$$\text{Grid convergence} = 0° \, 20' \, 18''$$
$$(t - T) \text{ correction} = 0° \, 00' \, 04''$$
$$\text{NG easting of baseline} = 500\,000 \text{ m}$$

Prior to the above observations, the gyro-theodolite was checked on a surface baseline of known azimuth. The following mean data were obtained.

$$\text{Known azimuth of surface base} = 140° \, 25' \, 54''$$
$$\text{Gyro azimuth of surface base} = 141° \, 30' \, 58''$$

Determine the national grid bearing of the underground baseline. (KU)

Refer to Chapter 10 for information on the gyro-theodolite.
Using Schuler's mean

$$N_1 = \tfrac{1}{4}(r_1 + 2r_2 + r_3) = 334° \, 33' \, 24''$$
$$N_2 = \tfrac{1}{4}(r_2 + 2r_3 + r_4) = 334° \, 29' \, 54''$$
$$\therefore N = (N_1 + N_2)/2 = 334° \, 31' \, 39''$$

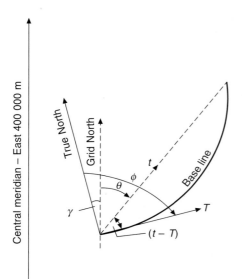

Fig. 5.36

Horizontal circle reading of the base = 28° 32′ 46″

∴ Gyro azimuth of the baseline = 28° 32′ 46″ − 334° 31′ 39″

= 54° 01′ 07″

However, observations on the surface base show the gyro-theodolite to be 'over-reading' by (141° 30′ 58″ − 140° 25′ 54″) = 1° 05′ 04″.

∴ True azimuth of baseline ϕ = gyro azimuth − Instrument constant

= 54° 01′ 07″ − 1° 05′ 04″

= 52° 56′ 03″

Now by reference to *Figure 5.36*, the sign of the correction to give the NG bearing can be seen, i.e.

Azimuth ϕ = 52° 56′ 03″

Grid convergence γ = −0° 20′ 18″

$(t - T)$ = −0° 00′ 04″

∴ NG bearing θ = 52° 35′ 41″

Exercises

(5.1) Explain the meaning of the term 'grid convergence'. Show how this factor has to be taken into account when running long survey lines by theodolite.

From a point *A* in latitude 53° N, longitude 2° W, a line is run at right angles to the initial meridian for a distance of 31 680 m in a westernly direction to point *B*.

Calculate the true bearing of the line at *B*, and the longitude of that point. Calculate also the bearing and distance from *B* of a point on the meridian of *B* at the same latitude as the starting point *A*. The radius of the Earth may be taken as 6273 km. (LU)

(Answer: 269° 37′ 00″; 2° 28′ 51″ W; 106.5 m)

(5.2) Two points, *A* and *B*, have the following coordinates:

	Latitude				Longitude			
	°	′	″		°	′	″	
A	52	21	14	N	93	48	50	E
B	52	24	18	N	93	42	30	E

Given the following values:

Latitude	1″ of latitude	1″ of longitude
52° 20′	30.423 45 m	18.638 16 m
52° 25′	30.423 87 m	18.603 12 m

find the azimuths of *B* from *A* and of *A* from *B*, also the distance *AB*.　　(LU)

(*Answer*: 308° 23′ 36″, 128° 18′ 35″, 9021.9 m)

(5.3) At a terminal station *A* in latitude N 47° 22′ 40″, longitude E 0° 41′ 10″, the azimuth of a line *AB* of length 29 623 m was 23° 44′ 00″,

　　Calculate the latitude and longitude of station *B* and the reverse azimuth of the line from station *B* to the nearest second.　　(LU)

Latitude	1″ of longitude	1″ of latitude
47° 30′	20.601 m	30.399
47° 35′	20.568 m	30.399

(*Answer*: N 47° 37′ 32″; E 0° 50′ 50″; 203° 51′ 08″)

REFERENCES

Ashkenazi, V. (1988) 'Co-ordinate Datums and Applications', Seminar on GPS, The University of Nottingham, April 1988.

Bomford, G. (1982) *Geodesy*, 4th edn, Oxford University Press, London.

Cross, P.A., Hollwey, J.R. and Small, L. (1985) 'Geodetic Appreciation', NELP Working Paper No. 2.

Jackson, J.E. (1982) *Sphere, Spheroid and Projections for Surveyors,* Granada Publishing Ltd, Herts.

Schofield, W. (1984) *Engineering Surveying*, Vol. 2, Butterworths, London.

6
Control surveys

A control survey provides a framework of survey points, whose relative positions, in two or three dimensions, are known to prescribed degrees of accuracy. The areas covered by these points may extend over a whole country and form the basis for the national maps of that country. Alternatively the area may be relatively small, encompassing a construction site for which a large-scale plan is required. Although the areas covered in construction are usually quite small, the accuracy may be required to a very high order. The types of engineering project envisaged are the construction of long tunnels and/or bridges, deformation surveys for dams and reservoirs, three-dimensional tectonic ground movement for landslide prediction, to name just a few. Hence control networks provide a reference framework of points for:

(1) Topographic mapping and large-scale plan production.
(2) Dimensional control of construction work.
(3) Deformation surveys for all manner of structures, both new and old.
(4) The extension and densification of existing control networks.

The methods of establishing the vertical control have already been discussed in Chapter 2, so only two-dimensional horizontal control will be dealt with here. Elements of geodetic surveying have been dealt with in Chapter 5 and so plane surveying for engineering control will be concentrated upon.

The methods used for control surveys are:

(1) Traversing.
(2) Triangulation.
(3) Trilateration.
(4) A combination of (2) and (3), sometimes referred to as triangulateration.
(5) Satellite position fixing (see Chapter 7).
(6) Inertial position fixing.

Whilst the above systems establish a network of points, single points may be fixed by intersection and/or resection.

6.1 TRAVERSING

Since the advent of EDM equipment, traversing has emerged as the most popular method of establishing control networks not only in engineering surveying but also in geodetic work. In underground mining it is the only method of control applicable whilst in civil engineering it lends itself ideally to surveys and dimensional control of route-type projects such as highway and pipeline construction.

Traverse networks are, to a large extent, free of the limitations imposed on the other systems and, compared with them, have the following advantages:

(1) Much less reconnaissance and organization required in establishing a single line of easily accessible stations compared with the laying out of well-conditioned geometric figures.
(2) In conjunction with (1), the limitations imposed on the other systems by topographic conditions do not apply to traversing.
(3) The extent of observations to only two stations at a time is relatively small and flexible compared with the extensive angular and/or linear observations at stations in the other systems. It is thus much easier to organize.
(4) Traverse networks are free of the strength of figure considerations so characteristic of triangular systems. Thus once again the organizational requirements are reduced.
(5) Scale error does not accrue as in triangulation, whilst the use of longer sides, easily measured with EDM equipment, reduces azimuth swing errors.
(6) Traverse stations can usually be chosen so as to be easily accessible, as well as convenient for the subsequent densification of lower order control.
(7) Traversing permits the control to closely follow the route of a highway, pipline or tunnel, etc., with the minimum number of stations.

From the accuracy point of view it has been shown (Chrzanowski and Konecny, 1965; Adler and Schmutter, 1971) that traverse is superior to triangulation and trilateration and, in some instances, even to triangulateration. However, it must be said that these findings are disputed by Phillips (1967). Nevertheless, it can be argued that, from the accuracy point of view, traversing compares more than favourably with the other methods.

Thus, from a consideration of all the above statements it is obvious that from the logistical point of view, traversing is far superior to the other methods and offers at least equivalent accuracy.

6.1.1 Types of traverse

Using the technique of traversing, the relative position of the control points is fixed by measuring the horizontal angle at each point, subtended by the adjacent stations, and the horizontal distance between consecutive pairs of stations.

The procedures for measuring angles have been dealt with in Chapter 4 and for measuring distance in Chapter 3. The majority of traverses carried out today would most probably capture the field data using a total station. Hence the distance is measured by EDM and the angles by digital processes. Occasionally, steel tapes may be used for distance.

The liability of a traverse to undetected error makes it essential that there should be some external check on its accuracy. To this end the traverse may commence from and connect into known points of greater accuracy than the traverse. In this way the error vector of misclose can be quantified and distributed throughout the network, to produce geometric correctness. Such a traverse is called a 'link' traverse.

Alternatively, the error vector can be obtained by completing the traverse back to its starting origin. Such a traverse is called a 'polygonal' or 'loop' traverse. Both the 'link' and 'polygonal' traverse are generally referred to as 'closed' traverses.

The third type of traverse is the 'free' or 'open' traverse, which does not close back onto any known point and which therefore has no way of detecting or quantifying the errors.

(1) Link traverse

Figure 6.1 illustrates a typical link traverse commencing from higher order point *Y* and closing onto

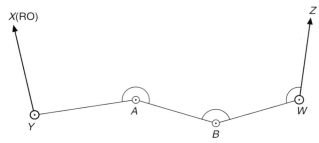

Fig. 6.1   *Link traverse*

point *W*, with terminal orienting bearing to points *X* and *Z*. Generally, points *X*, *Y*, *W* and *Z* would be part of a higher order control network, although this may not always be the case. It may be that when tying surveys into the OSNG, due to the use of very precise EDM equipment the intervening traverse is more precise than the relative positions of the NG stations. This is purely a problem of scale arising from a lack of knowledge, on the behalf of the surveyor, of the positional accuracy of the grid points. In such a case, adjustment of the traverse to the NG could result in distortion of the intervening traverse.

The usual form of an adjustment generally adopted in the case of a link traverse is to hold points *Y* and *W* fixed whilst distributing the error throughout the intervening points. This implies that points *Y* and *W* are free from error and is tantamount to allocating a weight of infinity to the length and bearing of line *YW*. It is thus both obvious and important that the control into which the traverse is linked should be of a higher order of precision than the connecting traverse.

The link traverse has certain advantages over the remaining types, in that systematic error in distance measurement and orientation are clearly revealed by the error vector.

(2) Polygonal traverse

Figures 6.2 and *6.3* illustrate the concept of a polygonal traverse. This type of network is quite popular and is used extensively for peripheral control on all types of engineering sites. If no orientation facility is available, the control can only be used for independent sites and plans and cannot be connected to other survey systems.

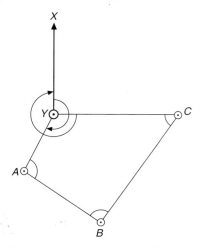

Fig. 6.2   *Loop traverse (oriented)*

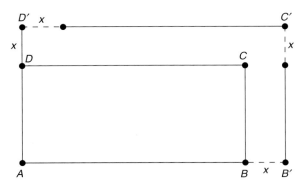

Fig. 6.3 *Loop traverse (independent)*

In this type of traverse the systematic errors of distance measurement are not eliminated and enter into the result with their full weight. Similarly, orientation error would simply cause the whole network to swing through the amount of error involved and would not be revealed in the angular misclosure.

This is illustrated in *Figures 6.4* and *6.5*. In the first instance a scale error equal to X is introduced into each line of a rectangular-shaped traverse *ABCD*. Then, assuming the angles are error free, the traverse appears to close perfectly back to *A*, regardless of the totally incorrect coordinates which would give the position of *B, C* and *D* as *B', C'* and *D'*.

Figure 6.5 shows the displacement of *B, C* and *D* to *B', C'* and *D'* caused by an orientation error (θ) for *AB*. This can occur when *AB* may be part of another network and the incorrect value is taken for its bearing. The traverse will still appear to close correctly, however. Fortunately, in this particular case, the coordinates of *B* would be known and would obviously indicate some form of mistake when compared with *B'*.

(3) Open (or free) traverse

Figure 6.6 illustrates the open traverse which does not close into any known point and therefore cannot provide any indication of the magnitude of measuring errors. In all surveying literature, this form of traversing is not recommended due to the lack of checks. Nevertheless, it is frequently utilized in mining and tunnelling work because of the physical restriction on closure.

Fig. 6.4

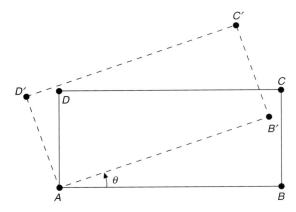

Fig. 6.5

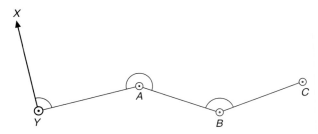

Fig. 6.6 *Open (or free) traverse*

6.1.2 Reconnaissance

Reconnaissance is a vitally important part of any survey project, as emphasized in Chapter 1. Its purpose is to decide the best location for the traverse points.

In the first instance the points should be intervisible from the point of view of traverse observations.

If the purpose of the control network is the location of topographic detail only, then they should be positioned to afford the best view of the terrain, thereby ensuring that the maximum amount of detail can be surveyed from each point.

If the traverse is to be used for setting out, say, the centre-line of a road, then the stations should be sited to afford the best positions for setting out the intersection points (IPs) and tangent points (TPs), to provide accurate location.

The distance between stations should be kept as long as possible to minimize effect of centring errors.

Finally, as cost is always important, the scheme should be one that can be completed in the minimum of time, with the minimum of personnel.

The type of survey station used will also be governed by the purpose of the traverse points. If they are required as control for a quick, one-off survey of a small area, then wooden pegs about 0.25 m long and driven down to ground level may suffice. A fine point on the top of the peg defines the control point. Alternatively, long-life stations may require construction or some form of commercially available station. *Figure 6.7* shows the type of survey station recommended by the Department of Transport (UK) for major road projects. They are recommended to be placed at 250-m intervals and remain stable for at least five years. *Figure 6.8* shows a type of station commercially produced by

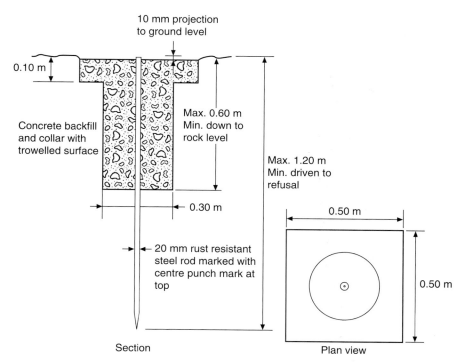

Fig. 6.7

Earth Anchors Ltd, England. Road, masonry or hilti nails may be used on paved or black-topped surfaces.

6.1.3 Sources of error

The sources of error in traversing are:

(1) Errors in the observation of horizontal and vertical angles (angular error).
(2) Errors in the measurement of distance (linear error).
(3) Errors in the accurate centring of the instrument and targets, directly over the survey point (centring error).

Linear and angular errors have been fully dealt with in Chapters 3 and 4 respectively.

Centring errors were dealt with in Chapter 4 also, but only insofar as they affected the measurement of a single angle. Their propagation effects through a traverse will now be examined.

In precise traversing the effect of centring errors will be greater than the effect of reading errors if appropriate procedures are not adopted to eliminate them. As already illustrated in Chapter 4, the effect on the angular measurements increases with decreasing lengths of the traverse legs (*Figure 6.9*).

The inclusion of short lines cannot be avoided in many engineering surveys, particularly in underground tunnelling work. In order, therefore, to minimize the propagational effect of centring error, a constrained centring system called the three-tripod system (TTS) is used.

The TTS uses interchangeable levelling heads or tribrachs and targets, and works much more efficiently with a fourth tripod (*Figures 6.10* and *6.11*).

Consider *Figure 6.12*. Tripods are set up at *A*, *B*, *C* and *D* with the detachable tribrachs carefully

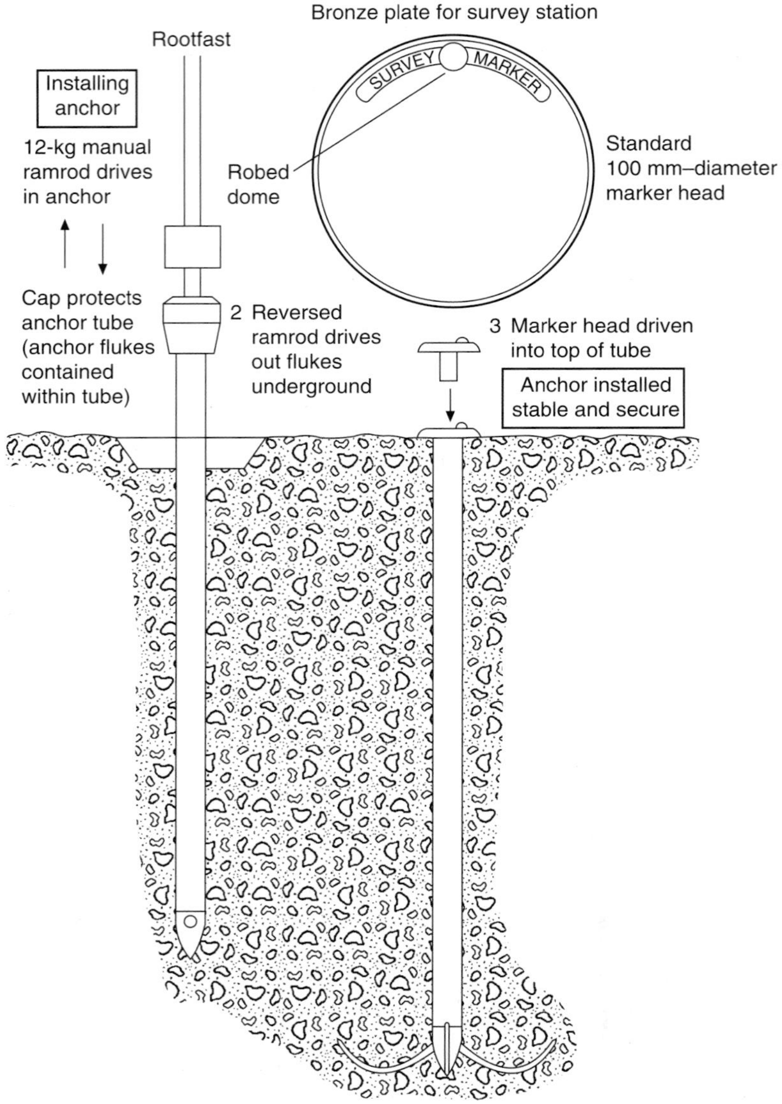

Fig. 6.8

levelled and centred over each station. Targets are clamped into the tribrachs at A and C, whilst the theodolite is clamped into the one at B. When the angle ABC has been measured, the target (T_1) is clamped into the tribrach at B, the theodolite into the tribrach at C and that target into the tribrach at D. Whilst the angle BCD is being measured, the tripod and tribrach are removed from A and set up at E in preparation for the next move forward. This technique not only produces maximum speed and efficiency, but also confines the centring error to the station at which it occurred. Indeed, the error in question here is not one of centring in the conventional sense, but one of knowing whether or not the central axes of the targets and theodolite, when moved forward, occupy exactly the same positions as did their previous occupants.

The confining of centring errors using the above system can be explained by reference to

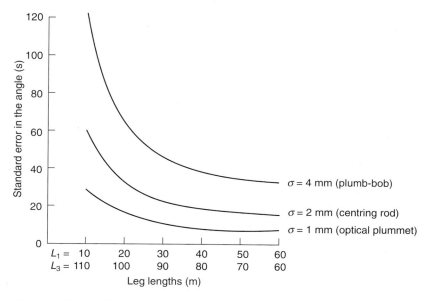

Fig. 6.9 *The combined effect of target and theodolite centring errors*

Fig. 6.10 *Interchangeable target and tribrach*

Figure 6.13. Consider first the use of the TTS. The target erected at *C*, 100 m from *B*, is badly centred, resulting in a displacement of 50 mm to *C′*. The angle measured at *B* would be *ABC′* in error by *e*. The error is 1 in 2000 ≈ 2 min. (*N.B.* If *BC* was 10 m long, then *e* = 20 min.)

The target is removed from *C′* and replaced by the theodolite, which measures the angle *BC′D*, thus bringing the survey back onto *D*. The only error would therefore be a coordinate error at *C* equal to the centring error and would obviously be much less than the exaggerated 50 mm used here.

Fig. 6.11 *Interchangeable theodolite and tribrach*

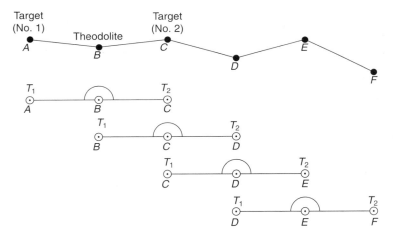

Fig. 6.12 *Conventional three-tripod system*

Consider now conventional equipment using one tripod and theodolite and sighting to ranging rods. Assume that the rod at C, due to bad centring or tilting, appears to be at C'; the wrong angle, ABC', would be measured. Now, when the *theodolite* is moved it would this time be correctly centred over the station at C and the correct angle BCD measured. However, this correct angle would be added to the previous computed bearing, which would be that of BC', giving the bearing $C'D'$. Thus the error e is propagated from the already incorrect position at C', producing a further error at D' of the traverse. Centring of the instrument and targets precisely over the survey stations is thus of paramount importance; hence the need for constrained centring systems in precise traversing.

It can be shown that if the theodolite and targets are re-centred with an error of ±0.3 mm, these

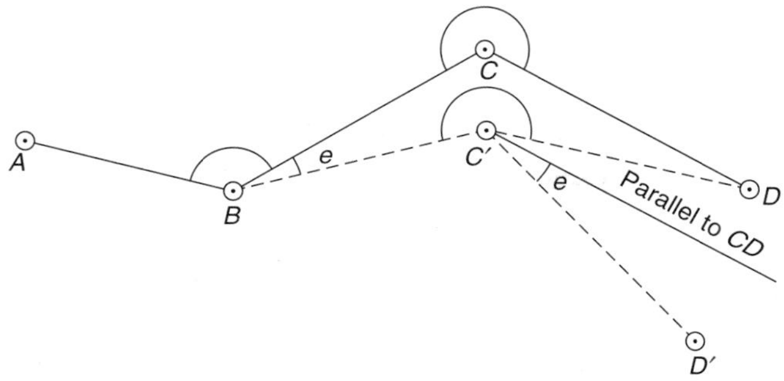

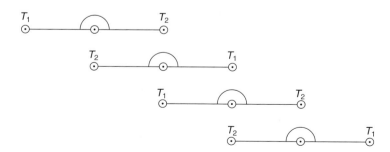

Fig. 6.13 *Propagation of centring error*

centring errors alone will produce an error of ±6″ in bearing after 1500 m of traversing with 100-m sights. If the final bearing is required to ±2″, the component caused by centring must be limited to about one-third of the total component, that is ±0.6″, and would therefore require centring errors in the region of ±0.03 mm. Thus in general a mean error ±0.1 mm would be compatible with a total mean error of ±6″ in the final bearing of the above traverse. This therefore imposes a very rigorous standard on constrained centring systems.

Many different systems of constrained centring are available which have been investigated (Berthon-Jones, 1972) to obtain a knowledge of the expected accuracy.

When considering the errors of constrained centring, the relevant criterion is the repeatability of the system and not the absolute accuracy of centring over a ground mark. One is concerned with the degree to which the vertical axis of the theodolite placed in a tribrach coincides with the vertical through the centre of the target previously occupying the tribrach.

The error sources identified were:

(1) The aim mark of the target, eccentric to the vertical axis.
(2) The vertical axis, eccentric to the centring axis as these are separate components.
(3) Variations in clamping pressures.
(4) Tolerance on fits, which is essentially a manufacturing problem.

Although the accuracy of replacement is generally quoted as ±0.1 mm, the above investigation showed errors greater than this in the majority of systems tested. A variation on the conventional tripod system is therefore recommended, as shown in *Figure 6.14*, which reduced the errors by a factor of *n*, equal to the number of traverse stations.

Fig. 6.14 *Suggested three-tripod system*

6.1.4 *Traverse computation*

The various steps in traverse computation will now be carried out, with reference to the traverse shown in *Figure 6.15*. The observed horizontal angles and distances are shown in columns 2 and 7 of *Table 6.1*.

A common practice is to assume coordinate values for a point in the traverse, usually the first station, and allocate an arbitrary bearing for the first line from that point. For instance, in *Figure 6.15*, point A has been allocated coordinates of E 0.00, N0.00, and line AB a bearing of 0°00′ 00″. This has the effect of establishing a plane rectangular grid and orientating the traverse on it. As shown, AB becomes the direction of the N-axis, with the E-axis at 90° and passing through the grid origin at A.

The computational steps, in the order in which they are carried out, are:

(1) Obtain the angular misclosure W, by comparing the sum of the observed angles (α) with the sum of error-free angles in a geometrically correct figure.
(2) Assess the acceptability or otherwise of W.
(3) If W is acceptable, distribute it throughout the traverse in equal amounts to each angle.
(4) From the corrected angles compute the whole circle bearing of the traverse lines relative to AB.
(5) Compute the coordinates (ΔE, ΔN) of each traverse line.
(6) Assess the coordinate misclosure ($\Delta'E$, $\Delta'N$).
(7) Balance the traverse by distributing the coordinate misclosure throughout the traverse lines.
(8) Compute the final coordinates (E, N) of each point in the traverse relative to A, using the balanced values of ΔE, ΔN per line.

The above steps will now be dealt with in detail.

(1) *Distribution of angular errror*

On the measurement of the horizontal angles of the traverse, the majority of the systematic errors are eliminated by repeated double-face observation. The remaining random errors are distributed equally around the network, as follows.

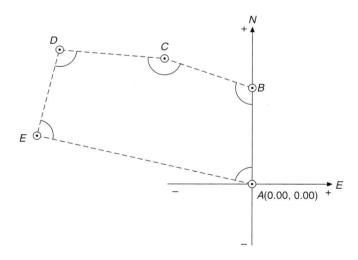

Fig. 6.15   *Polygonal traverse*

Table 6.1 Bowditch Adjustment

Angle	Observed horizontal angle °	'	"	Corr. "	Corrected horizontal angle °	'	"	Line	W.C.B. °	'	"	Horiz. Length m	Difference in coordinates ΔE	ΔN	δE	δN	Corrected Values ΔE	ΔN	Final Values E	N	Pt.
(1)	(2)			(3)	(4)			(5)	(6)			(7)	(8)	(9)	(10)	(11)	(12)	(13)	(14)	(15)	(16)
ABC	120	25	50	+10	120	26	00	AB	000	00	00 (Assumed)	155.00	0.00	155.00	0.07	0.10	0.07	155.10	0.00	0.00	A
BCD	149	33	50	+10	149	34	00	BC	300	26	00	200.00	−172.44	101.31	0.09	0.13	−172.35	101.44	0.07	155.10	B
CDE	095	41	50	+10	095	42	00	CD	270	00	00	249.00	−249.00	0.00	0.11	0.17	−248.89	0.17	−172.28	256.54	C
DEA	093	05	50	+10	093	06	00	DE	185	42	00	190.00	−18.87	−189.06	0.08	0.13	−18.79	−188.93	−421.17	256.71	D
EAB	081	11	50	+10	081	12	00	EA	098	48	00	445.00	439.76	−68.08	0.20	0.30	439.96	−67.78	−439.96	67.78	E
								AB	000	00	00								0.00	0.00	A
Sum (2n−4)90°	539 / 540	59 / 00	10 / 00	+50	540	00	00					ΣL = 1239.00	−0.55	−0.83	0.55	0.83	0.00 Sum	0.00 Sum			
Error	−50											Correction =	+0.55 ΔE	+0.83 ΔN							

Error Vector = $(0.55^2 + 0.83^2)^{\frac{1}{2}} = 0.99\,(213°\ 32')$

Accuracy = 1/1252

In a polygon the sum of the *internal* angles should equal $(2n - 4)90°$, the sum of the *external* angles should equal $(2n + 4)90°$.

$$\therefore \text{Angular misclosure} = W = \sum_{i=1}^{n} \alpha_i - (2n \pm 4)90° = -50'' \text{ (Table 6.1)}$$

where α = mean observed angle
$\quad\quad n$ = number of angles in the traverse

The angular misclosure W is now distributed by equal amounts on each angle, thus:

$$\text{Correction per angle} = W/n = +10'' \text{ (Table 6.1)}$$

However, before the angles are corrected, the angular misclosure W must be considered to be acceptable. If W was too great, and therefore indicative of poor observations, the whole traverse may need to be re-measured. A method of assessing the acceptability or otherwise of W is given in *Section 2*.

(2) Acceptable angular misclosure

The following procedure may be adopted provided that there is evidence of the variance of the mean observed angles, i.e.

$$\sigma_w^2 = \sigma_{a1}^2 + \sigma_{a2}^2 + \ldots + \sigma_{an}^2$$

where σ_{an}^2 = variance of the mean observed angle
$\quad\quad \sigma_w^2$ = variance of the sum of the angles of the traverse

Assuming that each angle is measured with equal precision:

$$\sigma_{a1}^2 = \sigma_{a2}^2 = \ldots = \sigma_{an}^2 = \sigma_A^2$$

then $\sigma_w^2 = n \cdot \sigma_A^2$ and

$$\sigma_w = n^{\frac{1}{2}} \cdot \sigma_A \tag{6.1}$$

$$\text{Angular misclosure} = W = \sum_{i=1}^{n} \alpha_i - [(2n \pm 4)90°]$$

where α = mean observed angle
$\quad\quad n$ = number of angles in traverse

then for 95% confidence

$$P(-1.96\sigma_w < W < +1.96\alpha_w) = 0.95 \tag{6.2}$$

and for 99.73% confidence

$$P(-3\sigma_w < W < +3\sigma_w) = 0.9973 \tag{6.3}$$

For example, consider a closed traverse of nine angles. Tests prior to the survey showed that the particular theodolite used had a standard error (σ_A) of $\pm 3''$. What would be considered an acceptable angular misclosure for the traverse?

$$\sigma_w = 9^{\frac{1}{2}} \cdot 3'' = \pm 9''$$

$$P(-1.96 \times 9'' < W < +1.96 \times 9'') = 0.95$$

$$P(-18'' < W < +18'') = 0.95$$

Similarly $P(-27'' < W < +27'') = 0.9973$

Thus, if the angular misclosure W is greater than $\pm18''$ there is evidence to suggest unacceptable error in the observed angles, provided the estimate for σ_A is reliable. If W exceeds $\pm27''$ there is definitely angular error present of such proportions as to be quite unacceptable.

Research has shown that a reasonable value for the standard error of the mean of a double face observation is about 2.5 times the least count of the instrument. Thus for a 1-sec theodolite:

$$\sigma_A = \pm2.5''$$

Assuming the theodolite used in the traverse of *Figure 6.15* had a least count of 10'':

$$\sigma_w = 5^{\frac{1}{2}} \times 25'' = \pm56''$$

95% confidence:

$$P(-110'' < W < 110'') = 0.95$$

Thus as the angular misclosure is less than $+110''$, the traverse computation may proceed and after the distribution of the angular error, the WCBs are computed.

(3) Whole circle bearings (WCB)

The concept of WCBs has been dealt with in Chapter 5 and should be referred to for the 'rule' that is adopted. The corrected angles will now be changed to WCBs relative to *AB* using that rule.

	Degree	Minute	Second
WCB *AB*	000	00	00
Angle *ABC*	120	26	00
Sum	120 +180	26	00
WCB *BC*	300	26	00
Angle *BCD*	149	34	00
Sum	450 −180	00	00
WCB *CD*	270	00	00
Angle *CDE*	95	42	00
Sum	365 −180	42	00
WCB *DE*	185	42	00
Angle *DEA*	93	06	00
Sum	278 −180	48	00
WCB *EA*	98	48	00
Angle *EAB*	81	12	00
Sum	180 −180	00	00
WCB *AB*	000	00	00 (Check)

(4) Plane rectangular coordinates

Using the observed distance, reduced to the horizontal, and the bearing of the line, transform this data (polar coordinates) to rectangular coordinates for each line of the traverse. This may be done using the basic formula (Chapter 5):

$$\Delta E = L \sin \text{WCB}$$

$$\Delta N = L \cos \text{WCB}$$

or the P → R keys on a pocket calculator. The results are shown in columns 8 and 9 of *Table 6.1*.

As the traverse is a closed polygonal, starting from and ending on point A, the respective algebraic sums of the ΔE and ΔN values would equal zero if there was no observational error present. However, as shown, the error in $\Delta E = -0.55$ m and in $\Delta N = -0.83$ m and is 'the coordinate misclosure'. As the correction is always of opposite sign to the error, i.e.

$$\text{Correction} = -\text{Error} \tag{6.4}$$

then the ΔE values must be corrected by $+0.55 = \Delta'E$ and the ΔN values by $+0.83 = \Delta'N$. The situation is as shown in *Figure 6.16*, where the resultant amount of misclosure AA' is called the 'error vector'. This value, when expressed in relation to the total length of the traverse, is used as a measure of the precision of the traverse.

For example:

$$\text{Error vector} = (\Delta'E^2 + \Delta'N^2)^{\frac{1}{2}} = 0.99 \text{ m}$$

$$\text{Accuracy of traverse} = 0.99/1239 = 1/1252$$

(The error vector can be computed using the R → P keys.)

(5) Balancing the traversing

Balancing the traverse, sometimes referred to as 'adjusting' the traverse, involves distributing $\Delta'E$ and $\Delta'N$ throughout the traverse in order to make it geometrically correct.

There is no ideal method of balancing and a large variety of procedures are available, ranging from the very elementary to the much more rigorous. Where a non-rigorous method is used, the most popular procedure is to use the *Bowditch rule*.

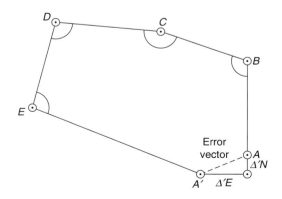

Fig. 6.16   *Coordinate misclosure*

The 'Bowditch rule' was devised by Nathaniel Bowditch, surveyor, navigator and mathematician, as a proposed solution to the problem of compass traverse adjustment, which was posed in the American journal *The Analyst* in 1807.

The Bowditch rule is as follows:

$$\delta E_i = \frac{\Delta'E}{\sum\limits_{i=1}^{n} L_i} \cdot L_i = K_1 \cdot L_i \qquad (6.5)$$

and

$$\delta N_i = \frac{\Delta'N}{\sum\limits_{i=1}^{n} L_i} \cdot L_i = K_2 \cdot L_i \qquad (6.6)$$

where δE_i, δN_i = the coordinate corrections

$\Delta'E$, $\Delta'N$ = the coordinate misclosure (constant)

$\sum\limits_{i=1}^{n} L_i$ = the sum of the lengths of the traverse (constant)

L_i = the horizontal length of the ith traverse leg

K_1, K_2 = the resultant constants

From equations (6.5) and (6.6), it can be seen that the corrections made are simply in proportion to the length of the line.

The correction for each length is now computed in order.

For the first line *AB*:

$$\delta E_1 = (\Delta'E/\Sigma L)L_1 = K_1 \cdot L_1$$

where $K_1 = +0.55/1239 = 4.4 \times 10^{-4}$

$\therefore \delta E_1 = (4.4 \times 10^{-4})\,155.00 = +0.07$

Similarly for the second line *BC*:

$$\delta E_2 = (4.4 \times 10^{-4})200.00 = +0.09$$

and so on:

$$\delta E_3 = (4.4 \times 10^{-4})\,249.00 = +0.11$$
$$\delta E_4 = (4.4 \times 10^{-4})\,190.00 = +0.08$$
$$\delta E_5 = (4.4 \times 10^{-4})\,445.00 = +0.20$$

$$\text{Sum} = +0.55 \text{ (Check)}$$

Similarly for the ΔN value of each line:

$$\delta N_1 = (\Delta'N/\Sigma L)L_1 = K_2 L_1$$

where $K_2 = +0.83/1239 = 6.7 \times 10^{-4}$

$\therefore \delta N_1 = (6.7 \times 10^{-4})155.00 = +0.10$

and so on for each line:

$$\delta N_2 = +0.13$$
$$\delta N_3 = +0.17$$
$$\delta N_4 = +0.13$$
$$\delta N_5 = +0.30$$

Sum   +0.83   (Check)

These corrections (as shown in columns 10 and 11 of *Table 6.1*) are added algebraically to the values ΔE, ΔN in columns 8 and 9 to produce the balanced values shown in columns 12 and 13.

The final step is to algebraically add the values in columns 12 and 13 to the coordinates of A to produce the coordinates of each point in turn, relative to A, the origin (as shown in the final three columns).

6.1.5 Link traverse adjustment

A link traverse (*Figure 6.17*) commences from known stations (*AB*) and connects into known stations (*CD*). Stations A, B, C and D are usually fixed to a higher order of accuracy, with values which remain unaltered in the subsequent computation. The method of computation and adjustments proceeds as follows:

(1) Angular adjustment

(1) Compute the WCB of *CD* through the traverse from *AB* and compare it with the known bearing of *CD*. The difference (Δ) of the two bearings is the angular misclosure.
(2) As a check on the value of Δ the following rule may be applied. Computed WCB of *CD* = (sum of observed angles + initial bearing (*AB*)) – $n \times 180°$ where n is the number of angles and is positive if even, negative if odd. If the result is negative, add 360°.
(3) The correction per angle would be Δ/n, which is distributed accumulatively over the WCB as shown in the example.

(2) Coordinate adjustment

(1) Compute the initial coordinates of C through the traverse from E as origin. Comparison with the known coordinates of C gives the coordinate misclosure $\Delta'E$, and $\Delta'N$.
(2) As the computed coordinates are total values, distribute the misclosure accumulatively over stations E_1 to C.

Now study the example given in *Table 6.2*.

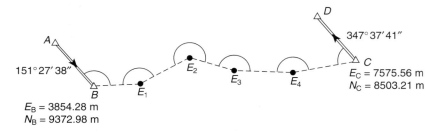

Fig. 6.17

6.1.6 *The effect of the balancing procedure*

The purpose of this section is to show that balancing a traverse does not in any way improve it; it simply makes the figure geometrically correct.

The survey stations set in the ground represent the 'true' traverse, which in practice is unknown. Observation of the angles and distances is an attempt to obtain the value of the true traverse. It is never achieved, due to observational error, and hence we have an 'observed' traverse, which may approximate very closely to the 'true', but is not geometrically correct, i.e. there is coordinate misclosure. Finally, we have the 'balanced' traverse after the application of the Bowditch rule. This traverse is now geometrically correct, but in the majority of cases will be significantly different from both the 'true' and 'observed' network.

As field data are generally captured to the highest accuracy possible, relative to the expertise of the surveyor and the instrumentation used, it could be argued that the best balancing process is that which alters the field data the least.

Basically the Bowditch rule adjusts the positions of the traverse stations, resulting in changes to the observed data. For instance, it can be shown that the changes to the angles will be equal to:

$$\delta\alpha_i = 2\cos\frac{\alpha_i}{2}\ (\Delta'E\cos\beta + \Delta'N\sin\beta)/\Sigma L \qquad (6.7a)$$

where β is the mean bearing of the lines subtending the angle α. This does not, however, apply to the first and last angle, where the corrections are

$$\delta\alpha_1 = -(\Delta'E\sin\beta_1 + \Delta'N\cos\beta_1)/\Sigma L \qquad (6.7b)$$
$$\delta\alpha_n = +(\Delta'E\sin\beta_n + \Delta'N\cos\beta_n)/\Sigma L \qquad (6.7c)$$

The Bowditch adjustment results in changes to the distances equal to

$$\delta_L = \frac{fL}{t}\ (\Delta'N\cos\beta + \Delta'E\sin\beta) \qquad (6.7d)$$

where f = the factor of proportion
t = the error vector.

It can be seen from equation (6.7a) that in a relatively straight traverse, where the angle (α) approximates to 180°, the corrections to the angles ($\delta\alpha$) will be zero for all but the first and last angles.

6.1.7 *Accuracy of traversing*

Computer simulation of all types of network has shown that traversing is generally more accurate than classical triangulation and trilateration. Due to the weak geometry of a traverse, it generally has only three degrees of freedom (that is three redundant observations), it is difficult to arrive at an estimate of accuracy. This, coupled with the effect of the balancing procedure, makes it virtually impossible. Although there have been many attempts to produce equations defining the accuracy of a traverse, at the present time the best approach is a strength analysis using variance–covariance matrices from a least squares adjustment.

6.1.8 *Blunders in the observed data*

Blunders or mistakes in the measurement of the angles, results in gross angular misclosure. Provided it is only a single blunder it can easily be located.

Table 6.2 Bowditch adjustment of a link traverse

Stns	Observed angles ° ′ ″	Line	WCB ° ′ ″	Corrn	Adjusted WCB ° ′ ″	Dist (m)	Unadjusted E	Unadjusted N	Corrn δE	Corrn δN	Adjusted E	Adjusted N	Stn
A		A–B	151 27 38		151 27 38		3854.28	9372.98			3854.28	9372.98	B
B	143 54 47	B–E1	115 22 25	−4	115 22 21	651.16	4442.63	9093.96	+0.03	−0.05	4442.66	9093.91	E.1
E.1	149 08 11	E1–E2	84 30 36	−8	84 30 28	870.92	5309.55	9177.31	+0.08	−0.11	5309.63	9177.20	E.2
E.2	224 07 32	E2–E3	128 38 08	−12	128 37 56	522.08	5171.38	8851.36	+0.11	−0.15	5171.49	8851.21	E.3
E.3	157 21 53	E3–E4	106 00 01	−16	105 59 45	1107.36	6781.87	8546.23	+0.17	−0.22	6782.04	8546.01	E.4
E.4	167 05 15	E4–C	93 05 16	−20	93 04 56	794.35	7575.35	8503.49	+0.21	−0.28	7575.56	8503.21	C
C	74 32 48	C–D	347 38 04	−23	347 37 41								
D		C–D	347 37 41										
		C–D Δ =	+23		Sum =	3945.87	7575.56	8503.21					
					ΔE, ΔN		−0.21	+0.28					

Angle misclosure:

	° ′ ″
Sum	916 10 26
Initial bearing	151 27 38
Total	1067 38 04
−6 × 180°	1080 00 00
	−12 21 56
	+360 00 00
CD (comp)	347 38 04
CD (known)	347 37 41
Δ =	+23

Error vector = $(0.21^2 + 0.28^2)^{\frac{1}{2}} = 0.35$

Proportional error = $\dfrac{0.35}{3946} = 1/11\,300$

Check

In the case of an angle, the traverse can be computed forward from X (*Figure 6.18*) and then backwards from Y. The point which has the same co-ordinates in each case, is where the blunder occurred and the angle must be reobserved. This process can be carried out by plotting using a protractor and scale. Alternatively the right angled bisector of the error vector YY' of the plotted traverse, will pass through the required point (*Figure 6.18*). The theory is that BYY' forms an equilateral triangle.

In the case of a blunder in measuring distance, the incorrect leg is the one whose bearing is similar to the bearing of the error vector. If there are several legs with similar bearings the method fails. Again the incorrect leg must be remeasured.

6.2 TRIANGULATION

Because, at one time, it was easier to measure angles than it was distance, triangulation was the preferred method of establishing the position of control points.

Many countries used triangulation as the basis of their national mapping system. The procedure was generally to establish primary triangulation networks, with triangles having sides ranging from 30 to 50 km in length. The primary trig points were fixed at the corners of these triangles and the sum of the measured angles was correct to $\pm 3''$. These points were usually established on the tops of mountains to afford long, uninterrupted sight lines. The primary network was then densified with points at closer intervals connected into the primary triangles. This secondary network had sides of 10–20 km with a reduction in observational accuracy. Finally, a third-order net, adjusted to the secondary control, was established at 3–5-km intervals and fourth-order points fixed by intersection. *Figure 6.19* illustrates such a triangulation system established by the Ordnance Survey of Great Britain and used as control for the production of national maps. The base line and check base line would be measured by invar tapes in catenary and connected into the triangulation by angular extension procedures. This approach is classical triangulation, which is now obsolete. The more modern approach would be to measure the base lines with EDM equipment and to include many more measured lines in the network, to afford greater control of scale error.

Although the areas involved in construction are relatively small compared with national surveys (resulting in the term 'microtriangulation') the accuracy required in establishing the control surveys is frequently of a very high order, e.g. long tunnels or dam deformation measurements.

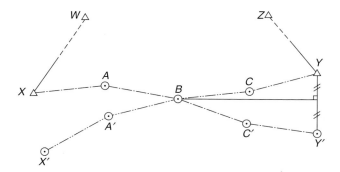

Fig. 6.18

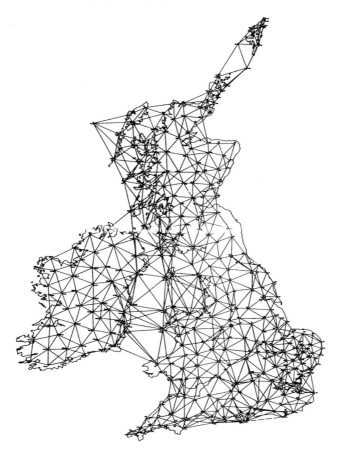

Fig. 6.19 *An example of a triangulation*

The principles of the method are illustrated by the typical basic figures shown in *Figure 6.20*. If all the angles are measured, then the scale of the network is obtained by the measurement of one side only, i.e. the base line. Any error, therefore, in the measurement of the base line will result in scale error throughout the network. Thus, in order to control this error, check base lines should be measured at intervals. The scale error is defined as the difference between the measured and computed check base. Using the base line and adjusted angles the remaining sides of the triangles may be found and subsequently the coordinates of the control stations.

Triangulation is best suited to open, hilly country, affording long sights well clear of intervening terrain. In urban areas, roof-top triangulation is used, in which the control stations are situated on the roofs of accessible buildings.

6.2.1 Shape of the triangle

The sides of the network are computed by the sine rule. From triangle ABC in *Figure 6.20(a)*:

$$\log b = \log c + \log \sin B_1 - \log \sin C_1$$

The effect on side b of errors in the measurement of angles B and C is found in the usual way. Consider an error δb in side b due to an angular error δB in the measurement of angle B; then

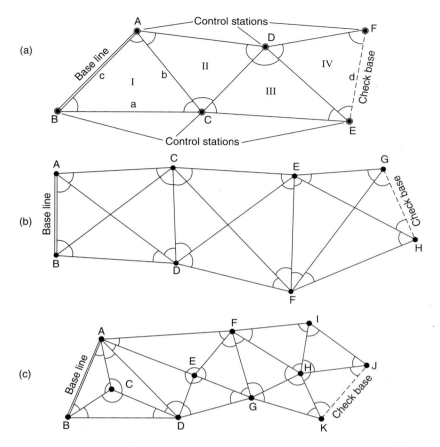

Fig. 6.20 *(a) Chain of simple triangles, (b) braced quadrilaterals and (c) polygons with central points*

$$\frac{\delta b}{b} = \delta B \cot B$$

Similarly for an error δC in angle C:

$$\frac{\delta b}{b} = \delta C \cot C$$

If we regard the above errors as standard errors and combine them, the result is

$$\frac{\sigma b}{b} = [(\sigma_B \cot B)^2 + (\sigma_C \cot C)^2]^{\frac{1}{2}}$$

Further, assuming equal angular errors, i.e. $\sigma_B = \sigma_C = \sigma$ rad, then

$$\frac{\sigma b}{b} = \sigma(\cot^2 B + \cot^2 C)^{\frac{1}{2}} \tag{6.8}$$

Equation (6.8) indicates that as angles B and C approach 90°, the effect of angular error on the computed side b will be a minimum. Thus the ideal network for *Figure 6.20(a)* would be to have very small angles opposite the sides which do not enter into the scale error computation, i.e. sides BC, AD, CE and DF. Such a network would not, however, be a practical proposition due to the very limited ground coverage, and the best compromise is the use of equilateral triangles where possible. If small angles are inevitable and cannot be fixed so as not to enter the scale computation, they should be measured with extra precision.

Assuming now that $B = C = 60°$ and $\sigma = \pm 1''$, then as $\cot 60° = 3^{-\frac{1}{2}}$ and σ rad $\approx 1/200\,000$:

$$\frac{\sigma b}{b} = 1/200\,000 \left(\frac{2}{3}\right)^{\frac{1}{2}} = 1/245\,000$$

After n triangles the error will be $n^{\frac{1}{2}}$ times the error in each triangle:

$$\therefore \frac{\sigma b}{b} = n^{\frac{1}{2}}/245\,000$$

Thus, after, say, nine triangles, the scale error would be approximately $1/82\,000$. This result indicates the need for maximum accuracy in the measurement of the base line and angles, as well as the need for regular check bases and well-conditioned triangles. It can be shown that when the angles are adjusted, equation (6.8) becomes

$$\frac{\sigma b}{b} = \sigma \left(\frac{2}{3} (\cot^2 B + \cot B \cot C + \cot^2 C)\right)^{\frac{1}{2}} \tag{6.9}$$

which theoretically shows no improvement in the scale error if $B = C$.

6.2.2 *General procedure*

(1) Reconnaissance of the area, to ensure the best possible positions for stations and base lines.
(2) Construction of the stations.
(3) Consideration of the type of target and instrument to be used and also the method of observation. All of these depend on the precision required and the length of sights involved.
(4) Observation of angles and base-line measurements.
(5) Computation: base line reduction, station and figural adjustment, coordinates of stations by direct methods.

A general introduction to triangulation has been presented, aspects of which will now be dealt with in detail.

(1) Reconnaissance is the most important aspect of any well-designed surveying project. Its main function is to ensure the best positions for the survey stations commensurate with well-conditioned figures, ease of access to the stations and economy of observation.

A careful study of all existing maps or plans of the area is essential. The best position for the survey stations can be drawn on the plan and the overall shape of the network studied. Whilst chains of single triangles are the most economic to observe, braced quadrilaterals provide many more conditions of adjustment and are at their strongest when square shaped. Using the contours of the plan, profiles between stations can be plotted to ensure intervisibility. Stereo-pairs of aerial photographs, giving a three-dimensional view of the terrain, are useful in this respect. Whilst every attempt should be made to ensure that there are no angles less than 25°, if a small angle cannot be avoided it should be situated opposite a side which does not enter into the scale computation.

When the paper triangulation is complete, the area should then be visited and the site of every station carefully investigated. With the aid of binoculars, intervisibility between stations should be checked and ground-grazing rays avoided. Since the advent of EDM, base-line siting is not so critical. Soil conditions should be studied to ensure that the ground is satisfactory for the construction of long-term survey stations. Finally, whilst the strength of the network is a function of its shape, the purpose of the survey stations should not be forgotten and their position located accordingly.

(2) Stations must be constructed for long-term stability and may take the form of those illustrated in *Figures 6.7* and *6.8*. A complete referencing of the station should then be carried out in order to ensure its location at a future date.

(3) As already stated, the type of target used will depend on the length of sight involved and the accuracy required. for highly precise networks, the observations may be carried out at night when refraction is minimal. In such a case, signal lamps would be the only type of target to use.

For short sights it may be possible to use the precise targets shown in *Figure 6.10*. Whatever form the target takes, the essential considerations are that it should be capable of being accurately centred over the survey point and afford the necessary size and shape for accurate bisection at the observation distances used.

(4) The observation of the angles has already been dealt with in Chapter 4. In triangulation the method of directions would inevitably be used and the horizon closed. An appropriate number of sets would be taken on each face. The base line and check base would most certainly be measured by EDM, with all the necessary corrections made to ensure high accuracy, as illustrated in Chapter 3.

(5) Since the use of computers is now well established, there is no reason why a least squares adjustment using the now standard variation of coordinates method should not be carried out.

Alternatively the angles may be balanced by simpler, less rigorous methods known as 'equal shifts'. On completion, the sides may be computed using the sine rule and finally the coordinates of each survey point obtained.

If the survey is to be connected to the national mapping system of the country, then all the base-line measurements must be reduced to MSL and multiplied by the local scale factor. As many of the national survey points as possible should be included in the scheme.

6.2.3 Figural adjustment by equal shifts

Whilst least squares methods permit the adjustment of the network as a whole, the simpler 'equal shifts' approach treats each figure in the network separately. The final values for the angles must, however, satisfy the conditions of adjustment of each figure.

(1) *Simple triangle.* The condition of adjustment of a plane triangle is that the sum of the angles should equal 180°. It is due to this minimum number of conditions that other figures, such as braced quadrilaterals, are favoured rather than triangles.

For large triangles on an ellipsoid of reference, with sides greater than 10 km, the angles should sum to 180° + E. E is the spherical excess of the triangle and may be computed from

$$E'' = \frac{\text{Area of triangle} \times 206\,265}{R} \tag{6.10}$$

where R = local radius of the Earth

For greater accuracy $R^2 = \rho v$ may be used, taking the latitude at the centre of the triangle to evaluate ρ and v.

After adjusting the angles to equal (180° + E), Legendre's theorem stipulates that if one-third of the spherical excess is deducted from each angle, the triangle may be treated as a plane triangle for

Table 6.3

Angle	Observed value			Correction	Ellipsoidal angles			E″/3	Plane angles		
	°	′	″	″	°	′	″		°	′	″
A	052	12	48.15	+ 1.11	052	12	49.26	−2.29	052	12	46.97
B	076	09	10.27	+ 1.11	076	09	11.38	−2.29	076	09	09.09
C	051	38	05.12	+ 1.11	051	38	06.23	−2.29	051	38	03.94
Sum	180	00	03.54	+ 3.33	180	00	06.87	−6.87	180	00	00.00

the computation of its side lengths. Error analysis shows that the computation of the area of the triangle is not critical and may even be scaled from a map. An example is illustrated in *Table 6.3*.

(a) The spherical excess was computed as $E'' = 6.87''$; therefore the observed angles are balanced by + 1.11″ per angle so that the sum equals $180° + E'' = 180° 00' 06.87''$.
(b) The ellipsoidal angles are then reduced to plane by deducting $E/3$ from each.
(c) The unknown sides of the triangle can now be computed using the sine rule in plane trigonometry.
(d) The azimuths of the sides are obtained using the ellipsoidal angles.
(e) The latitudes and longitudes can now be computed using the mid-latitude rule, for example.

(2) In a braced quadrilateral (*Figure 6.21*), the final balanced angles should fulfil the following conditions of adjustment, if the figure is to be geometrically correct.
   Conditions of adjustment (*Figure 6.21*):

$$\text{Angles } 1 + 2 + 3 + 4 + 5 + 6 + 7 + 8 = 360°$$
$$1 + 2 + 3 + 4 = 180°$$
$$3 + 4 + 5 + 6 = 180°$$
$$5 + 6 + 7 + 8 = 180°$$
$$7 + 8 + 1 + 2 = 180°$$
$$1 + 2 = 5 + 6$$
$$3 + 4 = 7 + 8$$

Side condition:

   Σ log sins of the odd angles = Σ log sins of the even angles

As many of the above conditions are dependent upon each other, only four are used in the actual

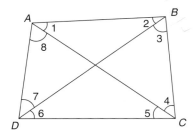

Fig. 6.21

adjustment. The 'method of adjustment' is: (a) adjust angles (1–8) to equal 360°; (b) adjust angles (1 + 2) to equal (5 + 6); (c) adjust angles (3 + 4) to equal (7 + 8); (d) side condition.

Proof of side condition:
  From *Figure 6.21* it is required to calculate length *CD* from base *AB*.
  This may be done via route *BC* or *AD* as follows:

  AB/sin 4 = BC/sin 1   ∴ $BC = AB$ sin 1/sin 4

  Now BC/sin 6 = DC/sin 3 $DC = BC$ sin 3/sin 6 $= \dfrac{AB \sin 1 \sin 3}{\sin 4 \sin 6}$

  Similarly via *AD* $DC = AB$ sin 2 sin 8/(sin 7 sin 5)

As there can only be one length for *DC*, then cancelling *AB* gives

  sin 1 sin 3/(sin 4 sin 6) = sin 2 sin 8/(sin 7 sin 5)

Cross-multiplying and taking logs:

  log sin 1 + log sin 3 + log sin 5 + log sin 7 = log sin 2 + log sin 4 + log sin 6 + log sin 8

This method of adjustment will now be illustrated using the following mean observed angles in *Figure 6.21*:

(a) The first step in the method of adjustment is clearly seen.
(b) The second step shows that the difference between angles (1 + 2) and (5 + 6) is 4″, i.e. 1″ per angle, which is added to the smaller sum and subtracted from the larger.
(c) The third step is identical to the above: the corrections of 2″ and 1″ have been arbitrarily made to prevent the introduction of decimals of a second (correction per angle = 1.5″).
    Three steps have produced corrected angles which satisfy the first seven conditions of adjustment. It is now necessary to find the log sin of these angles and to compare their sums. This can be done very quickly on a pocket calculator. (Table 6.5)

  Adjustment = (15/217) × 10″ = 0.7″ ≈ 1″

(d) Column 5 represents the change in the log sin of the angles for a change of 10″ in the angle. These values are easily obtained by increasing the value of the angle by 10″ and then finding its log sin on the pocket calculator. The difference of the two log sin values is the difference for 10″ change in the angle.

Table 6.4

Number	Observed angles				1st correction							2nd correction		
	°	′	″	″	°	′	″	°	′	″	″	°	′	″
1	50	42	27	−1	50	42	26⎫	117	30	19	1	50	42	27
2	66	47	54	−1	66	47	53⎭				1	66	47	54
3	41	24	32	−1	41	24	31⎫	62	29	36	2	41	24	33
4	21	05	06	−1	21	05	05⎭				1	21	05	06
5	74	13	36	−1	74	13	35⎫	117	30	23	−1	74	13	34
6	43	16	49	−1	43	16	48⎭				−1	43	16	47
7	18	36	14	−1	18	36	13⎫	62	29	42	−1	18	36	12
8	43	53	30	−1	43	53	29⎭				−2	43	53	27
	360	00	08	−8	360	00	00				0	360	00	00

Table 6.5

1	2 Angles			3 Log sin (odd)	4 Log sin (even)	5 Difference for 10″ arc	6	7 Final values		
	°	′	″					°	′	″
1	50	42	27	1.888698		0.000 017	1″	50	42	28
2	66	47	54		1.963374	9	−1″	66	47	53
3	41	24	33	1.820485		24	1″	41	24	34
4	21	05	06		1.556004	55	−1″	21	05	05
5	74	13	34	1.983329		6	1″	74	13	35
6	43	16	47		1.836046	22	−1″	43	16	46
7	18	36	12	1.503810		62	1″	18	36	13
8	43	53	27		1.840913	22	−1″	43	53	26
				1.196322	1.196337	0.000 217		360	00	00
					1.196322					
					0.000015					

Normally the difference for 1″ of arc is used, but in this case 10″ differences are used in order to facilitate understanding of the principles.

(e) Summing columns 3 and 4 shows a difference of 15 (0.000 015) which must be adjusted. The necessary angular correction (0.7″) is obtained by dividing 15 by the sum of columns 5, i.e. 217 (0.000 217), as shown. This may be explained as follows: if one altered all the angles by 10″, the total change in the log sins would be 0.000 217. However, the change required is only 0.000 015, which by proportion represents an angular change of $15/217 \times 10″ = 0.7″$.

(g) If any angle is greater than 90°, then a positive correction to the angle would require a negative correction to its log sin. Thus the difference value in column 5 should have a negative sign which is applied in the summing of this column and throughout.

(h) It is worth noting that the accuracy of a triangulation figure is expressed by the magnitude of the difference in the sum of log sins, i.e. 0.000 015. Compensating errors can occur in angles, tending to indicate excellent closure; such errors would, however, substantially unbalance the side equation.

Although the above method can be done easily on a pocket calculator, the following approach (Smith, 1982) has been produced specifically for a pocket calculator.

The method precludes the use of logarithms and differences for 1″ or 10″, and is as follows: in the side condition assume v is the correction per angle; then

$$\sin(1+v)\sin(3+v)\sin(5+v)\sin(7+v) = \sin(2+v)\sin(4+v)\sin(6+v)\sin(8+v)$$

Now $\sin(1+v) = \sin 1 \cos v + \cos 1 \sin v$, which, as v is very small, $= \sin 1 + \cos 1\, v$

$$\frac{(\sin 1 + \cos 1v)(\sin 3 + \cos 3v)(\sin 5 + \cos 5v)(\sin 7 + \cos 7v)}{(\sin 2 + \cos 2v)(\sin 4 + \cos 4v)(\sin 6 + \cos 6v)(\sin 8 + \cos 8v)} = 1$$

Expanding to the first order only:

$$\frac{(\sin 1 \sin 3 + \sin 1 \cos 3v + \cos 1 \sin 3v)(\sin 5 \sin 7 + \sin 5 \cos 7v + \cos 5 \sin 7v)}{(\sin 2 \sin 4 + \sin 2 \cos 4v + \cos 2 \sin 4v)(\sin 6 \sin 8 + \sin 6 \cos 8v + \cos 6 \sin 8v)}$$

$$= \frac{\sin 1 \sin 3 \sin 5 \sin 7 + \sin 1 \sin 3 \sin 5 \sin 7v (\cot 1 + \cot 3 + \cot 5 + \cot 7)}{\sin 2 \sin 4 \sin 6 \sin 8 + \sin 2 \sin 4 \sin 6 \sin 8v (\cot 2 + \cot 4 + \cot 6 + \cot 8)}$$

Let sin 1 sin 3 sin 5 sin 7 = A   cot 1 cot 3 cot 5 cot 7 = B
    sin 2 sin 4 sin 6 sin 8 = C   cot 2 cot 4 cot 6 cot 8 = D

Then the above expression can be rearranged and expressed thus:

$$v'' = \frac{206\,265(A - C)}{AB + CD}$$

If v'' is positive then $A > C$ and v'' is subtracted from the odd angles and added to the even.

All the digits as displayed on the pocket calculator are significant and should be carried through the computation.

The previous example is now re-worked using this method for the side condition, and it is shown in *Table 6.6*.

(3) Polygon with central point. The basic triangulation figures are shown in *Figure 6.22*.

Conditions of adjustment:

(a) Each triangle to equal 180°, i.e. I, II . . . V in *Figure 6.22(c)*.
(b) Central angles to equal 360°.
(c) Side condition using base angles only, i.e. 1, 2 . . . 10 in *Figure 6.22(c)*.

Method of adjustment:

(a) Adjust each triangle to 180°.
(b)   (i) Adjust the central angles to 360°.
      (ii) Readjust the triangles to 180° using the two base angles in each triangle.
(c) Side condition adjustment using the base angles only.

Steps (b) (i) and (ii) are in fact one step, for correction of say, +10″ to each of the central angles would automatically give a correction of, say, −5″ to each base angle of the triangle. The side condition would then be carried out in exactly the same manner already described, in each case excluding the angles at the centre point.

6.2.4 Satellite stations

In *Figure 6.23* it is required to find the angles measured to A, B and C from D, or alternatively the bearings DA, DB and DC. If D is an 'up-station', e.g. church spire, lightning conductor or tall structure, or the lines of sight are blocked by natural or man-made obstacles, then it is necessary to establish a satellite station S nearby, from which angles to A, B, C and D are measured. These measured angles about S are then reduced to their equivalent about D. This is illustrated as follows: If the line SD is assumed to be due north, then it can be seen that the bearing of DB is greater than that of SB by the amount δ_B. Thus the measured bearing SB is increased by δ_B to give the required bearing DB.

If working directly in angles, then regarding $ABSD$ as a crossed quadrilateral, it can be seen that

$$A\hat{D}B = A\hat{S}B + \delta_B - \delta_A \text{ (with } S \text{ due south of } D)$$

Students should draw the following and verify for themselves:

S due west of D	$A\hat{D}B = A\hat{S}B - \delta_B - \delta_A$
S due east of D	$A\hat{D}B = A\hat{S}B + \delta_B + \delta_A$
S due north of D	$A\hat{D}B = A\hat{S}B - \delta_B + \delta_A$

The method of solving the problem is determined largely by the data supplied. If the angles at A

Table 6.6 Adjustment of braced quadrilateral by equal shifts (using pocket calculator)

Angle no.	Observed angle ° ' "	1st corr'n	1st corr'd angle ° ' "	1+2=5+6 3+4=7+8 ° ' "	2nd corr'n	2nd corr'd angle ° ' "	Sin odd ∠$_s$ product	Cot odd ∠$_s$ sum	Sin even ∠$_s$ product	Cot even ∠$_s$ sum	3rd corr'n	Final corr'd angle ° ' "
1	50 42 27	–1"	50 42 26	117 30 19	+1"	50 42 27	0.773 923	0.818 270 3			+1"	50 42 28
2	66 47 54	–1"	66 47 53		+1"	66 47 54	×	+	0.919 124	0.428 634 4	–1"	66 47 53
3	41 24 32	–1"	41 24 31	62 29 36	+2"	41 24 33	0.661 432	1.133 911 5	×	+	+1"	41 24 34
4	21 05 06	–1"	21 05 05		+1"	21 05 06	×	+	0.359 753	2.593 582 9	–1"	21 05 05
5	74 13 36	–1"	74 13 35	117 30 23	–1"	74 13 34	0.962 342	0.282 479 4	×	+	+1"	74 13 35
6	43 16 49	–1"	43 16 48		–1"	43 16 47	×	+	0.685 561	1.061 927 3	–1"	43 16 46
7	18 36 14	–1"	18 36 13	62 29 42	–1"	18 36 12	0.319 014	2.970 867 6	×	+	+1"	18 36 13
8	43 53 30	–1"	43 53 29		–2"	43 53 27			0.693 287	1.039 486 9	–1"	43 53 26
Σ	360 00 08	–8"	360 00 00		00	360 00 00	0.157 152 8	5.205 528 8	0.157 158 4	5.123 6315	00	360 00 00
							A	B	C	D		

$$\text{3rd correction} = \frac{206\,265\,(A - C)}{AB + CD}$$
$$= 0.7'' \approx 1''$$

If A > C then add to even ∠$_s$ and subtract from odd ∠$_s$, and *vice versa*

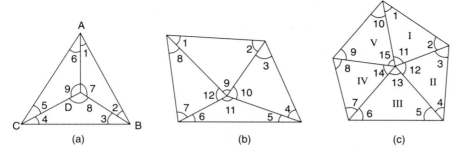

Fig. 6.22

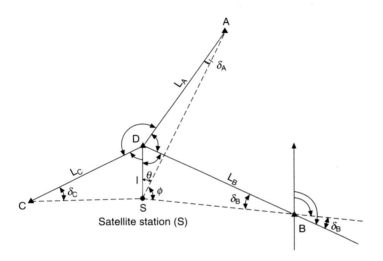

Fig. 6.23

and B to D are given, then one can find an approximate value for $A\hat{D}B$ from $(180° - D\hat{A}B - D\hat{B}A)$, and then use the sine rule with length AB to find L_A and L_B. Then by the sine rule in ΔDAS

$$\delta_A'' = \frac{l \sin \theta}{L_A} \times 206\,265 \tag{6.10}$$

To assess the effect of errors in the measured quantities on δ_A, differentiate with respect to each in turn

$$\frac{\delta(\delta_A)}{\delta_A} = \frac{\delta l}{l} = \frac{\delta L}{L} = \cot \theta \; \delta\theta$$

This indicates:

(1) That the fractional error in δ_A is directly proportional to the fractional error in l and L. Thus if $\delta_A = 600'' \pm 1''$, $l = 10$ m and $L = 10$ km, then l needs only be measured to the nearest 0.017 m and L to 17 m, i.e. 1 in 600.
(2) That the error in δ_A is proportional to $\cot \theta \; \delta\theta \; \delta_A$, thus the angle θ should be as large as possible and angle δ_A as small as possible, making l as small as possible. The accuracy to which one

measures θ, i.e. $\delta\theta$, varies with the value of θ. If it is very large, then cot θ is very small and θ need be measured with only normal accuracy.

The sum effect of the standard errors is

$$\frac{\delta(\delta_A)}{\delta_A} = \pm\left\{\left(\frac{\delta l}{l}\right)^2 + \left(\frac{\delta L}{l}\right)^2 + (\cot\theta\ \delta\theta)^2\right\}^{\frac{1}{2}}$$

6.2.5 Resection and intersection

Using these techniques, one can establish the coordinates of a point P, by observations to at least three known points. These techniques are useful for obtaining the position of single points, to provide control for setting out or detail survey in better positions than the existing control may be.

(1) Intersection

This involves sighting in to P from known positions (*Figure 6.24*). If the bearings of the rays are used, then using the rays in combinations of two, the coordinates of P are obtained as follows:

In *Figure 6.25* it is required to find the coordinates of P, using the *bearings* α and β to P from known points A and B whose coordinates are E_A, N_A and E_B, N_B.

$$PL = E_P - E_A \qquad AL = N_P - N_A$$
$$PM = E_P - E_B \qquad M_B = N_P - N_B$$

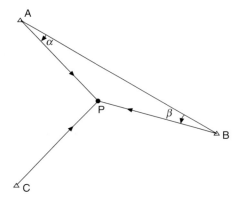

Fig. 6.24

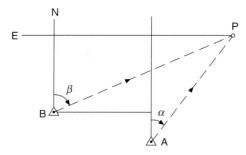

Fig. 6.25

Now as $\qquad PL = AL \tan \alpha$ $\qquad\qquad\qquad\qquad\qquad\qquad\qquad$ (1)

then $\qquad E_P - E_A = (N_P - N_A) \tan \alpha$

Similarly $\qquad PM = MB \tan \beta$

then $\qquad E_P - E_B = (N_P - N_B) \tan \beta$ $\qquad\qquad\qquad\qquad\qquad$ (2)

Subtracting (*1*) from (*2*) gives

$$E_B - E_A = (N_P - N_A) \tan \alpha - (N_P - N_B) \tan \beta$$
$$= N_P \tan \alpha - N_A \tan \alpha - N_P \tan \beta + N_B \tan \beta$$
$$\therefore N_P (\tan \alpha - \tan \beta) = E_B - E_A + N_A \tan \alpha - N_B \tan \beta$$

Thus $\qquad\qquad\qquad N_P = \dfrac{E_B - E_A + N_A \tan \alpha - N_B \tan \beta}{\tan \alpha - \tan \beta}$ $\qquad\qquad\qquad$ (6.12a)

Similarly $\qquad N_P - N_A = (E_P - E_A) \cot \alpha$

$$N_P - N_B = (E_P - E_B) \cot \beta$$

Subtracting $\qquad N_B - N_A = (E_P - E_A) \cot \alpha - (E_P - E_B) \cot \beta$

Thus $\qquad\qquad\qquad E_P = \dfrac{N_B - N_A + E_A \cot \alpha - E_B \cot \beta}{\cot \alpha - \cot \beta}$ $\qquad\qquad\qquad$ (6.12b)

Using equations (6.12a) and (6.12b) the coordinates of *P* are computed. It is assumed that *P* is always to the right of A $\rightarrow$ B, in the equations.

If the observed angles into *P* are used (*Figure 6.24*) the equations become

$$E_P = \frac{N_B - N_A + E_A \cot \beta + E_B \cot \alpha}{\cot \alpha + \cot \beta} \qquad\qquad\qquad (6.13a)$$

$$N_p = \frac{E_A - E_B + N_A \cot \beta + N_B \cot \alpha}{\cot \alpha + \cot \beta} \qquad\qquad\qquad (6.13b)$$

The above equations are also used in the direct solution of triangulation. The inclusion of an additional ray from *C*, affords a check on the observations and the computation.

(2) Resection

This involves the angular measurement from *P* out to the known points A, B, C (*Figure 6.26*). As only *P* is occupied in this technique, it is considered to provide a weaker solution than intersection. It is, however, an extremely useful technique for quickly fixing position where it is best required for setting-out purposes. Where only three known points are used a variety of analytical methods is available for the solution of *P*.

The following approach is referred to as the 'analytical method' (from *Figure 6.26*).

Let $BAP = \theta$, then $\qquad BCP = (360° - \alpha - \beta - \phi) - \theta = S - \theta$

where ϕ is computed from the coordinates of stations *A*, *B* and *C*; thus *S* is known.

From $\triangle PAB \qquad PB = BA \sin \theta / \sin \alpha$ $\qquad\qquad\qquad\qquad\qquad\qquad$ (1)

From $\triangle PBC \qquad PB = BC \sin(S - \theta) / \sin \beta$ $\qquad\qquad\qquad\qquad\qquad$ (2)

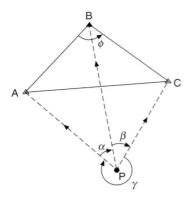

Fig. 6.26

Equating (*1*) and (*2*) $\dfrac{\sin (S - \theta)}{\sin \theta} = \dfrac{BA \sin \beta}{BC \sin \alpha} = Q$ (known)

then $(\sin S \cos \theta - \cos S \sin \theta)/\sin \theta = Q$

$\quad \sin S \cot \theta - \cos S = Q$

$\therefore \cot \theta = (Q + \cos S)/\sin S$ (6.14)

Thus, knowing θ and $(S - \theta)$, the triangles can be solved for lengths and bearings *AP*, *BP* and *CP*, and three values for the coordinates of *P* obtained if necessary.

The method fails, as do all three-point resections, if *P* lies on the circumference of a circle passing through *A*, *B* and *C* and thereby has an infinite number of positions.

Worked example

Example 6.1. The coordinates of *A*, *B* and *C* (*Figure 6.26*) are:

E_A 1234.96 m N_A 17594.48 m
E_B 7994.42 m N_B 24343.45 m
E_C 17913.83 m N_C 21364.73 m

Observed angles are:

$APB = \alpha = 61°41'46.6''$

$BPC = \beta = 74°14'58.1''$

Find the coordinates of *P*.

(1) From the coordinates of *A* and *B*:

$\Delta E_{AB} = 6759.46, \; \Delta N_{AB} = 6748.97$

$\therefore$ Horizontal distance $AB = (\Delta E^2 + \Delta N^2)^{\frac{1}{2}} = 9551.91$ m

Bearing $AB = \tan^{-1}(\Delta E/\Delta N) = 45°02'40.2''$

(or use the R → P keys on pocket calculator)

(2) Similarly from the coordinates of *B* and *C*:

$\Delta E_{BC} = 9919.41$ m,　$\Delta N_{BC} = -2978.72$ m

∴ Horizontal distance $BC = 10357.00$ m

Bearing $BC = 106°42'52.6''$

From the bearings of *AB* and *BC*:

$C\hat{B}A = \phi = 180°19'47.6''$

(3)　　$S = (360° - \alpha - \beta - \phi) = 105.724352°$

and　$Q = AB \sin \beta/BC \sin \alpha = 1.008167$

∴ $\cot \theta = (Q + \cos S)/\sin S$, from which

$\theta = 52.554505°$

(4) $BP = AB \sin \theta/\sin \alpha = 8613.32$ m

$BP = BC \sin (S - \theta)/\sin \beta = 8613.32$ m (Check)

Angle $CBP = 180° - [\beta + (S - \theta)] = 52.580681°$

∴ Bearing $BP =$ Bearing $BC + C\hat{B}P = 159.29529° = \delta$

Now using length and bearing of *BP*, transform to rectangular coordinate by formulae or P → R keys.

$\Delta E_{BP} = BP \sin \delta = 3045.25$ m

$\Delta N_{BP} = BP \cos \delta = -8057.03$ m

$E_P = E_B + \Delta E_{BP} = 11039.67$ m

$N_P = N_B + \Delta N_{BP} = 16286.43$ m

Checks on the observations and the computation can be had by computing the coordinates of *P* using the length and bearing of *AP* and *CP*.

In order to illustrate the diversity of methods available, the following method is proposed by Dr T.L. Thomas of Imperial College, London.

$E_P = E_A + ZV/(V^2 + W^2)$

$N_P = N_A + ZW/(V^2 + W^2)$

where:

$V = \Delta E_1 \cot \alpha - \Delta E_2 \cot(\alpha + \beta) + (N_C - N_B)$

$W = \Delta N_1 \cot \alpha - \Delta N_2 \cot(\alpha + \beta) + (E_B - E_C)$

$Z = X \cot \alpha - X \cot(\alpha + \beta) + Y + Y\cot \alpha\cot(\alpha + \beta)$

$X = \Delta E_1 \cdot \Delta E_2 + \Delta N_1 \cdot \Delta N_2$

$Y = \Delta E_1 \cdot \Delta N_2 - \Delta N_1 \cdot \Delta E_2$

$\Delta E_1 = E_B - E_A, \Delta E_2 = E_C - E_A, \Delta N_1 = N_B - N_A, \Delta N_2 = N_C - N_A$

Using the data in the previous example:

$\Delta E_1 = 6759.46, \Delta E_2 = 16678.87, \Delta N_1 = 6748.77, \Delta N_2 = 3770.25$

giving

$V = 17900.221$

$W = -2388.080$

$X = 1.3818546 \times 10^9$

$$Y = -870\,803\,392$$
$$Z = 1.786\,303\,1 \times 10^9$$

producing values for P of

$$E_P = 11\,039.67, \; N_P = 16\,286.43$$

A further approach is illustrated below:

A, B and C (*Figure 6.27*) are fixed points whose coordinates are known, and the coordinates of the circle centres O_1 and O_2 are

$$E_1 = \tfrac{1}{2}[E_A + E_B + (N_A - N_B)\cot\alpha]$$
$$N_1 = \tfrac{1}{2}[N_A + N_B - (E_A - E_B)\cot\alpha]$$
$$E_2 = \tfrac{1}{2}[E_B + E_C + (N_B - N_C)\cot\beta]$$
$$N_2 = \tfrac{1}{2}[N_B + N_C - (E_B - E_C)\cot\beta]$$

Thus the bearing δ of $O_1 \rightarrow O_2$ is obtained in the usual way, i.e.

$$\delta = \tan^{-1}[(E_2 - E_1)/(N_2 - N_1)]$$

then $E_P = E_B + 2[(E_B - E_1)\sin\delta - (N_B - N_1)\cos\delta]\sin\delta$ (6.15a)

$N_P = N_B + 2[(E_B - E_1)\sin\delta - (N_B - N_1)\cos\delta]\cos\delta$ (6.15b)

Intersection and resection can also be carried out using observed distances.

Although there are a large number of methods for the solution of a three-point resection, all of them fail if A, B, C and P lie on the circumference of a circle. Many of the methods also give dubious results if A, B and C lie in a straight line. Care should be exercised in the method of computation adopted and independent checks used wherever possible. Field configurations should be used which will clearly eliminate either of the above situations occurring; for example, siting P within a triangle formed by A, B and C, is an obvious solution.

6.3 TRILATERATION

Trilateration, based exclusively on measured horizontal distances, has gained acceptance because

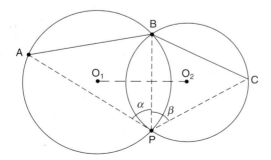

Fig. 6.27

of the advent of EDM instrumentation. The geometric figures used are similar to those employed in triangulation, although not as standardized due to greater control of scale error. It was originally considered that trilateration would supersede triangulation as a method of control due to the scale error factor. However, subsequent results have shown that the system is liable to a rapid accumulation of azimuth error, ethereby requiring a dense system of control points.

The fact that there is no horizontal angle measurement required in trilateration would appear to make it more rapid and thus, at first glance, more economical than triangulation. However, much depends on the length of line involved and the accuracy requirements.

All EDM equipment measures slope distance, which therefore needs to be reduced to the horizontal at some datum level. This requires then not only the measurement of slope length, but also the relative levels of the control points and instrument heights, or the measurement of vertical angles.

EDM instruments are calibrated for the velocity of electromagnetic waves under certain standard meteorological conditions. Thus actual meteorological conditions along the measured path need to be known in order to correct the measured distance. At the present time this is not a practical proposition and one has to be content with the measurement of temperature and pressure at each end of the line being measured. For the best possible results under these conditions one requires carefully calibrated thermometers and barometers hung as high as possible by the instruments and read at the same instant of measurement. In order to comply with this latter requirement some form of intercommunication is necessary.

Similar precautions are also required when measuring the vertical angles. In order to achieve the accuracy required, one needs to use highly precise theodolites, preferably with automatic vertical circle indexing. Ideally, simultaneous reciprocal observations are necessary. If vertical angles are possible at only one end of the line, then corrections for curvature and refraction must be applied. Also, depending on the terrain and accuracy requirement, it may be necessary to consider the effect of 'deviation of the vertical' on the angles measured.

It would appear, therefore, that not only is trilateration possibly less economical than triangulation but on consideration of the above error sources (Chrzanowski and Wilson, 1967) it may also prove less accurate. There appears to be conflicting evidence on this point (Burke, 1971), although Hodges *et al.* (1967) has shown conclusively that angles computed through a trilateration are as accurate as those measured with a 1″ theodolite on the same control net.

A further reason why trilateration has not superseded triangulation must be in the superior internal checks given by triangulation. For instance, a triangle with three angles measured has an angle check whereas with three sides measured there is none; a braced quadrilateral with angles observed has four conditions (three angles, one side) to be satisfied, whereas with the sides there is only the single condition that the computed total angle at one corner equals the sum of the two computed component angles.

Network design is therefore especially critical in trilateration. In order to obtain sufficient redundancy for checks on the accuracy, the geometric figures become quite complicated. For instance, to obtain the same redundancy as a triangulation braced quadrilateral, a pentagon with all ten sides measured would have to be used. Indeed, experts in trilateration analysis have proposed the use of the hexagon, with all sides measured (20 giving 10 checks) as the basic network figure. However, from the practical viewpoint, pentagons and hexagons with all stations intervisible are difficult to establish in the field. Thus, from the logistic viewpoint, trilateration would require as much organization as triangulation. However, all trilateration must include the measurement of some of the angles, to increase the number of redundant measurements. It follows from this that the modern practice is to combine trilateration and triangulation, thereby producing very strong networks. The network may be computed by the method of variation of coordinates or the following less rigorous approaches may be used.

In the figures formed, use the reduced lengths to obtain the angles. This may be done using the half-angle equation, i.e.

$$\tan A/2 = \{S(S-a)/[(S-b)(S-c)]\}^{\frac{1}{2}} \tag{6.16}$$

where $2S = (a + b + c)$, the sum of the three sides

The network may then be treated as a triangulation network; the angles adjusted, the lengths computed from the adjusted angles and the point coordinates obtained from the length and bearing of each line.

Alternatively, direct coordination of the points may be obtained. To find the coordinates of C, given the coordinates of A and B, and the length of the sides a, b, c of the triangle:

$$E_c = \tfrac{1}{2}(E_A + E_B) + \frac{a^2 - b^2}{2c^2}(E_A - E_B) - \frac{2\Delta}{c^2}(N_A - N_B) \tag{6.17a}$$

and

$$N_c = \tfrac{1}{2}(N_A + N_B) + \frac{a^2 - b^2}{2c^2}(N_A - N_B) + \frac{2\Delta}{c^2}(E_A - E_B) \tag{6.17b}$$

where A, B and C are in clockwise order, and

$$\Delta = \{s(s-a)(s-b)(s-c)\}^{\frac{1}{2}}$$

If the survey is to be tied into the national grid, the local scale factor would need to be found from 'provisional coordinates' and applied to the ellipsoidal lengths to give the grid lengths. These latter lengths are then used in the formula to give the grid coordinates.

Dr. T.L. Thomas (1971) offers the following alternative equations for trilateration computation:

$$\Delta E = E_B - E_A \qquad \Delta N = N_B - N_A \qquad c^2 = \Delta E^2 + \Delta N^2$$

$$p = \frac{\Delta E}{c} \qquad q = \frac{\Delta N}{c} \qquad k = \frac{(b^2 + c^2 - a^2)}{2c} \qquad h = (b^2 - k^2)^{\frac{1}{2}}$$

Then $E_c = E_A + pk - qh \qquad N_c = N_A + qk + ph$

Checks $a^2 = (E_C - E_B)^2 + (N_C - N_B)^2$

$$b^2 = (E_C - E_A)^2 + (N_C - N_A)^2$$

It is assumed in the above that C is to the left of $\overrightarrow{AB}$.

6.4 TRIANGULATERATION

As its name implies, triangulateration is simply the combining of triangulation and trilateration to produce a control system in which all the angles and sides are measured.

From the accuracy point of view, the system should be very strong, possessing all the advantages of both systems from which it is derived. The improvement in the redundancy checks for a braced quadrilateral and central point pentagon are shown on the next page:

	Triangulation	*Trilateration*	*Triangulateration*
Quadrilateral			
No. of directions	12	0	12
No. of sides	1	6	6
No. of checks	4	1	9
Pentagon			
No. of directions	20	0	20
No. of sides	1	10	10
No. of checks	6	4	15

Whilst it is generally acknowledged that triangulateration is more accurate than the previously mentioned systems, one must consider whether or not it is economically justified. The logistics of the system will certainly not be equal to the sum of the previous two methods, for once set up at the observation station and targets/reflectors have been established on the stations to be observed, a skilled surveyor could acquire all the necessary field data with little extra time and effort. The use of electronic 'total stations' makes the prospect even more viable and may justify the initial high capital expenditure involved. Further, as there would be little or no accumulation of scale and azimuth error, ill-conditioned figures could be utilized, thereby reducing the reconnaissance time.

It should be possible through pre-survey analysis to optimize the system so that every station in the network need not be occupied, thus further improving the viability. The adjustment of such a network containing dissimilar quantities presents no difficulty if computer facilities are available. Using the variation of coordinates method, all the data can be adjusted *en masse* to produce the corrected coordinates of the network plus a complete error analysis and a posterior weighting of the field data.

It is thus evident that triangulateration is to be preferred over the use of classical triangulation or trilateration and this seems to be modern practice. However, it is unlikely to supersede traversing because of the basic difference between the two systems and the accuracy/economy factor.

6.5 INERTIAL SURVEYING

The inertial surveying system (ISS) provides three-dimensional positioning without any of the problems that beset most position-fixing procedures. It is unaffected by atmospheric refraction, it does not require intervisibility between points, it can be operated day or night regardless of weather conditions and it can progress at the speed of the vehicle in which it is housed.

The system consists basically of an inertial platform with three accelerometers held in three mutually orthogonal axes by three similarly orientated precise gyroscopes (*Figure 6.28*).

The system is housed in a vehicle or helicopter, and, commencing from a point of known coordinates the acceleration components in the direction of the axes are sampled at microscopic intervals. The change in position from the starting point is obtained by double integration of the acceleration with respect to time.

An essential part of the system is a computer which is used for monitoring and controlling the initial alignment of the system in an Earth-fixed coordinate system and computing relative position in real time.

The total system is therefore composed of three accelerometers, three gyroscopes, computer, power supply and units for the control, storage and display of field data. The components measured

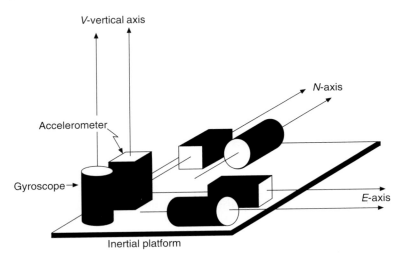

Fig. 6.28 *Schematic inertial platform*

by each of the three accelerometers are accelerations and times, which, when processed and corrected for errors, provide spatial position.

6.5.1 Measurement of acceleration

Acceleration is the rate of change of velocity with respect to time. An acceleration of 4 m/s^2 produces a velocity of $4 \times 1 = 4$ m/s over a 1-s interval; as velocity is rate of change of distance with time, then in 1 s the incremental distance is $4 \times 1 = 4$ m. This double numerical integration process is continually applied, so that at the next survey position the distance components in all three axes relative to the initial starting point are known. As the axes are maintained parallel in space to the coordinate system chosen as the starting point, the coordinates of the next point are known.

Accelerators measure the acceleration in the line of the accelerator. The principle can be explained by considering a pendulum contained within a case (*Figure 6.29*). When the case is at rest, or moving at a constant velocity, the pendulum hangs vertically in its null position. Movement of the case to the right, in the direction of the sensitive axis, would cause the pendulum to swing to the left. The detector senses the movement and sends a signal to the torquing system to direct a force (F) to the pendulum to keep it from swinging. The signal sent is directly proportional to the force exerted and can be amplified and measured. Then, knowing the force and mass of the pendulum, acceleration is obtained from Newton's first law that force equals mass × acceleration. This form of accelerometer is referred to as a 'torque-rebalanced pendulus accelerometer'. The digitized electrical signal, when multiplied by the time interval, produces the incremental velocity component.

The characteristics of the gyros and their application to direction orientation are dealt with in the final chapter. The function of the gyroscopes is to stabilize the triad of accelerometers parallel to the initial three-dimensional coordinate system adopted. The system therefore resists rotation but allows measurement of translation from point to point. Misalignment of the accelerators will result in changes in the measurement of acceleration. For instance, if the accelerometer in the east axis is misaligned by an angle of α, the acceleration a_E becomes $a_E \cos \alpha$, whilst the north axis acceleration a_N becomes $a_N \sin \alpha$. In the vertical axis the accelerometer not only measures vertical acceleration but also acceleration due to gravity. This must be subtracted from the output in order to compute difference in elevation.

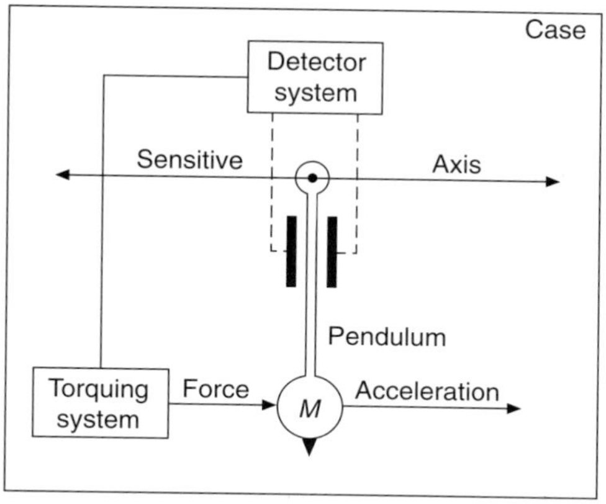

Fig. 6.29 *Principle of the accelerator*

6.5.2 Survey procedure

The first step in the survey procedure is the calibration of the system. This largely comprises a sequence of tests controlled by the on-board computer. Gyro and accelerometer parameters are tested to see if they are within the specified ranges and are recorded for later use in data analysis. All appropriate components are monitored for operational stability. The process is repeated several times and may take up to two hours to complete.

The next step is the alignment of the accelerometer coordinate system to the geodetic coordinate system to be used. This is referred to as a local vertical or local north system. Throughout the survey, the accelerometer system is rotated into the geodetic system by complex mechanical procedures controlled by the computer. Regardless of this, displacement from the initial orientation does occur, resulting in coordinate errors of the points surveyed. In addition, other errors affect the accelerations measured and steps must be taken to reduce their cumulative effect. This is done by stopping the vehicle every 3–5 min and informing the computer that the vehicle has stopped and all three velocity values should be zero. If they are not zero, the drift of each accelerometer and gyroscope is recorded. These velocity errors may be put into a Kalman filter and a new set of relations between the errors and the system is calculated to enable corrections to be made to the position of that point and all previous points. A Kalman filter is a series of mathematical models which statistically relates the different sources of error to one another, and computes the most likely estimates for the corrections in real time. The overall procedure is called a 'zero velocity update' or ZUPT. (If post-processing is to be used then the raw data, without statistical filtering are preferred.) In this way, the error effect is kept to a minimum and the coordinate differences obtained are applied to the input coordinates of the starting point, to give real-time coordinate values. A ZUPT takes less than a minute to complete.

When the ISS reaches a survey point whose coordinate position is required, a station-marking procedure is initiated. The procedure is similar to a ZUPT but in addition to the gyro and accelerometer data, the preliminary coordinates of the point are recorded. Also, as the point of reference for the ISS cannot be located over the station point, the appropriate measurements taken to relate the ISS to the station are also logged in the computer. Depending on the location of the survey point relative to the ISS, the measurements may be simple taped offsets or length and bearing using a total station. Whatever method is used, the measurements must be to an accuracy commensurate with that of the ISS system.

At the completion of the ISS traverse onto a known survey point, the computed coordinates and elevation are compared with the known values and the difference used in an appropriate adjustment or balancing procedure. The balancing may be carried out in proportion to travel time or travel distance (Bowditch) or a combination of both. If post-processing is employed, a more rigorous analysis of the error sources may enable a more accurate distribution of errors throughout the traverse.

To ensure the best results, the following procedures should be adopted:

(1) The route between survey points should be driven in a straight line and at a uniform time rate. This will optimize the Kalman filtering, as orientation errors would increase with route changes, rapid accelerations and bumpy roads.
(2) ZUPTs must be performed every 3–5 min to provide systems control.
(3) The system must be recalibrated every 5–6 hours.
(4) Known points should be included for reference purposes every 1–2 hours.
(5) All the known control points should be identified in advance to ensure easy access.
(6) The traverse must close on a known point and the survey run back to the beginning.

6.5.3 Accuracies and applications

At the present time accuracies in the region of 200 mm in plan and 100 mm in elevation have been quoted. This accuracy may be further enhanced by shortening the interval between ZUPTs and reducing the total survey time to less than two hours. In such cases, accuracies in the region of 10 mm or less have been obtained.

ISS provides a flexible and accurate surveying system which can provide point positioning at high speeds. It is independent of weather conditions and so completely computerized that it is virtually free from human error. The one negative aspect of ISS is its very high cost, which makes it viable only for the type of organization with sufficient project work to make it cost effective. In the UK it has been used for road inventory surveys, where many kilometres of road have been surveyed in record time. It has been estimated that the rate of point positioning is as much as 20 times quicker than conventional methods. It should be remembered than an ISS survey must start from and connect into known points. Hence the system must always be integrated with an established system of points. Basically, any project requiring the fixing of a large number of points over a large area might best be carried out using ISS.

Worked examples

Example 6.2. The following table gives the coordinates of the sides of a traverse *ABCDEFA*.

Side	ΔE (m)	ΔN (m)
AB	−76.35	−138.26
BC	145.12	−67.91
CD	20.97	109.82
DE	187.06	31.73
EF	−162.73	77.36
FA	−87.14	−25.24

It is apparent from these values that an error of 30 m has occurred, and is most likely to be in either *BC* or *EF*. Explain the reasons for these statements.

Tacheometric readings were taken from *A* to a vertical staff at *D*. The telescope angle was 24°

below horizontal and stadia readings of 1.737, 2.530 and 3.322 m were recorded. Use these readings to decide which length should be re-measured and also find the difference in level between stations *A* and *D* if the instrument height was 1.463 m above the station at *A*. (LU)

Summing the above coordinates gives an error of + 26.93 (*E*), – 12.5 (*N*), the error vector being $(26.93^2 + 12.5^2)^{\frac{1}{2}} = 30$ m.

Thus, inspection of the above coordinates indicates the lines *BC* or *EF* as being the only possible sources of the error.

$$\text{Bearing of error vector} = \tan^{-1}\frac{26.9}{12.5} \approx \frac{2}{1}$$

$$\text{Bearing of } BC = \tan^{-1}\frac{145.12}{67.91} \approx \frac{2}{1}$$

$$\text{Bearing of } EF = \tan^{-1}\frac{162.73}{77.36} \approx \frac{2}{1}$$

Thus, the error could lie in either line as they are both parallel to the error vector. One must therefore utilize the tacheometric data as follows in order to isolate the line in question.

$$\text{Distance } AD = 100S \cos^2 \theta$$
$$= 100 \times 1.585 \cos^2 24° = 132.3 \text{ m}$$

Distance *AD* from coordinates $(96.35^2 + 89.74^2)^{\frac{1}{2}} = 131.7$ m

Thus, the error of 30 m cannot be in the line *BC* and must be in *EF*. An inspection of the co-ordinates indicates that *EF* should be increased by 30 m.

Vertical height by tacheometry = 132.3 tan 24° = 58.90 m

∴ Difference in level of *A* and *D* = 1.463 – 58.90 – 2.530 = 59.97 m

Example 6.3. The following survey was carried out from the bottom of a shaft at *A*, along an existing tunnel to the bottom of a shaft at *E*.

Line	WCB °	′	″	Measured distance (m)	Remarks
AB	70	30	00	150.00	Rising 1 in 10
BC	0	00	00	200.50	Level
CD	154	12	00	250.00	Level
DE	90	00	00	400.56	Falling 1 in 30

If the two shafts are to be connected by a straight tunnel, calculate the bearing *A* to *E* and the grade.

If a theodolite is set up at *A* and backsighted to *B*, what is the value of the clockwise angle to be turned off, to give the line of the new tunnel? (KU)

$$\text{Horizontal distance } AB = \frac{150}{(101)^{\frac{1}{2}}} \times 10 = 149.25 \text{ m}$$

$$\text{Rise from } A \text{ to } B = 150/(101)^{\frac{1}{2}} = 14.92 \text{ m}$$

$$\text{Fall from } D \text{ to } E = \frac{400.56}{(901)^{\frac{1}{2}}} = 13.34 \text{ m}$$

$$\text{Horizontal distance } DE = \frac{400.56}{(901)^{\frac{1}{2}}} \times 30 = 400.34 \text{ m}$$

Coordinates (ΔE, ΔN)	0	0	A
149.25 $\begin{matrix}\sin\\\cos\end{matrix}$ 70° 30′ 00″	140.69	49.82	B
200.50 due N	0	200.50	C
200.00 $\begin{matrix}\sin\\\cos\end{matrix}$ 154° 12′ 00″	108.81	−225.08	D
400.34 due E	400.34	0	E
Total coords of *E*	(E) 649.84	(N) 25.24	

∴ Tunnel is rising from *A* to *E* by (14.92 − 13.34) = 1.58 m

∴ Bearing $AE = \tan^{-1}\dfrac{+649.84}{+25.24} = 87° \, 47$

Length = 649.84/sin 87° 47′ = 652.33 m

Grade = 1.58 in 652.33 = 1 in 413

Angle turned off = *BAE* = (87° 47′ − 70° 30′) = 17° 17′ 00″

Example 6.4. A level railway is to be constructed from *A* to *D* in a straight line, passing through a large hill situated between *A* and *D*. In order to speed the work, the tunnel is to be driven from both sides of the hill (*Figure 6.30*).

The centre-line has been established from *A* to the foot of the hill at *B* where the tunnel will commence, and it is now required to establish the centre-line on the other side of the hill at *C*, from which the tunnel will be driven back towards *B*.

To provide this data the following traverse was carried out around the hill:

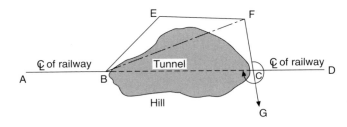

Fig. 6.30

Side	Bearing			Horizontal distance (m)	Remarks
	°	′	″		
AB	88	00	00	–	Centre line of railway
BE	46	30	00	495.8 m	
EF	90	00	00	350.0 m	
FG	174	12	00	–	Long sight past hill

Calculate:

(1) The horizontal distance from *F* along *FG* to establish point *C*.
(2) The clockwise angle turned off from *CF* to give the line of the reverse tunnel drivage.
(3) The horizontal length of tunnel to be driven. (KU)

Find total coordinates of F relative to B

	ΔE (m)	ΔN (m)	Station
$495.8 \genfrac{}{}{0pt}{}{\sin}{\cos} 46°30'00''$ →	359.6	341.3	BE
$350.0 - 90°00'00''$ →	350.0	–	EF
Total coordinates of *F*	E 709.6	N 341.3	F

$$\text{WCB of } BF = \tan^{-1}\frac{709.60}{341.30} = 64°18'48''$$

Distance $BF = 709.60/\sin 64°18'48'' = 787.42$ m

Solve triangle *BFC* for the required data.
The bearings of all three sides of the triangle are known, from which the following values for the angles are obtained:

FBC = 23°41′12″
BCF = 86°12′00″
CFB = 70°06′48″
―――――――
180°00′00″ (Check)

By sine rule:

(a) $FC = \dfrac{BF \sin FBC}{\sin BCF} = \dfrac{787.42 \sin 23°41'12''}{\sin 86°12'00''} = 317.03$ m

(b) $360° - BCF = 273°48'00''$

(c) $BC = \dfrac{BF \sin CFB}{\sin BCF} = \dfrac{787.42 \sin 70°06'48''}{\sin 86°12'00''} = 742.10$ m

Example 6.5.

Table 6.7   *Details of a traverse* ABCDEFA:

Line	Length (m)	WCB	ΔE (m)	ΔN (m)
AB	560.5		0	−560.5
BC	901.5		795.4	−424.3
CD	557.0		−243.0	501.2
DE	639.8		488.7	412.9
EF	679.5	293°59′		
FA	467.2	244°42′		

Adjust the traverse by the Bowditch method and determine the coordinates of the stations relative to *A* (0.0). What are the length and bearing of the line *BE*? (LU)

Complete the above table of coordinates:

	Line	ΔE (m)	ΔN (m)
$679.5 {\sin \atop \cos} 293°59′ \quad \rightarrow$	EF	−620.8	+276.2
$467.2 {\sin \atop \cos} 244°42′ \quad \rightarrow$	FA	−422.4	−199.7

Table 6.8

Line	Lengths (m)	ΔE (m)	ΔN (m)	Corrected ΔE	Corrected ΔN	E	N	Stns
A						0.0	0.0	A
AB	560.5	0	−560.5	0.3	−561.3	0.3	−561.3	B
BC	901.5	795.4	−424.3	795.5	−425.7	796.2	−987.0	C
CD	557.0	−243.0	501.2	−242.7	500.3	553.5	−486.7	D
DE	639.8	488.7	412.9	489.0	411.9	1042.5	−74.8	E
EF	679.5	−620.8	276.2	−620.4	275.2	422.1	200.4	F
FA	467.2	−422.4	−199.7	−422.1	−200.4	0.0	0.0	A
						Check	Check	
Sum	3805.5	−2.1	5.8	0.0	0.0			
Correction to coordinates		2.1	−5.8					

The Bowditch corrections (δE, δN) are computed as follows, and added algebraically to the co-ordinate differences, as shown in *Table 6.8*.

Line	δE (m)	δN (m)
	$\dfrac{2.1}{3805.5} \times 560.5$ giving	$\dfrac{-5.8}{3805.5} \times 560.5$ giving
AB	$K_2 \times 560.5 = 0.3$	$K_1 \times 560.5 = -0.8$
BC	$K_2 \times 901.5 = 0.5$	$K_1 \times 901.5 = -1.4$
CD	$K_2 \times 557.0 = 0.3$	$K_1 \times 557.0 = -0.9$
DE	$K_2 \times 639.8 = 0.3$	$K_1 \times 639.8 = -1.0$
EF	$K_2 \times 679.5 = 0.4$	$K_1 \times 679.5 = -1.0$
FA	$K_2 \times 467.2 = 0.3$	$K_1 \times 467.2 = -0.7$
	Sum = 2.1	Sum = -5.8

To find the length and bearing of *BE*:

$$\Delta E = 1042.2, \quad \Delta N = 486.5$$

$$\therefore \text{ Bearing } BE = \tan^{-1}\frac{1042.2}{486.5} = 64°59'$$

Length $BE = 1042.2/\sin 64°59' = 1150.1$ m

Example 6.6. In a quadrilateral *ABCD* (*Figure 6.31*), the coordinates of the points, in metres, are as follows:

Point	E	N
A	0	0
B	0	−893.8
C	634.8	−728.8
D	1068.4	699.3

Find the area of the figure by calculation.

If *E* is the mid-point of *AB*, find, either graphically or by calculation, the coordinates of a point *F* on the line *CD*, such that the area *AEFD* equals the area *EBCF*. (LU)

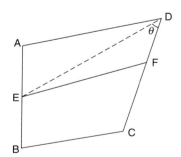

Fig. 6.31

The above coordinates are total coordinates and, therefore the appropriate rule is used.

Stn	E	N	Difference of E	Sum of N	Double +	Area −
A	0	0	0	−893.8		
B	0	−893.8	−634.8	−1622.6	1030026	
C	634.8	−728.8	−433.6	−29.5	12791	
D	1068.4	699.3	1068.4	699.3	747132	
A	0	0				
				Σ	1789949	
				Area	894974	m²

Rounding off the above values to the correct number of significant figures gives 895000 m².

To find the coordinates of *F* by calculation:
From coordinate geometry it is easily shown that the coordinates of *E* are the mean of *A* and *B*.

Stn	E	N	Difference of E	Sum of N	Double +	Area −
A	0	0	0	−446.9		
E	0	−446.9	−1068.6	252.4		269700
D	1068.4	699.3	1068.4	699.3	747100	
				Σ	477400	
				Area	238700 m²	

By coordinates, as above, area of triangle *AED* is found.

$$\therefore \text{ Area of triangle } EDF = \frac{895000}{2} - 238700 = 208800 \text{ m}^2$$

From coordinates

$$\text{Bearing } ED = \tan^{-1} \frac{+1068.4}{+1146.2} = 42°59'$$

$$\text{Length} = 1146.2 \cos 42°59' = 1567.0 \text{ m}$$

$$\text{Bearing } DC = \tan^{-1} \frac{-433.6}{-1428.1} = 196°54'$$

$$\therefore \theta = (42°59' - 16°54') = 26°05°$$

Now: Area of triangle $EDF = \frac{1}{2} DE \times DF \sin \theta = 208800 \text{ m}^2$

$$\therefore DF = 208800/(0.5 \times 1567 \times \sin 26°05') = 606 \text{ m}$$

Thus coordinates of *F* relative to *D* are

$$606 \frac{\sin}{\cos} 196° 54' = -176.2(\Delta E) - 579.9(\Delta N)$$

∴ Total coordinates of $F = 892.2$ E119.4 N

Example 6.7. The mean values of the angles *A*, *B* and *C* of a triangle as measured in a major triangulation were as follows, with the weights shown: *A* 50°22′32.5″, 5; *B* 65°40′47.5″, 3; *C* 63°56′ 46.5″, 6. The length of the side *BC* was 37.5 km and the radius of the Earth 6267 km. Calculate: (a) the spherical excess; (b) the probable values of the spherical angles. (LU)

(a) Spherical excess $\quad E'' = \dfrac{\frac{1}{2}ab \sin \hat{C} \times 206\,265}{R^2}$

From the sine rule $\quad b = a \sin B/\sin A$

$$\therefore E'' = \frac{a^2 \sin B \sin C}{2R^2 \sin A} \times 206\,265 = 39''$$

(b) Sum of adjusted spheroidal angles should equal $180° + E''$, i.e. $180°00'03.9''$.

Angle	Mean value ° ′ ″	Weight	Reciprocal weight	Correction angles	Corrected angles ° ′ ″
A	50 22 32.5	5	$\frac{1}{5} \times 30 = 6$	$\dfrac{-2.6 \times 6}{21} = -0.7''$	50 22 31.8
B	65 40 47.5	3	$\frac{1}{3} \times 30 = 10$	$\dfrac{-2.6 \times 10}{21} = -1.3''$	65 40 46.2
C	63 56 46.5	6	$\frac{1}{6} \times 30 = 5$	$\dfrac{-2.6 \times 5}{21} = -0.6''$	63 56 45.9
Sum	180 00 06.5		Sum = 21	Sum = −2.6″	180 00 03.9

03.9

∴ Correction $= -2.6''$

Example 6.8. Four triangulation stations are in the form of a triangle *ABC*, within which lies the fourth station *D*. The measured angles with the log sins of the outer angles are given below. Adjust the angles to the nearest second by the method of equal shifts.

Number		Measured angle ° ′ ″			Log sin	Difference in LS for 1″
1	BAD	26	31	32	$\bar{1}$.649 915 6	0.000 004 2
2	ABD	20	57	35	$\bar{1}$.553 532 9	55
3	DBC	35	05	09	$\bar{1}$.759 519 0	32
4	BCD	30	28	41	$\bar{1}$.705 186 3	36
5	ACD	26	59	46	$\bar{1}$.656 989 0	41
6	CAD	39	57	26	$\bar{1}$.807 680 7	25
7	ADB	132	30	50		
8	BDC	114	26	04		
9	CDA	113	03	06		

Refer to *Figure 6.22(a)* and use the method outlined in *Section 6.2.3* (page 279)   (LU)

Δ	Number	Angle °	′	″	First corr'n ″	Corrected angles ″	Central angles °	′	″	Second corr'n ″	Corrected angles ″
	1	26	31	32	1	33				−0.5	32.5
ABD	2	20	57	35	1	36				−0.5	35.5
	7	132	30	50	1	51	132	30	51	1	52
Sum		179	59	57							
	3	35	05	09	2	11				−0.5	10.5
BCD	4	30	28	41	2	43				−0.5	42.5
	8	114	26	04	2	06	114	26	06	1	7
Sum		179	59	54							
	5	26	59	46	−6	40				−0.5	39.5
CAD	6	39	57	26	−6	20				−0.5	19.5
	9	113	03	06	−6	00	113	03	00	1	1
Sum		180	00	18			359	59	57		

As the correction to the central angles is 1″, this automatically gives a correction of −0.5″ to each of the base angles of the triangles to restore them to 180°.

Number	Side condition Log sin (odd)	Number	Side condition Log sin (even)	Difference 1″	Correction ″	Final angles °	′	″
1	$\overline{1}$.649 915 6			42	−1	26	31	31.5
		2	$\overline{1}$.553 532 9	55	1	20	57	36.5
3	$\overline{1}$.759 519 0			32	−1	35	05	09.5
		4	$\overline{1}$.705 186 3	36	1	30	28	43.5
5	$\overline{1}$.656 989 0			41	−1	26	59	38.5
		6	$\overline{1}$.807 680 7	25	1	39	57	20.5
Sum	$\overline{1}$.066 423 6		$\overline{1}$.066 399 9	231				
	399 9							

Difference = 237   But $\frac{237}{231}$ of 1″ ≈ 1″

The central angles are as shown at the end of the second correction. The final angles shown may now be rounded off to the nearest second.

Example 6.9. In the triangulation network shown in *Figure 6.32*, all the angles have been observed and the sides *DH* and *GC* measured as base and check base respectively, with the following results:

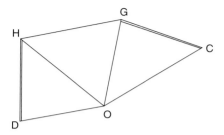

Fig. 6.32

	ΔDHO				ΔHGO				ΔGCO		
	°	′	″		°	′	″		°	′	″
$\hat{D}$ = 79		47	05	$\hat{H}$ = 77		28	58	$\hat{G}$ = 82		22	17
$\hat{H}$ = 58		32	35	$\hat{G}$ = 36		02	38	$\hat{C}$ = 71		29	47
$\hat{O}$ = 41		40	05	$\hat{O}$ = 66		28	48	$\hat{O}$ = 26		08	17
DH = 426.58 m								GC = 486.83 m			

Adjust the observed angles by 'equal shifts' to give a consistent figure.   (ICE)

The requirement in this question is that the figure should be adjusted so that the 'computed' value of the check base equals the 'measured' value.

First, adjust each triangle. Summing the angles of each triangle gives: *DHO* = 179°59′45″, *HGO* = 180°00′24″ and *GCO* = 180°00′21″. There is thus a correction per angle of 5″, –8″ and –7″ per triangle, respectively. The corrected angles are now as follows:

	ΔDHO				ΔHGO				ΔGCO		
	°	′	″		°	′	″		°	′	″
$\hat{D}$ = 79		47	10	$\hat{H}$ = 77		28	50	$\hat{G}$ = 82		22	10
$\hat{H}$ = 58		32	40	$\hat{G}$ = 36		02	30	$\hat{C}$ = 71		29	40
$\hat{O}$ = 41		40	10	$\hat{O}$ = 66		28	40	$\hat{O}$ = 26		08	10

By the sine rule through *Figure 6.32*, the computed value for

$$GC = \frac{HD \sin H\hat{D}O \sin G\hat{H}O \sin G\hat{O}C}{\sin H\hat{O}D \sin O\hat{G}H \sin O\hat{C}G}$$

Taking logs	*Difference for 10″*	° ′ ″	*Difference for 10″*
log 426.58 = 2.630 001			
log sin 79°47′10″ = $\bar{1}$.993 063	3.7	log sin 41 40 10 = $\bar{1}$.822 712	23.7
log sin 77°28′50″ = $\bar{1}$.989 548	4.7	log sin 36 02 30 = $\bar{1}$.769 653	29.0
log sin 26°08′10″ = $\bar{1}$.643 951	42.8	log sin 71 29 40 = $\bar{1}$.976 943	7.2
Σ = 2.256 563		Σ = $\bar{1}$.569 308	

$\therefore$ Log $GC = 2.687255 = 486.69$ m   (computed)
　　Log $GC = 2.687378 = 486.83$ m   (measured)
　　　　　　　　　　　‾‾‾‾‾‾‾‾‾‾‾
Difference $= 0.000\,123$

This difference must be adjusted among the six angles used in the computation so that the final log value of *GC* (computed) would equal that of *GC* (measured).

Sum of differences for $10'' = 111.1$

$\therefore$ Correction per angle $= \left(\dfrac{123}{111}\right) \times 10'' = 11''$

As the final log value of *GC* (computed) needs to be increased, then inspection of the log computation shows that angles *HDO*, *GHO* and *GOC* would be adjusted by $+11''$ each, whilst *HOD*, *OGH* and *OCG* are adjusted by $-11''$ each. The three angles not used in the computation remain as shown in the first correction.

Example 6.10. In a triangle *ABC*, $AB = 5205.0$m, $AC = 5113.8$ m and the angles *B* and *C* were $55°01'05''$ and $62°04'20''$, respectively. Station *A* could not be occupied and observations were taken from satellite station *P*, 11.1 m from *A* and inside the triangle. Instrument readings at *P* were: on *A*, $0°00'00''$; on *C*, $148°28'40''$; on *B* $211°31'10''$. Calculate the angular error in the triangle.
(LU)

As the theodolite is a clockwise-measuring instrument, the instrument readings at *P* serve to fix the relative positions of *A*, *B* and *C* (*Figure 6.33*), as well as the following angular values:

$A\hat{P}C = 148°28'40'', \quad C\hat{P}B = 63°02'30'', \quad B\hat{P}A = 148°28'50''.$

By the sine rule in $\triangle APC$

$\alpha'' = \dfrac{AP \sin A\hat{P}C}{AC} \times 206\,265 = \dfrac{11.1 \sin 148°28'40''}{5113.8} \times 206\,265$

$= 234'' = 0°03'54''$

Similarly in $\triangle APB$

$\theta'' = \dfrac{AP \sin B\hat{P}A}{AB} \times 206\,265 = \dfrac{11.1 \sin 148°28'50''}{5205.0} \times 206\,265$

$= 230'' = 0°03'50''$

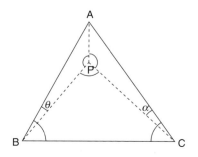

Fig. 6.33

$$\therefore \quad C\hat{A}B = C\hat{P}B - \alpha'' - \theta''$$

$$= 63°02'30'' - 03'54'' - 03'50'' = 62°54'46''$$

$$\therefore \text{ Angular error } = 180° - (\hat{A} + \hat{B} + \hat{C})$$

$$= 180° - (62°54'46'' + 55°01'05'' + 62°04'20'')$$

$$= +11''$$

Exercises

(*6.1*) In a closed traverse *ABCDEFA* the angles and lengths of sides were measured and, after the angles had been adjusted, the traverse sheet shown below was prepared.

It became apparent on checking through the sheet that it contained mistakes. Rectify the sheet where necessary and then correct the coordinates by Bowditch's method. Hence, determine the co-ordinates of all the stations. The coordinates of *A* are E − 235.5, N + 1070.0.

Line	Length (m)	WCB °	′	″	Reduced bearing °	′	″		ΔE (m)	ΔN (m)
AB	355.52	58	30	00	N 58	30	00	E	303.13	185.75
BC	476.65	185	12	30	S 84	47	30	W	−474.70	−43.27
CD	809.08	259	32	40	S 79	32	40	W	−795.68	−146.82
DE	671.18	344	35	40	N 15	24	20	W	−647.08	178.30
EF	502.20	92	30	30	S 87	30	30	E	501.72	−21.83
FA	287.25	131	22	00	S 48	38	00	E	215.58	−189.84

(*Answer*: Mistakes Bearing *BC* to S 5°12′30″ W, hence ΔE and ΔN interchange. ΔE and ΔN of *DE* interchanged. Bearing *EF* to S 87°29′30″ E, giving new ΔN of −21.97 m. Coordinates (*B*) E 67.27, N 1255.18; (*C*) E 23.51, N 781.19; (*D*) E −773.00, N 634.50; (*E*) E −951.99, N 1281.69; (*F*) E −450.78, N 1259.80)

(*6.2*) In a traverse *ABCDEFG*, the line *BA* is taken as the reference meridian. The coordinates of the sides *AB*, *BC*, *CD*, *DE* and *EF* are:

Line	AB	BC	CD	DE	EF
ΔN	− 1190.0	−565.3	590.5	606.9	1017.2
ΔE	0	736.4	796.8	−468.0	370.4

If the bearing of *FG* is 284°13′ and its length is 896.0 m, find the length and bearing of *GA*

(LU)

(*Answer*: 947.8 m, 216°45′)

(*6.3*) The following measurements were obtained when surveying a closed traverse *ABCDEA*:

Line	EA	AB	BC	
Length (m)	793.7	1512.1	863.7	

Included angles	DEA	EAB	ABC	BCD
	93°14′	112°36′	131°42′	95°43′

It was not possible to occupy *D*, but it could be observed from *C* and *E*. Calculate the angle *CDE* and the lengths *CD* and *DE*, taking *DE* as the datum, and assuming all the observations to be correct.   (LU)

(*Answer*: *CDE* = 96°45′, *DE* = 1847.8 m, *CD* = 1502.0 m)

(*6.4*) An open traverse was run from *A* to *E* in order to obtain the length and bearing of the line *AE* which could not be measured direct, with the following results:

Line	AB	BC	CD	DE
Length (m)	1025	1087	925	1250
WCB	261°41′	09°06′	282°22′	71°31′

Find, by calculation, the required information.   (LU)

(*Answer*: 1620.0 m, 339°46′)

(*6.5*) A traverse *ACDB* was surveyed by theodolite and chain. The lengths and bearings of the lines *AC*, *CD* and *DB* are given below:

Line	AC	CD	DB
Length (m)	480.6	292.0	448.1
Bearing	25°19′	37°53′	301°00′

If the coordinates of *A* are *x* = 0, *y* = 0 and those of *B* are *x* = 0, *y* = 897.05, adjust the traverse and determine the coordinates of *C* and *D*. The coordinates of *A* and *B* must not be altered.

(*Answer*: Coordinate error: *x* = 0.71, *y* = 1.41. (*C*) *x* = 205.2*y* = 434.9, (*D*) *x* = 179.1, *y* = 230.8)

(*6.6*) A polygon *ABCDEA* with a central station *O* forms part of a triangulation scheme. The angles in each of the figures which form the complete network are being adjusted, and in this case the angles in each of the triangles *DOE* and *EOA* have already been adjusted and need no further correction.

Making use of the information given in the table below, use the method of equal shifts to determine the correction that must be applied to each of the remaining angles.   (ICE).

Triangle	Angle	Observed value			Log sin	Log sin *difference* for 1″
		°	′	″		
AOB	OAB	40	17	57	$\overline{1}$.810 755 7	25
	OBA	64	11	20	$\overline{1}$.954 355 6	10
	AOB	75	30	52		
BOC	OBC	37	22	27	$\overline{1}$.783 201 4	28
	OCB	71	10	50	$\overline{1}$.976 139 0	7

	BOC	71	26	22		46
COB	OCD	24	51	25	$\bar{1}$.623 615 4	
	ODC	51	48	47	$\bar{1}$.895 421 4	17
	COD	103	19	33		

		Adjusted values			
		°	′	″	
DOE	ODE	67	18	59	$\bar{1}$.965 036 2
	OED	51	02	00	$\bar{1}$.890 707 1
	DOE	61	39	01	
EOA	OEA	116	47	40	$\bar{1}$.950 671 4
	OAE	15	08	02	$\bar{1}$.416 766 2
	EOA	48	04	18	

(*Answer*: OAB 4.8″; OBA −5.8″; AOB −8.0″; OBC 14.8″; OCB 4.2″; BOC 2.0″; OCD 12.8″; ODC 2.2″; COD 0″).

(*6.7*) A bridge is to be built across a river where it is approximately 1.5 km wide and a survey station has been established on each bank to mark the centre line.

Excluding the use of electronic devices, describe how the distance between these two stations can be determined to a high degree of accuracy. Outline the calculations involved and quote the relevant equations at each stage. (ICE).

(*Answer*: Triangulation; braced quadrilateral; base line; figural adjustment)

(*6.8*) In order to demonstrate how a triangulation is adjusted by the method of equal shifts, consider a figure which consists of a triangle *ABC* with a central (internal) point *D* and in which the following fictitious angles are given as 'observed angles': *BAD* = *ABD* = *CBD* = *BCD* = *ACD* = 30°00′; *ADB* = *BDC* = *CDA* = 120°00′; *CAD* = 33°00′.

REFERENCES AND FURTHER READING

Adler, R.K. and Schmutter, B. (1971) 'Precise Traverses in Major Geodetic Networks', *Canadian Surveyor*, March.

Berthon-Jones, P. (1972) 'A Comparison of the Precision of Traverses Adjusted by Bowditch Rule and Least Squares', *Survey Review*, April, No. 164.

Burke, K.F. (1971) 'Why Compare Triangulation and Trilateration?', *Proc. ASCE, Journal of the Surveying and Mapping Division,* October.

Chrzanowski, A. and Konecny, G. (1965) 'Theoretical Comparison of Triangulation, Trilateration and Traversing', *Canadian Surveyor*, Vol. XIX, No. 4, September.

Chrzanowski, A. and Wilson, P. (1967) 'Pre-Analysis of Networks for Precise Engineering Surveys', *Proc. Third S. African Nat. Surv. Conf.*

Curl, S.J. (1977) 'The Effects of Refraction on Engineering Survey Measurements', Ph.D. Thesis, The University of Nottingham.

Hodges, D.J. (1975) 'Calibration and Testing of Electromagnetic Distance-Measuring Instruments', *Colliery Guardian*, No. 11, November.

Hodges, D.J. (1980) 'Electro-Optical Distance Measurement', *Conf. Assoc. of Surveyors in Civil Eng.*, April.

Hodges, D.J., Skellern, P. and Morley, J.A. (1967) 'Trials with a Model 6 Geodimeter for Surface Surveys', *The Mining Engineer*, No. 84, September.

Leahy, F.J. (1977) 'Bowditch Revisited', *Australian Surveyor*, December, Vol. 28, No. 8.

Murphy, B.T. (1974) 'The Adjustment of Single Traverses', *Australian Surveyor*, December, Vol. 26, No. 4.

Ordnance Survey (1950) *Constants, Formulae and Methods Used in Traverse Mercator Projection*, HMSO, London.

Phillips, J.O. (1967) 'Electronic Traverse versus Triangulation', *Proc. ASCE, Journal of the Surveying and Mapping Division*, October.

Schofield, W. (1973) 'Engineering Surveys on the National Grid', *Journal of the Institution of Highway Engineers*, Vol. XX, No. 10, October.

Schwendener, H.R. (1972) 'Electronic Distancers for Short Ranges: Accuracy and Checking Procedures', *Survey Review*, Vol. XXI, NO. 164, April.

Smith, J.R. (1982) 'Equal Shifts by Pocket Calculator', *Civil Engineering Surveyor*, Vol. VII, Issue 5, June 1982.

Thomas, T.L. (1971) 'Desk Computers in Surveying', *Chartered Surveyor*, No. 11.

7

Satellite positioning

7.1 INTRODUCTION

Before commencing this chapter, the reader should have studied Chapter 5 and acquired a knowledge of local and global geoids and ellipsoids, transformations and heights, i.e. Sections 5.1 to 5.8.

The concept of satellite position fixing commenced with the launch of the first Sputnik satellite by the USSR in October 1957. This was rapidly followed by the development of the Navy Navigation Satellite System (NNSS) by the US navy. This system, commonly referred to as the Transit system, was to provide world-wide navigation capability for the US Polaris submarine fleet. The Transit system was made available for civilian use in 1967. However, as it required very long observation periods and had a rather low accuracy, its application was limited to geodetic and navigation uses.

In 1973, the US Department of Defense commenced the development of NAVSTAR (Navigation System with Time and Ranging) global positioning system (GPS), and the first satellites were launched in 1978. These satellites were essentially experimental, with the operational system scheduled for 1987. Now that GPS is fully operational, relative positioning to several millimetres, with extremely short observation periods of a few minutes, has been achieved. For distances in excess of 5 km GPS has been shown to be more accurate than EDM traversing. It therefore has a wide application in engineering surveying, with an effect even greater than the advent of EDM. Apart from the high accuracies attainable, GPS offers the following significant advantages:

(1) Position is determined directly in an X, Y, Z coordinate system.
(2) Intervisibility between ground stations is unnecessary.
(3) As each point is fixed discretely, there is no error propagation as in networks.
(4) Survey points may therefore be selected according to their required function, rather than to produce a well-conditioned network configuration.
(5) Low skill required by the operator.
(6) Position may be fixed on land, at sea or in the air. This latter facility may have a profound effect in aerial photogrammetry.
(7) Measurement may be carried out, day or night, anywhere in the world, at any time and in any type of weather.
(8) Continuous measurement may be carried out, resulting in greatly improved deformation monitoring.

7.2 GPS SEGMENTS

The GPS system can be broadly divided into three segments: the *space* segment, the *control* segment and the *user* segment. The *space* segment is composed of satellites weighing about 400 kg and powered by means of two solar panels with three back-up, nickel-cadmium batteries (*Figure 7.1*). The operational phase consists of 28 satellites, at the present time, with three spares. They are in near-circular orbits, at a height of 20 200 km above the Earth, with an orbit time of 12 hours (11 h 58 min). The six equally spaced orbital planes (*Figure 7.2*), are inclined at 55° to the equator, resulting in five hours above the horizon. The system therefore guarantees that at least four satellites will always be in view.

Each satellite has a fundamental frequency of 10.23 MHz and transmits two L-band radio signals. Signal L_1 has a frequency of 1575.42 MHz (10.23×154) and L_2 a frequency of 1227.60 MHz (10.23×120). Modulated onto these signals are a Coarse Acquisition (C/A) code, now referred to as the Standard S-code, and a Precise P-code. The L_1 frequency has both the P- and S-codes, whereas the L_2 has only the P-code. The codes are pseudo-random binary sequences transmitted at frequencies of 1.023 MHz (S-code) and 10.23 MHz (P-code) (*Figure 7.3*). The P-code provides what is termed the precise positioning service (PPS) and the S-code the standard positioning service (SPS). The SPS will provide absolute point position to an accuracy of 100–300 m; the PPS to an accuracy of 5–10 m.

The codes are, in effect, time marks linked to ultra-accurate clocks (oscillators) on board the satellites. Each satellite carries three rubidium or caesium clocks having a precision in the region of 10^{-13}s. In addition, both L_1 and L_2 carry a formatted data message, transmitted at a rate of 50 bits per second, containing satellite identification, satellite ephemeris, clock information, ionospheric data, etc.

The *control* segment has the task of supervising the satellite timing system, the orbits and the mechanical condition of the individual satellites. Neither the timing system nor the orbits are sufficiently stable to be left unchecked for any great period of time.

The satellites are currently tracked by five monitor stations, situated in Kwajalein, Hawaii, Ascension and Diego Garcia, with the master control in Colorado Springs.

As the basic principle of position fixing using GPS is that of a resection, using distances to three

Fig. 7.1 *GPS satellite*

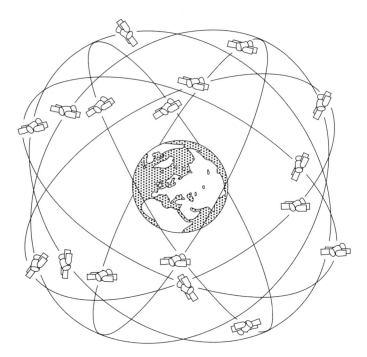

Fig. 7.2 *The GPS satellite constellation: 24 satellites, 6 orbital planes, 55° inclination, 20 200 km altitude 12-hour orbits. (Courtesy Wild Heerbrugg)*

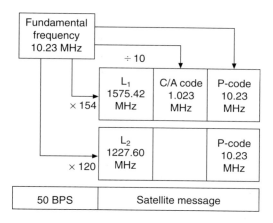

Fig. 7.3 *(Courtesy of Leical)*

known points (satellites), the position of the satellites (in a known coordinate system) is critical. The position of the satellite is obtained from data broadcast by the satellite and called the 'broadcast ephemeris'. The positional data from all the tracking stations are sent to the master control for processing. These data, combined with the satellite's positions on previous orbits, make it possible to predict the satellite's position for several hours ahead. This information is uploaded to the satellite, for subsequent transmission to the user, every eight hours. Orbital positioning is currently accurate to about 10 m, but would degrade if not continuously updated.

The master control is also connected to the time standard of the US Naval Observatory in Washington, DC. In this way, satellite time can be synchronized and data relating it to Universal Time transmitted. Other data regularly updated are the parameters defining the ionosphere, to facilitate the computation of refraction corrections to the distances measured. The *user* segment consists essentially of a portable receiver/processor with power supply and an omnidirectional antenna (*Figure 7.4*). The processor is basically a microcomputer containing all the software for processing the field data.

7.3 GPS RECEIVERS

Basically, a receiver obtains pseudo-range or carrier phase data to at least four satellites. As GPS receiver technology is developing so rapidly, it is only possible to deal with some of the basic operational characteristics. The type of receiver used (*Figure 7.5*) will depend largely upon the requirements of the user. For instance, if GPS is to be used for absolute as well as relative positioning, then it is necessary to use pseudo-ranges. If high-accuracy relative positioning is the requirement, then the carrier phase would be the observable involved. From this initial consideration it can be seen that, for real-time pseudo-range positioning, the user would need access to the navigation message (Broadcast Ephemerides). If carrier waves are to be used, the data are post-processed and an external precise ephemeris may be used. Thus, where the navigation message is essential, a code-correlating receiver would be used. If carrier phase and post-processing are the requirement, a codeless receiver may be preferred.

Fig. 7.4   *GPS antenna and receiver (Courtesy of Aga Geotronics)*

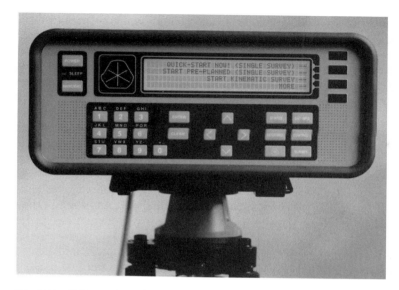

Fig. 7.5   *GPS receiver (Courtesy of Aga Geotronics)*

A receiver generally has one or more channels. A channel consists of the hardware and software necessary to track a satellite on the code and/or carrier phase measurement, continuously. Receivers can therefore be multichannel in order to track a number of satellites. Multiplexing enables a single channel to rapidly sequence the signals from a number of satellites at a rate of about 50 per second. The sequencing is so rapid that continuous pseudo-ranges can be measured to all the satellites being tracked, plus capture of all the navigation messages. Their great limitation from the engineering surveying viewpoint is that they cannot track carrier phase. Sequencing through satellites using one, two or three channels is the process used by some receivers. Where there is only one channel, the satellite is tracked and the data acquired before moving on to the next satellite. Where there are two or three channels, the extra channels may be used to locate the next satellite and update the ephemeris, thereby speeding up the process. A maximum of four satellites can be sequenced at a rate of about one every five seconds.

When using the carrier phase observable, it is necessary to remove the modulations. The code-correlation-type receiver uses a delay lock loop to maintain alignment between the incoming, satellite-generated signal. The incoming signal is multiplied by its equivalent part of the generated signal, which has the effect of removing the codes. It does still retain the navigation message and can therefore utilize the broadcast ephemeris.

The codeless receiver uses a squaring channel and multiplies the received signal by itself, thereby doubling the frequency and removing the code modulation. This process, whilst reducing the signal-to-noise ratio, loses the navigation message and necessitates the use of an external ephemeris service.

Each different type of instrument has its own peculiar advantages and disadvantages. For instance, code-correlating receivers need access to the P-code if tracking on L_2 frequency. As the P-code may be changed to the Y-code and made unavailable to civilian users, L_2 tracking would be eliminated. However, these receivers are capable of tracking satellites at lower elevations than the codeless type.

The receivers used for navigation purposes generally track up to six satellites obtaining L_1 pseudo-range data and, for the majority of harbour entrancing, need to be able to accept differential corrections from an on-shore reference receiver.

Geodetic receivers used in engineering surveying may be single or dual frequency, with from 12 to 24 channels in order to track all the satellites available. Some 24-channel receivers have allocated 12 to GPS and 12 to Glonass, the Russian equivalent system.

All modern receivers can acquire the L_1 pseudo-range observable using a code correlation process illustrated later. When the pseudo-range is computed using the S-code (or C/A-code as it is sometimes referred to), it can be removed from the signal in order to access the L_1 carrier phase and the navigation message. These two signals could be classified as civilian data. Dual frequency receivers also use code correlation to access the P-code pseudo-range data and the L_2 carrier phase. However, this is only possible with the 'permission' of the US military who can prevent access to the P-code. This process is called Anti-Spoofing (AS) and is dealt with later. When AS is operative, a signal squaring technique may be used to access the L_2 code. The process has problems which have been mentioned earlier. Some manufacturers use a code correlation squaring process, which gives half-wavelength L_2 carrier phase data. Two other approaches used by different manufacturers and called 'cross-correlation' and 'PW code tracking' are capable of producing full wavelength L_2 carrier phase data.

No doubt receiver technology will continue to develop smaller and yet more sophisticated instrumentation. Indeed, at the time of writing (May 2000) a wrist watch incorporating GPS has been developed.

7.4 SATELLITE ORBITS

The German astronomer Johannes Kepler (1571–1630) established three laws defining the movement of planets around the Sun, which have been applied to the movement of satellites around the Earth:

(1) Satellites move around the Earth in elliptical orbits, with the centre of mass of the Earth situated at one of the focal points G (*Figure 7.6*). The other focus G' is unused.
(2) The radius vector from the Earth's centre to the satellite sweeps out equal areas at equal time intervals (*Figure 7.7*).
(3) The square of the orbital period is proportional to the cube of the semi-major axis a, i.e. $T^2 = a^3 \times$ constant.

These laws therefore define the geometry of the orbit, the velocity variation of the satellite along its orbital path, and the time taken to complete an orbit.

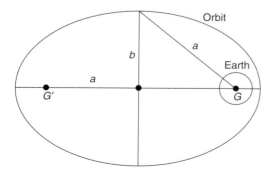

Fig. 7.6

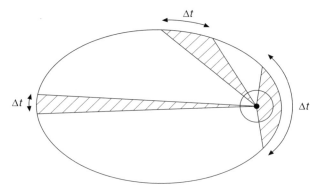

Fig. 7.7

Whilst a and e (the eccentricity) define the shape of the ellipse (see Chapter 5), its orientation in space must be specified by three angles defined with respect to a space-fixed reference coordinate system. The spatial orientation of the orbital ellipse is shown in *Figure 7.8* where:

(1) Angle Ω is the right ascension (RA) of the ascending node of the orbital path, measured on the equator, eastward from the vernal equinox (γ).
(2) i is the inclination of the orbital plane to the equatorial plane.
(3) ω is the argument of perigee, measured in the plane of the orbit from the ascending node.

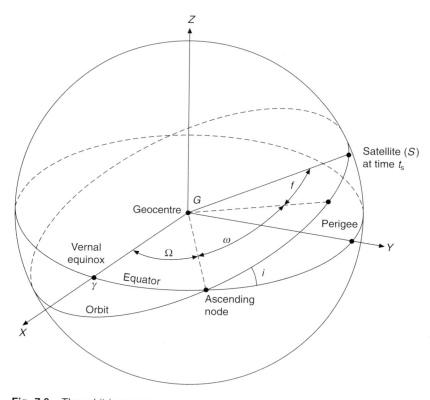

Fig. 7.8 *The orbit in space*

Thus, having defined the orbit in space, the satellite is located relative to the perigee using the angle *f*, called the 'true anomaly' at the time when it passed through the perigee.

The 'perigee' is the point when the satellite is closest to the Earth, and the 'apogee' when it is furthest away. A line joining these two points is called the 'line of apsides' and is the *X*-axis of the orbital space coordinate system. The *Y*-axis is in the mean orbital plane at right angles to the *X*-axis. The *Z*-axis is normal to the orbital plane and will be used to represent small perturbations from the mean orbit. The *XYZ* space coordinate system has its origin at *G*. It can be seen from *Figure 7.9* that the space coordinates of the satellite at time *t* are:

$$X_0 = r \cos f$$
$$Y_0 = r \sin f$$
$$Z_0 = 0 \text{ (in a pure Keplerian orbit)}$$

where *r* = the distance from the Earth's centre to the satellite.

The space coordinates can easily be computed using the information contained in the broadcast ephemeris. The procedure is as follows:

(1) Compute *T*, which is the orbital period of the satellite, i.e. the time it takes to complete its orbit. Using Kepler's third law:

$$T = 2\pi a \, (a/\mu)^{1/2} \tag{7.1}$$

μ is the Earth's gravitational constant and is equal to 398 601 km³ s⁻².

(2) Compute the 'mean anomaly' *M*, which is the angle swept out by the satellite in the time interval $(t_s - t_p)$ from

$$M = 2\pi(t_s - t_p)/T \tag{7.2}$$

where t_s = the time of the satellite signal transmission (observed) and
t_p = the time of the satellite's passage through perigee (obtained from the broadcast ephemeris).

M defines the position of the satellite in orbit but only for ellipses with *e* = *0*, i.e. circles. To correct for this it is necessary to obtain the 'eccentric anomaly' *E* and the 'true anomaly' *f* (*Figure 7.10*) for the near-circular GPS orbits.

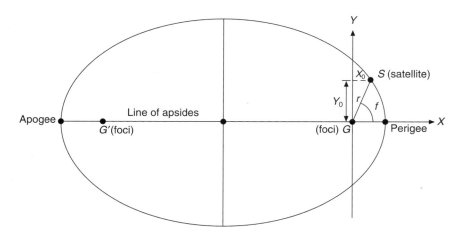

Fig. 7.9 *Orbital coordinate system*

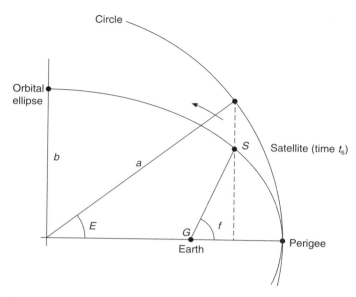

Fig. 7.10 *The orbital ellipse*

(3) From Kepler's equation:

$$E - e \sin E = M \tag{7.3}$$

which is solved iteratively for E. e is the well-known eccentricity of an ellipse, calculated from

$$e = (1 - b^2/a^2)^{1/2}$$

Now the 'true anomaly' f is computed from

$$\cos f = (\cos E - e)/(1 - e \cos E) \tag{7.4}$$

or

$$\tan (f/2) = [(1 + e)/(1 - e)]^{1/2} \tan (E/2) \tag{7.5}$$

(4) Finally, the distance from the centre of the Earth to the satellite (GS), equal to r, is calculated from

$$r = a(1 - e^2)/1 - e \cos f \tag{7.6}$$

and as first indicated

$$
\begin{aligned}
X_0 &= r \cos f \\
Y_0 &= r \sin f \\
Z_0 &= 0
\end{aligned} \tag{7.7}
$$

Thus the position of the satellite is defined in the mathematically pure Keplerian orbit at the time (t_s) of observation.

The actual orbit of the satellite departs from the Keplerian orbit due to the effects of

(1) the non-uniformity of the Earth's gravity field $= q_1$;

(2) the attraction of the moon and sun = q_2;
(3) atmospheric drag = q_3;
(4) direct and reflected solar radiation pressure = q_4 and q_5;
(5) earth tides = q_6;
(6) ocean tides = q_7.

These forces produce orbital perturbations, the total effect ($q_t = q_1 + q_2 + \ldots q_7$) of which must be mathematically modelled to produce a precise position for the satellite at the time of observation.

As already illustrated, the pure, smooth Keplerian orbit is obtained from the elements:

a – semi-major axis
e – eccentricity
    which give the size and shape of the orbit.
i – inclination
Ω – right ascension of the ascending node
    which orient the orbital plane in space with respect to the Earth.
ω – argument of perigee
t_p – ephemeris reference time
    which fixes the position of the satellite.

Additional parameters given in the Broadcast Ephemeris describe the deviations of the satellite motion from the pure Keplerian form. There are two ephemerides available: the Broadcast, shown below, and the Precise.

M_0 = Mean anomaly
Δn = Mean motion difference
e = Eccentricity
$\sqrt{a}$ = Square root of semi-major axis
Ω = Right ascension
i_0 = Inclination
ω = Argument of perigee
$\dot{\Omega}$ = Rate of right ascension
$\dot{i}$ = Rate of inclination
C_{uc}, C_{us} = Correction terms to argument of latitude
C_{rc}, C_{rs} = Correction terms to orbital radius
C_{ic}, C_{is} = Correction terms to inclination
t_p = Ephemeris reference time

Using the Broadcast Ephemeris, plus two additional values from the WGS 84 geopotential model, namely:

ω_e – the angular velocity of the Earth ($7\,292\,115 \times 10^{-11}$ rad s^{-1})
μ – the gravitational/mass constant of the Earth ($3\,986\,005 \times 10^{8}$ m^3 s^{-2})

The Cartesian coordinates in a perturbed satellite orbit can be computed using:

u – the argument of latitude (the angle in the orbital plane, from the ascending node to the satellite)
r – the geocentric radius, as follows

$$X_0 = r \cos u$$

$$Y_0 = r \sin u \quad\quad\quad\quad\quad\quad\quad\quad (7.8)$$

$$Z_0 = 0$$

where

$$r = a(1 - e \cos E) + C_{rc} \cos 2 (\omega + f) + C_{rs} \sin 2 (\omega + f)$$
$$u = \omega + f + C_{uc} \cos 2 (\omega + f) + C_{us} \sin 2 (\omega + f)$$

where $a(1 - e \cos E)$ is the elliptical radius, C_{rc} and C_{rs} the *cosine* and *sine* correction terms of the geocentric radius and C_{uc}, C_{us} the correction terms for u.

It is now necessary to rotate the orbital plane about the X_0 axis, through the inclination i, to make the orbital plane coincide with the equatorial plane and the Z_0 axis coincide with the Z axis of the Earth fixed system (IRP). Thus:

$$X_E = X_0$$
$$Y_E = Y_0 \cos i \tag{7.9}$$
$$Z_E = Y_0 \sin i$$

where

$$i = i_0 + \dot{i}t + C_{ic} \cos 2 (\omega + f) + C_{is} \sin 2 (\omega + f)$$

and i_0 is the inclination of the orbit plane at reference time t_p.

         $\dot{i}$ is the linear change in inclination since the reference time,

     C_{ic}, C_{is} are the amplitude of the cosine and sine correction terms of the inclination of the orbital plane.

Finally, although the Z_E axis is now correct, the X_E axis aligns with the First Point of Aries and requires a rotation about Z towards the Zero Meridian (IRM) usually referred to as the Greenwich Meridian. The required angle of rotation is Greenwich Apparent Sidereal Time (GAST) and is in effect the longitude of the ascending node of the orbital plane (λ_0) at the time of observation t_s.

To compute λ_0 we use the right ascension parameter Ω_0, the change in GAST using the Earth's rotation rate ω_e during the time interval ($t_s - t_p$) and change in longitude since the reference time, thus:

$$\lambda_0 = \Omega_0 + (\dot{\Omega} - \omega_e)(t_s - t_p) - \omega_e t_p$$

and:

$$X = X_E \cos \lambda_0 - Y_E \sin \lambda_0$$
$$Y = X_E \sin \lambda_0 - Y_E \cos \lambda_0 \tag{7.10}$$
$$Z = Z_E$$

The accuracy of the orbit deduced from the Broadcast Ephemeris is about 10 m at best and is directly reflected in the *absolute* position of points. Whilst this may be adequate for some applications, such as navigation, it would not be acceptable for most engineering surveying purposes. Fortunately, differential procedures and the fact that engineering generally requires *relative* positioning using carrier phase, eliminates the effect of orbital error. However, relative positioning accuracies better than 0.1 ppm of the length of the baseline can only be achieved using a Precise Ephemeris.

The GPS satellite coordinates are defined on an ellipsoid of reference called the World Global System 1984 (WGS 84). As stated, the system has its centre coinciding with the centre of mass of the Earth and orientated to coincide with the IERS axes, as described in Chapter 5. Its size and shape is the one that best fits the geoid over the whole Earth, and is identical to the Ellipsoid GRS80 with $a = 6378137.0$ m and $1/f = 298.257223563$.

In addition to being a coordinate system, other values, such as a description of the Earth's gravity field, the velocity of light and the Earth's angular velocity, are also supplied. Consequently, the velocity of light as quoted for the WGS 84 model must be used to compute ranges from observer to satellite, and the subsequent position, based on all the relevant parameters supplied.

The WGS 84 reference system was based on TRANSIT doppler measurements at about 1600 sites, combined with data from satellite laser ranging (SLR) and very long baseline interferometry (VBLI). The result of this is that, as the position of these points defining the global reference system change due to movement of the land masses, the coordinate system will need to be regularly revised. The ability to measure this movement to a very high accuracy by use of the satellites themselves now results in four-dimensional positioning, possibly including rate of change of position.

The final stage, then, of the positioning process is the transformation of the WGS 84 coordinates to local geodetic or plane rectangular coordinates and height. This is usually done using the Helmert transformation outlined in Chapter 5. The translation, scale and rotational parameters between GPS and national mapping coordinate systems have been published. The practical problems involved have already been mentioned in Chapter 5. If the parameters are unavailable, they can be obtained by obtaining the WGS 84 coordinates of points whose local coordinates are known. A least squares solution will produce the parameters required. (The transformation processes are dealt with later in the chapter.)

It must be remembered that the height obtained from satellites is the ellipsoidal height and will require accurate knowledge of the geoid–ellipsoid separation (N) to change it to orthometric.

7.5 BASIC PRINCIPLE OF POSITION FIXING

As previously stated, the principle involves the measurement of distance (or range) to at least three satellites whose X, Y and Z position is known, in order to define the user's X_p, Y_p and Z_p position.

In its simplest form, the satellite transmits a signal on which the time of its departure (t_D)from the satellite is modulated. The receiver in turn notes the time of arrival (t_A) of this time mark. Then the time which it took the signal to go from satellite to receiver is ($t_A - t_D$) $= \Delta t$ (called the delay time). The measured range R is obtained from

$$R_1 = (t_A - t_D)\,C = \Delta t\,C \tag{7.11}$$

where c = the velocity of light.

Whilst the above describes the basic principle of range measurement, to achieve it one would require the receiver to have a clock as accurate as the satellite's and perfectly synchronized with it. As this would render the receiver impossibly expensive, a correlation procedure, using the pseudo-random binary codes (P or S), usually 'S', is adopted. The signal from the satellite arrives at the receiver and triggers the receiver to commence generating the S-code. The receiver-generated code is cross-correlated with the satellite code (*Figure 7.11*). The ground receiver is then able to determine the time delay (Δt) since it generated the same portion of the code received from the satellite. However, whilst this eliminates the problem of an expensive receiver clock, it does not eliminate the problem of exact synchronization of the two clocks. Thus, the time difference between the two clocks, termed clock bias, results in an incorrect assessment of Δt. The distances computed are therefore called 'pseudo-ranges'. The effect of clock bias, however, can be eliminated by the use of four satellites rather than three.

A line in space is defined by its difference in coordinates in an X, Y and Z system:

$$R = (\Delta X^2 + \Delta Y^2 + \Delta Z^2)^{1/2}$$

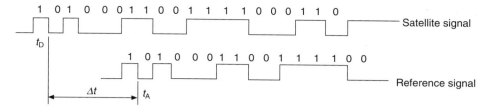

Fig. 7.11   *Correlation of the pseudo-binary codes*

If the error in R, due to clock bias, is δR and is constant throughout, then:

$$R_1 + \delta R = [(X_1 - X_p)^2 + (Y_1 - Y_p)^2 + (Z_1 - Z_p)^2]^{1/2}$$
$$R_2 + \delta R = [(X_2 - X_p)^2 + (Y_2 - Y_p)^2 + (Z_2 - Z_p)^2]^{1/2}$$
$$R_3 + \delta R = [(X_3 - X_p)^2 + (Y_3 - Y_p)^2 + (Z_3 - Z_p)^2]^{1/2}$$
$$R_4 + \delta R = [(X_4 - X_p)^2 + (Y_4 - Y_p)^2 + (Z_4 - Z_p)^2]^{1/2}$$

where X_n, Y_n, Z_n = the coordinates of satellites 1, 2, 3 and 4 (n = 1 to 4)
X_p, Y_p, Z_p = the coordinates required for point P
R_n = the measured ranges to the satellites

Solving the four equations for the four unknowns X_p, Y_p, Z_p and δR eliminates the error due to clock bias.

Whilst the use of pseudo-range is sufficient for navigational purposes and constitutes the fundamental approach for which the system was designed, a much more accurate measurement of range is required for positioning in engineering surveying. This is done by measuring phase difference by means of the carrier wave in a manner analogous to EDM measurement. As observational resolution is about 1% of the signal wavelength λ, the following table shows the reason for using the carrier waves; this is referred to as the carrier phase observable.

GPS signal	*Wavelength λ*	*1% of λ*
S-code	300 m	3 m
P-code	30 m	0.3 m
Carrier	200 mm	2 mm

Carrier phase is the difference between the incoming satellite carrier signal and the phase of the constant-frequency signal generated by the receiver. It should be noted that the satellite carrier signal when it arrives at the receiver is different from that initially transmitted, because of the relative velocity between transmitter and receiver; this is the well-known Doppler effect. The carrier phase therefore changes according to the continuously integrated Doppler shift of the incoming signal. This observable is biased by the unknown offset between the satellite and receiver clocks and represents the difference in range to the satellite at different times or epochs. The carrier phase movement, although analogous to EDM measurement, is a one-way measuring system, and thus the number of whole wavelengths (N) at lock-on is missing; this is referred to as the integer or phase ambiguity. The value of N can be obtained from GPS network adjustment or from double differencing or eliminated by triple differencing.

7.6 DIFFERENCING DATA

Whilst the system was essentially designed to use pseudo-range for navigation purposes, it is the carrier phase observable which is used in engineering surveying to produce high accuracy relative positioning. Carrier phase measurement is similar to the measuring process used in EDM. However, as it is not a two-way process, as in EDM, the observations are ambiguous because of the unknown integer number of cycles between the satellite and receiver at lock-on. Once the satellite signals have been acquired by the receiver, the number of cycles can be tracked and counted (carrier phase) with the initial integer number of cycles, known as the integer ambiguity, still unknown. However, the integer ambiguity will be the same throughout the survey and can be represented by a single bias term (N) (*Figure 7.12*). The integer ambiguity will change only if the receiver loses lock on the satellite. This is known as cycle slip. Cycle slips occur when there is a temporary obstruction in the line of sight between the receiver and satellite. For a moment the tracking stops and only recommences when the line of sight is repaired. Whilst the fractional phase measurement is the same as prior to the break in the line of sight, the integer number of cycles is different. With dual-frequency receivers, cycle slips can occur on either frequency, thereby complicating the problem of their detection even more.

The carrier phase observation equation comprises:

- carrier phase from satellite to receiver in the receiver time frame t_r, and consisting of the fractional part of the wavelength plus the number of wavelengths different from those of the integer ambiguity N, i.e. $\Phi\ t_r$);

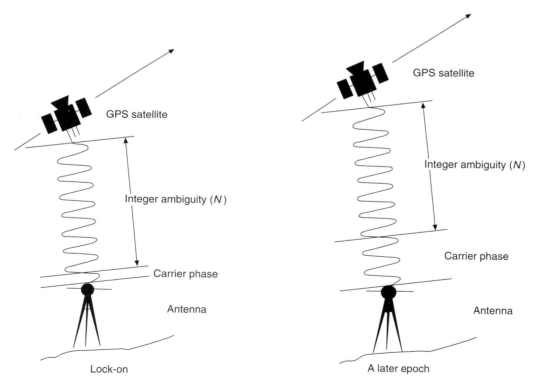

Fig. 7.12 *Integer ambiguity*

- the integer ambiguity N, comprising the integer number of wavelengths from satellite to receiver when the receiver first locks on to the satellite;
- the frequency of the carrier wave (f);
- the speed of light in vacuo (c);
- the geometric range from receiver to satellite (R), in which the coordinates of the satellite and receiver are implicit, i.e. $R = [(X_s - X_r)^2 = (Y_s - Y_r)^2 = (Z_s - Z_r)^2]^{1/2}$;
- the error due to atmospheric refraction through the ionosphere (e_{ION});
- the error due to atmospheric error through the troposphere (e_{TRO}).

It must also be remembered that we are dealing with different time frames, namely:

t_r = the receiver time
t^s = the satellite time
T = the ideal GPS time, and
$T_r = t_r + \delta t_r$
$T_s = t^s + \delta t^s$

where δt_r is the receiver clock offset and δt^s the satellite clock offset.

The pure phase observation equation may be expressed as:

$$\Phi(t_r) = \frac{f}{c} R(t_s, T_r) - f[\delta t_r(t_r) - \delta t^s(t^s)] + N + e_{ION} + e_{TRO} \qquad (7.11)$$

By differencing these phase equations in a variety of ways, one can obtain the position of one point relative to another to a much greater accuracy from that obtained from 'stand alone' data.

7.6.1 *Single differencing* (Figure 7.13)

A single difference is the difference in phase of simultaneous measurements between one satellite position and two ground stations. As shown in *Figure 7.13*, consider two ground stations A and B observing to one satellite S_1. The result would be two equations of the above form for Φ_A and Φ_B, which, when differenced, give:

$$\Phi_{AB}(t_A, t_B) = \frac{f}{c} R_{AB}(T_s, T_A, T_B) - f[\delta t_{AB}(t_A, t_B)] + N_{AB} + e_{ION} + e_{TRO} \qquad (7.12)$$

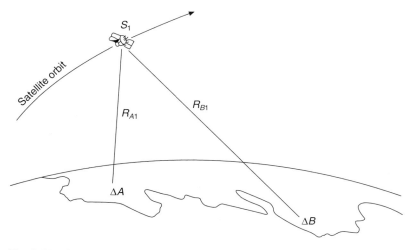

Fig. 7.13 *Single difference*

From this it can be seen that the satellite clock offset δt^s has been eliminated. Orbital and atmospheric errors are also virtually eliminated in relative positioning, as the errors may be assumed identical. Baselines up to 50 km in length would be regarded as short compared with the height of the satellites (20 200 km). Thus it could be argued that the signals to each end of the baseline would pass through the same column of atmosphere, resulting in equal errors cancelling each other out. Similarly with orbital errors and, indeed, with the effects of selective availability (SA).

The above differencing procedure is sometimes referred to as the 'between-station' difference and is the basis of differential GPS. Similarly, simultaneous observations from one ground station to two satellites (between-satellites difference) eliminates receiver clock error (δt_r).

7.6.2 Double differencing *(Figure 7.14)*

A double difference is the difference between two single differences using two ground stations A and B observing simultaneously to two satellites S_1 and S_2. This is the process used in almost all GPS software, for, not only does it remove all the errors as stated for the single difference, but it also removes receiver clock offsets (δt_r). A further asset of this procedure is that it retains and therefore enables the integer ambiguity (N) to be resolved.

7.6.3 Triple differencing *(Figure 7.15)*

A triple difference is the difference of two double differences. The same satellite/receiver combinations are used, but at different epochs. In addition to all the errors removed by double differencing, the integer ambiguity is also removed.

In engineering surveying, single differencing is little used, although it could be used with permanent active stations over long baselines for orbit determination. Triple differencing reduces the number of observations and creates a high noise level. It can, however, be useful in the first state of data editing, particularly the location of cycle slips and their subsequent correction. The magnitude of a cycle slip is the difference between the initial integer ambiguity and the subsequent one, after signal loss. It generally shows up as a 'jump' or 'gap' in the residual output from a least squares

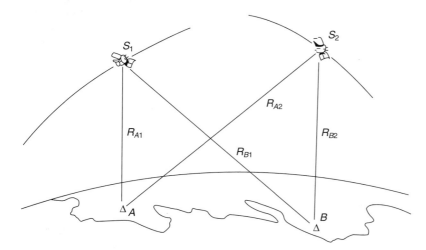

Fig. 7.14   *Double difference*

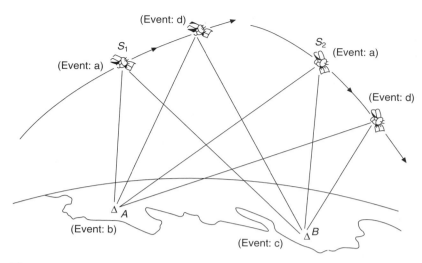

(Event: d)

S_1
(Event: a)

S_2
(Event: a)

(Event: d)

Δ A
(Event: b)

Δ B
(Event: c)

Fig. 7.15 *Triple difference*

adjustment. Graphical output of the residuals in single, double and triple differencing clearly illustrates the cycle slip and its magnitude (*Figure 7.16*).

However, once the data has been filtered and pre-processed and a final value for the integer ambiguity obtained using double and triple differencing, the double difference algorithm is used with the resolved integer ambiguity to produce a least squares value for the vector of the baseline, i.e. the difference in coordinates.

7.7 GPS FIELD PROCEDURES

The use of GPS for positioning to varying degrees of accuracy, in situations ranging from dynamic (navigation) to static (control networks), has resulted in a wide variety of different field procedures using one or other of the basic observables. Generally pseudo-range measurements are used for navigation, whilst the higher precision necessary in engineering surveys requires carrier frequency phase measurements. The basic measuring unit of the S-code (C/A) used in navigation is about 30 m, whilst the L_1 carrier is 19 cm, with range measurement to millimetres.

The basic point positioning method used in navigation gave the *X, Y, Z* position to an accuracy better than 30 m by observation to four satellites. However, the introduction of SA degraded this accuracy to 100 m or more and led to the development of the more accurate differential technique. In this technique the vector between two receivers (baseline) is obtained, i.e. the difference in coordinates (ΔX, ΔY, ΔZ). If one of the receivers is set up over a fixed station whose coordinates are known, then comparison with the observed coordinates enables the differences to be transmitted as corrections to the second receiver (rover). In this way, all the various GPS errors are lumped together in a single correction. The corrections transmitted may be in a simple coordinate format, i.e. δX, δY, δZ, which are easy to apply. Alternatively, the difference in coordinate position of the fixed station may be used to derive corrections to the ranges to the various satellites used. The rover then applies those corrections to its own observations before computing its position. The fundamental assumption in Differential GPS (DGPS) is that the errors within the area of survey would be

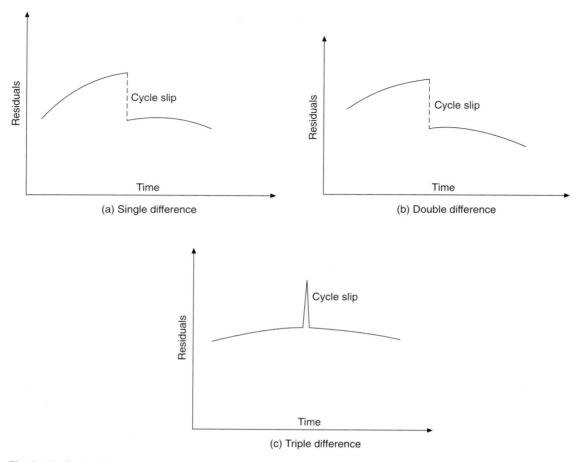

Fig. 7.16 *Cycle slips*

identical. This assumption is acceptable for most engineering surveying where the areas involved are small compared with the distance to the satellites.

Where the area of survey becomes extensive this argument may not hold and a slightly different approach is used called Wide Area Differential GPS.

It can now be seen that, using differential GPS, the position of a roving receiver can be found relative to a fixed master or base station without the effect of errors in satellite and receiver clocks, ionospheric and tropospheric refraction and even ephemeris error. This idea has been expanded to the concept of having permanent base stations established throughout a wide area or even a whole country.

As GPS is essentially a military product, the US Department of Defense has the facility to reduce the accuracy of the system by interfering with the satellite clocks and the ephemeris of the satellite. This is known as selective availability (SA) of the precise positioning service (PPS). There is also a possibility that the P-code could be altered to a Y-code, to prevent imitation of the PPS by hostile forces, and made unavailable to civilian users. This is known as anti-spoofing (AS). However, the carrier wave is not affected and differential methods should correct for most SA effects.

Using the carrier phase observable in the differential mode produces accuracies of 1 ppm of the baseline length. The need to resolve for the integer ambiguity, however, results in post-processing.

Whilst this, depending on the software, can result in even greater accuracies than 1 ppm (up to 0.01 ppm), it precludes real-time positioning. However, the development of Kinematic GPS and 'On-the-Fly' ambiguity resolution has made real-time positioning possible and greatly reduced the observing times.

The following methods are based on the use of carrier phase measurement for relative positioning using two receivers.

7.7.1 Static positioning

This method is used to give high precision over long baselines such as are used in geodetic control surveys. One receiver is set up over a station of known X, Y, Z coordinates, preferably in the WGS 84 reference system, whilst a second receiver occupies the station whose coordinates are required. Observation times may vary from 45 min to several hours. This long observational time is necessary to allow a change in the relative receiver/satellite geometry in order to calculate the initial integer ambiguity terms. Accuracies in the order of 5 mm ± 1 ppm of the baseline are achievable as the majority of error in GPS, such as clock, orbital, atmospheric error and SA, are eliminated or substantially reduced by the differential process. The use of permanent active GPS networks established by a government agency or private company could result in a further increase in accuracy for static positioning.

Apart from establishing high precision control networks, it is used in control densification using a leap-frog technique; measuring plate movement in crustal dynamics and oil rig monitoring.

7.7.2 Rapid static

Rapid static surveying is ideal for many engineering surveys and is halfway between static and kinematic procedures. The 'master' receiver is set up on a reference point and continuously tracks all visible satellites throughout the duration of the survey. The 'roving' receiver visits each of the remaining points to be surveyed, but stays for just a few minutes, typically 2–15 min.

Using double difference algorithms, the integer ambiguity terms are quickly resolved and position, relative to the reference point, obtained to sub-centimetre accuracy. Each point is treated independently and as it is not necessary to maintain lock on the satellites, the roving receiver may be switched off whilst travelling between stations. Apart from a saving in power, the necessity to maintain lock, which is very onerous in urban surveys, is removed.

This method is accurate and economic where there are a great many points to be surveyed. It is ideally suited for short baselines where systematic errors such as atmospheric, orbital, etc., may be regarded as equal at all points and so differenced out. It can be used on large lines (> 10 km) but may require longer observing periods due to the erratic behaviour of the ionosphere, i.e. up to 30 min of dual frequency observation. These times can be halved if the observations are carried out at night when the ionosphere is more stable.

7.7.3 Reoccupation

This technique is regarded as a third form of static surveying or as a pseudo-kinematic procedure. It is based on repeating the survey after a time gap of one or two hours in order to make use of the change in receiver/satellite geometry to resolve the integer ambiguities.

The master receiver is once again positioned over a known point, whilst the roving receiver visits the unknown points for a few minutes only. After one or two hours, the roving receiver returns to the first unknown point and repeats the survey. There is no need to track the satellites whilst moving from point to point. This technique therefore makes use of the first few epochs of data and the last few epochs which reflect the relative change in receiver/satellite geometry and so permit the ambiguities and coordinate differences to be resolved.

Using dual frequency data gives values comparable with the rapid static technique. Due to the method of changing the receiver/satellite geometry, it can be used with cheaper single-frequency receivers (although extended measuring times are recommended) and a poorer satellite constellation.

7.7.4 Kinematic positioning

The major problem with static GPS is the time required for an appreciable change in the satellite/receiver geometry so that the initial integer ambiguities can be resolved. However, if the integer ambiguities could be resolved (and constrained in a least squares solution) prior to the survey, then a single epoch of data would be sufficient to obtain relative positioning to sub-centimetre accuracy. This concept is the basis of kinematic surveying. It can be seen from this that, if the integer ambiguities are resolved initially and quickly, it will be necessary to keep lock on these satellites whilst moving the antenna.

7.7.4.1 Resolving the integer ambiguities

The process of resolving the integer ambiguities is called 'initialization' and may be done by setting-up both receivers at each end of a baseline whose coordinates are accurately known. In subsequent data processing, the coordinates are held fixed and the integers determined using only a single epoch of data. These values are now held fixed throughout the duration of the survey and coordinates estimated every epoch, provided there are no cycle slips.

The initial baseline may comprise points of known coordinates fixed from previous surveys, by static GPS just prior to the survey, or by transformation of points in a local coordinate system to WGS 84.

An alternative approach is called the 'antenna swap' method. An antenna is placed at each end of a short base (5–10 m) and observations taken over a short period of time. The antennae are interchanged, lock maintained, and observations continued. This results in a massive change in the relative receiver/satellite geometry and, consequently, rapid determination of the integers. The antennae are returned to their original position prior to the surveys.

It should be realized that the whole survey will be invalidated if a cycle slip occurs. Thus, reconnaissance of the area is still of vital importance, otherwise re-initialization will be necessary. A further help in this matter is to observe to many more satellites than the minimum four required.

7.7.4.2 Traditional kinematic surveying

Assuming the ambiguities have been resolved, a master receiver is positioned over a reference point of known coordinates and the roving receiver commences its movement along the route required. As the movement is continuous, the observations take place at pre-set time intervals (less than 1 s). Lock must be maintained to at least four satellites, or re-established when lost.

In this technique it is the trajectory of the rover that is surveyed, hence linear detail such as roads, rivers, railways, etc., can be rapidly surveyed. Antennae can be fitted to the roofs of cars, which can be driven at a slow speed along a road to obtain a three-dimensional profile.

7.7.4.3 Stop and go surveying

As the name implies, this kinematic technique is practically identical to the previous one, only in this case the rover stops at the point of detail or position required (*Figure 7.17*). The accent is

Fig. 7.17 *The roving antenna*

therefore on individual points rather than a trajectory route, so data is collected only at those points. Lock must be maintained, though the data observed when moving is not necessarily recorded. This method is ideal for engineering and topographic surveys.

7.7.4.4 Real-time kinematic (RTK)

The previous methods described all require post-processing of the results. However, RTK provides the relative position to be determined instantaneously as the roving receiver occupies a position. The essential difference is the use of mobile data communication to transmit information from the reference point to the rover. Indeed, it is this procedure which imposes limitation due to the range over which the communication system can operate. Also, the effect of SA on the reference data may have changed by the time the data is communicated to the rover, resulting in small positional errors. With the removal of SA (1 May 2000) this is no longer a problem.

The system requires two receivers with only one positioned over a known point. The base station transmits code and carrier phase data to the rover. On-board data processing resolves ambiguities and solves for a change in coordinate differences between roving and reference receivers. This technique can use single or dual frequency receivers. Loss of lock can be regained by remaining static for a short time over a point of known position.

The great advantage of this method for the engineer is that GPS can be used for setting-out on site. With on-board application software and palm sized processor, the setting-out coordinates can be keyed in, and graphical output indicates the direction and distance through which the pole-antenna must be moved. The point to be set-out is shown as a dot with a central cross representing the antenna. When the two coincide, the point of the pole-antenna is at the setting-out position.

Throughout all the procedures described above, it can be seen that initialization or re-initialization can only be done with the receiver static. This may be impossible in high accuracy hydrographic surveys or road profiling in a moving vehicle. With this in mind, Leica Ltd of Heerbrugg, Switzerland, have developed, in conjunction with the Astronomical Institute of the University of Berne, a Fast

Ambiguity Resolution Algorithm that enables ambiguity resolution whilst the receiver is moving. The acronym for the algorithm is FARA, used in a technique called Ambiguity Resolution On the Fly (AROF). The technique requires L_1 and L_2 observations from at least five satellites with a good GDOP. Depending on the level of ionospheric disturbances, the maximum range from the reference to the rover for resolving ambiguities whilst the rover is in motion is 10 km, with an accuracy of 10–20 mm. This relatively recent development will undoubtedly improve the process and role of kinematic GPS surveying.

7.8 ERROR SOURCES

The final position of the survey station is influenced by:

(1) The error in the range measurement.
(2) The satellite–receiver geometry.
(3) The accuracy of the satellite ephemerides.
(4) The effect of atmospheric refraction.
(5) The processing software used.

It is necessary, therefore, to consider the various errors involved, many of which have already been mentioned.

The majority of the error sources are eliminated or substantially reduced if relative positioning is used, rather than single-point positioning. This fact is common to many aspects of surveying. For instance, in simple levelling it is generally the difference in elevation between points that is required. Therefore, if we consider two points A and B whose heights H_A and H_B were obtained at the same time, by the same observer, using the same equipment, the errors would be identical, i.e. $\delta H_A = \delta H_B = \delta H$, then:

$$\Delta H_{AB} = (H_A + \delta H) - (H_B + \delta H)$$

with the result that δH is differenced out and difference in height is much more accurate than the individual heights. Thus, if the absolute position of point A fixed by GPS was 10 m in error, the same would apply to point B, so their relative position would be almost error free. Then, knowing the actual coordinates of A would bring B to its correct relative position. This should be borne in mind when examining the error sources in GPS.

7.8.1 Receiver clock error

This error is a result of the receiver clock not being compatible and in the same time system as the satellite clock. Range measurement is thus contaminated (pseudo-range). As the speed of light is approximately 300 000 km s^{-1}, then an error of 0.01 s results in a range error of about 3000 km.

As already shown, this error can be obtained using four satellites or cancelled using differencing software.

7.8.2 Satellite clock error

Excessive temperature variations in space may result in the variation of the satellite clock from GPS time. Careful monitoring allows the amount of drift to be assessed and included in the broadcast message and therefore eliminated if the user is using the same data. Differential procedures eliminate this error.

7.8.3 Satellite ephemeris error

Orbital data has already been discussed in detail with reference to Broadcast and Precise Ephemeris, and the effect of selective availability (SA). Nevertheless even excluding SA, which at present is no longer applied, errors are still present and influence baseline measurement in the ratio:

$$\delta b/b = \delta S/R_s$$

where

δb = error in baseline b
δS = error in satellite orbit
R_s = satellite range

The specification for GPS is that orbital errors should not exceed 3.7 m, but this is not always possible. Error in the range of 10–20 m may occur using the Broadcast Ephemeris. Thus, for an orbital error of 10 m on a 10 km baseline with a range of 20 000 km, the error in the baseline would be 5 mm. This error is eliminated over moderate length baselines using differential techniques.

7.8.4 Atmospheric refraction

Atmospheric refraction error is usually dealt with in two parts, namely ionospheric and tropospheric. The effects are substantially reduced by DGPS compared with single-point positioning. Comparable figures are:

	Single point	Differential
Ionosphere	15–20 m	2–3 m
Troposphere	3–4 m	1 m

If it was identical at each end of a small baseline, then the total effect would cancel using DGPS.

The ionosphere is the region of the atmosphere from 50 to 1000 km in altitude in which ultraviolet radiation has ionized a fraction of the gas molecules, thereby releasing free electrons. GPS signals are slowed down and refracted from their true path when passing through this medium. The effect on range measurement can vary from 5 to 150 m. As the ionospheric effect is frequency dependent, carrier wave measurement using the different L-band frequencies, i.e. L_1 and L_2, can be processed to eliminate the ionospheric correction. However, it is considered in some circles that using frequency doubling to achieve the elimination will not yield the accuracies required, due to the resultant noise.

If the ionosphere was of constant thickness and electron density, then DGPS, as already mentioned, would eliminate its effect. This, unfortunately, is not so and residual effects remain. Positional and temporal variation in the electron density makes complete elimination over longer baselines impossible and may require complex software modelling.

The troposphere is even more variable than the ionosphere and is not frequency dependent. However, being closer to the ground, it can be easily measured and modelled. If conditions are identical at each end of a baseline, then its effect is completely eliminated by DGPS. Over longer baselines measurements can be taken and used in an appropriate model to reduce the error by as much as 95%.

7.8.5 Multipath error *(Figure 7.18)*

This is caused by the satellite signals being reflected off local surfaces, resulting in a time delay and consequently a greater range. At the frequencies used in GPS they can be of considerable amplitude, due to the fact that the antenna must be designed to track several satellites and cannot therefore be more directional. Antenna design cannot preclude this effect. The only solution at this stage of GPS is the careful siting of the survey station, clear of any reflecting surfaces. In built-up areas, multipaths may present insurmountable problems unless the position of the satellites, with reference to the ground stations, is very carefully planned. It cannot be eliminated by DGPS.

7.8.6 Geometric dilution of precision (GDOP)

As with a distance resection to survey stations on the ground, the geometric relationship of the stations to the resected point will have an effect on the accuracy of the point positioning. Exactly the same situation exists in GPS, where the position of the satellites will affect the three-dimensional angles of intersection. When the satellites are close together or in a straight line, a low-accuracy fix is obtained. When they are wide apart, almost forming a square, high accuracy is obtainable (*Figure 7.19*). The satellite configuration geometry with respect to the ground station is called the GDOP. The GDOP number is small for good configuration and large for poor configuration. Other DOP parameters are:

    VDOP = Vertical vector – one dimension
    HDOP = Horizontal vector – two dimensions
    PDOP = Position vector – three dimensions
    TDOP = Time vector
    GDOP = Geometric position and time vector – four dimensions

Observations should be avoided when large DOP values prevail. Fifty per cent of the time

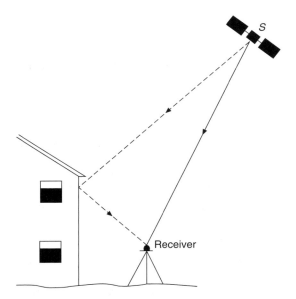

Fig. 7.18 *Multipath effect*

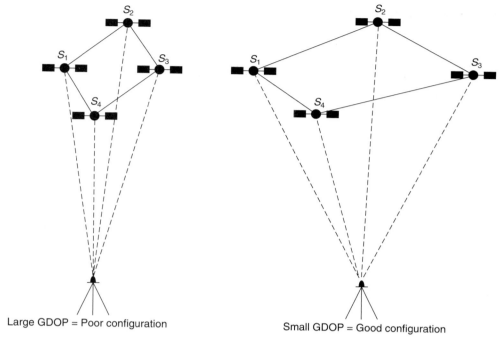

Large GDOP = Poor configuration
Small GDOP = Good configuration

Fig. 7.19 *Satellite configuration*

HDOP < 1.4 and VDOP < 2.0; 90% of the time the values are 1.7 and 2.8 respectively. The GPS receiver searches for and uses the best GDOP satellites during observation.

The DOP values can be used in the following relationship:

$$\sigma_P = \chi \, DOP \times \sigma_R$$

where σ_P = standard deviation of positional accuracy
σ_R = standard deviation of the range

Thus, for a VDOP = 2.0, HDOP = 1.5, and $\sigma_R = \pm 5$ m, then $\sigma_P = \pm 10$ m for the vertical position and ± 7.5 m for the horizontal.

7.8.7 Selective availability (SA)

Static positioning with the P-code is accurate to 5–10 m and is therefore denied access to by civilian users by encryption of the code. This is referred to as anti-snooping (AS).

It was anticipated that use of the S-code or, as it was originally called, the coarse acquisition code C/A, would result in very much worse positional accuracies. This was not the case, and accuracies in the region of 30 m were attained. This gave the American government cause for concern as to its use by an enemy in time of war, and a decision was made to degrade pseudo-range measurement. This process was called selective availability (SA) and comprised:

- Epsilon, which was a corruption of the Broadcast Ephemeris on the S-code, resulting in incorrect positioning of the satellites.
- Dither, which was a corruption of the rate at which the satellite clocks function, resulting in further degrading of observed pseudo-ranges to an accuracy no greater than 30 m.

Whilst absolute position would be affected, the errors at each end of an inter-station vector would be identical and completely cancelled by DGPS.

At the time of writing (May 2000), the American government has turned off SA, resulting in navigational accuracies improving from 100 m to about 10 m. With the exception of multipath errors, DGPS used over the relatively small areas encountered in the majority of engineering surveys produces a practically error free inter-station vector. Thus turning off SA does not improve the precise positioning required in engineering. Over distances greater than 100 km, atmospheric errors are a cause for concern.

7.9 GPS SURVEY PLANNING

Planning of a GPS survey is much more critical than for a conventional survey as the present high cost of GPS hardware and software would make it economically prohibitive to have to repeat a survey.

In the first instance, the points to be surveyed are plotted on a plan and the lengths of the baselines noted. The position of existing horizontal and vertical control should also be shown. Where any baseline is excessively long, the accuracy may be improved by splitting it into two shorter lengths.

Next, using satellite visibility software, determine the window available for the area involved, at the time planned (*Figure 7.20*). Satellite sky plots (*Figure 7.21*) should be generated to show the satellite configuration when at least 25° above the horizon of the observer. All this information is necessary for designing the observing schedule for tracking only those satellites with the best GDOP. It also ensures that all the receivers used will be programmed to observe the same satellites.

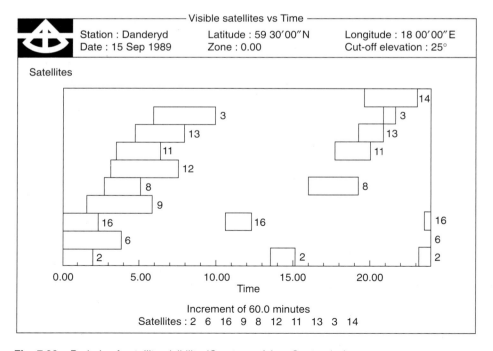

Fig. 7.20   *Periods of satellite visibility (Courtesy of Aga Geotronics)*

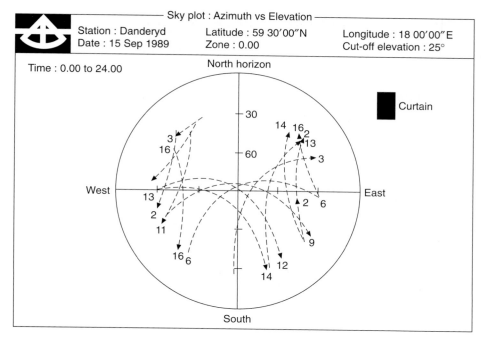

Fig. 7.21 *Computer-generated sky plots (Courtesy of Aga Geotronics)*

Using this information, computer simulation of σ_P calculations can be carried out to estimate attainable accuracy.

Armed with a detailed observation schedule, the reconnaissance should now be carried out. Points should be positioned to avoid multipath problems and to be easily accessible: this is particularly important in wooded or urban areas. Knowing which stations are critical, because of obstructions or limited visibility time, the observation schedule can be drawn up. The distance between points and accessibility determines the rate of work and hence the future observation patterns. The reference point should always be sited where there are no obstructions.

During the field work, the data is stored in the receiver's memory for post-processing by the computer (*Figure 7.22*) using differencing techniques. Finally, network adjustment of the reduced data provides the final positions of the ground points. All the polygons formed by the chains of baselines should be closed and redundancy introduced to improve the accuracy.

7.10 TRANSFORMATION BETWEEN REFERENCE SYSTEMS

As all geodetic systems are theoretically parallel, it would appear that the transformation of the Cartesian coordinates of a point in one system (WGS84) to that in another system (OSGB36), for instance, would simply involve a 3-D translation of the origin of one coordinate system to the other, i.e. ΔX, ΔY, ΔZ (*Figure 7.23*). However, due to observational errors, the orientation of the co-ordinate axis of both systems may not be parallel and must therefore be made so by rotations θ_x, θ_y, θ_z about the X, Y, Z axis. The size and shape of the reference ellipsoid is not relevant when working in 3-D Cartesian coordinates, hence six parameters should provide the transformation necessary.

Fig. 7.22 *Computer processing (sky plots)*

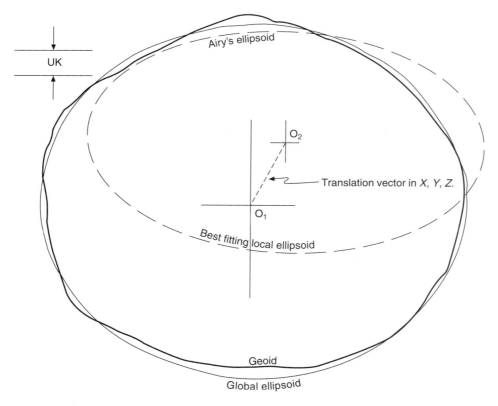

Fig. 7.23 *Relationship of local to global ellipsoid*

However, it is usual to include a seventh parameter S which allows the scale of the axes to vary between the two coordinate systems.

A clockwise rotation about the X-axis (θ_x) has been shown in Chapter 5, Section 5.6 to be:

$$\begin{bmatrix} X \\ Y \\ Z \end{bmatrix}_{\theta_x} = \begin{bmatrix} 1 & 0 & 0 \\ 0 & \cos\theta_x & -\sin\theta_x \\ 0 & \sin\theta_x & \cos\theta_x \end{bmatrix} \begin{bmatrix} X \\ Y \\ Z \end{bmatrix}_{WGS\,84} = R_{\theta_x}\, x_{WGS\,84} \qquad (7.13)$$

Similarly, rotation about the Y-axis (θ_y) will give:

$$\begin{bmatrix} X \\ Y \\ Z \end{bmatrix}_{\theta_{xy}} = \begin{bmatrix} \cos\theta_y & 0 & -\sin\theta_y \\ 0 & 1 & 0 \\ \sin\theta_y & 0 & \cos\theta_y \end{bmatrix} \begin{bmatrix} X \\ Y \\ Z \end{bmatrix}_{\theta_x} = R_{\theta_x}\, R_{\theta_y}\, x_{WGS\,84} \qquad (7.14)$$

Finally, rotation about the Z-axis (θ_z) gives:

$$\begin{bmatrix} X \\ Y \\ Z \end{bmatrix}_{\theta_{xyz}} = \begin{bmatrix} \cos\theta_z & \sin\theta_z & 0 \\ -\sin\theta_z & \cos\theta_z & 0 \\ 0 & 0 & 1 \end{bmatrix} \begin{bmatrix} X \\ Y \\ Z \end{bmatrix}_{\theta_{xy}} = R_{\theta_x}\, R_{\theta_y}\, R_{\theta_z}\, x_{WGS\,84} \qquad (7.15)$$

where R are the rotation matrices. Combining the rotations, including the translation to the origin and applying the scale factor S, gives:

$$\begin{bmatrix} X \\ Y \\ Z \end{bmatrix} = \begin{bmatrix} \Delta X \\ \Delta Y \\ \Delta Z \end{bmatrix} + (1+S) \begin{bmatrix} r_{11} & r_{12} & r_{13} \\ r_{21} & r_{22} & r_{23} \\ r_{31} & r_{32} & r_{33} \end{bmatrix} \begin{bmatrix} X \\ Y \\ Z \end{bmatrix}_{WGS\,84} \qquad (7.16)$$

where: $r_{11} = \cos\theta_y \cos\theta_z$
$\quad r_{12} = \cos\theta_x \sin\theta_z + \sin\theta_x \sin\theta_y \cos\theta_z$
$\quad r_{13} = \sin\theta_x \sin\theta_z - \cos\theta_x \sin\theta_y \cos\theta_z$
$\quad r_{21} = -\cos\theta_y \sin\theta_z$
$\quad r_{22} = \cos\theta_x \cos\theta_z - \sin\theta_x \sin\theta_y \cos\theta_z$
$\quad r_{23} = \sin\theta_x \cos\theta_z + \cos\theta_x \sin\theta_y \sin\theta_z$
$\quad r_{31} = \sin\theta_y$
$\quad r_{32} = -\sin\theta_x \cos\theta_y$
$\quad r_{33} = \cos\theta_x \cos\theta_y$

In matrix form:

$$x = \Delta + S \cdot R \cdot \bar{x} \qquad (7.17)$$

where $x =$ vector of 3-D Cartesian coordinates in a local system
$\quad \Delta =$ is the 3-D shift vector of the origins $(\Delta x, \Delta y\ \Delta z)$
$\quad S =$ scale factor
$\quad R =$ is the orthogonal matrix of the three successive rotation matrices, $\theta_x\ \theta_y\ \theta_z$
$\quad \bar{x} =$ vector of 3-D Cartesian coordinates in the GPS satellite system, WGS84

This seven-parameter transformation is called the Helmert transformation and, whilst mathematically rigorous, is entirely dependent on the rigour of the parameters used. In practice, these parameters are computed from the inclusion of at least three *known* points in the networks. However, the co-ordinates of the known points will contain observational error which, in turn, will affect the transformation parameters. Thus, the output coordinates will contain error. It follows that any transformation in the 'real' world can only be a 'best estimate' and should contain a statistical measure of its quality.

As all geodetic systems are theoretically aligned with the International Reference Pole (IRP) and International Reference Meridian (IRM), which is approximately Greenwich, the rotation parameters are usually less than 5 seconds of arc. In which case, cos $\theta \approx 1$ and sin $\theta = \theta$ rads making the Helmert transformation linear, as follows:

$$\begin{bmatrix} X \\ Y \\ Z \end{bmatrix} = \begin{bmatrix} \Delta X \\ \Delta Y \\ \Delta Z \end{bmatrix} + \begin{bmatrix} 1 + S & -\theta_z & \theta_y \\ \theta_z & 1 + S & -\theta_x \\ -\theta_y & \theta_x & 1 + S \end{bmatrix} \begin{bmatrix} X \\ Y \\ Z \end{bmatrix}_{\text{WGS 84}} \tag{7.18}$$

The rotations θ are in radians, the scale factor S is unitless and, as it is usually expressed in ppm, must be divided by a million.

When solving for the transformation parameters from a minimum of three known points, the XYZ_{LOCAL} and XYZ_{WGS84} are known for each point. The difference in their values would give ΔX, ΔY and ΔZ, which would probably vary slightly for each point. Thus a least squares estimate is taken, the three points give nine observation equations from which the seven transformation parameters are obtained.

It is not always necessary to use a seven-parameter transformation. Five-parameter transformations are quite common, comprising three translations, a longitude rotation and scale change. For small areas involved in construction, small rotations can be described by translations (3), and including scale factors gives a four-parameter transformation. The linear formula (7.18) can still be used by simply setting the unused parameters to zero.

It is important to realize that the Helmert transformation is designed to transform between two datums and cannot consider the scale errors and distortions that exist throughout the Terrestrial Reference Framework of points that exist in most countries. For example, in Great Britain a single set of transformation parameters to relate WGS 84 to OSGB36 would give errors in some parts of the country as high as 4 m. It should be noted that the Molodensky datum transformation (Chapter 5, Section 5.6) deals only with ellipsoidal coordinates (ϕ, λ, h), their translation of origin and changes in reference ellipsoid size and shape. Orientation of the ellipsoid axes is not catered for. However, the advantage of Molodensky is that it provides a single-stage procedure between data.

The next step in the transformation process is to convert the X, Y, Z Cartesian coordinates in a local system to corresponding ellipsoidal coordinates' latitude (ϕ), longitude (λ) and height above the local ellipsoid of reference (h) whose size and shape are defined by its semi-major axis a and eccentricity e. The coordinate axes of both systems are coincident and so the Cartesian to ellipsoidal conversion formulae can be used as given in Chapter 5, equations (5.4) to (5.7), i.e.

$$\tan \lambda = Y/X \tag{7.19}$$

$$\tan \phi = (Z + e^2 v \sin \phi)/(X^2 + Y^2)^{1/2} \tag{7.20}$$

$$h = [X/(\cos \phi \cos \lambda)] - v \tag{7.21}$$

where $v = a/(1 - e^2 \sin^2 \phi)^{1/2}$
       $e = (a^2 - b^2)^{1/2}/a$

An iterative solution is required in (7.20), although a direct formula exists as shown in Chapter 5 equation (5.9).

The final stage is the transformation from the ellipsoidal coordinates ϕ, λ and h to plane projection coordinates and height above mean sea level (MSL). In Great Britain this would constitute grid eastings and northings on the Transverse Mercator projection of Airy's Ellipsoid and height above MSL, as defined by continuous tidal observations from 1915 to 1921 at Newlyn in Cornwall, i.e. E, N and H.

An example of the basic transformation formula is shown in Chapter 5, Section 5.7, equations (5.46) to (5.48). The Ordnance Survey offer the following approach:

N_0 = northing of true origin $(-100\ 000\ \text{m})$

E_0 = easting of true origin $(400\ 000\ \text{m})$

F_0 = scale factor of central meridian $(0.999\ 601\ 27\ 17)$

ϕ_0 = latitude of true origin $(49°\text{N})$

λ_0 = longitude of true origin $(2°\text{W})$

a = semi-major axis $(6\ 377\ 563.396\ \text{m})$

b = semi-minor axis $(6\ 356\ 256.910\ \text{m})$

e^2 = eccentricity squared = $(a^2 - b^2)/a^2$

$n = (a - b)/(a + b)$

$v = aF_0(1 - e^2 \sin^2 \phi)^{-1/2}$

$\rho = aF_0(1 - e^2)(1 - e^2 \sin^2 \phi)^{-3/2}$

$\eta = \dfrac{v}{\rho} - 1$

$$M = b F_0 \left[\begin{array}{l} \left(1 + n + \dfrac{5}{4}n^2 + \dfrac{5}{4}n^3\right)(\phi - \phi_0) - \left(3n + 3n^2 + \dfrac{21}{8}n^3\right)\sin(\phi - \phi_0)\cos(\phi + \phi_0) \\[2mm] + \left(\dfrac{15}{8}n^2 + \dfrac{15}{8}n^3\right)\sin(2(\phi - \phi_0))\cos(2(\phi + \phi_0)) \\[2mm] - \dfrac{35}{24}n^3 \sin(3(\phi - \phi_0))\cos(3(\phi + \phi_0)) \end{array} \right]$$

$$(7.22)$$

$\text{I} = M + N_0$

$\text{II} = \dfrac{v}{2} \sin \phi \cos \phi$

$\text{III} = \dfrac{v}{24} \sin \phi \cos^3 \phi (5 - \tan^2 \phi + 9\eta^2)$

$\text{IIIA} = \dfrac{v}{720} \sin \phi \cos^5 \phi (61 - 58 \tan^2 \phi + \tan^4 \phi)$

$\text{IV} = v \cos \phi$

$\text{V} = \dfrac{v}{6} \cos^3 \phi \left(\dfrac{v}{\rho} - \tan^2 \phi\right)$

$\text{VI} = \dfrac{v}{120} \cos^5 \phi (5 - 18 \tan^2 \phi + \tan^4 \phi + 14\eta^2 - 58 (\tan^2 \phi)\eta^2)$

then: $N = \text{I} + \text{II} (\lambda - \lambda_0)^2 + \text{III} (\lambda - \lambda_0)^4 + \text{IIIA} (\lambda - \lambda_0)^6$

$E = E_0 + \text{IV} (\lambda - \lambda_0) + \text{V} (\lambda - \lambda_0)^3 + \text{VI} (\lambda - \lambda_0)^5$

The computation must be done using double-precision arithmetic with angles in radians.

As shown in Chapter 5, Section 5.3.5, *Figure 5.10*, it can be seen that the ellipsoidal height *h* is the linear distance, measured along the normal, from the ellipsoid to a point above or below the ellipsoid. These heights are not relative to gravity and so cannot indicate flow in water, for instance.

The orthometric height *H* of a point is the linear distance from that point, measured along the gravity vector, to the equipotential surface of the Earth that approximates to MSL. The difference between the two heights is called the geoid-ellipsoid separation, or the geoid height and is denoted by *N*, thus:

$$h = N + H \tag{7.23}$$

In relatively small areas, generally encountered in construction, GPS heights can be obtained on several benchmarks (BM) surrounding and within the area. The difference between the two sets of values gives the value of *N* at each benchmark. The geoid can be regarded as a plane between these points or a contouring program used, thus providing corrections for further GPS heighting within the area. Accuracies relative to tertiary levelling are achievable.

It is worth noting that if, within a small area, *height differences* are required and the geoid–ellipsoid separations are constant, then the value of *N* can be ignored and ellipsoidal heights only used.

The importance of orthometric heights (*H*) relative to ellipsoidal heights (*h*) cannot be over-emphasized. As *Figure 7.24* clearly illustrates, the orthometric heights, relative to the geoid, indicate that the lake is level, i.e. $H_A = H_B$. However, the ellipsoidal heights would indicate water flowing from B to A, i.e. $h_B > h_A$. As the engineer generally requires difference in height (ΔH), then from GPS ellipsoidal heights the following would be needed:

$$\Delta H_{AB} = \Delta h_{AB} - \Delta N_{AB} \tag{7.24}$$

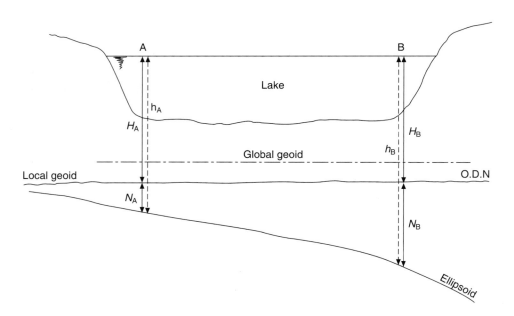

Fig. 7.24 *Orthometric (H) and ellipsoidal (h) heights O.D. Newlyn geoid lies approximately 800 m below the global geoid*

An approximate method of obtaining N on a small site has already been mentioned. On a national basis an accurate national geoid model is required.

In Great Britain the complex, irregular surface of the geoid was established by a combination of astrogeodetic, gravimetric and satellite observations (OSGM91) to such an accuracy that precise GPS heights can be transformed to orthometric with the same accuracy achievable as with precise spirit levelling. However, over distances greater than 5 km, standard GPS heights (accurate to 20–50 mm), when transformed using OSGM91 (accurate to 3 mm$(k)^{1/2}$, where k is the distance in kilometres between points), will produce relative orthometric values as good as those achieved by standard (tertiary) levelling. This means, in effect, that the National GPS Network of points can also be treated as benchmarks.

To summarize, the transformation process when using GPS is:

$(X, Y, Z)_{GPS}$ Using Keplerian elements and time parameters
$\downarrow$

$(X, Y, Z)_{LOCAL}$ Using transformation parameters
$\downarrow$

$(\phi, \lambda, h)_{LOCAL}$ Using ellipsoidal conversion formulae
$\downarrow$

$(E, N, H)_{LOCAL}$ Using projection parameters and geoid–ellipsoid separation

7.11 DATUMS

As already mentioned, the Broadcast Ephemeris is sufficiently accurate for relative positioning over a limited area. For instance, an error of 20 m in the satellite position would produce an error of only 10 mm in a 10 km baseline. However, to achieve an accuracy of 1 mm would require satellite positioning accurate to 2 m and so require the use of a Precise Ephemeris.

Whilst the Broadcast Ephemeris is computed from only five monitoring stations, the Precise is computed from 24 stations situated throughout the world and is available about five days after the event it describes. Precise Ephemerides are available from a variety of government, commercial and academic sources, as most individual users would find the complex computation a daunting task.

An alternative approach known as 'orbit relaxation' can be used, in which the broadcast ephemeris forms a reference orbit which can be corrected for the various error sources involved. However, the method is dependent upon the coordinates of the many tracking stations being known to a high degree of accuracy. A network of such points, established by VLBI (very long baseline interferometry) or SLR (satellite laser ranging) is called a fiducial network. The position of points in a network or terrestrial framework are the basis of reference datums in which the satellite ephemerides describe position.

7.11.1 Global datums

Modern engineering surveying uses GPS in an increasing number of situations. Indeed, in the very near future it will be the primary method of survey. What may not be so apparent to the user is the fact that they will be using a global positioning system for the most local of operations.

Global datums are established by assigning Cartesian coordinates to various positions throughout the world. Observational errors in these positions will obviously be reflected in the datum.

The WGS84 was established from the coordinate position of about 1600 points around the globe, fixed largely by TRANSIT satellite observations. At the present time its origin is geocentric (i.e. the centre of mass of the whole Earth) and its axes virtually coincide with the International Reference Pole and International Reference Meridian. Designed to best fit the global geoid as a whole means it does not fit many of the local ellipsoids in use by many countries. In Great Britain, for instance, it lies about 50 m below the geoid and slopes from east to west, resulting in the geoid–ellipsoid separation being 10 m greater in the west than in the east.

It is also worth noting that the axes are stationary with respect to the *average* motions of this dynamically changing Earth. For instance, tectonic plate movement causes continents to move relative to each other by about 10 cm per year. Local movements caused by tides, pressure weather systems, etc., can result in movement of several centimetres. The result is that the WGS84 datum appears to move relative to the various countries. In Great Britain, the latitudes and longitudes are changing at a rate of 2.5 cm per year in a north-easterly direction. In time, this effect will be noticeable in large-scale mapping.

It can be seen from the above statements that constant monitoring of the WGS84 system is necessary to maintain its validity. In 1997, 13 tracking stations situated throughout the globe had their positional accuracies redefined to an accuracy better than 5 cm, thereby bringing the origin, orientation and scale of the system to within the accuracy of its theoretical specification. Until recently this accuracy was not available to the civilian user due to the application of selective availability (SA). Thus, using a single receiver, absolute positional accuracy was no better than 100 m. However, on 1 May 2000 the US government removed SA, which will no doubt have a significant effect on some areas of GPS use.

Another global datum almost identical to the WGS84 Reference System is the International Terrestrial Reference Frame (ITRF) produced by the International Earth Rotation Service (IERS) in Paris, France. The system was produced from the positional coordinates of over 500 stations throughout the world, fixed by a variety of geodetic space positioning techniques such as satellite laser ranging (SLR), very long baseline interferometry (VLBI), lunar laser ranging (LLR), Doppler ranging integrated on satellite (DORIS) and GPS. Combined with the constant monitoring of Earth rotation, crustal plate movement and polar motion, the IERS have established a very precise terrestrial reference frame, the latest version of which is the ITRF96. This TRF has been established by the civil GPS community, not the US military. It comprises a list of Cartesian coordinates (X, Y, Z) as of the date 1996, with the change in position (dX, dY, dZ) in metres per year for each station. The ITRF96 is available as a SINEX format text file from the IERS internet web site. The ITRF is the most accurate global TRF and for all purposes is identical to the WGS84 TRF.

The ITRF96 is important to GPS users in that it generates a precise ephemeris and stations that can be used in a fiducial network, all of which is supplied free on the internet by the International GPS Service (IGS). The precise ephemeris, used in conjunction with any of the 200 IGS tracking stations and the dual-frequency GPS data from those stations, enables single receiver positional accuracy to a few millimetres.

7.11.2 Local datums

Historically, the majority of local datums were made accessible to the user by means of a TRF of points coordinated by triangulation. These points gave horizontal position only and the triangulation pillars were situated on hilltops. A vertical TRF of benchmarks, established by spirit levelling, required low-lying, easily traversed routes. Hence, there were two entirely different systems.

A TRF established by GPS gives a single three-dimensional system of easily accessible points which can be transformed to give more accurate position in the local system. As WGS84 is continually changing position due to tectonic movement, the local system must be based on a certain time in WGS84. Thus, we have local datums like the North American Datum *1983* and the European Terrestrial Reference System *1989* (ETRS89). In 1989 a high precision European Three-Dimensional Reference Frame (EUREF) was established by GPS observations on 93 stations throughout Europe (ETRF89). The datum used (ETRS89) was consistent with WGS84/1TRF96 and extends into Great Britain, where it forms the datum for the Ordnance Survey National GPS Network.

The OS system will be briefly described here as it illustrates a representative model that will be of benefit to the everyday user of GPS. At the present time (May 2000), the National GPS network TRF comprises two types of GPS station consisting of:

- Active layer: this is a primary network of 30 continuously observing, permanent GPS receivers whose precise coordinates are known. Using a single dual frequency receiver and data downloaded from these stations, which are located within 100 km of any point in Britain, precise positioning can be achieved to accuracies of 10 mm.
- Passive layer: this is a secondary network of about 900 easily accessible stations at a density of 20–35 km. These stations can be used in kinematic form using two receivers to obtain real-time positioning to accuracies of 50–100 mm.

The coordinates obtained by the user are ETRS89 and can be transformed to and from WGS84/1TRF96 by a six-parameter transformation published by IERS on their internet site. Of more interest to local users would be the transformation from ETRS89 to OSGB36, which is the basic mapping system for the country. The establishment of the OSGB36 TRF by triangulation has resulted in variable scale changes throughout the framework, which renders the use of a single Helmert transformation unacceptable. The OS have therefore developed a 'rubber-sheet' transformation called OSTN97, which copes not only with a change of datum but also with the scale distortions and removed the need to compute local transformation parameters by the inclusion of at least three known points within the survey. As OSTN97 works in National Grid eastings and northings, the ETRS89 coordinates must be converted to National Grid coordinates using the map projection formulae already illustrated, but on the global ellipsoid GRS80 rather than the Airy ellipsoid. Transformation back to Airy can then be done if necessary.

Using the National GPS Network in static, post-processing mode, can produce horizontal accuracy of a few millimetres. The same mode of operation using 'active' stations and one hour of data would give accuracies of about 20 mm. Heighting accuracies would typically be twice these values. However, it must be understood that these accuracies apply to position computed in ETRS89 datum on a suitable projection. Transformation to the National Grid using OSTN97 would degrade the above accuracy, and positional errors in the region of 200 mm have been quoted. The final accuracy would, nevertheless, be greater than the OS base-map, where errors of 500 mm in the position of detail at the 1/1250 scale are the norm. Thus, transformation to the National Grid should only be used when integration to the OS base-map is required.

Thus, from the practical user's point of view, the National GPS Network can be used to establish further control, which can then be transformed to National Grid eastings and northings and heights above Newlyn MSL using the OS National Grid Transformation (OSTN97) and the OS National Geoid Model (OSGM91) respectively, supplied free on the OS web site.

As a measure of the high regard with which this procedure is held, work is in hand to improve and thereby replace both OSTN97 and OSGM91.

The evolution of GPS technology and future trends have been illustrated by a description of the OS National GPS Network and its application. Other countries are also establishing continuously operating GPS reference station networks. For instance, in Japan a network of about 1000 stations are deployed throughout the country with a spacing of about 20–30 km (GEONET).

Many innovative techniques have been used in GPS surveying to resolve the problem of ambiguity resolution (AR), reduce observation time and increase accuracy. They involve the development of sophisticated AR algorithms to reduce the time for static surveying, or carry out 'on-the-fly' carrier phase AR whilst the antenna is continually moving. Other techniques such as 'stop and go' and 'rapid static' techniques were also developed. They all, however, require the use of at least two, multi-channel, dual frequency receivers, and there are limitations on the length of baseline observed (< 15 km). However, downloading the data from only one of the 'active stations' in the above network results in the following benefit:

1. For many engineering surveying operations the use of only one single-frequency receiver, thereby reducing costs. However this may result in antenna phase centre variation due to the use of different autenuae.
2. Baseline lengths greatly extended for rapid static and kinematic procedures.
3. 'On-the-fly' techniques can be used with ambiguity resolution algorithms from a single epoch.

For many users the use of continuously operating reference stations is still in its early stages (May, 2000) and many problems may still need to be resolved, particularly for real-time users. There may be communication problems if the distances to the active network stations are too great. This may require the establishment of smaller, higher density networks within the area of use, supplied by professional organizations and offered on a 'fee-for-service' basis. Although the mobile phone may resolve any such problems. Other problems may result from the use of existing, non-compatible software and turnkey systems, requiring a concerted effort on the part of all involved to use all available technology to produce a thoroughly integrated, user friendly service.

7.12 OTHER SATELLITE SYSTEMS

There are many systems available, mostly based on a small number of satellites in a geosynchronous orbit and used for communications as well as positioning. Such systems are Starfix, Geostar, Locstar, etc., which, due to their geosychronous orbits, do not provide a good geometric configuration with the receiver and cannot compete with GPS.

One system, however, whose planned constellation was similar to GPS, is the Russian GLONASS, GLObal NAvigation Satellite System. The constellation comprises 24 satellites, with eight in each of three orbital planes inclined at $64.8°$ and orbiting at a height of 19 100 km. The satellite signals consist of two L-band carrier frequencies at 1250 MHZ (L_2) and 1600 MHZ (L_1). A precise code with a bandwidth of 5.11 MHZ is on both L_1 and L_2, whilst the coarser code of 0.511 MHz is on L_1 only. Thus, the satellites permit pseudo-range and carrier phase measurement.

Two differences that exist between GPS and GLONASS are that SA is not possible on GLONASS and the coordinate datum is different, i.e. SGS-85 datum compared with WGS-84. However, both the UK and Russia are keen to make both systems compatible with each other, although doubts are cast on the ability of Russia, at the present time, to maintain and progress their system as there has been a gradual decline in the number of satellites available. Using a receiver capable of utilizing both systems would offer the user twice the number of satellites, increased accuracy, savings in time and an independent check on each system.

Whilst the future of GPS looks very bright compared with GLONASS, the European community are also considering a Global Navigation Satellite System (GNSS) as outlined in a Commission communication (February 1999) entitled 'GALILEO'.

The American government wish to promote GPS as a global standard within the overriding consideration of national security requirements.

At the present time (May 2000) SA has been removed, which must be of great benefit to the Standard Positioning Service (SPS) user, and there is a declared intention to continue to provide

SPS free on a world-wide basis. The US is prepared to co-operate with any government, as evidenced with GLONASS, to ensure the needs of the civil, commercial and scientific community are met, along with international security. A GPS Executive Board has been established to manage the services offered by GPS and, at the present time, have evidenced goodwill by giving all users access to the L_2 carrier and agreeing to provide a fully coded navigation signal on that frequency, i.e. a C/A code. There is also a proposal to provide a third civil signal to be called L_5, with a carrier frequency similar to the L_2 signal, and studies are in hand to discuss the possibility of civil and military users sharing common frequencies without impeding security measures. A greater number of satellites are planned, with new operational systems which allow the satellite to maintain its own ephemeris and clock data by ranging to and communicating with other satellites.

In spite of all this obvious goodwill and intention on behalf of the US, the GPS system is fundamentally a military system in complete control of the Department of Defense at times of national security. This reason, combined with commercial concerns and the possibility of future user charges, has caused the European Commission to consider their own GNSS. Basically, there are two possibilities. The first is a system of 21 satellites used in conjunction with GPS to meet European user requirements. The second is a system of 36 satellites which would not be in conjunction with GPS. Whatever decisions are arrived at, the use of Global Navigation Satellite Systems seems assured.

7.13 APPLICATIONS

The previous pages have already indicated the basic application of GPS in engineering surveying – that is: the establishment of control surveys, topographic surveys and setting-out on site. Indeed, any three-dimensional spatial data normally captured using conventional surveying techniques with a total station can be done by GPS, even during the night, provided sufficient satellites are visible.

On a national scale, horizontal and, to a certain extent, vertical control, used for mapping purposes and established by classical triangulation with all its built-in scale error, are being replaced by three-dimensional GPS networks. In relation to Great Britain and the Ordnance Survey, this has been dealt with in previous pages. The great advantage of this to the engineering surveyor is that, when using GPS on a local level, there is no requirement for coordinate transformation and the resultant plans are more consistent. Also, in mapping at a local level there is, in effect, no need to establish a control network as it already exists in the form of the orbiting satellite. Thus, time and money are saved. Kinematic methods can be used for rapid detailing, and real-time kinematic (RTK) for setting-out.

Whilst the above constitutes the main area of interest for the engineering surveyor, other applications will be briefly mentioned to illustrate the power and versatility of GPS.

7.13.1 *Machine guidance* (Figures 7.25 and 7.26)

Earthmoving and grading plant are now being controlled in three dimensions using GPS in the real-time kinematic (RTK) mode. Tests carried out at the IESSG (Nottingham University) using one antenna on the cab and one at each end of the bulldozer blade gave accuracies of a few millimetres. Combined with in-vehicle digital ground and design models, the machine operator can complete the design without reliance on extraneous equipment such as sight rails, batter boards or lasers.

The advantage of GPS over existing systems that use total stations or lasers is that full three-dimensional information is supplied to the operator permitting alignment positioning as well as

Fig. 7.25   *'Site vision GPS' machine guidance by* Trimble, *showing 2 GPS antennae and the in-cab control system*

depth excavation or grading. Also, the system permits several items of plant to work simultaneously and over distances of 10 km from the base station. The plant is no longer dependent on the use of stakes, profile boards, strings, etc., and so does not suffer from downtime waiting when 'wood' has been disturbed and needs replacing. In this way the dozers, etc., can be kept running continuously with resultant productivity gains.

7.13.2 Global mean sea level

The determination of MSL on a global scale requires a network of tide gauges throughout the world connected to a single global reference frame. GPS is being used for this purpose within an international programme called GLOSS (global level of the sea surface) established by the Intergovernmental Oceanographic Commission. GPS is also involved in a similar exercise on a European basis.

7.13.3 Plate tectonics

Plate tectonics is centred around the theories of continental drift and is the most widely accepted model describing crustal movement. GPS is being used on a local and regional basis to measure three-dimensional movement. On the local basis, inter-station vectors across faults are being continually monitored to millimetre accuracy, whilst, on a regional basis, GPS networks have been established

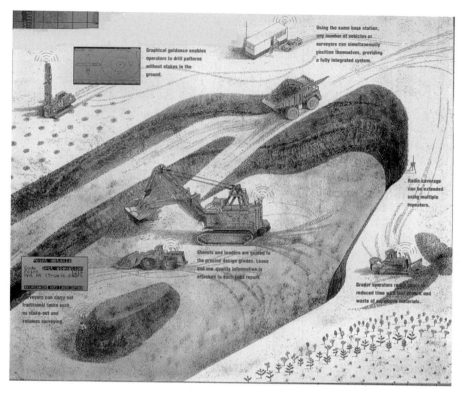

Fig. 7.26 *Various applications of machine guidance by GPS*

on all continental plate boundaries. The information obtained adds greatly to the study of earthquake prediction, volcanoes and plate motion.

In a secondary way, it is also linked to global reference systems defined by coordinate points on the Earth's surface. As the plates continue to move, coordinate points will alter position and global reference systems will need to be redefined.

7.13.4 Geographical information systems (GIS)

GIS has already been defined in the earlier chapters. Such is the growth in this area of spatial information management that GPS systems have been specifically designed with GIS in mind. Facilities management covers such a vast area of varying requirements that it is not possible to be specific. Suffice to say that the speed and accuracy of GPS, particularly in the kinematic mode, render it ideal for data collection and real-time coding of that date. The relatively new Leica GS50 GIS/GPS system has been designed with rapid data collection and accurate attribute description in mind. Its very production is indicative of the importance attached to GPS in GIS. The palm-size computer contains GIS Data PRO Office Software that converts GPS position information to vector GIS format, and a coding facility that permits smooth transfer of data to plan or computer database.

7.13.5 Navigation

GPS is now used in all aspects of navigation.

GPS voice navigation systems are now built in to several models of car. Simply typing the required destination into the on-board computer results in a graphical display of the route, along with voiced directions. Similar systems are used by private boats and aeroplanes, whilst hand-held receivers are now standard equipment for walkers and cyclists. A wristwatch produced by Casio contains a GPS system giving position and route information to the wearer.

GPS can be used for fleet management when the position and status of vehicles can be transmitted to a central control, thereby permitting better management of the vehicles, whilst the driver can use it as an aid to route location.

It is used by surveying ships for major offshore hydrographic surveys. Ocean-going liners use it for navigation purposes, whilst most harbours have a DGPS system to enable precise docking.

At the present time, aircraft landing and navigation are controlled by a variety of disparate systems. GPS is gradually being introduced and will eventually provide a single system for all aircraft operations.

The uses to which GPS can be put are limited only by the imagination of the user. They can range from the complexities of measuring gravity waves to the simplicity of spreading fertilizer in precision farming, and include such areas of study as meteorology oceanography, geophysics and in-depth analysis on a local and global basis. As GPS equipment and procedures improve, its applications will continue to grow.

REFERENCES

Ackroyd, N. (1999) 'The Application of GPS to Machine Guidance', *Journal of the ICES*, August.

Ashkenazi, V. (1988) 'Global Positioning System' Seminar on GPS, The University of Nottingham.

Ashkenazi, V. and Suterfield, P.J. (1989) 'Rapid Static and Kinematic GPS Surveying', *Land and Minerals Surveying*, October.

Ashkenazi, V. (1989) 'Positioning by Satellites. Principles, Achievements and Prospects.' The University of Nottingham.

Ashkenazi, V. *et al.* (1992) 'Wide Area Differential GPS.' Institute of Engineering Surveying Space Geodesy (IESSG), The University of Nottingham.

Bingley, R. (1994) 'GPS Observables and Algorithms.' IESSG, The University of Nottingham.

Bingley, R. (1994) 'Surveying with GPS.' IESSG, The University of Nottingham.

Cross, P.A. (1988) 'Geophysical Applications of GPS Surveying.' Seminar on GPS, The University of Nottingham.

Cross, P. A. (2000) 'Prospects for GPS – new systems, new applications, new techniques', *Engineering Surveying Showcase 2000*, Issue One.

Davies, P. (1999) 'Improving access to the National Coordinate System.' *Surveying World*, **7**(4).

Davies, P. (2000) 'Information Paper 1/2000 Coordinate Positioning – Ordnance Survey Policy and Strategy', *Journal of ICES*, May.

Decker, B. (1984) 'World Geodetic System 1984.' Defence Mapping Agency, Aerospace Centre, Missouri, USA.

de la Fuente, C. (1988) 'Kinematic GPS Surveying.' Seminar on GPS, The University of Nottingham.

Moore, T. (1987) 'The Computation of GPS Satellite Orbits.' Seminar on GPS, The University of Nottingham.

Moore, T. (1994) 'Coordinate Systems, Frames and Datums.' IESSG, The University of Nottingham.

Moore, T. 'GPS Orbit Determination and Fiducial Networks.' IESSG, The University of Nottingham.

Moore, T. (1994) 'An Introduction to Differential GPS.' IESSG, The University of Nottingham.

Ordnance Survey (2000) 'A Guide to Coordinate Systems in Great Britain'. O.S. Southampton.

Pichot, G. (2000) 'GPS on site – new prospects for machine guidance'. *Engineering Surveying Showcase 2000*, Issue One.

Roberts, G.W. and Dodson, A.H. (1999) 'Using RTK GPS to control construction plant', *Journal of the ICES*, November.

Stanford, N.M. and Cross, P.A. (1994) 'A Review of some current and future applications of GPS', Dept of Geomatics, The University of Newcastle.

8
Curves

In the geometric design of motorways, railways, pipelines, etc., the design and setting out of curves is an important aspect of the engineer's work.

The initial design is usually based on a series of straight sections whose positions are defined largely by the topography of the area. The intersections of pairs of straights are then connected by horizontal curves (see *Section 8.2*). In the vertical design, intersecting gradients are connected by curves in the vertical plane.

Curves can be listed under three main headings, as follows:

(1) Circular curves of constant radius.
(2) Transition curves of varying radius (spirals).
(3) Vertical curves of parabolic form.

8.1 CIRCULAR CURVES

Two straights, D_1T_1 and D_2T_2 in *Figure 8.1*, are connected by a circular curve of radius R:

(1) The straights when projected forward, meet at I: the *intersection point*.
(2) The angle Δ at I is called the *angle of intersection* or the *deflection angle*, and equals the angle T_1OT_2 subtended at the centre of the curve 0.
(3) The angle ϕ at I is called the *apex angle*, but is little used in curve computations.
(4) The curve commences from T_1 and ends at T_2; these points are called the *tangent points*.
(5) Distances T_1I and T_2I are the tangent lengths and are equal to $R \tan \Delta/2$.
(6) The length of curve T_1AT_2 is obtained from:

> Curve length $= R\Delta$ where Δ is expressed in radians, or

> Curve length $= \dfrac{\Delta° \cdot 100}{D°}$ where *degree of curve* (D) is used (see *Section 8.1.1*)

(7) Distance T_1T_2 is called the *main chord* (C), and from *Figure 8.1*.

$$\sin \frac{\Delta}{2} = \frac{T_1B}{T_10} = \frac{\frac{1}{2} \text{ chord } (C)}{R} \qquad \therefore C = 2R \sin \frac{\Delta}{2}$$

(8) IA is called the *apex distance* and equals

> $I0 - R = R \sec \Delta/2 - R = R (\sec \Delta/2 - 1)$

(9) AB is the *rise* and equals $R - 0B = R - R \cos \Delta/2$

> $\therefore AB = R (1 - \cos \Delta/2)$

These equations should be deduced using a curve diagram (*Figure 8.1*).

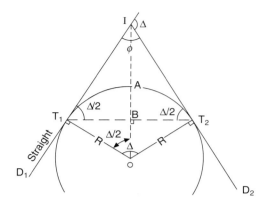

Fig. 8.1

8.1.1 Curve designation

Curves are designated either by their *radius* (R) or their *degree of curvature* ($D°$). The degree of curvature is defined as the angle subtended at the centre of a circle by an *arc* of 100 m (*Figure 8.2*).

Thus
$$R = \frac{100 \text{ m}}{D \text{ rad}} = \frac{100 \times 180°}{D° \times \pi}$$

$$\therefore R = \frac{5729.578}{D°} \text{ m} \tag{8.1}$$

Thus a 10° curve has a radius of 572.9578 m.

8.1.2 Through chainage

Through chainage is the horizontal distance from the start of a scheme for route construction.

Consider *Figure 8.3*. If the distance from the start of the route (Chn 0.00 m) to the tangent point T_1 is 2115.50 m, then it is said that the chainage of T_1 is 2115.50 m, written as (Chn 2115.50 m).

If the route centre-line is being staked out at 20-m chord intervals, then the peg immediately prior to T_1 must have a chainage of 2100 m (an integer number of 20 m intervals). The next peg on the centre-line must therefore have a chainage of 2120 m. It follows that the length of the first sub-chord on the curve from T_1 must be (2120–2115.50) = 4.50 m.

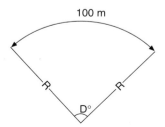

Fig. 8.2

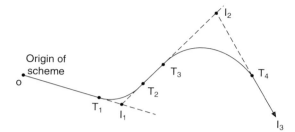

Fig. 8.3

Similarly, if the chord interval had been 30 m, the peg chainage prior to T_1 must be 2100 m and the next peg (on the curve) 2130 m, thus the first sub-chord will be (2130 − 2115.50) = 14.50 m.

A further point to note in regard to chainage is that if the chainage at I_1 is known, then the chainage at T_1 = Chn I_1 − distance $I_1 T_1$, the tangent length. However the chainage at T_2 = Chn T_1 + curve length, as chainage is measured along the route under construction.

8.2 SETTING OUT CURVES

This is the process of establishing the centre-line of the curve on the ground by means of pegs at 10-m to 30-m intervals. In order to do this the tangent and intersection points must first be fixed in the ground, in their correct positions.

Consider *Figure 8.3*. The straights $0I_1$, I_1I_2, I_2I_3, etc., will have been designed on the plan in the first instance. Using railway curves, appropriate curves will now be designed to connect the straights. The tangent points of these curves will then be fixed, making sure that the tangent lengths are equal, i.e. $T_1I_1 = T_2I_1$ and $T_3I_2 = T_4I_2$. The coordinates of the origin, point 0, and all the intersection points only will now be carefully scaled from the plan. Using these coordinates, the bearings of the straights are computed and, using the tangent lengths on these bearings, the coordinates of the tangent points are also computed. The difference of the bearings of the straights provides the deflection angles (Δ) of the curves which, combined with the tangent length, enables computation of the curve radius, through chainage and all setting-out data. Now the tangent and intersection points are set out from existing control survey stations and the curves ranged between them using the methods detailed below.

8.2.1 Setting out with theodolite and tape

The following method of setting out curves is the most popular and it is called *Rankine's deflection or tangential angle method*, the latter term being more definitive.

In *Figure 8.4* the curve is established by a series of chords T_1X, XY, etc. Thus, peg 1 at X is fixed by sighting to I with the theodolite reading zero, turning off the angle δ_1 and measuring out the chord length T_1X along this line. Setting the instrument to read the second deflection angle gives the direction T_1Y, and peg 2 is fixed by measuring the chord length XY from X until it intersects at Y. The procedure is now continued, the angles being set out from T_1I, and the chords measured from the previous station.

It is thus necessary to be able to calculate the setting-out angles δ as follows:
Assume $0A$ bisects the chord T_1X at right angles; then

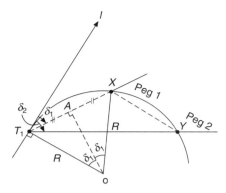

Fig. 8.4

$$A\hat{T}_1 0 = 90° - \delta_1, \quad \text{but} \quad I\hat{T}_1 0 = 90°$$

$$\therefore \quad I\hat{T}_1 A = \delta_1$$

By radians, arc length $T_1X = R2\delta_1$

$$\therefore \delta_1 \text{ rad} = \frac{\text{arc } T_1X}{2R} \approx \frac{\text{chord } T_1X}{2R}$$

$$\therefore \delta_1° = \frac{\text{chord } T_1X \times 180°}{2R \cdot \pi} = 28.6479 \frac{\text{chord}}{R} = 28.6479 \frac{C}{R} \tag{8.2a}$$

$$\text{or} \quad \delta° = \frac{D° \times \text{chord}}{200} \text{ where } \textit{degree of curve} \text{ is used} \tag{8.2b}$$

(Using equation (8.2a) the angle is obtained in degree and decimals of a degree; a single key operation converts it to degrees, minutes, seconds.)

An example will now be worked to illustrate these principles.

The centre-line of two straights is projected forward to meet at I, the deflection angle being 30°. If the straights are to be connected by a circular curve of radius 200 m, tabulate all the setting-out data, assuming 20-m chords on a through chainage basis, the chainage of I being 2259.59 m.

Tangent length = $R \tan \Delta/2 = 200 \tan 15° = 53.59$ m

$\therefore$ Chainage of $T_1 = 2259.59 - 53.59 = 2206$ m

$\therefore$ 1st sub-chord 14 m

Length of circular arc = $R\Delta = 200(30° \cdot \pi/180)$ m = 104.72 m
From which the number of chords may now be deduced

i.e. 1st sub-chord = 14 m

2nd, 3rd, 4th, 5th chords = 20 m each

Final sub-chord = 10.72 m

Total = 104.72 m (*Check*)

$\therefore$ Chainage of $T_2 = 2206$ m + 104.72 m = 2310.72 m

Deflection angles:

$$\text{For 1st sub-chord} = 28.6479 \cdot \frac{14}{200} = 2°\,00'\,19''$$

$$\text{Standard chord} = 28.6479 \cdot \frac{20}{200} = 2°\,51'\,53''$$

$$\text{Final sub-chord} = 28.6479 \cdot \frac{10.72}{200} = 1°\,32'\,08''$$

Check: The sum of the deflection angles = $\Delta/2 = 14°\,59'\,59'' \approx 15°$

Table 8.1

Chord number	*Chord length* (m)	*Chainage* (m)	*Deflection angle* °	'	"	*Setting-out angle* °	'	"	*Remarks*
1	14	2220.00	2	00	19	2	00	19	peg 1
2	20	2240.00	2	51	53	4	52	12	peg 2
3	20	2260.00	2	51	53	7	44	05	peg 3
4	20	2280.00	2	51	53	10	35	58	peg 4
5	20	2300.00	2	51	53	13	27	51	peg 5
6	10.72	2310.72	1	32	08	14	59	59	peg 6

The error of 1″ is, in this case, due to the rounding-off of the angles to the nearest second and is negligible.

8.2.2 *Setting out with two theodolites*

Where chord taping is impossible, the curve may be set out using two theodolites at T_1 and T_2 respectively, the intersection of the lines of sight giving the position of the curve pegs.

The method is explained by reference to *Figure 8.5*. Set out the deflection angles from T_1I in the usual way. From T_2, set out the same angles from the main chord T_2T_1. The intersection of the corresponding angles gives the peg position.

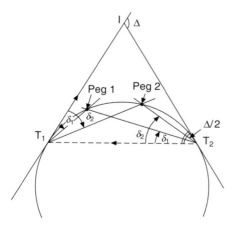

Fig. 8.5

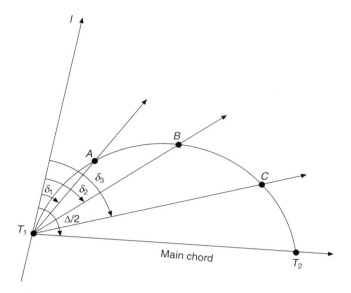

Fig. 8.6   *Setting-out by EDM*

If T_1 cannot be seen from T_2, sight to I and turn off the corresponding angles $\Delta/2 - \delta_1$, $\Delta/2 - \delta_2$, etc.

8.2.3 Setting-out using EDM

When setting-out by EDM, the total distance from T_1 to the peg is set out, i.e. distances $T_1 A$, $T_1 B$, and $T_1 C$ etc. in *Figure 8.6*. However, the chord and sub-chord distances are still required in the usual way, plus the setting-out angles for those chords. Thus all the data and setting-out computation as shown in *Table 8.1* must first be carried out prior to computing the distances to the pegs direct from T_1. These distances are computed using equation 7 in 8.1, i.e.

$T_1 A = 2R \sin \delta_1 = 2R \sin 2°00'19'' = 14.00$ m.

$T_1 B = 2R \sin \delta_2 = 2R \sin 4°52'12'' = 33.96$ m.

$T_1 C = 2R \sin \delta_3 = 2R \sin 7°44'05'' = 53.83$ m.

$T_1 T_2 = 2R \sin (\Delta/2) = 2R \sin 15°00'00'' = 103.53$ m.

In this way the curve is set-out by measuring the distances direct from T_1 and turning off the necessary direction in the manner already described.

8.2.4 Setting-out using coordinates

In this procedure the coordinates along the centre-line of the curve are computed relative to the existing control points. Consider *Figure 8.7*:

(1) From the design process, the coordinates of the tangent and intersection point are obtained.
(2) The chord intervals are decided in the usual way and the setting-out angles δ_1, δ_2 . . . δ_n, computed in the usual way (*Section 8.2.1.*).
(3) From the known coordinates of T_1 and I, the bearing $T_1 I$ is computed.

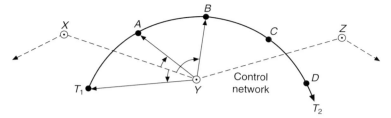

Fig. 8.7 *Setting-out using coordinates*

(4) Using the setting-out angles, the bearings of the rays $T_1 A$, $T_1 B$, $T_1 C$, etc. are computed relative to $T_1 I$. The distances are obtained as in *Section 8.2.3*.

(5) Using the bearings and distances in (4) the coordinates of the curve points A, B, C, etc. are obtained.

(6) These points can now be set out from the nearest control points either by 'polars' or by 'intersection', as follows:

(7) Using the coordinates, compute the bearing and distance from, say, station Y to T_1, A and B.

(8) Set up theodolite at Y and backsight to X; set the horizontal circle to the known bearing YX.

(9) Now turn the instrument until it reads the computed bearing YT_1 and set out the computed distance in that direction to fix the position of T_1. Repeat the process for A and B. The ideal instrument for this is a total station, many of which will have onboard software to carry out the computation in real time. However, provided that the ground conditions are suitable and the distances within, say, a 50 m tape length, a theodolite and steel tape would suffice.

Other points around the curve are set out in the same way from appropriate control points.

Intersection may be used, thereby precluding distance measurement, by computing the bearings to the curve points from *two* control stations. For instance, the theodolites are set up at Y and Z respectively. Instrument Y is orientated to Z and the bearing YZ set on the horizontal circle. Repeat from Z to Y. The instruments are set to bearings YB and ZB respectively, intersecting at peg B. The process is repeated around the curve.

Using coordinates eliminates many of the problems encountered in curve ranging and does not require the initial establishment of tangent and intersection points.

8.2.5 Setting out with two tapes (method of offsets)

Theoretically this method is exact, but in practice errors of measurement propagate round the curve. It is therefore generally used for minor curves.

In *Figure 8.8*, line OE bisects chord $T_1 A$ at right-angles, then $ET_1 O = 90° − \delta$, $\therefore CT_1 A = \delta$, and triangles $CT_1 A$ and $ET_1 O$ are similar, thus

$$\frac{CA}{T_1 A} = \frac{T_1 E}{T_1 O} \qquad \therefore CA = \frac{T_1 E}{T_1 O} \times T_1 A$$

i.e. offset $\quad CA = \dfrac{\frac{1}{2} \text{ chord} \times \text{chord}}{\text{radius}} = \dfrac{\text{chord}^2}{2R}$ $\qquad\qquad$ (8.3)

From *Figure 8.8*, assuming lengths $T_1 A = AB = AD$

then angle $DAB = 2\delta$, $\quad$ and so offset $\quad DB = 2CA = \dfrac{\text{chord}^2}{R}$ $\qquad\qquad$ (8.4)

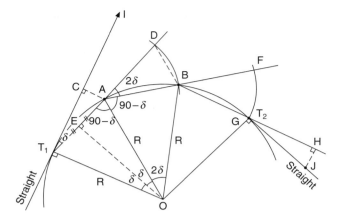

Fig. 8.8

The remaining offsets round the curve to T_2 are all equal to DB whilst, if required, the offset HJ to fix the line of the straight from T_2, equals CA.

The method of setting out is as follows:

It is sufficient to approximate distance T_1C to the chord length T_1A and measure this distance along the tangent to fix C. From C a right-angled offset CA fixed the first peg at A. Extend T_1A to D so that AD equals a chord length; peg B is then fixed by pulling out offset length from D and chord length from A, and where they meet is the position B. This process is continued to T_2.

The above assumes equal chords. When the first or last chords are sub-chords, the following (*Section 8.2.6*) should be noted.

8.2.6 Setting out by offsets with sub-chords

In *Figure 8.9* assume T_1A is a sub-chord of length x; from equation (8.3) the offset $CA = O_1 = x^2/2R$.

As the normal chord AB differs in length from T_1A, the angle subtended at the centre will be 2θ not 2δ. Thus, as shown in *Figure 8.8*, the offset DB will not in this case equal $2CA$.

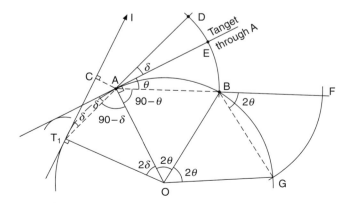

Fig. 8.9

Construct a tangent through point A, then from the figure it is obvious that angle $EAB = \theta$, and if chord $AB = y$, then offset $EB = y^2/2R$.

Angle $DAE = \delta$, therefore offset DE will be directly proportional to the chord length, thus:

$$DE = \frac{O_1}{x}\,y = \frac{x^2}{2R}\,\frac{y}{x} = \frac{xy}{2R}$$

Thus the total offset $DB = DE + EB$

$$= \frac{y}{2R}\,(x + y) \tag{8.5}$$

$$\text{i.e.} \ = \frac{\text{chord}}{2R}\,(\text{sub-chord} + \text{chord})$$

Thus having fixed B, the remaining offsets to T_2 are calculated as y^2/R and set out in the usual way. If the final chord is a sub-chord of length x_1, however, then the offset will be

$$\frac{x_1}{2R}\,(x_1 + y) \tag{8.6}$$

Students should note the difference between equations (8.5) and (8.6).

A more practical approach to this problem is actually to establish the tangent through A in the field. This is done by swinging an arc of radius equal to CA, i.e. $x^2/2R$ from T_1. A line tangential to the arc and passing through peg A will then be the required tangent from which offset EB, i.e. $y^2/2R$, may be set off.

8.2.7 Setting out with inaccessible intersection point

In *Figure 8.10* it is required to fix T_1 and T_2, and obtain the angle Δ, when I is inaccessible.

Project the straights forward as far as possible and establish two points A and B on them. Measure distance AB and angles BAC and DBA then:

angle $IAB = 180° - B\hat{A}C$ and angle $IBA = 180° - D\hat{B}A$, from which angle BIA is deduced and angle Δ. The triangle AIB can now be solved for lengths IA and IB. These lengths, when subtracted from the computed tangent lengths ($R \tan \Delta/2$), give AT_1 and BT_2, which are set off along the straight to give positions T_1 and T_2 respectively.

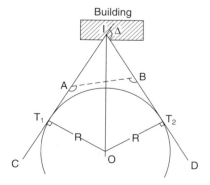

Fig. 8.10

8.2.8 Setting out with theodolite at an intermediate point on the curve

Due to an obstruction on the line of sight (*Figure 8.11*) or difficult communications and visibility on long curves, it may be necessary to continue the curve by ranging from a point on the curve. Assume that the setting-out angle to fix peg 4 is obstructed. The theodolite is moved to peg 3, backsighted to T_1 with the instrument reading 180°, and then turned to read 0°, thus giving the direction 3 – T. The setting-out angle for peg 4, δ_4, is turned off and the chord distance measured from 3. The remainder of the curve is now set off in the usual way, that is, δ_5 is set on the theodolite and the chord distance measured from 4 to 5.

The proof of this method is easily seen by constructing a tangent through peg 3, then angle $A3T_1$ = $AT_13 = \delta_3 = T3B$. If peg 4 was fixed by turning off δ from this tangent, then the required angle from $3T$ would be $\delta_3 + \delta = \delta_4$.

8.2.9 Setting out with an obstruction on the curve

In this case (*Figure 8.12*) an obstruction on the curve prevents the chaining of the chord from 3 to 4. One may either

(1) Set out the curve from T_2 to the obstacle.
(2) Set out the chord length $T_14 = 2R \sin \delta_4$ (EDM).
(3) Set out using intersection from theodolites at T_1 and T_2.
(4) Use coordinate method.

8.2.10 Passing a curve through a given point

In *Figure 8.13*, it is required to find the radius of a curve which will pass through a point P, the position of which is defined by the distance IP at an angle of ϕ to the tangent.

Consider triangle IPO:

$$\text{angle } \beta = 90° - \Delta/2 - \phi \text{ (right-angled triangle } IT_2O\text{)}$$

by sine rule: $\sin \alpha = \dfrac{IO}{PO} \sin \beta$ but $IO = R \sec \dfrac{\Delta}{2}$

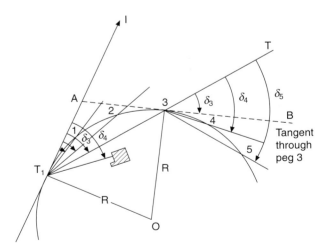

Fig. 8.11

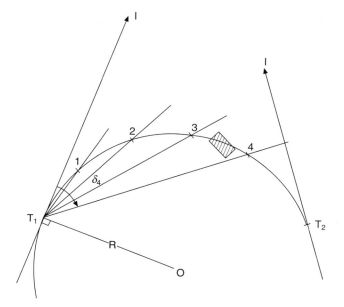

Fig. 8.12

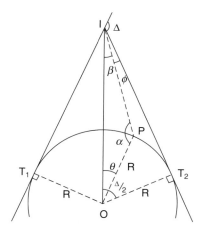

Fig. 8.13

$$\therefore \sin \alpha = \sin \beta \frac{R \sec \Delta/2}{R} = \sin \beta \sec \frac{\Delta}{2}$$

then $\theta = 180° - \alpha - \beta$,   and by the sine rule: $R = IP \dfrac{\sin \beta}{\sin \theta}$

8.3 COMPOUND AND REVERSE CURVES

Although equations are available which solve compound curves (*Figure 8.14*) and reverse curves (*Figure 8.15*), they are difficult to remember and students are advised to treat the problem as two simple curves with a common tangent point *t*.

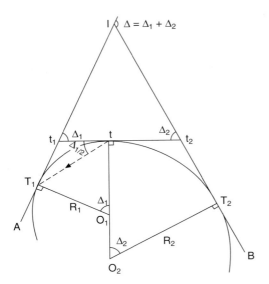

Fig. 8.14

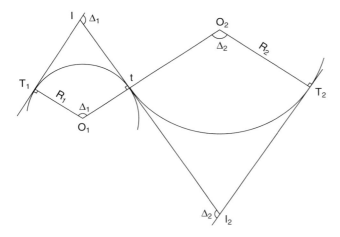

Fig. 8.15

In the case of the compound curve, the total tangent lengths T_1I and T_2I are found as follows:

$$R_1 \tan \Delta_1/2 = T_1t_1 = t_1t \qquad \text{and} \qquad R_2 \tan \Delta_2/2 = T_2t_2 = t_2t, \text{ as } t_1t_2 = t_1t + t_2t$$

then triangle t_1It_2 may be solved for lengths t_1I and t_2I which, if added to the known lengths T_1t_1 and T_2t_2 respectively, give the total tangent lengths.

In setting out this curve, the first curve R_1 is set out in the usual way to point t. The theodolite is moved to t and backsighted to T_1, with the horizontal circle reading $(180° - \Delta_1/2)$. Set the instrument to read zero and it will then be pointing to t_2. Thus the instrument is now oriented and reading zero, prior to setting out curve R_2.

In the case of the reverse curve, both arcs can be set out from the common point t.

8.4 SHORT AND/OR SMALL-RADIUS CURVES

Short and/or small-radius curves such as for kerb lines, bay windows or for the construction of large templates may be set out by the following methods.

8.4.1 *Offsets from the tangent*

The position of the curve (in *Figure 8.16*) is located by right-angled offsets Y set out from distances X, measured along each tangent, thereby fixing half the curve from each side.

The offsets may be calculated as follows for a given distance X. Consider offset Y_3, for example.

In $\triangle ABO$, $AO^2 = OB^2 - AB^2$ ∴ $(R - Y_3)^2 = R^2 - X_3^2$ and $Y_3 = R - (R^2 - X_3^2)^{\frac{1}{2}}$

Thus for any offset Y_i at distance X_i along the tangent

$$Y_i = R - (R^2 - X_i^2)^{\frac{1}{2}}$$

(8.7)

8.4.2 *Offsets from the long chord*

In this case (*Figure 8.17*) the right-angled offsets Y are set off from the long chord C, at distances X to each side of the centre offset Y_0.

An examination of *Figure 8.17*, shows the central offset Y_0 equivalent to the distance T_1A on *Figure 8.16*; thus:

$$Y_0 = R - [R^2 - (C/2)^2]^{\frac{1}{2}}$$

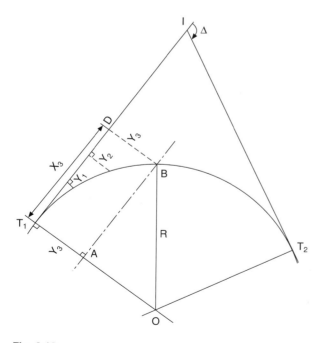

Fig. 8.16

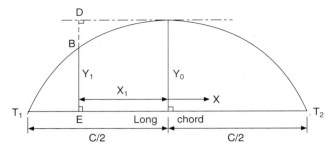

Fig. 8.17

Similarly, DB is equivalent to DB on *Figure 8.16*, thus: $DB = R - (R^2 - X_1^2)^{\frac{1}{2}}$

and offset $Y_1 = Y_0 - DB$ $\therefore$ $Y_1 = Y_0 - [R - (R^2 - X_1^2)^{\frac{1}{2}}]$

and for any offset Y_i at distance X_i each side of the mid-point of T_1T_2:

$$\text{mid-point of } T_1T_2: \qquad Y_i = Y_0 - [R - (R^2 - X^2)^{\frac{1}{2}}] \qquad\qquad (8.8)$$

Therefore, after computation of the central offset, further offsets at distances X_i, each side of Y_0, can be found.

8.4.3 Halving and quartering

Referring to *Figure 8.18*:

(1) Join T_1 and T_2 to form the long chord. Compute and set out the central offset Y_0 to A from B (assume $Y_0 = 20$ m), as in *Section 8.4.2*.
(2) Join T_1 and A, and now halve this chord and quarter the offset. That is, from mid-point E set out offset $Y_1 = 20/4 = 5$ m to D.
(3) Repeat to give chords T_1D and DA; the mid-offsets FG will be equal to $Y_1/4 = 1.25$ m.

Repeat as often as necessary on both sides of the long chord.

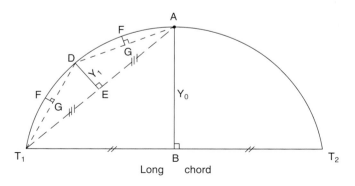

Fig. 8.18

Worked examples

Example 8.1. The tangent length of a simple curve was 202.12 m and the deflection angle for a 30-m chord 2°18′.

Calculate the radius, the total deflection angle, the length of curve and the final deflection angle.

(LU)

$$2°18′ = 2.3° = 28.6479 \cdot \frac{30}{R} \qquad \therefore R = 373.67 \text{ m}$$

$$202.12 = R \tan \Delta/2 = 373.67 \tan \Delta/2 \qquad \therefore \Delta = 56°49′06″$$

Length of curve = $R\Delta$ rad = 373.67 × 0.991 667 rad = 370.56 m
Using 30-m chords, the final sub-chord = 10.56 m

$$\therefore \text{ final deflection angle } = \frac{138′ \times 10.56}{30} = 48.58′ = 0°48′35″$$

Exmple 8.2. The straight lines *ABI* and *CDI* are tangents to a proposed circular curve of radius 1600 m. The lengths *AB* and *CD* are each 1200 m. The intersection point is inaccessible so that it is not possible directly to measure the deflection angle; but the angles at *B* and *D* are measured as

$$A\hat{B}D = 123°48′, \quad B\hat{D}C = 126°12′ \text{ and the length } BD \text{ is } 1485 \text{ m}$$

Calculate the distances from *A* and *C* of the tangent points on their respective straights and calculate the deflection angles for setting out 30-m chords from one of the tangent points. (LU)

Referring to *Figure 8.19*:

$$\Delta_1 = 180° - 123°48′ = 56°12′, \qquad \Delta_2 = 180° - 126°12′ = 53°48′$$

$$\therefore \Delta = \Delta_1 + \Delta_2 = 110°$$

$$\phi = 180° - \Delta = 70°$$

Tangent lengths IT_1 and $IT_2 = R \tan \Delta/2 = 1600 \tan 55° = 2285$ m

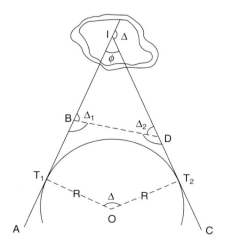

Fig. 8.19

By sine rule in triangle *BID*:

$$BI = \frac{BD \sin \Delta_2}{\sin \phi} = \frac{1484 \sin 53° \, 48'}{\sin 70°} = 1275.2 \text{ m}$$

$$ID = \frac{BD \sin \Delta_1}{\sin \phi} = \frac{1485 \sin 56°15'}{\sin 70°} \ 1314 \text{ m}$$

Thus $AI = AB + BI = 1200 + 1275.2 = 2475.2$ m

$CI = CD + ID = 1200 + 1314 = 2514$ m

$\therefore AT_1 = AI - IT_1 = 2475.2 - 2285 = 190.2$ m

$CT_2 = CI - IT_2 = 2514 - 2285 = 229$ m

Deflection angle for 30-m chord $= 28.6479 \times 30/1600 = 0.537148°$

$$= 0°32'14''$$

Example 8.3. A circular curve of 800 m radius has been set out connecting two straights with a deflection angle of 42°. It is decided, for construction reasons, that the mid-point of the curve must be moved 4 m towards the centre, i.e. away from the intersection point. The alignment of the straights is to remain unaltered.

Calculate:

(1) The radius of the new curve.
(2) The distances from the intersection point to the new tangent points.
(3) The deflection angles required for setting out 30-m chords of the new curve.
(4) The length of the final sub-chord.   (LU)

Referring to *Figure 8.20*

$$IA = R_1 (\sec \Delta/2 - 1) = 800(\sec 21° - 1) = 56.92 \text{ m}$$

$$\therefore IB = IA + 4 \text{ m} = 60.92 \text{ m}$$

(1) Thus, $60.92 = R_2(\sec 21° - 1)$, from which $R_2 = 856$ m
(2) Tangent length $= IT_1 = R_2 \tan \Delta/2 = 856 \tan 21° = 328.6$ m

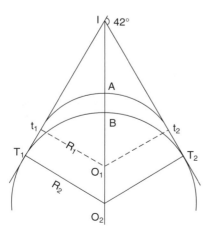

Fig. 8.20

(3) Deflection angle for 30-m chord = $28.6479 \cdot C/R = 28.6479 \cdot \dfrac{30}{856} = 1°\,00'14''$

(4) Curve length = $R\Delta$ rad $= \dfrac{856 \times 42° \times 3600}{206\,265} = 627.5$ m

∴ Length of final sub-chord = 27.5 m

Example 8.4. The centre-line of a new railway is to be set out along a valley. The first straight *AI* bears 75°, whilst the connecting straight *IB* bears 120°. Due to site conditions it has been decided to join the straights with a compound curve.

The first curve of 500 m radius commences at T_1, situated 300 m from *I* on straight *AI*, and deflects through an angle of 25° before joining the second curve.

Calculate the radius of the second curve and the distance of the tangent point T_2 from *I* on the straight *IB*.

(KU)

Referring to *Figure 8.14*:

$\Delta = 45°$, $\Delta_1 = 25°$ ∴ $\Delta_2 = 20°$

Tangent length $T_1t_1 = R_1 \tan \Delta_1/2 = 500 \tan 12°30' = 110.8$ m. In triangle t_1It_2

Angle $t_2It_1 = 180° - \Delta = 135°$

Length $It_1 = T_1I - T_1t_1 = 300 - 110.8 = 189.2$ m

By sine rule

$$t_1t_2 = \frac{It_1 \sin t_2It_1}{\sin \Delta_2} = \frac{189.2 \sin 135°}{\sin 20°} = 391.2 \text{ m}$$

$$It_2 = \frac{It_1 \sin \Delta_1}{\sin \Delta_2} = \frac{189.2 \sin 25°}{\sin 20°} = 233.8 \text{ m}$$

∴ $tt_2 = t_1t_2 - T_1t_1 = 391.2 - 110.8 = 280.4$ m

∴ $280.4 = R_2 \tan \Delta_2/2 = R_2 \tan 10°;$ ∴ $R_2 = 1590$ m

Distance $IT_2 = It_2 + t_2T_2 = 233.8 + 280.4$

$= 514.2$ m

Example 8.5. Two straights intersecting at a point *B* have the following bearings, *BA* 270°, *BC* 110°. They are to be joined by a circular curve which must pass through a point *D* which is 150 m from *B* and the bearing of *BD* is 260°.

Find the required radius, tangent lengths, length of curve and setting-out angle for a 30-m chord.

(LU)

Referring to *Figure 8.2*:

From the bearings, the apex angle = $(270° - 110°) = 160°$

∴ $\Delta = 20°$

and angle $DBA = 10°$ (from bearings)

∴ $OBD = \beta = 70°$

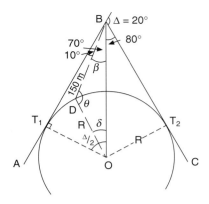

Fig. 8.21

In triangle *BDO* by sine rule

$$\sin \theta = \frac{OB}{OD} \sin \beta = \frac{R \sec \Delta/2}{R} \sin \beta = \sec \frac{\Delta}{2} \sin \beta$$

$$\therefore \ \sin \theta = \sec 10° \times \sin 70°$$

$$\theta = \sin^{-1} 0.954\,190 = 72°35'25'' \ \text{or}$$

$$(180° - 72°35'25'') = 107°24'35''$$

An examination of the figure shows that δ must be less than 10°,

$$\therefore \ \theta = 107°24'35''$$

$$\delta = 180° - (\theta + \beta) = 2°35'25''$$

By sine rule $\ DO = R = \dfrac{DB \sin \beta}{\sin \delta} = \dfrac{150 \sin 70°}{\sin 2°35'25''}$

$$\therefore \ R = 3119 \text{ m}$$

Tangent length $= R \tan \Delta/2 = 3119 \tan 10° = 550 \text{ m}$

Length of curve $= R\Delta$ rad $= \dfrac{3119 \times 20° \times 3600}{206\,265} = 1089 \text{ m}$

Deflection angle for 30-m chord $= 28.6479 \times \dfrac{30}{3119} = 0°16'32''$

Example 8.6. Two straights *AEI* and *CFI*, whose bearings are respectively 35° and 335°, are connected by a straight from *E* to *F*. The coordinates of *E* and *F* in metres are:

E	E 600.36	N 341.45
F	E 850.06	N 466.85

Calculate the radius of a connecting curve which shall be tangential to each of the lines *AE*, *EF* and *CF*. Determine also the coordinates of *I*, T_1, and T_2, the intersection and tangent points respectively.

(KU)

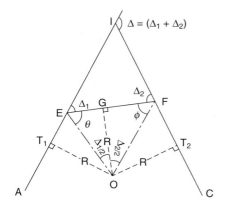

Fig. 8.22

Referring to *Figure 8.22*:

Bearing $AI = 35°$,   bearing $IC = (335° - 180°) = 155°$

$$\therefore \Delta = 155° - 35° = 120°$$

By coordinates

Bearing $EF = \tan^{-1}\dfrac{+\,249.70\ \text{N}}{+\,125.40\ \text{N}} = 63°20'$

Length $EF = 249.7/\sin 63°20' = 279.42$ m

From bearings AI and EF, angle $IEF = \Delta_1 = (63°20' - 35°)$

$$= 28°20'$$

From bearings CI and EF, angle $IFE = \Delta_2 = (155° - 63°20')$

$$= 91°\,40', \text{ check } (\Delta_1 + \Delta_2) = \Delta = 120°$$

In triangle EFO

Angle $FEO = (90° - \Delta_1/2) = \theta = 75°50'$

Angle $EFO = (90° - \Delta_2/2) = \phi = 44°10'$

$$EG = GO \cot \theta = R\cot \theta$$

$$GF = GO \cot \phi = R\cot \phi$$

$$\therefore EG + GF = EF = R\,(\cot \theta + \cot \phi)$$

$$\therefore R = \frac{EF}{(\cot \theta + \cot \phi)} = \frac{279.42}{\cot 75°\,50' + \cot 44°10'}$$

$$= 217.97 \text{ m}$$

$$ET_1 = R\tan \Delta_1/2 = 217.97 \tan 14°10' = 55.02 \text{ m}$$

$$FT_2 = R\tan \Delta_2/2 = 217.97 \tan 45°50' = 224.4 \text{ m}$$

bearing $ET_1 = 215°$

bearing $FT_2 = 155°$

$\therefore$ Coordinates of $T_1 = $ 55.02 sin 215° = −31.56 E,

55.02 cos 215° = −45.07 N.

$\therefore$ Total coordinates of T_1 = E 600.36 − 31.56 = E 568.80 m

= N 341.45 − 45.07 = N 296.38 m

Similarly

Coordinates of T_2 = 224.4 sin 155° = +98.84 E,

224.4 cos 155° = −203.38 N

$\therefore$ Total coordinates of T_2 = E 850.06 + 94.84 = E 944.90 m

= N 466.85 − 203.38 = N 263.47 m

$T_1 I = R \tan \Delta/2 = 217.97 \tan 60° = 377.54$ m

Bearing of $T_1 I = 35°$

$\therefore$ Coordinates of I = 377.54 sin 35° = +216.55 E,

377.54 cos 35° = +309.26 N.

$\therefore$ Total coordinates of I = E 586.20 + 216.55 = E 802.75

= N 321.23 + 309.26 = N 630.49

The coordinates of I can be checked via $T_2 I$.

Example 8.7. The coordinates in metres of two points B and C with respect to A are:

B 470 E   500 N
C 770 E   550 N

Calculate the radius of a circular curve passing through the three points, and the coordinates of the intersection point I, assuming that A and C are tangent points.   (KU)

Referring to *Figure 8.23*:
By coordinates

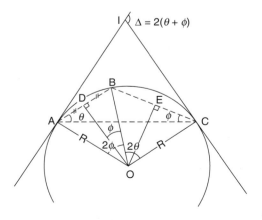

Fig. 8.23

Bearing $\quad AB = \tan^{-1}\dfrac{+470 \text{ E}}{+500 \text{ N}} = 43°14'$

Bearing $\quad AC = \tan^{-1}\dfrac{+770 \text{ E}}{+550 \text{ N}} = 54°28'$

Bearing $\quad BC = \tan^{-1}\dfrac{+330 \text{ E}}{+50 \text{ N}} = 80°32'$

Distance $\quad AB = 500/\cos 43°14' = 686$ m

From bearings of AB and AC, angle $BAC = \theta = 11°14'$
From bearings of CA and CB, angle $BCA = \phi = 26°04'$

As a check, the remaining angle, calculated from the bearings of BA and $BC = 142°42'$, summing to 180°.

In right-angled triangle DOB:

$$OB = R = \frac{DB}{\sin \phi} = \frac{343}{\sin 26°04'} = 781 \text{ m}$$

This result could now be checked through triangle OEC.

$$\Delta = 2(\phi + \theta) = 74°36'$$

$$\therefore AI = R \tan \Delta/2 = 781 \tan 37°18' = 595 \text{ m}$$

Bearing AI = bearing $AC - \Delta/2 = 54°82' - 37°18'$

$$= 17°10'$$

$\therefore$ Coordinates of I equal:

595 sin 17°10′ = + 176 E

595 cos 17°10′ = + 569 N

Exercises

(8.1) In a town planning scheme, a road 9 m wide is to intersect another road 12 m wide at 60°, both being straight. The kerbs forming the acute angle are to be joined by a circular curve of 30 m radius and those forming the obtuse angle by one of 120 m radius.

Calculate the distances required for setting out the four tangent points.

Describe how to set out the larger curve by the deflection angle method and tabulate the angles for 15-m chords. (LU)

(Answer: 75, 62, 72, 62 m. $\delta = 3°35'$)

(8.2) A straight BC deflects 24° right from a straight AB. These are to be joined by a circular curve which passes through a point P, 200 m from B and 50 m from AB.

Calculate the tangent length, length of curve and deflection angle for a 30-m chord. (LU)

(Answer: $R = 3754$ m, $IT = 798$ m, curve length = 1572 m, 0°14′)

(8.3) A reverse curve is to start at a point A and end at C with a change of curvature at B. The chord lengths AB and BC are respectively 661.54 m and 725.76 m and the radii likewise 1200 and 1500 m.

Due to irregular ground the curves are to be set out using two theodolites and no tape or chain.

Calculate the data for setting out and describe the procedure in the field.   (LU)

(*Answer*: Tangent lengths: 344.09, 373.99; curve length: 670.2, 733, per 30-m chords: $\delta_1 = 0°42'54''$, $\delta_2 = 0°34''30''$)

(*8.4*) Two straights intersect making a deflection angle of 59° 24′, the chainage at the intersection point being 880 m. The straights are to be joined by a simple curve commencing from chainage 708 m.

If the curve is to be set out using 30-m chords on a through chainage basis, by the method of offsets from the chord produced, determine the first three offsets.

Find also the chainage of the second tangent point, and with the aid of sketches, describe the method of setting out.   (KU)

(*Answer*: 0.066, 1.806, 2.985 m, 864.3 m)

(*8.5*) A circular curve of radius 250 m is to connect two straights, but in the initial setting out it soon becomes apparent that the intersection point is in an inaccessible position. Describe how it is possible in this case to determine by what angle one straight deflects from the other, and how the two tangent points may be accurately located and their through chainages calculated.

On the assumption that the chainages of the two tangent points are 502.2 m and 728.4 m, describe the procedure to be adopted in setting out the first three pegs on the curve by a theodolite (reading to 20″) and a steel tape from the first tangent point at 30-m intervals of through chainage, and show the necessary calculations.

If it is found to be impossible to set out any more pegs on the curve from the first tangent point because of an obstruction between it and the pegs, describe a procedure (without using the second tangent point) for accurately locating the fourth and succeeding pegs. No further calculations are required.   (ICE)

(*Answer*: 03° 11′ 10″, 06° 37′ 20″, 10° 03′ 40″)

8.5 TRANSITION CURVES

The *transition curve* is a curve of constantly changing radius. If used to connect a straight to a curve of radius R, then the commencing radius of the transition will be the same as the straight (∞), and the final radius will be that of the curve R (see *Figure 8.28*).

Consider a vehicle travelling at speed (V) along a straight. The forces acting on the vehicle will be its weight W, acting vertically down, and an equal and opposite force acting vertically up through the wheels. When the vehicle enters the curve of radius R at tangent point T_1, an additional centrifugal force (P) acts on the vehicle, as shown in *Figures 8.24* and *8.25*. If P is large the vehicle will be forced to the outside of the curve and may skid or overturn. In *Figure 8.25* the resultant of the two forces is shown as N, and if the road is super-elevated normal to this force, there will be no tendency for the vehicle to skid. It should be noted that as

$$P = WV^2/Rg \tag{8.9}$$

super-elevation will be maximum at minimum radius R.

It therefore requires a length of spiral curve to permit the gradual introduction of superelevation, from zero at the start of the transition to maximum at the end, where the radius is the minimum safe radius R.

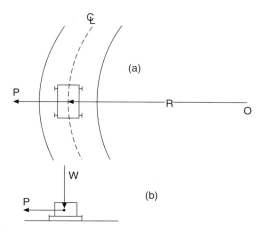

Fig. 8.24

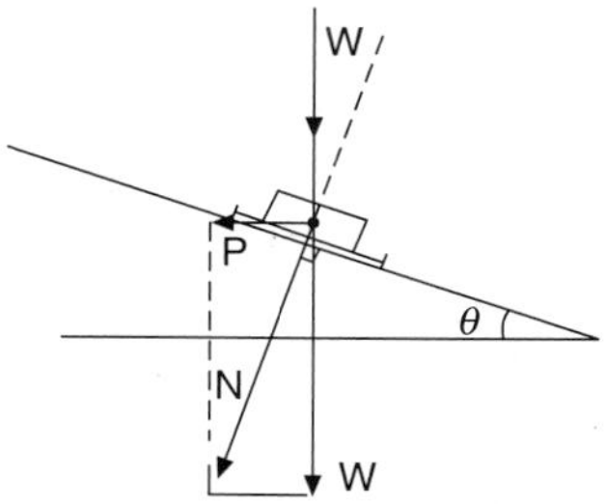

Fig. 8.25

8.5.1 Principle of the transition

The purpose then of a transition curve is to:

(1) Achieve a gradual change of direction from the straight (radius ∞) to the curve (radius R).
(2) Permit the gradual application of super-elevation to counteract centrifugal force and minimize passenger discomfort.

Since P cannot be eliminated, it is allowed for by permitting it to increase uniformly along the curve. From equation (8.9), as P is inversely proportional to R, the basic requirement of the ideal transition curve is that its radius should decrease uniformly with distance along it. This requirement also permits the uniform application of super-elevation; thus at distance l along the transition the radius is r and $rl = c$(constant):

$$\therefore\ l/c = 1/r$$

From *Figure 8.26*, tt_1 is an infinitely small portion of a transition δl of radius r; thus:

$$\delta l = r\ \delta\phi$$

$$\therefore\ 1/r = \delta\phi/\delta l \quad \text{which on substitution above gives}$$

$$l/c = \delta\phi/\delta l$$

integrating: $\phi = l^2/2c$ $\therefore\ l = (2c\phi)^{\frac{1}{2}}$

putting $a = (2c)^{\frac{1}{2}}$

$$l = a(\phi)^{\frac{1}{2}} \tag{8.10}$$

when $c = RL$, $a = (2RL)^{\frac{1}{2}}$ and (8.10) may be written:

$$l = (2RL\phi)^{\frac{1}{2}} \tag{8.11}$$

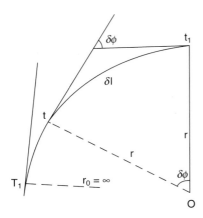

Fig. 8.26

The above expressions are for the *clothoid curve*, sometimes called the *Euler spiral*, which is the one most used in road design.

8.5.2 Curve design

The basic requirements in the design of transition curves are:

(1) The value of the minimum safe radius (R), and
(2) The length (L) of the transition curve. *Sections 8.5.5 or 8.5.6*

The value R may be found using either of the approaches *Sections 8.5.3 or 8.5.4*.

8.5.3 Centrifugal ratio

Centrifugal force is defined as $P = WV^2/Rg$; however, this 'overturning force' is counteracted by the mass (W) of the vehicle, and may be expressed as P/W, termed the *centrifugal ratio*. Thus, centrifugal ratio:

$$P/W = V^2/Rg \tag{8.12}$$

where V is the design speed in m/s, g is acceleration due to gravity in m/s^2 and R is the minimum safe radius in metres.

When V is expressed in km/h, the expression becomes

$$P/W = V^2/127R \tag{8.13}$$

Commonly used values for centrifugal ratio are

0.21 to 0.25 on roads,   0.125 on railways

Thus, if a value of $P/W = 0.25$ is adopted for a design speed of $V = 50$ km/h, then

$$R = \frac{50^2}{127 \times 0.25} = 79 \text{ m}$$

The minimum safe radius R may be set either equal to or greater than this value.

8.5.4 Coefficient of friction

The alternative approach to find R is based on Road Research Laboratory (RRL) values for the coefficient of friction between the car tyres and the road surface.

Figure 8.27(a) illustrates a vehicle passing around a correctly super-elevated curve. The resultant of the two forces is N. The force F acting towards the centre of the curve is the friction applied by the car tyres to the road surface. These forces are shown in greater detail in *Figure 8.27(b)* from which it can be seen that

$$F_2 = \frac{WV^2}{Rg} \cos \theta, \text{ and } F_1 = W \cos (90 - \theta) = W \sin \theta$$

$$\therefore F = F_2 - F_1 = \frac{WV^2}{Rg} \cos \theta - W \sin \theta$$

Similarly $N_2 = \dfrac{WV^2}{Rg} \sin \theta$, and $N_1 = W \cos \theta$

$$\therefore N = N_2 + N_1 = \frac{WV^2}{Rg} \sin \theta + W \cos \theta$$

Then
$$\frac{F}{N} = \frac{\dfrac{WV^2}{Rg} \cos \theta - W \sin \theta}{\dfrac{WV^2}{Rg} \sin \theta + W \cos \theta} = \frac{\dfrac{V^2}{Rg} - \tan \theta}{\dfrac{V^2}{Rg} \tan \theta + 1}$$

For Department of Transport (DTp) requirements, the maximum value for $\tan \theta = 1$ in $14.5 = 0.069$, and as V^2/Rg cannot exceed 0.25 the term in the denominator can be ignored and

$$\frac{F}{N} = \frac{V^2}{Rg} - \tan \theta = \frac{V^2}{127R} - \tan \theta \tag{8.14}$$

(a)

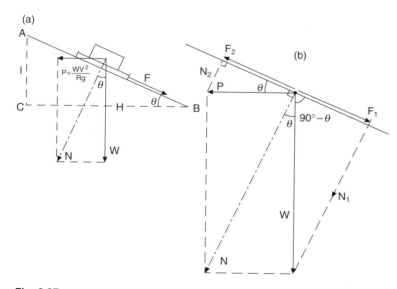

Fig. 8.27

To prevent vehicles slipping sideways, F/N must be greater than the coefficient of friction μ between tyre and road. The RRL quote value for μ of 0.15, whilst 0.18 may be used up to 50 km/h, thus:

$$V^2/127R \not> \tan \theta + \mu \tag{8.15}$$

For example, if the design speed is to be 100 km/h, super-elevation limited to 1 in 14.5 (0.069) and $\mu = 0.15$, then

$$\frac{100^2}{127R} = 0.069 + 0.15$$

$$\therefore R = 360 \text{ m}$$

In the UK, the geometric parameters used in design are normally related to design speed. *Table 8.3* (p. 398) shows typical desirable and absolute minimum values for horizontal and vertical curvature; there is also an additional lower level designated 'limiting radius', specific of horizontal curvature.

Designs for new roads should aim to achieve the desirable values for each design parameter. However, absolute minimum values can be used wherever substantial saving in construction or environmental costs can be achieved.

DTp Technical Standard TD9/93 advises that in the design of new roads, the use of radii tighter than the limiting values is undesirable and not recommended.

8.5.5 *Rate of application of super-elevation*

It is recommended that on motorways super-elevation should be applied at a rate of 1 in 200, on all-purpose roads at 1 in 100, and on railways at 1 in 480. Thus, if on a motorway the super-elevation is computed as 0.5 m, then 100 m of transition curve would be required to accommodate 0.5 m at the required rate of 1 in 200, i.e. 0.5. in 100 = 1 in 200. In this way the length L of the transition is found.

The amount of super-elevation is obtained as follows:
From the triangle of forces in *Figure 8.27(a)*

$$\tan \theta = V^2/Rg = 1/H = 1 \text{ in } H$$
$$\text{thus } H = Rg/V^2 = 127R/V^2, \quad \text{where } V \text{ is in km/h.}$$

In the UK, the design speed V is replaced by the 'average speed' which is 63.6% of V,

$$\text{and } H = 127R/(0.636V)^2 = 314R/V^2$$
$$\therefore \text{ super-elevation} = 1 \text{ in } 314R/V^2 \tag{8.16}$$

The percentage super-elevation (or crossfall), S may be found from:

$$S = V^2/2.828R \tag{8.17}$$

The DTp recommend the crossfall should never be greater than 1 in 14.5, or less than 1 in 48, to allow rainwater to run off the road surface.

It is further recommended that adverse camber should be replaced by a favourable crossfall of 2.5% when the value of V^2/R is greater than 5 and less than 7 (see *Table 8.3*).

Driver studies have shown that whilst super-elevation is instrumental to driver comfort and safety, it need not be applied too rigidly. Thus for sharp curves in urban areas with at-grade junctions and side access, super-elevation should be limited to 5%.

The rate of crossfall, combined with the road width, allows the amount of super-elevation to be calculated. Its application at the given rate produces the length L of transition required.

8.5.6 *Rate of change of radial acceleration*

An alternative approach to finding the length of the transition is to use values for 'rate of change of radial acceleration' which would be unnoticeable to passengers when travelling by train. The appropriate values were obtained empirically by W.H. Shortt, an engineer working for the railways; hence it is usually referred to as *Shortt's Factor*.

Radial acceleration $= V^2/R$

Thus, as radial acceleration is inversely proportional to R it will change at a rate proportional to the rate of change in R. The transition curve must therefore be long enough to ensure that the rate of change of radius, and hence radial acceleration, is unnoticeable to passengers.

Acceptable values for rate of change of radial acceleration (q) are 0.3 m/s^3, 0.45 m/s^3 and 0.6 m/s^3.

Now, as radial acceleration is V^2/R and the time taken to travel the length L of the transition curve is L/V, then

Rate of change of radial acceleration $= q = \dfrac{V^2}{R} \div \dfrac{L}{V} = \dfrac{V^3}{RL}$

$$\therefore L = \frac{V^3}{Rq} = \frac{V^3}{3 \cdot 6^3 \cdot R \cdot q} \tag{8.18}$$

where the design speed (V) is expressed in km/h.

Although this method was originally devised for railway practice, it is also applied to road design. q should normally not be less than 0.3 m/s^3 for unrestricted design, although in urban areas it may be necessary to increase it to 0.6 m/s^3 or even higher, for sharp curves in tight locations.

8.6 SETTING-OUT DATA

Figure 8.28 indicates the usual situation of two straights projected forward to intersect at I with a clothoid transition curve commencing from tangent point T_1 and joining the circular arc at t_1. The second equal transition commences at t_2 and joins at T_2. Thus the composite curve from T_1 to T_2 consists of a circular arc with transitions at entry and exit.

(1) *Fixing the tangent points T_1 and T_2*

In order to fix T_1 and T_2 the tangent lengths T_1I and T_2I are measured from I back down the straights, or they are set out direct by coordinates.

$$T_1I = T_2I = (R + S) \tan \Delta/2 + C \tag{8.19}$$

where S = shift $= L^2/24R - L^4/3! \times 7 \times 8 \times 2^3R^3 + L^6/5! \times 11 \times 12 \times 2^5R^5 - L^8/7! \times 15 \times 16 \times 2^7R^7 \dots$

and $\quad C = L/2 - L^3/2! \times 5 \times 6 \times 2^2R^2 + L^5/4! \times 9 \times 10 \times 2^4R^4 - L^7/6! \times 13 \times 14 \times 2^6R^6 \dots$

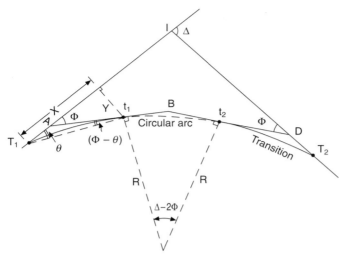

Fig. 8.28

The values of *S* and *C* are abstracted from the *Highway Transition Curve Tables (Metric)* (see *Table 8.2*).

(2) Setting out the transitions

Referring to *Figure 8.29*:

The theodolite is set at T_1 and oriented to *I* with the horizontal circle reading zero. The transition is then pegged out using deflection angles (θ) and chords (Rankine's method) in exactly the same way as for a simple curve.

The data are calculated as follows:

(a) The length of transition *L* is calculated (see design factors in *Section 8.5.5* and *8.5.6*), assume $L = 100$ m.

(b) It is then split into, say, 10 arcs, each 10 m in length (ignoring through chainage), the equivalent chord lengths being obtained from:

$$A - \frac{A^3}{24R^2} + \frac{A^5}{1920R^4}, \text{ where } A \text{ is the arc length}$$

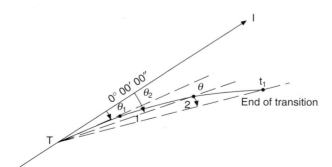

Fig. 8.29

(c) The setting-out angles $\theta_1, \theta_2 \ldots \theta_n$ are obtained as follows:

Basic formula for clothoid: $l = (2RL\phi)^{\frac{1}{2}}$

$$\therefore \Phi = \frac{l^2}{2RL} = \frac{L}{2R} \qquad \text{when} \quad l = L \tag{8.20}$$

(l is any distance along the transition other than total distance L)

then $\qquad \theta = \Phi/3 - 8\Phi^3/2835 - 32\Phi^5/467\,775 \ldots$ (8.21)

$= \Phi/3 - N$, where N is taken from Tables and ranges in value from 0.1″ when $\Phi = 3°$, to 34′ 41.3″ when $\Phi = 86°$ (see *Table 8.1*).

Now $\qquad \dfrac{\phi_1}{\Phi} = \dfrac{l_1^2}{L^2}$ (8.22)

$\therefore \phi_1 = \Phi \dfrac{l_1^2}{L^2}$ where l_1 = chord length = 10 m, say

and $\qquad \theta_1 = \phi_1/3 - N_1$ (where N_1 is the value relative to ϕ_1 in *Table 8.1*)

Similarly $\phi_2 = \Phi \dfrac{l_2^2}{L^2}$ where $l_2 = 20$ m

and $\qquad \theta_2 = \phi_2/3 - N_2$ and so on.

Students should note:

(a) The values for l_1, l_2, etc. are accumulative.

Table 8.1 *Interpolated deflection angles*
For any point on a spiral where angle consumed = ϕ, the true deflection is $\phi/3$ minus the correction tabled below. The back angle is $2\phi/3$ plus the same correction.

Angle consumed	$\phi/3$		Deduct N −		Deflection angle			Angle consumed	$\phi/3$		Deduct N −		Deflection angle		
ϕ °	°	′	′	″	°	′	″	ϕ °	°	′	′	″	°	′	″
			NIL												
2	0	40		0.1		40	0	45	15	00	4	46.2	14	55	13.8
3	1	00		0.2		59	59.9	46	15	20	5	6.0	15	14	54.0
4	1	20			1	19	59.8	47	15	40	5	26.6	15	34	33.4
5	1	40		0.4	1	39	59.6	48	16	00	5	48.1	15	54	11.9
6	2	00		0.7	1	59	59.3	49	16	20	6	10.6	16	13	49.4
7	2	20		1.0	2	19	59.0	50	16	40	6	34.1	6	33	25.9
				Continued at 1° intervals of ϕ											
41	13	40	3	35.9	13	36	24.1	84	28	00	32	14.4	27	27	45.6
42	14	00	3	52.1	13	56	7.7	85	28	20	33	26.9	27	46	33.1
43	14	20	4	9.4	14	15	50.6	86	28	40	34	41.3	28	5	18.7
44	14	40	4	27.4	14	35	32.6								

Reproduced with permission of the County Surveyors' Society

(b) Thus the values obtained for θ_1, θ_2, etc. are the final setting-out angles and are obviously not to be summed.
(c) Although the chord length used is accumulative, the method of setting out is still the same as for the simple curve.

(3) Setting-out circular arc $t_1 t_2$

In order to set out the circular arc it is first necessary to establish the direction of the tangent $t_1 B$ (*Figure 8.28*). The theodolite is set at t_1 and backsighted to T_1 with the horizontal circle reading $[180° - (\Phi - \theta)]$. setting the instrument to zero will now orient it in the direction $t_1 B$ with the circle reading zero, prior to setting-out the simple circular arc. The angle $(\Phi - \theta)$ is called the *back-angle to the origin* and may be expressed as follows:

$$\theta = \Phi/3 - N$$
$$\therefore (\Phi - \theta) = \Phi - (\Phi/3 - N) = 2/3\Phi + N \tag{8.23}$$

and is obtained direct from tables.
The remaining setting-out data are obtained as follows:

(a) As each transition absorbs an angle Φ, then the angle subtending the circular arc = $(\Delta - 2\Phi)$.
(b) Length of circular arc = $R(\Delta - 2\Phi)$, which is then split into the required chord lengths C.
(c) The deflection angles $\delta° = 28.6479 \cdot C/R$ are then set out from the tangent $t_1 B$ in the usual way.

The second transition is best set out from T_2 to t_2. Setting-out from t_2 to T_2 involves the 'osculating-circle' technique (see *Section 8.9*).
The preceding formulae for clothoid transitions are specified in accordance with the latest *Highway Transition Curve Tables* (*Metric*) compiled by the County Surveyors' Society. As the equations involved in the setting-out data are complex, the information is generally taken straight from tables. However, approximation of the formulae produces two further transition curves, the *cubic spiral* and the *cubic parabola* (see *Section 8.7*).
In the case of the clothoid, *Figure 8.28* indicates an offset Y at the end of the transition, distance X along the straight, where

$$X = L - L^3/5 \times 4 \times 2!R^2 + L^5/9 \times 4^2 \times 4!R^4 - L^7/13 \times 4^3 \times 6!R^6 + \ldots \tag{8.24}$$
$$Y = L^2/3 \times 2R - L^4/7 \times 3! \times 2^3R^3 + L^6/11 \times 5! \times 2^5R^5 - L^8/15 \times 7! \times 2^7R^7 + \ldots \tag{8.25}$$

The clothoid is always set out by deflection angles, but the values for X and Y are useful in the large-scale plotting of such curves, and are taken from tables.
Refer to the end of the chapter for derivation of clothoid formulae.

8.6.1 Highway transition curve tables (metric)

An examination of the complex equations defining the clothoid transition spiral indicates the obvious need for tables of prepared data to facilitate the design and setting out of such curves. These tables have been produced by the County Surveyors' Society under the title *Highway Transition Curve Tables* (*Metric*), and contain a great deal of valuable information relating to the geometric design of highways. A very brief sample of the Tables is given here simply to convey some idea of the format and information contained therein (*Table 8.2*).
As shown in *Section 8.6(3)*, $\theta = \phi/3 - N$ and the 'back-angle' is $2\phi/3 + N$, all this information for various values of ϕ is supplied in *Table 8.1* and clearly shows that for large values of ϕ, N cannot be ignored.

Table 8.2

Gain of Accn. m/s³	0.30	0.45	0.60
Speed Value km/h	84.4	96.6	106.3

Increase in Degree of Curve per Metre = D/L = 0° 8' 0.0"
RL Constant = 42971.835

Degree of Curvature Based on 100 m. Standard Arc

Radius R	Degree of Curve D			Spiral Length L	Φ Angle Consumed			Shift S	R + S R	C	Long Chord	Coordinates X	Coordinates Y	Θ Deflection Angle from Origin			Back Angle to Origin		
Metres	°	'	"	Metres	°	'	"	Metres	Metres	Metres	Metres	Metres	Metres	°	'	"	°	'	"
8594.3669	0	40	0.0	5.00	0	1	0.0	0.0001	8594.3670	2.5000	5.0000	5.0000	0.0005	0	0	20.0	0	0	40.0
4297.1835	1	20	0.0	10.00	0	4	0.0	0.0010	4297.1844	5.0000	10.0000	10.0000	0.0039	0	1	20.0	0	2	40.0
2864.7890	2	0	0.0	15.00	0	9	0.0	0.0033	2864.7922	7.5000	15.0000	15.0000	0.0131	0	3	0.0	0	6	0.0
2148.5917	2	40	0.0	20.00	0	16	0.0	0.0078	2148.5995	10.0000	20.0000	20.0000	0.0310	0	5	20.0	0	10	40.0
1718.8734	3	20	0.0	25.00	0	25	0.0	0.0152	1718.8885	12.5000	24.9999	24.9999	0.0606	0	8	20.0	0	16	40.0
1432.3945	4	0	0.0	30.00	0	36	0.0	0.0262	1432.4207	14.9999	29.9999	29.9997	0.1047	0	12	0.0	0	24	0.0
1227.7667	4	40	0.0	35.00	0	49	0.0	0.0416	1227.8083	17.4999	34.9997	34.9993	0.1663	0	16	20.0	0	32	40.0
1074.2959	5	20	0.0	40.00	1	4	0.0	0.0621	1074.3579	19.9998	39.9994	39.9986	0.2482	0	21	20.0	0	42	40.0
954.9297	6	0	0.0	45.00	1	21	0.0	0.0884	955.0180	22.4996	44.9989	44.9975	0.3534	0	27	0.0	0	54	0.0
859.4367	6	40	0.0	50.00	1	40	0.0	0.1212	859.5579	24.9993	49.9981	49.9958	0.4848	0	33	20.0	1	6	40.0
781.3061	7	20	0.0	55.00	2	1	0.0	0.1613	781.4674	27.4989	54.9970	54.9932	0.6452	0	40	20.0	1	20	40.0
716.1972	8	0	0.0	60.00	2	24	0.0	0.2094	716.4067	29.9982	59.9953	59.9895	0.8377	0	48	0.0	1	36	0.0
661.1051	8	40	0.0	65.00	2	49	0.0	0.2663	661.3714	32.4974	64.9930	64.9843	1.0650	0	56	19.9	1	52	40.1
613.8834	9	20	0.0	70.00	3	16	0.0	0.3325	614.2159	34.9962	69.9899	69.9772	1.3300	1	5	19.9	2	10	40.1
572.9578	10	0	0.0	75.00	3	45	0.0	0.4090	573.3668	37.4946	74.9857	74.9679	1.6357	1	14	59.8	2	30	0.2
537.1479	10	40	0.0	80.00	4	16	0.0	0.4964	537.6443	39.9926	79.9803	79.9556	1.9850	1	25	19.8	2	50	40.2
505.5510	11	20	0.0	85.00	4	49	0.0	0.5953	506.1463	42.4900	84.9733	84.9399	2.3807	1	36	19.7	3	12	40.3
477.4648	12	0	0.0	90.00	5	24	0.0	0.7066	478.1715	44.9867	89.9645	89.9201	2.8256	1	47	59.5	3	36	0.5
452.3351	12	40	0.0	95.00	6	1	0.0	0.8310	453.1661	47.4825	94.9534	94.8953	3.3227	2	0	19.3	4	0	40.7
429.7183	13	20	0.0	100.00	6	40	0.0	0.9692	430.6875	49.9774	99.9398	99.8647	3.8748	2	13	19.1	4	26	40.9
409.2556	14	0	0.0	105.00	7	21	0.0	1.1218	410.3774	52.4712	104.9232	104.8273	4.4846	2	26	58.8	4	54	1.2
390.6530	14	40	0.0	110.00	8	4	0.0	1.2897	391.9427	54.9637	109.9031	109.7822	5.1550	2	41	18.4	5	22	41.6
373.6681	15	20	0.0	115.00	8	49	0.0	1.4734	375.1416	57.4546	114.8790	114.7280	5.8888	2	56	17.9	5	52	42.1
358.0986	16	0	0.0	120.00	9	36	0.0	1.6738	359.7725	59.9439	119.8503	119.6636	6.6886	3	11	57.3	6	24	2.7

Part only of *Table 8.2* is shown and it is the many tables like this that provide the bulk of the design data. Much of the information and its application to setting out should be easily understood by the student, so only a brief description of its use will be given here.

Use of tables

(1) Check the angle of intersection of the straights (Δ) by direct measurement in the field.
(2) Compare Δ with 2Φ, if $\Delta \leqslant 2\Phi$, then the curve is wholly transitional.
(3) Abstract $(R + S)$ and C in order to calculate the tangent lengths $= (R + S) \tan \Delta/2 + C$.
(4) Take Φ from tables and calculate length of circular arc using $R(\Delta - 2\Phi)$, or, if working in 'degree of curvature' D, use

$$\frac{100(\Delta - 2\Phi)}{D}$$

(5) Derive chainages at the beginning and end of both transitions.
(6) Compute the setting-out angles for the transition $\theta_1 \ldots \theta$ from $\phi_1/\Phi = l_1^2/L^2$ from which $\theta_1 = \phi_1/3 - N_1$, and so on, for accumulative values of l.
(7) As control for the setting out, the end point of the transition can be fixed first by turning off from T_1 (the start of the transition) the 'deflection angle from the origin' θ and laying out the 'long chord' as given in the Tables. Alternatively, the right-angled offset Y distance X along the tangent may be used.
(8) When the first transition is set out, set up the theodolite at the end point and with the theodolite reading $(180° - (\frac{2}{3}\Phi + N))$, backsight to T_1. Turn the theodolite to read $0°$ when it will be pointing in a direction tangential to the start of the circular curve prior to its setting out. This process has already been described.
(9) As a check on the setting out of the circular curve take $(R + S)$ and S from the Tables to calculate the apex distance $= (R + S) (\sec \Delta/2 - 1) + S$, from the intersection point I of the straights to the centre of the circular curve.
(10) The constants RL and D/L are given at the head of the Tables and can be used as follows:

(a) Radius at any point P on the transition $= r_p = RL/l_p$
(b) Degree of curve at $P = D_p = (D/L) \times l_p$, where l_p is the distance to P measured along the curve from T_1.
Similarly:

(c) Angle consumed at $P = \phi_p = \dfrac{l_p^2}{2RL}$ or $\dfrac{l_p^2}{200} \times \dfrac{D}{L}$

(d) Setting-out angle from T_1 to $P = \theta_p = \dfrac{\phi_p}{3} - N_p$ or

$$\frac{l^2}{600} \times \frac{D}{L} - N_p$$

Worked examples

Example 8.8. It is required to connect two straights ($\Delta = 50°$) with a composite curve, comprising entry transition, circular arc and exit transition. The initial design parameters are $V = 105$ km/h, $q = 0.6$ m/s³, $\mu = 0.15$ and crossfall 1 in 14.5 (tan $\theta = 0.07$).

(1) $V^2/127R = \tan \theta + \mu$

$105^2/127R = 0.07 + 0.15 = 0.22$

$R = 105^2/127 \times 0.22 = 394.596$ m

The nearest greater value is 409.2556 m for a value of $V = 106.3$ km/h (*Table 8.2*).

(2) From the table selected, the length L of transition is given as 105 m. Purely to illustrate the application of formulae:

$$(L = V^3/3.6^3 \times R \times q = 106.3^3/3.6^3 \times 409.2556 \times 0.6 = 104.8 \text{ m})$$

(3) From the Table, $\Phi = 7°21'00''$

Check $\Phi = L/2R = 105/2 \times 409.2556 = 0.12828169$ rads $= 7°21'00''$

Now as $2\Phi < \Delta$, there is obviously a portion of circular arc subtended by angle of $(\Delta - 2\Phi)$.

(4) From the Table, find the tangent length:

$(R + S) \tan \Delta/2 + C$

where $(R + S) = 410.3374$ ⎫
$\quad\quad\quad C = 52.4721$ ⎬ see *Table 8.2*
$\quad\quad\quad\quad\quad\quad\quad$ ⎭

$T_1I = T_2I = 410.3774 \tan 25° + 52.4712 = 243.8333$ m.

(5) The chainage of the intersection point $I = 5468.399$ m (see *Figure 8.28*)

Chn $T_1 = 5468.399 - 243.833 = 5224.566$ m

Chn $t_1 = 5224.566 + 105 = 5329.566$ m

Length of circular arc $= R(\Delta - 2\Phi) = 252.1429$ m,

or using 'degree of curve' $(D) = 100(\Delta - 2\Phi)/D = 252.1429$ m (check)

Chn $t_2 = 5329.566 + 252.143 = 5581.709$ m

Chn $T_2 = 5581.709 + 105 = 5686.709$ m

(6) As a check, the ends of the transition could be established using the offset $Y = 4.485$ m at distance $X = 104.827$ m along the tangent (see Table) or by the 'long chord' $= 104.923$ m at an angle θ to the tangent of $2°26'58.8''$. This procedure will establish t_1 and t_2 relative to T_1 and T_2 respectively. Now the setting-out angles ($\theta_1, \theta_2 \ldots \theta$) are computed. If the curve was to be set out by 10-m chords and there was no though chainage, their values could simply be abstracted from *Table 8.2* as shown:

10 m $= 0°01'20.0''$

20 m $= 0°05'20.0''$, and so on.

These values would be subtracted from 360°, for use in setting-out the 2nd transition from T_2 back to t_2.

However, the usual way is to set out on a through chainage basis.

Hence using 10-m standard chords, as the chainage at $T_1 = 5224.566$, the first sub-chord $= 10 - 4.566 = 5.434$ m. The setting-out angles are computed from equations (8.21) and (8.22), i.e.

(a) $\quad\quad\quad \Phi = 7°21'00''$ $\quad\quad$ from *Table 8.2*

and $\quad\quad\quad \theta = 2°26'58.8''$

as, $\theta_1/\theta = (l_1/L)^2$

then $\theta_1 = 2°26'58.8''\ (5.434/105)^2 = 0°00'23.6'' - N_1$

(b) Alternatively using 'degree of curve' D:

$$\theta_1 = \frac{l_1^2}{600} - \frac{D}{L} - N_1$$

D/L is given in the Table $= 0°08'00''$

$$\therefore \frac{D}{600 \cdot L} = 0.8''\text{ (a constant)}$$

and $\theta_1 = 0.8''\ (5.434)^2 = 0°00'23.6'' - N_1$

(It can be seen from *Table 8.1* that the deduction N_1 is negligible and will have a maximum value for the formal setting-out angle θ of just over 1''.)

Using either of the above approaches and remembering that the chord lengths are accumulative in the computation, the setting-out angles are:

Entry Transition

Chord m	Chainage m	Setting-out Angle (θ)	Deduction N	Remarks
0	5224.566	–	–	T_1 – Start of spiral
5.434	5230	0°00'23.6''	–	Peg. 1.
10	5240	0°03'10.6''	–	Peg. 2.
10	5250	0°08'37.6''	–	Peg. 3.
		etc.		

When computing the setting-out angles for the exit transition from T_2 back to t_2 one starts with the length of the final sub-chord at T_2.

(7) When the entry transition is set out to t_1, the theodolite is moved to t_1 and backsighted to T_1, with the horizontal circle reading:

$180° -$ (back-angle to the origin)

$= 180° - (\Phi - \theta)$

$= 180° - 4°54'01.2''$ (taken from final column of *Table 8.2*)

$= 175°05'58.8$

Rotating the upper circle to read $0°00'00''$ will establish the line of sight tangential to the route of the circular arc, ready for setting out.

(8) As the chainage of $t_1 = 5329.566$, the first sub-chord $= 0.434$ m, thereby putting Peg 1 on chainage 5330 m.

The setting out angles are now calculated in the usual way as shown in *Section 8.2.1*.

If it is decided to set out the exit transition from the end of the circular arc, then the theory of the osculating circle is used (see *Section 8.9.2*).

8.7 CUBIC SPIRAL AND CUBIC PARABOLA

Approximation of the clothoid formula produces the cubic spiral and cubic parabola, the latter being used on railway and tunnelling work because of the ease in setting out by offsets. The cubic spiral can be used for minor roads, as guide for excavation prior to the clothoid being set out, or as check on clothoid computation.

$Y = L^2/6R,$ which when $L = l$, $Y = y$, becomes

$y = l^3/6RL$ (8.26)

Approximating equation (8.24) gives

$X = L,$ thus $x = l$

$\therefore y = x^3/6RL$ (the equation for a cubic parabola)   (8.27)

In both cases:

Tangent length $T_1I = (R + S) \tan \Delta/2 + C$

where $S = L^2/24R$ (8.28)

and $C = L/2$ (8.29)

             $\Phi = L/2R = l^2/2RL$ (8.30)

and $\theta = \Phi/3$ (8.31)

The deflection angles for these curves may be obtained as follows (the value of N being ignored):

$\theta_1/\theta = l_1^2/L^2$, where l is the chord/arc length   (8.32)

When the value of $\Phi \approx 24°$, the radius of these curves starts to increase again, which makes them useless as transitions.

Refer to *worked examples* for application of the above equations. and note 'back-angle to origin' would be $(\Phi - \theta)$ or $\frac{2}{3}\Phi$.

8.8 CURVE TRANSITIONAL THROUGHOUT

A curve transitional throughout (*Figure 8.30*) comprises two transitions meeting at a common tangent point t.

Tangent length $T_1I = X + Y \tan \Phi$ (8.33)

where X and Y are obtained from equations (8.24) and (8.25) and $\Phi = \Delta/2 = L/2R$.

$\therefore \Delta = L/R$ (8.34)

8.9 THE OSCULATING CIRCLE

Figure 8.31 illustrates a transition curve T_1PE. Through P; where the transition radius is r, a simple curve of the same radius is drawn and called the *osculating circle*.

At T_1 the transition has the same radius as the straight T_1I, that is, ∞, but diverges from it at a

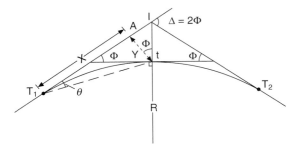

Fig. 8.30

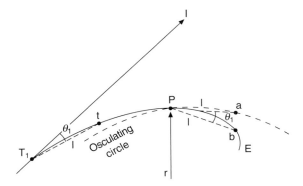

Fig. 8.31

constant rate. Exactly the same condition exists at P with the osculating circle, that is, the transition has the same radius as the osculating circle, r, but diverges from it at a constant rate. Thus if chords $T_1t = Pa = Pb = l$, then

angle $IT_1t = aPb = \theta_1$

This is the theory of the osculating circle, and its application is described in the following sections.

8.9.1 Setting out with the theodolite at an intermediate point along the transition curve

Figure 8.32 illustrates the situation where the transition has been set out from T_1 to P_3 in the normal way. The sight T_1P_4 is obstructed and the theodolite must be moved to P_3 where the remainder of the transition will be set out. The direction of the tangent P_3E is first required from the back-angle $(\phi_3 - \theta_3)$.

From the *Figure* it can be seen that the angle from the tangent to the chord P_3P_4' on the osculating circle is $\delta_1^\circ = 28.6479 \times l/r_3$. The angle between the chord on the osculating circle and that on the transition is $P_4'P_3P_4 = \theta_1$, thus the setting-out angle from the tangent to $P_4 = (\delta_1 + \theta_1)$, to $P_5 = (\delta_2 + \theta_2)$ and to $P_6 = (\delta_3 + \theta_3)$, etc.

For example, assuming $\Delta = 60°$, $L = 60$ m, $l = $ chord $= 10$ m, $R = 100$ m and $T_1P_3 = 30$ m, calculate the setting-out angles for the remainder of the transition from P_3.

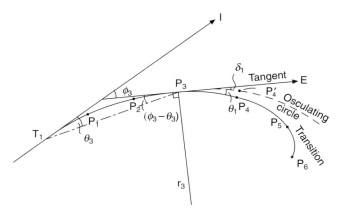

Fig. 8.32

From basic formula:

$$\phi_3 = \frac{l_3^2}{2\,RL} = \frac{30^2}{2 \times 100 \times 60} = 4°17'50'' \ (-N_3, \text{ if clothoid})$$

or, if curve is defined by its 'degree of curvature' $= D$, then

$$\phi_3 = \frac{l_3^2}{200} \cdot \frac{D}{L} \ (-N_3, \text{ if clothoid})$$

Thus the back-angle to the origin is found $(2/3)\,\phi_3$, and the tangent established as already shown.

Now from $\Phi = L/2R$ and $\theta = \Phi/3$, the angles θ_1, θ_2, and θ_3 are found as normal. In practice these angles would already be available, having been used to set out the first 30 m of the transition.

Before the angles to the osculating circle can be found, the value of r_3 must be known, thus, from $rl = RL$:

$$r_3 = RL/l_3 = 100 \times 60/30 = 200 \text{ m}$$

or 'degree of curvature' at P_3 30 m from $T_1 = \dfrac{D}{L} \times l_3$

$$\therefore \ \delta_1^° = 28.6479 \left(\frac{10}{200} \right) = 1°25'57''$$

$\delta_2 = 2\delta_1$ and $\delta_3 = 3\delta_1$ as for a simple curve

The setting-out angles are then $(\delta_1 + \theta_1)$, $(\delta_2 + \theta_2)$, $(\delta_3 + \theta_3)$.

8.9.2 Setting out transition from the circular arc

Figure 8.33 indicates the second transition in *Figure 8.28* to be set out from t_2 to T_2. The tangent t_2D would be established by backsighting to t_1 with the instrument reading $[180° - (\Delta - 2\Phi)/2]$, setting to zero to fix direction t_2D. It can now be seen that the setting-out angles here would be $(\delta_1 - \theta_1)$, $(\delta_2 - \theta_2)$, etc. computed in the usual way.

Consider the previous example, where the chainage at the end of the circular arc (t_2) was 5581.709 m, then the first sub-chord = 8.291 m.

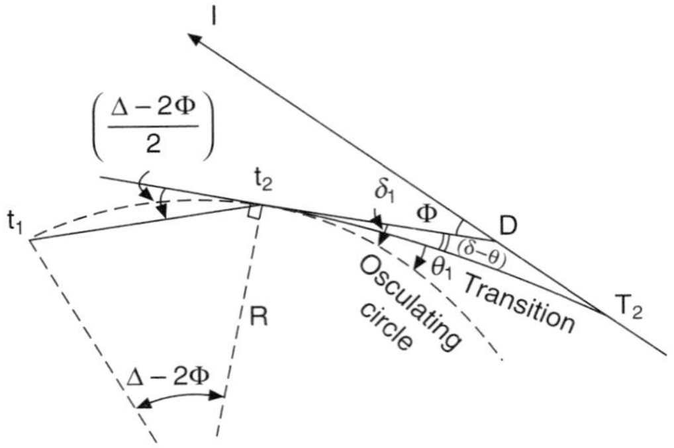

Fig. 8.33

The setting-out angles for the transition are computed in the usual way. Use, say, $D/600L = 0.8''$, then:

$$\theta_1 = 0.8''(8.291)^2 = 0°00'55.0''$$
$$\theta_2 = 0.8''(18.291)^2 = 0°04'27.6''$$
$$\theta_3 = 0.8''(28.291)^2 = 0°10'40.3''$$

and so on to the end of the transition.

Now compute the setting-out angles for a circular curve in the usual way:

$$\delta° = 28.6479 \times (C/R)$$
$$= 28.6479 \times (8.291/409.2556) = 0°34'49.3''$$

Also $\quad \delta°_{10} = 28.6479(10/409.2556) = 0°42'00.0''$

Remembering that the angles (δ) for a circular arc are accumulative, the setting-out angles for the transition are:

Chord (m)	Chainage (m)	δ			θ			Setting-out angle $\delta - \theta$			Remarks
		°	′	″	°	′	″	°	′	″	
8.291	5990	0	34	49.3	0	00	55.0	0	33	54.3	Peg 1
10	5600	01	16	49.3	0	04	27.6	01	12	22.3	Peg 2
10	5610	01	58	49.3	0	10	40.3	01	48	09.0	Peg 3
etc.	etc.	etc.			etc.			etc.			etc.

8.9.3 *Transitions joining arcs of different radii (compound curves)*

Figure 8.14 indicates a compound curve requiring transitions at T_1, t and T_2. To permit the entry of the transitions the circular arcs must be shifted forward as indicated in *Figure 8.34* where

$$S_1 = L_1^2/24R_1 \quad \text{and} \quad S_2 = L_2^2/24R_2$$

The lengths of transition at entry (L_1) and exit (L_2) are found in the normal way, whilst the transition connecting the compound arcs is

$$bc = L = (L_1 - L_2)$$

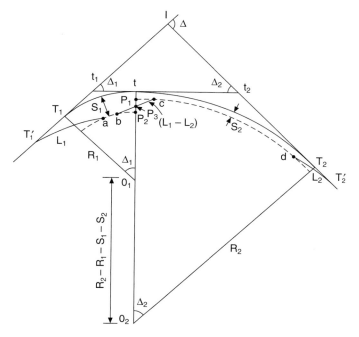

Fig. 8.34

The distance $P_1P_2 = (S_1 - S_2)$ is bisected by the transition curve at P_3. The curve itself is bisected and length $bP_3 = P_3c$. As the curves at entry and exit are set out in the normal way, only the fixing of their tangent points T_1' and T_2' will be considered. In triangle t_1It_2

$$t_1t_2 = t_1t + tt_2 = (R_1 + S_1)\tan \Delta_1/2 + (R_2 + S_2)\tan \Delta_2/2$$

from which the triangle may be solved for t_1I and t_2I.

Tangent length $T_1'I = T_1't_1 + t_1I = (R_1 + S_1)\tan \Delta_1/2 + L_1/2 + t_1I$

and $T_2'I = T_2't_2 + t_2I = (R_2 + S_2)\tan \Delta_2/2 + L_2/2 + t_2I$

The curve bc is drawn enlarged in *Figure 8.35* from which the method of setting out, using the osculating circle, may be seen.

Setting-out from b, the tangent is established from which the setting-out angles would be $(\delta_1 - \theta_1)$, $(\delta_2 - \theta_2)$, etc. as before, where δ_1, the angle to the osculating circle, is calculated using R_1.

If setting out from C, the angles are obviously $(\delta_1 + \theta_1)$, etc., where δ_1 is calculated using R_2.

Alternatively, the curve may be established by right-angled offsets from chords on the osculating circle, using the following equation:

$$y = \frac{x^3}{6RL} = \frac{x^3}{L^3}\frac{L^2}{6R} \quad \text{where} \quad \frac{L^2}{6R} = 4S$$

$$\therefore y = \frac{4x^3}{L^3}(S_1 - S_2) \tag{8.35}$$

It should be noted that the osculating circle provides only an approximate solution, but as the

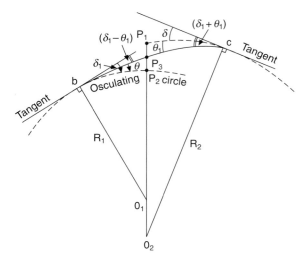

Fig. 8.35

transition is usually short, it may be satisfactory in practice. In the case of a reverse compound curve (*Figure 8.36*).

$$S = (S_1 + S_2), \qquad L = (L_1 + L_2) \quad \text{and} \quad y = \frac{4x^3}{L^3} (S_1 + S_2) \tag{8.36}$$

otherwise it may be regarded as two separate curves.

8.9.4 Coordinates on the Transition Spiral (Figure 8.37)

The setting-out of curves by traditional methods of angles and chords has been dealt with. However, these methods are frequently superseded by the use of coordinates (as indicated in section 8.2.4) of circular curves. The method of calculating coordinates along the centre-line of a transition curve is probably best illustrated with a worked example.

Consider 'Worked Example 8.8' dealing with the traditional computation of a clothoid spiral using Highway Transition Curve Tables. As with circular curves, it is necessary to calculate the traditional setting-out data first, as an aid to calculating coordinates.

In Example 8.8 the first three setting-out angles are:

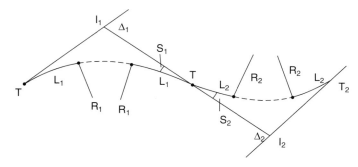

Fig. 8.36

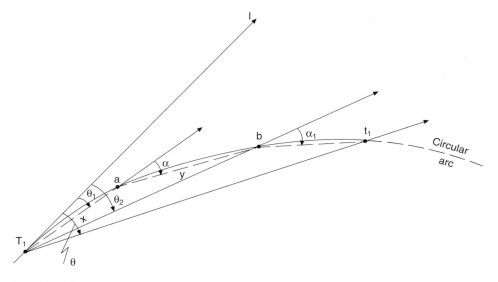

Fig. 8.37   *Coordinates along a transition curve*

$\theta_1 = 0°00'23.6''$, sub chord = 5.434 m (x)
$\theta_2 = 0°03'10.6''$, standard chord = 10 m (y)
$\theta_3 = 0°08'37.6''$, standard chord = 10 m (y)

If working in coordinates, the coordinates of T_1, the tangent point and I, the intersection point, would be known, say.

T_1 = E 500.000, N 800.000

And calculated using the coordinates of I, the WCB of $T_1I = 10°25'35.0''$

Now WCB T_1a = WCB $T_1I + \theta_1 = 10°25'58.6''$

And a sub-chord length $T_1a = x = 5.434$ m
Using the P and R keys, or the traditional formula the coordinates of the line T_1a are calculated, thus

$\Delta E = 0.9840$ m, $\Delta N = 5.3442$ m
$E_a = 500.000 + 0.9840 = 500.9840$ m
$N_a = 800.000 + 5.3442 = 805.3442$ m

In triangle T_1ab:

$aT_1b = (\theta_2 - \theta_1) = 0°02'47.0''$

Using the sine rule, the angle at b can be calculated

$$\frac{T_1a}{\sin T_1ba} = \frac{ab}{\sin aT_1b}$$

$$\sin T_1ba = \frac{T_1a \sin aT_1b}{ab} = \frac{5.434 \sin 0°02'47.0''}{10}$$

$T_1ba = 00°01'30.8''$, and
$$\alpha = aT_1b + T_1ba = 00°04'17.8''$$
and WCB ab = WCB $T_1a + \alpha = 10°30'16.4''$

dist. $ab = 10$ m

$\therefore \Delta E_{ab} = 1.8231, \quad \Delta N_{ab} = 9.8324$

and

$$E_b = E_a + \Delta E_{ab} = 500.9840 + 1.8231 = 502.8071 \text{ m}$$
$$N_b = N_a + \Delta N_{ab} = 805.3442 + 9.8324 = 815.1766 \text{ m}$$

This procedure is now repeated to the end of the spiral, as follows:
In triangle T_1bt_1,

(1) calculate distance T_1b from the coordinates of T_1 and b
(2) calculate angle T_1t_1b by sine rule, i.e.

$$\frac{T_1b}{\sin T_1t_1b} = \frac{bt_1}{\sin bT_1t_1}$$

where bt_1 is a known chord or sub-chord length and angle $bT_1t_1 = (\theta - \theta_2)$ in this instance.

(3) calculate $\alpha_1 = bT_1t_1 + T_1t_1b$
(4) WCB T_1b = WCB $T_1l + \theta_2$
(5) WCB bt_1 = WCB $T_1b + \alpha_1$ and length is known
(6) Computed ΔE, ΔN of the line bt_1 and add then algebraically to the E_b and N_b respectively, to give E_{t_1}, N_{t_1}

Using the 'back angle to the origin' $(\Phi - \theta)$ at the end of the spiral (t_1), the WCB of the tangent to the circular arc can be found and the coordinates along the centre-line calculated as shown in 8.2.4.

Worked examples

Example 8.9. Consider Figure 8.35 in which $R_1 = 700$ m, $R_2 = 1500$ m, $q = 0.3$ m/s^3 and $V = 100$ km/h

$L_1 = V^3/3.6^3R_1q = 102$ m
$L_2 = V^3/3.6^3R_2q = 48$ m
$L = (L_1 - L_2) = 54$ m

Setting-out from b:

$$\Phi = L/2R_1 = 54/2 \times 700 = 2°12'36''$$
Check: $\Phi = R_1L/2R_1^2 = 2°12'36''$
as $\theta = \Phi/3 = 0°44'12''$

Then, assuming 10 m chords and no through chainage:

$\theta_1 = 0°44'12'' (10/54)^2 = 0°01'31''$
$\theta_2 = 0°44'12'' (20/54)^2 = 0°06'04''$, etc.

For the osculating circle (R_1):

$$\delta^{\circ}_{10} = 28.6479(10/700) = 0°24'33''$$

Thus the setting-out angles for the 'suspended' transition are

Chord (m)	δ °	'	''	θ °	'	''	$\delta - \theta$ °	'	''	Remarks
10	0	24	33	0	01	31	0	23	02	Peg 1
10	0	49	06	0	06	04	0	43	02	Peg 2
etc.	etc.			etc.			etc.			etc.

Example 8.10. Part of a motorway scheme involves the design and setting out of a simple curve with cubic spiral transitions at each end. The transitions are to be designed such that the centrifugal ratio is 0.197, whilst the rate of change of centripetal acceleration is 0.45 m/s^3 at a design speed of 100 km/h.

If the chainage of the intersection to the straights is 2154.22 m and the angle of deflection 50°, calculate:

(a) The length of transition to the nearest 10 m.
(b) The chainage at the beginning and the end of the total composite curve.
(c) The setting-out angles for the first three 10-m chords on a through chainage basis.

Briefly state where and how you would orient the theodolite in order to set out the circular arc.
(KU)

Referring to *Figure 8.28*

Centrifugal ratio $P/W = V^2/127R$

$$\therefore R = \frac{100^2}{127 \times 0.197} = 400 \text{ m}$$

Rate of change of centripetal acceleration $= q = \dfrac{V^3}{3.6^3\,RL}$

(a) $\therefore L = \dfrac{100^3}{3.6^3 \times 400 \times 0.45} = 120 \text{ m}$

(b) To calculate chainage:

$$S = \frac{L^2}{24R} = \frac{120^2}{24 \times 400} = 1.5 \text{ m}$$

Tangent length $= (R + S) \tan \Delta/2 + L/2$

$$= (400 + 1.5) \tan 25° + 60 = 247.22 \text{ m}$$

$\therefore$ Chainage at $T_1 = 2154.22 - 247.22 = 1907 \text{ m}$

To find length of circular arc:

length of circular arc $= R(\Delta - 2\Phi)$ where $\Phi = L/2R$

thus $\qquad\qquad 2\Phi = \dfrac{L}{R} = \dfrac{120}{400} = 0.3 \text{ rad}$

and $$\Delta = 50° = 0.872\ 665\ \text{rad}$$

$$\therefore R(\Delta - 2\Phi) = 400(0.872\ 665 - 0.3) = 229.07\ \text{m}$$

Chainage at $$T_2 = 1907.00 + 2 \times 120 + 229.07 = 2376.07\ \text{m}$$

(c) To find setting-out angles from equation (8.32) $\theta_1/\theta = l_1^2/L^2$

$$\theta = \frac{\Phi}{3} = \frac{L}{6R} = \frac{120}{6 \times 400}\ \text{rad}$$

$$\theta'' = \frac{120 \times 206\ 265}{6 \times 400} = 10\ 313''$$

As the chainage of $T_1 = 1907$, then the first chord will be 3 m long to give a round chainage of 1910 m.

$$\therefore \theta_1 = \theta \frac{l_1^2}{L^2} = 10\ 313'' \times \frac{3^2}{120^2} = 0° 00' 06.5''$$

$$\theta_2 = 10\ 313'' \times \frac{13^2}{120^2} = 0° 02' 01''$$

$$\theta_3 = 10\ 313'' \times \frac{23^2}{120^2} = 0° 06' 19''$$

For final part of answer refer to *Sections 8.6(3)* and *8.7*.

Example 8.11. A transition curve of the cubic parabola type is to be set out from a straight centre-line. It must pass through a point which is 6 m away from the straight, measured at right angles from a point on the straight produced 60 m from the start of the curve.
   Tabulate the data for setting out a 120-m length of curve at 15-m intervals.
   Calculate the rate of change of radial acceleration for a speed of 50 km/h.   (LU)

   The above question may be read to assume that 120 m is only a part of the total transition length and thus L is unknown.

From expression for a cubic parabola: $y = x^3/6RL = cx^3$

$$y = 6\ \text{m}, \quad x = 60\ \text{m} \quad \therefore c = \frac{1}{36\ 000} = \frac{1}{6RL}$$

The offsets are now calculated using this constant:

$$y_1 = \frac{15^3}{36\ 00} = 0.094\ \text{m}$$

$$y_2 = \frac{30^3}{36\ 000} = 0.750\ \text{m}$$

$$y_3 = \frac{45^3}{36\ 000} = 2.531\ \text{m and so on}$$

Rate of change of radial acceleration = $q = V^3/3.6^3RL$

now $$\frac{1}{6RL} = \frac{1}{36\ 000} \quad \therefore \frac{1}{RL} = \frac{1}{6000}$$

$$\therefore q = \frac{50^3}{3.6^3 \times 6000} = 0.45 \text{ m/s}^3$$

Example 8.12. Two straights of a railway track of gauge 1.435 m have a deflection angle of 24° to the right. The straights are to be joined by a circular curve having cubic parabola transition spirals at entry and exit. The ratio of super-elevation to track gauge is not to exceed 1 in 12 on the combined curve, and the rate of increase/decrease of super-elevation on the spirals is not to exceed 1 cm in 6 m. If the through chainage of the intersection point of the two straights is 1488.8 m and the maximum allowable speed on the combined curve is to be 80 km/h, determine:

(a) The chainages of the four tangent points.
(b) The necessary deflection angles (to the nearest 20″) for setting out the first four pegs past the first tangent point, given that pegs are to be set out at the 30-m points of the through chainage.
(c) The rate of change of radial acceleration on the curve when trains are travelling at the maximum permissible speed.
                                                                                                    (ICE)

(a) Referring to *Figure 8.28*, the four tangent points are T_1, t_1, t_2, T_2.
   Referring to *Figure 8.27(a)*, as the super-elevation on railways is limited to 0.152 m, then $AB = 1.435 \text{ m} \approx CB$.

$$\therefore \text{ Super-elevation} = AC = \frac{1.435}{12} = 0.12 \text{ m} = 12 \text{ cm}$$

Rate of application = 1 cm in 6 m

$$\therefore \text{ Length of transition} = L = 6 \times 12 = 72 \text{ m}$$

From *Section 8.5.5* $\tan \theta = \dfrac{V^2}{127R} = \dfrac{1}{12}$

$$\therefore \frac{80^2}{127R} = \frac{1}{12} \quad \therefore R = 604.72 \text{ m}$$

$$\text{Shift} = S = \frac{L^2}{24R} = \frac{72^2}{24 \times 604.72} = 0.357 \text{ m}$$

$$\text{Tangent length} = (R + S) \tan \Delta/2 + L/2 = 605.077 \tan 12° + 36$$

$$= 164.6 \text{ m}$$

$$\therefore \text{ Chainage } T_1 = 1488.8 - 164.6 = 1324.2 \text{ m}$$

$$\text{Chainage } t_1 = 1324.2 + 72 = 1396.2 \text{ m (end of transition)}$$

To find length of circular curve:

$$2\Phi = \frac{L}{R} = \frac{72}{604.72} = 0.119\,063 \text{ rad}$$

$$\Delta = 24° = 0.418\,879 \text{ rad}$$

$$\therefore \text{ Length of curve} = R(\Delta - 2\Phi) = 604.72(0.418\,879 - 0.119\,063)$$

$$= 181.3 \text{ m}$$

$$\therefore \text{ Chainage } t_2 = 1396.2 + 181.3 = 1577.5 \text{ m}$$

$$\text{Chainage } T_2 = 1577.5 + 72 = 1649.5 \text{ m}$$

(b) From chainage of T_1 the first chord = 5.8 m

$$\theta = \frac{L}{6R} \times 206\,265 = \frac{72 \times 206\,265}{6 \times 604.72} = 4093''$$

$$\therefore \theta_1 = \theta \frac{l_1^2}{L^2} = 4093'' \times \frac{5.8^2}{72^2} = 27'' = 0°\,00'\,27'' \quad \text{peg 1}$$

$$\theta_2 = 4093 \times \frac{35.8^2}{72^2} = 1012'' = 0°\,16'\,52'' \qquad \text{peg 2}$$

$$\theta_3 = 4093 \times \frac{65.8^2}{72^2} = 3418'' = 0°\,56'\,58'' \qquad \text{per 3}$$

$$\theta_4 = 4093'' = \theta \text{ (end of transition)} = 1°\,08'\,10'' \quad \text{Peg 4}$$

(c) $$q = \frac{V^3}{3.6^3\,RL} = \frac{80^3}{3.6^3 \times 604.72 \times 72} = 0.25 \text{ m/s}^3$$

Example 8.13. A compound curve *AB, BC* is to be replaced by a single arc with transition curves 100 m long at each end. The chord lengths *AB* and *BC* are respectively 661.54 and 725.76 m and radii 1200 m and 1500 m. Calculate the single arc radius:

(a) If *A* is used as the first tangent point.
(b) If *C* is used as the last tangent point.   (LU)

Referring to *Figure 8.14* assume $T_1 = A$, $t = B$, $T_2 = C$, $R_1 = 1200$ m and $R_2 = 1500$ m. The requirements in this question are the tangent lengths *AI* and *CI*.

$$\text{chord } AB = 2R_1 \sin \frac{\Delta_1}{2}$$

$$\therefore \sin \frac{\Delta_1}{2} = \frac{661.54}{2 \times 1200}$$

$$\therefore \Delta_1 = 32°$$

Similarly, $\sin \dfrac{\Delta_2}{2} = \dfrac{725.76}{3000}$

$$\therefore \Delta_2 = 28°$$

Distance $At_1 = t_1B = R_1 \tan \dfrac{\Delta_1}{2} = 1200 \tan 16° = 344$ m

and $$Bt_2 = t_2C = R_2 \tan \frac{\Delta_2}{2} = 1500 \tan 14° = 374 \text{ m}$$

$$\therefore t_1t_2 = 718 \text{ m}$$

By sine rule in triangle t_1It_2

$$t_1I = \frac{718 \sin 28°}{\sin 120°} = 389 \text{ m}$$

and $$t_2I = \frac{718 \sin 32°}{\sin 120°} = 439 \text{ m}$$

$$\therefore AI = At_1 + t_1I = 733 \text{ m}$$
$$CI = Ct_2 + t_2I = 813 \text{ m}$$

To find single arc radius:

(a) From tangent point A

$$AI = (R + S) \tan \Delta/2 + L/2$$

where $\qquad S = L^2/24R$ and $\Delta = \Delta_1 + \Delta_2 = 60°$, $L = 100$ m

then $\qquad 733 = \left(R + \dfrac{L^2}{24R} \right) \tan 30° + 50$ from which

$$R = 1182 \text{ m}$$

(b) From tangent point C

$$CI = (R + S) \tan \Delta/2 + L/2$$
$$813 = (R + L^2/24R) \tan 30° + 50 \text{ from which}$$
$$R = 1321 \text{ m}$$

Example 8.14. Two straights with a deviation angle of 32° are to be joined by two transition curves of the form $\lambda = a(\Phi)^{\frac{1}{2}}$ where λ is the distance along the curve, Φ the angle made by the tangent with the original straight and a is a constant.
   The curves are to allow for a final 150 mm cant on a 1.435 m track, the straights being horizontal and the gradient from straight to full cant being 1 in 500.
   Tabulate the data for setting out the curve at 15-m intervals if the ratio of chord to curve for 16° is 0.9872. Find the design for this curve. (LU)

Referring to *Figure 8.30*:

   Cant = 0.15 m, rate of application = 1 in 500,
   therefore $L = 500 \times 0.15 = 75$ m

As the curve is wholly transitional $\Phi = \Delta/2 = 16°$

$$\therefore \text{ from } \Phi = \frac{L}{2R}, \qquad R = 134.3 \text{ m}$$

From ratio of chord to curve:

Chord $\qquad T_1 t = 75 \times 0.9872 = 74$ m
$$\therefore X = T_1 t \cos \theta = 73.7 \text{ m} \quad (\theta = \Phi/3)$$
$$Y = T_1 t \sin \theta = 6.9 \text{ m}$$
$$\therefore \text{ Tangent length} = X + Y \tan \Phi = 73.7 + 6.9 \tan 16° = 75.7 \text{ m}$$

Setting-out angles:

$$\theta_1 = (5°20') \frac{15^2}{75^2} = 12'48''$$

$$\theta_2 = (5°20') \frac{30^2}{75^2} = 51'12''$$

and so on to θ_5

Design speed:

$$\text{From } \textit{Figure 8.2(a)} \quad \tan \theta \approx \frac{AC}{CB} = \frac{0.15}{1.435} = \frac{V^2}{Rg}$$

from which $\qquad\qquad V = 11.8 \text{ m/s} = 42 \text{ km/h}$

Exercises

(*8.6*) The centre-line of a new road is being set out through a built-up area. The two straights of the road T_1I and T_2I meet giving a deflection angle of 45°, and are to be joined by a circular arc with spiral transitions 100 m long at each end. The spiral from T_1 must pass between two buildings, the position of the pass point being 70 m along the spiral from T_1 and 1 m from the straight measured at right angles.

Calculate all the necessary data for setting out the first spiral at 30-m intervals; thereafter find:

(a) The first three angles for setting out the circular arc, if it is to be set out by 10 equal chords.
(b) The design speed and rate of change of centripetal acceleration, given a centrifugal ratio of 0.1.
(c) The maximum super-elevation for a road width of 10 m. (KU)

(data $R = 572$ m, $T_1I = 237.23$ m, $\theta_1 = 9'01''$, $\theta_2 = 36'37''$, $\theta_3 = 1°40'10''$)

(*Answer*: (a) $1°44'53''$, $3°29'46''$, $5°14'39''$, (b) 85 km/h, 0.23 m/s³, (c) 1 m)

(*8.7*) A circular curve of 1800 m radius leaves a straight at through chainage 2468 m, joins a second circular curve of 1500 m radius at chainage 3976.5 m, and terminates on a second straight at chainage 4553 m. The compound curve is to be replaced by one of 2200 m radius with transition curves 100 m long at each end.

Calculate the chainages of the two new tangent points and the quarter point offsets of the transition curves. (LU)

(*Answer*: 2114.3 m, 4803.54 m; 0.012, 0.095, 0.32, 0.758 m)

(*8.8*) A circular curve must pass through a point P which is 70.23 m from I, the intersection point and on the bisector of the internal angle of the two straights AI, IB. Transition curves 200 m long are to be applied at each end and one of these must pass through a point whose coordinates are 167 m from the first tangent point along AI and 3.2 m at right angles from this straight. IB deflects $37°54'$ right from AI produced.

Calculate the radius and tabulate the data for setting out a complete curve. (LU)

(*Answer*: $R = 1200$ m, $AI = IB = 512.5$ m, setting-out angles or offsets calculated in usual way)

(*8.9*) The limiting speed around a circular curve of 667 m radius calls for a super-elevation of 1/24 across the 10-m carriageway. Adopting the Department of Transport recommendations of a rate of 1 in 200 for the application of super-elevation along the transition curve leading from the straight to the circular curve, calculate the tangential angles for setting out the transition curve with pegs at 15-m intervals from the tangent point with the straight. (ICE)

(*Answer*: $L = 83$ m, $2'20''$, $9'19''$, $20'58''$, $37'16''$, $58'13''$, $1°11'18''$)

(*8.10*) A circular curve of 610 m radius deflects through an angle of $40°30'$. This curve is to be replaced by one of smaller radius so as to admit transitions 107 m long at each end. The deviation of the new curve from the old at their mid-points is 0.46 m towards the intersection point.

Determine the amended radius assuming that the shift can be calculated with sufficient accuracy on the old radius. Calculate the lengths of track to be lifted and of new track to be laid.　　(LU)

(*Answer*: R = 590 m, new track = 521 m, old track = 524 m)

(*8.11*) The curve connecting two straights is to be wholly transitional without intermediate circular arc, and the junction of the two transitions is to be 5 m from the intersection point of the straights which deflects through an angle of 18°.

Calculate the tangent distances and the minimum radius of curvature. If the super-elevation is limited to 1 vertical to 16 horizontal, determine the correct velocity for the curve and the rate of gain of radial acceleration.　　(LU)

(*Answer*: 95 m, 602 m, 68 km/h, 0.06 m/s^3)

8.10 VERTICAL CURVES

Vertical curves (VC) are used to connect intersecting gradients in the vertical plane. Thus, in route design they are provided at all changes of gradient. They should be of sufficiently large curvature to provide comfort to the driver, that is, they should have a low 'rate of change of grade'. In addition, they should afford adequate 'sight distances' for safe stopping at a given design speed.

The type of curve generally used to connect the intersecting gradients g_1 and g_2 is the simple parabola. Its use as a *sag* or *crest* curve is illustrated in *Figure 8.38*.

8.10.1 Gradients

In vertical curve design the gradients are expressed as percentages, with a negative for a downgrade and a positive for an upgrade.

e.g.　A downgrade of 1 in 20 = 5 in 100 = −5% = −g_1%

An upgrade of 1 in 25 = 4 in 100 = +4% = +g_2%

The angle of deflection of the two intersecting gradients is called the *grade angle* and equals A in *Figure 8.38*. The grade angle simply represents the change of grade through which the vertical curve deflects and is the algebraic difference of the two gradients

$A\% = (g_1\% - g_2\%)$

In the above example $A\% = (-5\% - 4\%) = -9\%$ (negative indicates a sag curve).

8.10.2 Permissible approximations in vertical curve computation

In civil engineering, road design is carried out in accordance with the following documents:

(1) Layouts of Roads in Rural Areas.
(2) Roads in Urban Areas.
(3) Motorway Design Memorandum.

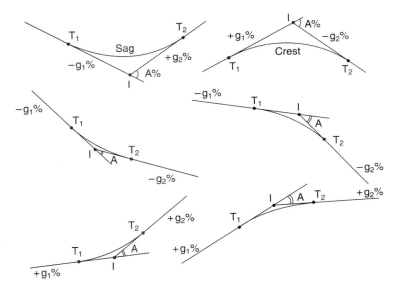

Fig. 8.38

However, practically all the geometric design in the above documents has been replaced by Department of Transport Standard TD 9/93, hereafter referred to simply as TD 9/93, with Advice Note TD 43/84.

In TD 9/93 the desirable maximum gradients for vertical curve design are:

Motorways	3%
Dual carriageways	4%
Single carriageways	6%

Due to the shallowness of these gradients, the following VC approximations are permissible, thereby resulting in simplified computation (*Figure 8.39*).

(1) Distance $T_1D = T_1BT_2 = T_1CT_2 = (T_1I + IT_2)$, without sensible error. This is very important and means that all distances may be regarded as horizontal in both the computation and setting out of vertical curves.
(2) The curve is of equal length each side of *I*. Thus $T_1C = CT_2 = T_1I = IT_2 = L/2$, without sensible error.
(3) The curve bisects *BI* at *C*, thus $BC = CI = Y$ (the mid-offset).
(4) From similar triangles T_1BI and T_1T_2J, if $BI = 2Y$, then $T_2J = 4Y$. $4Y$ represents the vertical divergence of the two gradients over half the curve length (*L/2*) and therefore equals *AL*/200.
(5) The basic equation for a simple parabola is

$$y = C \cdot l^2$$

where *y* is the *vertical* offset from gradient to curve, distance *l* from the start of the curve, and *C* is a constant. Thus, as the offsets are proportional to distance squared, the following equation is used to compute them:

$$\frac{y_1}{Y} = \frac{l_1^2}{(L/2)^2} \tag{8.37}$$

where *Y* = the mid-offset = *AL*/800 (see *Section 8.10.7*).

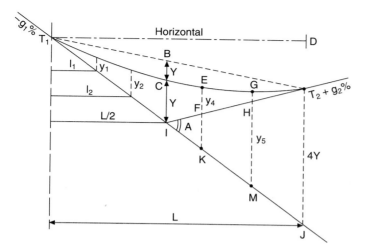

Fig. 8.39

8.10.3 *Vertical curve design*

In order to set out a vertical curve in the field, one requires levels along the curve at given chainage intervals. Before the levels can be computed, one must know the length L of the curve. The value of L is obtained from parameters supplied in *Table 3* of TD 9/93 (reproduced below as *Table 8.3*) and the appropriate parameters are K-values for specific design speeds and sight distances; then

$$L = KA \qquad (8.38)$$

where A = the difference between the two gradients (grade angle)
   K = the design speed related coefficient (*Table 8.3*)
 e.g. A + 4% gradient is linked to a −3% gradient by a crest curve. What length of curve is required for a design speed of 100 km/h?

$$A = (4\% - (-3\%)) = +7\% \text{ (positive for crest)}$$

From *Table 8.3*

 C1 Desirable minimum crest K-value = 100
 C2 One step below desirable minimum crest K-value = 55

   ∴ from $L = KA$

Desirable minimum length = $L = 100 \times 7 = 700$ m.

One step below desirable minimum length = $L = 55 \times 7 = 385$ m.

  Wherever possible the vertical and horizontal curves in the design process should be coordinated so that the sight distances are correlated and a more efficient the overtaking provision is ensured.
 The various design factors will now be dealt with in more detail.

8.10.3.1 *K-value*

Rate of change of gradient (r) is the rate at which the curve passes from one gradient ($g_1\%$) to the next ($g_2\%$) and is similar in concept to rate of change of radial acceleration in horizontal transitions.

When linked to design speed it is termed *rate of vertical acceleration* and should never exceed 0.3 m/s^2.

A typical example of a badly designed vertical curve with a high rate of change of grade is a hump-backed bridge where usually the two approaching gradients are quite steep and connected by a very short length of vertical curve. Thus one passes through a large grade angle A in a very short time, with the result that often a vehicle will leave the ground and/or cause great discomfort to its passengers.

Commonly-used design values for r are

> 3%   on crest curves
>
> 1.5%   on sag curves

thereby affording much larger curves to prevent rapid change of grade and provide adequate sight distances.

Working from first principles if $g_1 = -2\%$ and $g_2 = +4\%$ (sag curve), then the change of grade from -2% to $+4\% = 6\%$ (A), the grade angle. Thus, to provide for a rate of change of grade of 1.5%,

Table 8.3

Design speed (kph)	120	100	85	70	60	50	V^2/R
A *Stopping sight distance*, m							
Desirable minimum	295	215	160	120	90	70	
One step below desirable minimum	215	160	120	90	70	50	
B *Horizontal curvature*, m							
Minimum R^* without elimination of adverse camber and transitions	2880	2040	1440	1020	720	510	5
Minimum R^* with super-elevation of 2.5%	2040	1440	1020	720	510	360	7.07
Minimum R^* with super-elevation of 3.5%	1440	1020	720	510	360	255	10
Desirable minimum R with super-elevation of 5%	1020	720	510	360	255	180	14.14
One step below desirable minimum R with super-elevation of 7%	720	510	360	255	180	127	20
Two steps below desirable minimum radius with super-elevation of 7%	510	360	255	180	127	90	28.28
C *Vertical curvature*							
C1 Desirable minimum* crest K-value	182	100	55	30	17	10	
C2 One step below desirable minimum crest K-value	100	55	30	17	10	6.5	
C3 Absolute minimum sag K-value	37	26	20	20	13	9	
Overtaking sight distances							
C4 Full overtaking sight distance FOSD, m	*	580	490	410	345	290	
C5 FOSD overtaking crest K-value	*	400	285	200	142	100	

*Not recommended for use in the design of single carriageways.

The V^2/R values shown in Table 8.3 above simply represent a convenient means of identifying the relative levels of design parameters, irrespective of design speed.

(Reproduced with permission of the Controller of Her Majesty's Stationery Office)

one would require 400 m (*L*) of curve. If the curve was a crest curve, then using 3% gives 200 m (*L*) of curve:

$$\therefore L = 100A/r \qquad (8.39)$$

Now, expressing rate of change of grade as a single number we have

$$K = 100/r \qquad (8.40)$$

and as shown previously, *L = KA*.

8.10.3.2 Sight distances

Sight distance is a safety design factor which is intrinsically linked to rate of change of grade, and hence to *K*-values.

Consider once again the hump-backed bridge. Drivers approaching from each side of this particular vertical curve cannot see each other until they arrive, simultaneously, almost on the crest; by which time it may be too late to prevent an accident. Had the curve been longer and flatter, thus resulting in a low rate of change of grade, the drivers would have had a longer sight distance and consequently more time in which to take avoiding action.

Thus, sight distance, i.e. the length of road ahead that is visible to the driver, is a safety factor, and it is obvious that the sight distance must be greater than the stopping distance in which the vehicle can be brought to rest.

Stopping distance is dependent upon:

(1) Speed of the vehicle.
(2) Braking efficiency.
(3) Gradient.
(4) Coefficient of friction between tyre and road.
(5) Road conditions.
(6) Driver's reaction time.

In order to cater for all the above variables, the height of the driver's eye above the road surface is taken as being only 1.05 m; a height applicable to sports cars whose braking efficiency is usually very high. Thus, other vehicles, such as lorries, with a much greater eye height, would have a much longer sight distance in which to stop.

8.10.3.3 Sight distances on crests

Sight distances are defined as follows:

(1) Stopping sight distance (SSD) (Figure 8.40)

The SSD is the sight distance required by a driver to stop a vehicle when faced with an unexpected obstruction on the carriageway. It comprises two elements:

(a) The *perception-reaction distance*, which is the distance travelled from the time the driver sees the obstruction to the time it is realized that the vehicle must stop; and
(b) The *braking distance*, which is the distance travelled before the vehicle halts short of the obstruction.

The above are a function of driver age and fatigue, road conditions, etc. and thus the design

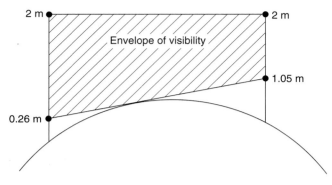

Fig. 8.40

parameters are based on average driver behaviour in wet conditions. *Table 8.3* provides values for desirable and absolute minimum SSD.

It has been shown that 95% of drivers' eye height is 1.05 m or above; the upper limit of 2 m represents large vehicles.

The height of the obstruction is between 0.26 m and 2.0 m. Forward visibility should be provided in both horizontal and vertical planes between points in the centre of the lane nearest the inside of the curve.

(2) Full overtaking sight distance (FOSD) (Figure 8.41)

On single carriageways, overtaking in the lane of the opposing traffic occurs. To do so in safety requires an adequate sight distance which will permit the driver to complete the normal overtaking procedure.

The FOSD consists of four elements:

(a) The *perception/reaction distance* travelled by the vehicle whilst the decision to overtake or not is made.
(b) The *overtaking distance* travelled by the vehicle to complete the overtaking manoeuvre.
(c) The *closing distance* travelled by the oncoming vehicle whilst overtaking is occurring.
(d) The *safety distance* required for clearance between the overtaking and oncoming vehicles at the instant the overtaking vehicle has returned to its own lane.

It has been shown that 85% of overtaking takes place in 10 seconds and *Table 8.3* gives appropriate FOSD values relative to design speed.

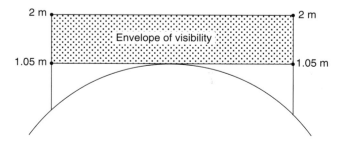

Fig. 8.41

It should be obvious from the concept of FOSD that it is used in the design of single carriageways only, where safety when overtaking is the prime consideration.

For instance, consider the design of a crest curve on a dual carriageway with a design speed of 100 km/h.

From *Table 8.3*:

C1 Desirable minimum K-value = 100
C2 One step below desirable minimum K-value = 55
C4 FOSD K-value = 400

As overtaking is not a safety hazard on a dual carriageway, FOSD is not necessary and one would use:

$L = 100$ A (desirable minimum)

or $L = 55$ A (one step below desirable minimum)

Had the above road been a single carriageway then FOSD would be required and:

$L = 400$ A

If this resulted in too long a curve, with excessive earthworks, then it might be decided to prohibit overtaking entirely, in which case:

$L = 55$ A

would be used.

Although equations are unnecessary when using design tables, they can be developed to calculate curve lengths L for given sight distances S, as follows:

(a) When $S < L$ (*Figure 8.42*)

From basic equation $y = Cl^2$
$Y = C(L/2)^2$, $h_1 = C(l_1)^2$ and $h_2 = C(l_2)^2$

then $$\frac{h_1}{Y} = \frac{l_1^2}{(L/2)^2} = \frac{4l_1^2}{L^2} \text{ and } \frac{h_2}{Y} = \frac{4l_2^2}{L^2}$$

thus $$l_1^2 = \frac{h_1 L^2}{4Y} \text{ but since } 4Y = \frac{AL}{200}$$

$$l_1^2 = \frac{200 h_1 L}{A}$$

and $$l_1 = (h_1)^{\frac{1}{2}} \left(\frac{200 L}{A}\right)^{\frac{1}{2}}$$

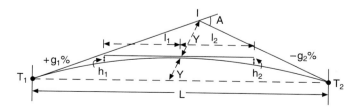

Fig. 8.42

Similarly $l_2 = (h_2)^{\frac{1}{2}} \left(\dfrac{200L}{A} \right)^{\frac{1}{2}}$

$$\therefore S = (l_1 + l_2) = [(h_1)^{\frac{1}{2}} + (h_2)^{\frac{1}{2}}] \left(\frac{200L}{A} \right)^{\frac{1}{2}} \tag{8.41}$$

and $$L = \frac{S^2 A}{200 \, [(h_1)^{\frac{1}{2}} + (h_2)^{\frac{1}{2}}]^2} \tag{8.42}$$

when $h_1 = h_2 = h$

$$L = \frac{S^2 A}{800h} \tag{8.43}$$

(b) When $S > L$ it can similarly be shown that

$$L = 2S - \frac{200}{A} \, [(h_1)^{\frac{1}{2}} + (h_2)^{\frac{1}{2}}]^2 \tag{8.44}$$

and when $h_1 = h_2 = h$

$$L = 2S - \frac{800h}{A} \tag{8.45}$$

When $S = L$, substituting in either of equations (8.43) or (8.45) will give the correct solution.

e.g. (8.43) $L = \dfrac{S^2 A}{800h} = \dfrac{L^2 A}{800h} = \dfrac{800h}{A}$

and (8.45) $L = 2S - \dfrac{800h}{A} = 2L - \dfrac{800h}{A} = \dfrac{800h}{A}$

N.B. If the relationship of S to L is not known then both cases must be considered; one of them will not fulfil the appropriate argument $S < L$ or $S > L$ and is therefore wrong.

8.10.3.4 *Sight distances on sags*

Visibility on sag curves is not obstructed as it is in the case of crests; thus sag curves are designed for at least absolute minimum comfort criteria of 0.3 m/sec^2. However, for design speeds of 70 km/h and below in unlit areas, sag curves are designed to ensure that headlamps illuminate the road surface for at least absolute minimum SSD. The relevant K values are given in C3, *Table 8.3*.

The headlight is generally considered as being 0.6 m above the road surface with its beam tilted up at 1° to the horizontal. As in the case of crests, equations can be developed if required.

Consider *Figure 8.43* where L is greater than S. From the equation for offsets:

$$\frac{BC}{T_2 D} = \frac{S^2}{L^2} \qquad \therefore BC = \frac{S^2 (T_2 D)}{L^2}$$

but $T_2 D$ is the vertical divergence of the gradients and equals

$$\frac{A \cdot L}{100 \cdot 2} \qquad \therefore BC = \frac{A \cdot S^2}{200L} \tag{a}$$

also $BC = h + S \tan x°$ (b)

Equating (a) and (b): $L = S^2 A(200h + 200S \tan x°)^{-1}$ (8.46)

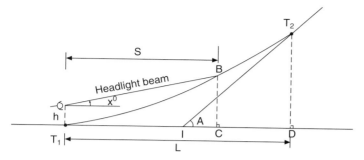

Fig. 8.43

putting $x° = 1°$ and $h = 0.6$ m

$$L = \frac{S^2 A}{120 + 3.5S}$$ (8.47)

Similarly, when S is greater than L (*Figure 8.44*)

$$BC = \frac{A}{100}\left(S - \frac{L}{2}\right) = h + S \tan x°$$

equating: $L = 2S - (200h + 200S \tan x°)/A$ (8.48)

when $x° = 1°$ and $h = 0.6$ m

$$L = 2S - (120 + 3.5S)/A$$ (8.49)

8.10.4 *Passing a curve through a point of known level*

In order to ensure sufficient clearance at a specific point along the curve it may be necessary to pass the curve through a point of known level. For example, if a bridge parapet or road furniture were likely to intrude into the envelope of visibility, it would be necessary to design the curve to prevent this.

This technique will be illustrated by the following example. A downgrade of 4% meets a rising grade of 5% in a sag curve. At the start of the curve the level is 123.06 m at chainage 3420 m, whilst at chainage 3620 m there is an overpass with an underside level of 127.06 m. If the designed curve is to afford a clearance of 5 m at this point, calculate the required length (*Figure 8.45*).

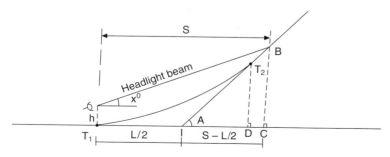

Fig. 8.44

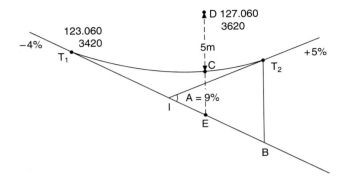

Fig. 8.45

To find the offset distance CE:
From chainage horizontal distance $T_1E = 200$ m at -4%

$$\therefore \text{ Level at } E = 123.06 - 8 = 115.06 \text{ m}$$

$$\text{Level at } C = 127.06 - 5 = 122.06 \text{ m}$$

$$\therefore \text{ Offset } CE = 7 \text{ m}$$

From offset equation $\dfrac{CE}{T_2B} = \dfrac{(T_1E)^2}{(T_1B)^2}$

but T_2B = the vertical divergence = $\dfrac{A}{100}\dfrac{L}{2}$, where $A = 9$

$$\therefore CE = \frac{AL}{200}\frac{200^2}{L^2} = \frac{1800}{L}$$

$$L = 257 \text{ m}$$

8.10.5 *To find the chainage of highest or lowest point on the curve*

The position and level of the highest or lowest point on the curve is frequently required for drainage design.

With reference to *Figure 8.45*, if one considers the curve as a series of straight lines, then at T_1 the grade of the line is -4% gradually changing throughout the length of the curve until at T_2 it is $+5\%$. There has thus been a change of grade of 9% in distance L. At the lowest point the grade will be horizontal, having just passed through -4% from T_1. Therefore, the chainage of the lowest point from the start of the curve is, by simple proportion,

$$D = \frac{L}{9\%} \times 4\% = \frac{L}{A} \times g_1 \qquad (8.50)$$

which in the previous example is $\dfrac{257}{9\%} \times 4\% = 114.24$ m from T_1.

Knowing the chainage, the offset and the curve level at that point may be found.
This simple approach suffices as the rate of change of grade is constant for a parabola, i.e. $y = Cl^2$, $\therefore d^2y/dl^2 = 2C$.

8.10.6 *Vertical curve radius*

Due to the very shallow gradients involved in VC design, the parabola may be approximated to a circular curve. In this way vertical curves may be expediently drawn on longitudinal sections using railway curves of a given radius, and vertical accelerations (V^2/R) easily assessed.

In circular curves (*Section 8.1*) the main chord from T_1 to $T_2 = 2R \sin \Delta/2$, where Δ is the deflection angle of the two straights. In vertical curves, the main chord may be approximated to the length (L) of the VC and the angle Δ to the grade angle A, i.e.

$$\Delta \approx A\%$$

$$\therefore \sin \Delta/2 \approx \Delta/2 \text{ rads} \approx A/200$$

$$\therefore L \approx 2RA/200 = AR/100 \tag{8.51}$$

and as $K = L/A = R/100$, then:

$$R = 100L/A = 100K \tag{8.52}$$

It is important to note that the reduced levels of VC must always be computed. Scaling levels from a longitudinal section, usually having a vertical scale different from the horizontal, will produce a curve that is neither parabolic nor circular. The use of railway curves is simply to indicate the position and extent of the curve on the section.

Thus, having obtained the radius R of the VC, it is required to know the number of the railway curve necessary to draw it on the longitudinal section.

If the horizontal scale of the section is 1 in H and the vertical scale is 1 in V, then the number of the railway curve required to draw the VC is:

Number of railway curve in mm = $R \cdot V/H^2$, with R in mm

e.g. If horizontal scale is 1 in 500, vertical scale 1 in 100 and curve radius is 200 m, then

Number of railway curve in mm = $200\,000 \times 100/500^2 = 80$

8.10.7 *Vertical curve computation*

The computation of a vertical curve will now be demonstrated using an example.

A '2nd difference' ($\delta^2 y/\delta l^2$) arithmetical check on the offset computation should automatically be applied. The check works on the principle that the change of grade of a parabola ($y = C \cdot l^2$) is constant, i.e. $\delta^2 y/\delta l^2 = 2C$. Thus, if the first and last chords are sub-chords of lengths different from the remaining standard chords, then the change of grade will be constant only for the equal-length chords.

For example, a 100-m curve is to connect a downgrade of 0.75% to an upgrade of 0.25%. If the level of the intersection point of the two grades is 150 m, calculate:

(1) Curve levels at 20-m intervals, showing the second difference ($d^2 y/dl^2$) check on the computations.
(2) The position and level of the lowest point on the curve.

Method

(a) Find the value of the central offset Y.
(b) Calculate offsets.
(c) Calculate levels along the gradients.
(d) Add/subtract (b) from (c) to get curve levels.

(a) *Referring to Figure 8.39:*
Grade angle $A = (-0.75 - 0.25) = -1\%$ (this is seen automatically).
$L/2 = 50$ m, thus as the grades IT_2, and IJ are diverging at the rate of 1% (1 m per 100 m) in 50 m, then

$$T_2J = 0.5 \text{ m} = 4Y \text{ and } Y = 0.125 \text{ m}$$

The computation can be quickly worked mentally by the student. Putting the above thinking into equation form gives

$$4Y = \frac{A}{100}\frac{L}{2} \qquad \therefore Y = \frac{AL}{800} = \frac{1 \times 100}{800} = 0.125 \tag{8.53}$$

(b) *Offsets from equation (8.37):*
There are two methods of approach.

(1) The offsets may be calculated from one gradient throughout; i.e. y_1, y_2, EK, GM, T_2J, from the grade T_1J.
(2) Calculate the offsets from one grade, say, T_1I, the offsets being equal on the other side from the other grade IT_2.

Method (1) is preferred due to the smaller risk of error when calculating curve levels at a constant interval and grade down T_1J.

From equation (8.37): $y_1 = Y \times \dfrac{l_1^2}{(L/2)^2}$

		1st diff.	2nd diff.
$T_1 =$	0 m		
		—— 0.02	
$y_1 = 0.125 \dfrac{20^2}{50^2} = 0.02$ m			—— 0.04
		—— 0.06	
$y_2 = 0.125 \dfrac{40^2}{50^2} = 0.08$ m			—— 0.04
		—— 0.10	
$y_3 = 0.125 \dfrac{60^2}{50^2} = 0.18$ m			—— 0.04
		—— 0.14	
$y_4 = 0.125 \dfrac{80^2}{50^2} = 0.32$ m			—— 0.04
		—— 0.18	

$$y_2 = T_2J = 4Y = 0.50 \text{ m}$$

The 2nd difference arithmetical check, which works only for equal chords, should be applied before any further computation.

(c) First find level at T_1 from known level at I:

Distance from I to $T_1 = 50$ m, grade $= 0.75\%$ (0.75 m per 100 m)

$\therefore$ Rise in level from I to $T_1 = \dfrac{0.75}{2} = 0.375$ m

Level at $T_1 = 150.000 + 0.375 = 150.375$ m

Levels are now calculated at 20-m intervals along T_1J, the fall being 0.15 m in 20 m. Thus, the following table may be made.

Chainage (m)	Gradient levels	Offsets	Curve levels	Remarks
0	150.375	0	150.375	Start of curve T_1
20	150.225	0.02	150.245	
40	150.075	0.08	150.155	
60	149.925	0.18	150.105	
80	149.775	0.32	150.095	
100	149.625	0.50	150.125	End of curve T_2

Position of lowest point on curve $= \dfrac{100 \text{ m}}{1\%} \times 0.75\% = 75$ m from T_1

$\therefore$ Offset at this point $= y_2 = 0.125 \times 75^2/50^2 = 0.281$ m

Tangent level 75 m from $T_1 = 150.375 - 0.563 = 149.812$ m

$\therefore$ Curve level $= 149.812 + 0.281 = 150.093$ m

8.10.8 Drawing-office practice

(1) Design

(a) Obtain grade angle (algebraic difference of the gradients) A.
(b) Extract the appropriate K-value from Design Table in TD 9/93.
(c) Length (L) of vertical curve $= KA$.
(d) Compute offsets and levels in the usual way.

(2) Drawing

To select the correct railway curve for drawing the vertical curve on a longitudinal section.

(a) Find equivalent radius (R) of vertical curve from $R = \dfrac{100 \cdot L}{A} = 100 \cdot K$.
(b) Number of railway curve in mm $= R$ mm $\times V/H^2$.

   If horizontal scale of section is say 1/500, then $H = 500$.
   If vertical scale of section is say 1/200, then $V = 200$.
   If the railway curves used are still in inches, simply express R in inches.

8.10.9 Computer-aided drawing and design (CADD)

The majority of survey packages now available contain a road design module, whilst highly specialized programs such as BIPS (British Integrated Program System) and MOSS (modelling systems) are standard tools in many road design offices.

Basically all the systems work from a digital ground model (DGM), established by ground survey methods or aerial photogrammetry. Thus not only is the road designed, but earthwork volumes, setting out data and costs are generated. The formation of DGMs and illustrations are shown in Chapter 1 (*Sections 1.5.4* and *1.5.5* and *Figure 1.19*).

In addition to road design using straights and standard curves, polynomial alignment procedures are available if required. The engineer is not eliminated from the CAD process; he must still specify such parameters as minimum permissible radius of curvature, maximum slope, minimum sight distances, coordination of horizontal and vertical alignment, along with any political, economic or aesthetic decisions. A series of gradients and curves can be input, until earthwork is minimized and balanced out. This is clearly illustrated by the generation of resultant mass-haul diagrams.

When a satisfactory road design has been arrived at, plans, longitudinal sections, cross-sections and mass-haul diagrams can be quickly produced. Three-dimensional views with colour shading are also available for environmental impact studies (Chapter 1). Bills of quantity, total costs and all setting-out data is provided as necessary.

Thus the computer provides a fast, flexible and highly economic method of road design, capable of generating contract drawings and schedules on request.

Worked examples

Example 8.15. An existing length of road consists of a rising gradient of 1 in 20, followed by a vertical parabolic crest curve 100 m long, and then a falling gradient of 1 in 40. The curve joins both gradients tangentially and the reduced level of the highest point on the curve is 173.07 m above datum.

Visibility is to be improved over this stretch of road by replacing this curve with another parabolic curve 200 m long.

Find the depth of excavation required at the mid-point of the curve. Tabulate the reduced levels of points at 30-m intervals on the new curve.

What will be the minimum visibility on the new curve for a driver whose eyes are 1.05 m above the road surface?   (ICE)

The first step here is to find the level of the start of the new curve; this can only be done from the information on the highest point P, (*Figure 8.45*).

Old curve $A = 7.5\%$, $L = 100$ m

Chainage of highest point P from $T_1 = \dfrac{100}{7.5\%} \times 5\% = 67$ m

Distance T_2C is the divergence of the grades (7.5 m per 100 m) over half the length of curve (50) = $7.5 \times 0.5 = 3.75$ m = $4Y$.

∴ Central offset $Y = 3.75/4 = 0.938$ m

Thus offset $PB = 0.938 \cdot \dfrac{67^2}{50^2} = 1.684$ m

Therefore the level of B on the tangent = $173.07 + 1.684 = 174.754$ m. This point is 17 m from I, and as the new curve is 200 m in length, it will be 177 m from the start of the new curve T_3.

∴ Fall from B to T_3 of new curve = $5 \times 1.17 = 5.85$ m

∴ Level of $T_3 = 174.754 - 5.85 = 168.904$ m

It can be seen that as the value of A is constant, when L is doubled, the value of Y, the central offset to the new curve, is doubled, giving 1.876 m.

∴ Amount of excavation at mid-point = 0.938 m

New curve offsets	*1st diff.*	*2nd diff.*
	— 0.169	
$y_1 = 1.876 \times \dfrac{30^2}{100^2} = 0.169$		— 0.337
	— 0.506	
$y_2 = 1.876 \times \dfrac{60^2}{100^2} = 0.675$		— 0.339
	— 0.845	
$y_3 = 1.876 \times \dfrac{90^2}{100^2} = 1.520$		— 0.336
	— 1.181	
$y_4 = 1.876 \times \dfrac{120^2}{100^2} = 2.701$		— 0.339
	— 1.520	
$y_5 = 1.876 \times \dfrac{150^2}{100^2} = 4.221$		— 0.337
	— 1.857	
$y_6 = 1.876 \times \dfrac{180^2}{100^2} = 6.078$		— 0.431*
	— 1.426	
$y_7 = 4y \qquad = 7.504$		

* Note change due to change in chord length from 30 m to 20 m.

Levels along the tangent $T_3 C$ are now obtained at 30-m intervals.

Chainage (m)	Tangent levels	Offsets	Curve levels	Remarks
0	168.904	0	168.904	T_3 of new curve
30	170.404	0.196	170.235	
60	171.904	0.675	171.229	
90	173.404	1.520	171.884	
120	174.904	2.701	172.203	
150	176.404	4.221	172.183	
180	177.904	6.078	171.826	
200	178.904	7.504	171.400	T_4 of new curve

From *Figure 8.46* it can be seen that the minimum visibility is half the sight distance and could thus be calculated from the necessary equation. However, if the driver's eye height of $h = 1.05$ m is taken as an offset then

$$\frac{h}{Y} = \frac{D^2}{(L/2)^2}, \text{ thus } \frac{1.05}{1.876} = \frac{D^2}{100^2}$$

∴ $D = 75$ m

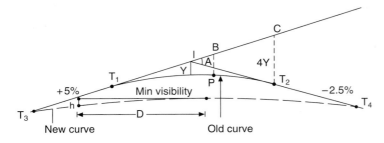

Fig. 8.46

Example 8.16. A rising gradient g_1 is followed by another rising gradient g_2 (g_2 less than g_1). The gradients are connected by a vertical curve having a constant rate of change of gradient. Show that at any point on the curve the height y above the first tangent point A is given by

$$y = g_1 x = - \frac{(g_1 - g_2)x^2}{2L}$$

where x is the horizontal distance of the point from A, and L is the horizontal distance between the two tangent points.

Draw up a table of heights above A for 100-m pegs from A when $g_1 = +5\%$, $g_2 = +2\%$ and $L = 1000$ m.

At what horizontal distance from A is the gradient $+3\%$? (ICE)

Figure 8.47 from equation for offsets. $\dfrac{BC}{Y} = \dfrac{x^2}{(L/2)^2}$

$$\therefore BC = Y \cdot \frac{4x^2}{L^2} \text{ but } Y = \frac{AL}{8} \text{ stations} = \frac{(g_1 - g_2)L}{8}$$

$$\therefore BC = \frac{(g_1 - g_2)L4x^2}{8L^2} = \frac{(g_1 - g_2)x^2}{2L}$$

Now $BD = g_1 x$

Thus, as $y = BD - BC = g_1 x - \dfrac{(g_1 - g_2)x^2}{2L}$

Using the above formula (which is correct only if horizontal distances x and L are expressed in stations, i.e. a station = 100 m)

$$y_1 = 5 - \frac{3 \times 1^2}{20} = 4.85 \text{ m}$$

$$y_2 = 10 - \frac{3 \times 2^2}{20} = 9.4 \text{ m}$$

$$y_3 = 15 - \frac{3 \times 3^2}{20} = 13.65 \text{ m and so on}$$

Grade angle = 3% in a 1000 m

Change of grade from 5% to 3% = 2%

$$\therefore \text{ Distance} = \frac{1000}{3\%} \times 2\% = 667 \text{ m}$$

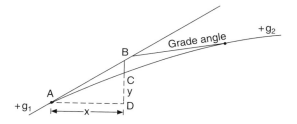

Fig. 8.47

Example 8.17. A falling gradient of 4% meets a rising gradient of 5% at chainage 2450 m and level 216.42 m. At chainage 2350 m the underside of a bridge has a level of 235.54 m. The two grades are to be joined by a vertical parabolic curve giving 14 m clearance under the bridge. List the levels at 50-m intervals along the curve.

(KU)

To find the offset to the curve at the bridge (Figure 8.48)

Level on gradient at chainage 2350 = 216.42 + 4   = 220.42 m

Level on curve at chainage 2350 = 235.54 − 14 = 221.54 m

∴ Offset at chainage 2350 = y_2 = 1.12 m

From equation for offsets $\dfrac{y_2}{Y} = \dfrac{(L/2 - 100)^2}{(L/2)^2}$

where $Y = \dfrac{AL}{800}$ and $A = 9\%$

$\dfrac{1.12 \times 800}{9 \times L} = \left(1 - \dfrac{200}{L}\right)^2$,   and putting $x = \dfrac{200}{L}$

$1.12 \times 4x = 9(1 - x)^2$

from which $x^2 - 2.5x + 1 = 0$, giving

$x = 2$ or   0.5   ∴ $L = 400$ m (as $x = 2$ is not possible)

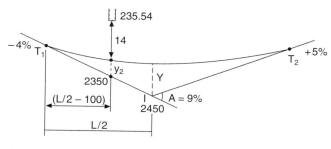

Fig. 8.48

Now $Y = \dfrac{9 \times 400}{800} = 4.5$ m from which the remaining offsets are found as follows:

at chainage   50 m offset $y_1 = 4.5 \dfrac{50^2}{200^2} = 0.28$ m

at chainage   100 m offset $y_2 = 4.5 \dfrac{100^2}{200^2} = 1.12$ m

at chainage   150 m offset $y_3 = 4.5 \dfrac{150^2}{200^2} = 2.52$ m

at chainage 200 m offset $Y = 4.50$ m

    To illustrate the alternative method, these offsets may be repeated on the other gradient at 250 m $= y_3$, 300 m $= y_2$, 350 m $= y_1$. The levels are now computed along each gradient from I to T_1 and T_2 respectively.

Chainage (m)	Gradient levels	Offsets	Curve levels	Remarks
0	224.42		224.42	Start of curve T
50	222.42	0.28	222.70	
100	220.42	1.12	221.54	
150	218.42	2.52	220.94	
200	216.42	4.50	220.92	Centre of curve I
250	218.92	2.52	221.44	
300	221.42	1.12	222.54	
350	223.92	0.28	224.20	
400	226.42		226.42	End of curve T_2

Example 8.18. A vertical parabolic curve 150 m in length connects an upward gradient of 1 in 100 to a downward gradient of 1 in 50. If the tangent point T_1 between the first gradient and the curve is taken as datum, calculate the levels of points at intervals of 25 m along the curve until it meets the second gradient at T_2. Calculate also the level of the summit giving the horizontal distance of this point from T_1.
    If an object 75 mm high is lying on the road between T_1 and T_2 at 3 m from T_2, and a car is approaching from the direction of T_1, calculate the position of the car when the driver first sees the object if his eye is 1.05 m above the road surface.   (LU)

To find offsets:

$$A = 3\% \qquad \therefore 4Y = \dfrac{L}{200} \times 3\% = 2.25 \text{ m}$$

and $$Y = 0.562 \text{ m}$$

$$\therefore y_1 = 0.562 \times \dfrac{25^2}{75^2} = 0.062 \qquad y_4 = 0.562 \times \dfrac{100^2}{75^2} = 1.000$$

$$y_2 = 0.562 \times \dfrac{50^2}{75^2} = 0.250 \qquad y_5 = 0.562 \times \dfrac{125^2}{75^2} = 1.562$$

$$y_3 = 0.562 \times \frac{75^2}{75^2} = 0.562 \qquad y_6 = 4y = 2.250$$

Second difference checks will verify these values.

With T_1 at datum, levels are now calculated at 25-m intervals for 150 m along the 1 in 100 (1%) gradient.

Chainage (m)	Gradient levels	Offsets	Curve levels	Remarks
0	100.00	0	100.000	Start of curve T_1
25	100.25	0.062	100.188	
50	100.75	0.250	100.250	
75	101.00	0.562	100.188	
100	101.25	1.000	100.000	
125	101.50	1.562	99.688	
150		2.250	99.250	End of curve T_2

Distance to highest point from $T_1 = \dfrac{150}{3\%} \times 1\% = 50$ m

Sight distance $(S < L)$
From expression (8.41)

$$S = ((h_1)^{\frac{1}{2}} + (h_2)^{\frac{1}{2}}\left(\frac{200L}{A}\right)^{\frac{1}{2}} \quad \text{when} \quad h_1 = 1.05 \text{ m}, h_2 = 0.075 \text{ m}$$

$\therefore$ $S = 130$ m, and the car is 17 m from T_1 and between T_1 and T_2

Example 8.19. A road gradient of 1 in 60 down is followed by an up-gradient of 1 in 30, the valley thus formed being smoothed by a circular curve of radius 1000 m in the vertical plane. The grades, if produced, would intersect at a point having a reduced level of 299.65 m and a chainage of 4020 m.

It is proposed to improve the road by introducing a longer curve, parabolic in form, and in order to limit the amount of filling it is decided that the level of the new road at chainage 4020 m shall be 3 m above the existing surface.

Determine:

(a) The length of new curve.
(b) The levels of the tangent points.
(c) The levels of the quarter points.
(d) The chainage of the lowest point on the new curve.   (LU)

To find central offset Y to new curve (Figure 8.49):
From simple curve data $\Delta = \cot 60 + \cot 30 = 2°51'51''$

Now $BI = R(\sec \Delta/2 - 1) = 0.312$ m

$\therefore$ Central offset $AI = Y = 3.312$ m and $T_2C = 4Y = 13.248$ m

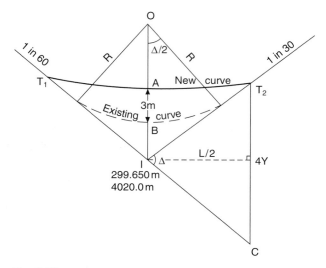

Fig. 8.49

To find length of new curve:
Grade 1 in 60 = 1.67%, 1 in 30 = 3.33%

∴ Grade angle Δ = 5%

(1) Then from T_2IC, $L/2 = \dfrac{13.248}{5} \times 100$

∴ $L = 530$ m

(2) Rise from I to T_1 = 1.67 × 2.65 = 4.426 m

∴ Level at T_1 = 299.65 + 4.426 = 304.076 m

Rise from I to T_2 = 3.33 × 2.65 = 8.824 m

∴ Level at T_2 = 299.65 + 8.824 = 308.474 m

(3) Levels at quarter points:

1st quarter point is 132.5 m from T_1

∴ Level on gradient = 304.076 − (1.67 × 1.325) = 301.863 m

Offset = $3.312 \times \dfrac{1^2}{2^2}$ = 0.828 m

∴ Curve level = 301.863 + 0.828 = 302.691 m

2nd quarter point is 397.5 m

∴ Level on gradient = 304.076 − (167 × 3.975) = 310.714 m

Offset = $3.312 \times \dfrac{3^2}{2^2}$ = 7.452 m

∴ Curve level = 310.714 + 7.452 = 318.166 m

(4) Position of lowest point on curve from $T_1 = \dfrac{530}{5\%} \times 1.67\% = 177$ m

Chainage at $T_1 = 4020 - 265 = 3755$ m

Chainage of lowest point $= 3755 + 177 = 3932$ m

Exercises

(*8.12*) A vertical curve 120 m long of the parabola type is to join a falling gradient of 1 in 200 to a rising gradient of 1 in 300. If the level of the intersection of the two gradients is 30.36 m give the levels at 15-m intervals along the curve.

If the headlamp of a car was 0.375 m above the road surface, at what distance will the beam strike the road surface when the car is at the start of the curve? Assume the beam is horizontal when the car is on a level surface. (LU)

(*Answer*: 30.660, 30.594, 30.541, 30.504, 30.486, 30.477, 30.489, 30.516, 30.588; 103.8 m)

(*8.13*) A road having an up-gradient of 1 in 15 is connected to a down-gradient of 1 in 20 by a vertical parabolic curve 120 m in length. Determine the visibility distance afforded by this curve for two approaching drivers whose eyes are 1.05 m above the road surface.

As part of a road improvement scheme a new vertical parabolic curve is to be set out to replace the original one so that the visibility distance is increased to 210 m for the same height of driver's eye.

Determine:

(a) The length of new curve.
(b) The horizontal distance between the old and new tangent points on the 1 in 5 gradient.
(c) The horizontal distance between the summits of the two curves. (ICE)

(*Answer*: 92.94 m, (a) 612 m, (b) 246 m, (c) 35.7 m)

(*8.14*) A vertical parabolic sag curve is to be designed to connect a down-gradient of 1 in 20 with an up-gradient of 1 in 15, the chainage and reduced level of the intersection point of the two gradients being 797.7 m and 83.544 m respectively.

In order to allow for necessary headroom, the reduced level of the curve at chainage 788.7 m on the down-gradient side of the intersection point is to be 85.044 m.

Calculate:

(a) The reduced levels and chainages of the tangent points and the lowest point on the curve.
(b) The reduced levels of the first two pegs on the curve, the pegs being set at the 30-m points of through chainage. (ICE)

(*Answer*: $T_1 = 745.24$ m, 86.166 m, $T_2 = 850.16$ m, 87.042 m, lowest pt $= 790.21$ m, 85.041 m, (b) 85.941 m, 85.104 m)

(*8.15*) The surface of a length of a proposed road consists of a rising gradient of 2% followed by a falling gradient of 4% with the two gradients joined by a vertical parabolic summit curve 120 m in length. The two gradients produced meet a reduced level of 28.5 m OD.

Compute the reduced levels of the curve at the ends, at 30-m intervals and at the highest point.

What is the minimum distance at which a driver, whose eyes are 1.125 m above the road surface, would be unable to see an obstruction 100 mm high? (ICE)

(*Answer*: 27.300, 27.675, 27.600, 27.075, 26.100 m; highest pt, 27.699 m, 87 m)

Derivation of Clothoid spiral formulae

AB is an infinitely small portion (δl) of the transition curve $T_1 t_1$. (*Figure B.1*)

$$\delta x/\delta l = \cos \phi = (1 - \phi^2/2! + \phi^4/4! - \phi^6/6! \ldots)$$

The basic equation for a clothoid curve is: $\phi = l^2/2RL$

$$\therefore \ \delta x/\delta l = \left(1 - \frac{(l^2/2RL)^2}{2!} + \frac{(l^2/2RL)^4}{4!} - \frac{(l^2/2RL)^6}{6!} \cdots\right)$$

Integrating: $x = l\left(1 - \dfrac{l^4}{40(RL)^2} + \dfrac{l^8}{3456(RL)^4} - \dfrac{l^{12}}{599\,040(RL)^6}\right)$

$$= l - \frac{l^5}{40(RL)^2} + \frac{l^9}{3456(RL)^4} - \frac{l^{13}}{599\,040(RL)^6} + \cdots$$

when $x = X$, $l = L$ and:

$$X = L - \frac{L^3}{5 \cdot 4 \cdot 2! R^2} + \frac{L^5}{9 \cdot 4^2 \cdot 4! R^4} - \frac{L^7}{13 \cdot 4^3 \cdot 6! R^6} + \cdots \qquad (8.54)$$

Similarly, $\delta y/\delta l = \sin \phi = (\phi - \phi^3/3! + \phi^5/5! \ldots)$
Substituting for ϕ as previously

$$\delta y/\delta l = \frac{l^2}{2RL} - \frac{(l^2/2RL)^3}{6} + \frac{(l^2/2RL)^5}{120} \cdots$$

Integrating: $y = l\left(\dfrac{l^2}{6RL} - \dfrac{l^6}{336(RL)^3} + \dfrac{l^{10}}{42\,240(RL)^5} \cdots\right)$

$$= \frac{l^3}{6RL} - \frac{l^7}{336\,(RL)^3} + \frac{l^{11}}{42\,240\,(RL)^5} \cdots$$

when $y = Y$, $l = L$, and

$$Y = \frac{L^2}{3 \cdot 2R} - \frac{L^4}{7 \cdot 3! 2^3 R^3} + \frac{L^6}{11 \cdot 5! 2^5 R^5} \cdots \qquad (8.55)$$

The basic equation for a clothoid is $l = a\,(\phi)^{\frac{1}{2}}$, where $l = L$, $\phi = \Phi$ and $L = a\,(\Phi)^{\frac{1}{2}}$, then squaring and dividing gives

$$L^2/l^2 = \Phi/\phi \qquad (8.56)$$

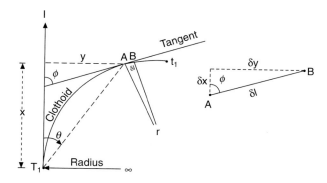

Fig. 8.50

From *Figure 8.50*

$$\tan \theta = \frac{y}{x} = \frac{l\,(l^2/6\,RL - l^6/336\,(RL)^3 + l^{10}/42\,240\,(RL)^5)}{l\,(1 - l^4/40\,(RL)^2 + l^8/3456\,(RL)^4)}$$

However, $l = (2\,RL\phi)^{\frac{1}{2}}$

$$\therefore\ \tan \theta = \frac{(2\,RL\phi)^{\frac{1}{2}}\,(\phi/3 - \phi^3/42 + \phi^5/1320)}{(2\,RL\phi)^{\frac{1}{2}}\,(1 - \phi^2/10 + \phi^4/216)}$$

$$= \left(\frac{\phi}{3} - \frac{\phi^3}{42} + \frac{\phi^5}{1320}\right)\left(1 - \frac{\phi^2}{10} + \frac{\phi^4}{216}\right)^{-1}$$

Let $x = -(\phi^2/10 - \phi^4/216)$ and expanding the second bracket binomially

i.e. $(1 + x)^{-1} = 1 - nx + \dfrac{n\,(n-1)x^2}{2!} \cdots$

$$= 1 + \frac{\phi^2}{10} - \frac{\phi^4}{216} + \frac{2}{2!}\left(\frac{\phi^2}{10} - \frac{\phi^4}{216}\right)^2$$

$$= 1 + \frac{\phi^2}{10} - \frac{\phi^4}{216} + \frac{\phi^4}{100} + \frac{\phi^8}{81 \cdot 4!4!} - \frac{2\phi^6}{45 \cdot 2!4!}$$

$$= 1 + \frac{\phi^2}{10} - \frac{\phi^4}{216} + \frac{\phi^4}{100}$$

$$\therefore\ \tan \theta = \left(\frac{\phi}{3} - \frac{\phi^3}{42} + \frac{\phi^5}{11.5!}\right)\left(1 + \frac{\phi^2}{10} - \frac{\phi^4}{216} + \frac{\phi^4}{100}\right)$$

$$= \frac{\phi}{3} + \frac{\phi^3}{105} + \frac{26\phi^5}{155\,925}$$

However, as $\theta = \tan \theta - \dfrac{1}{3}\tan^3 \theta + \dfrac{1}{5}\tan^5 \theta$

then
$$\theta = \frac{\phi}{3} - \frac{8\phi^3}{2835} - \frac{32\phi^5}{467\,775} \cdots \tag{8.57}$$

$$= \frac{\phi}{3} - N$$

When θ = maximum, $\phi = \Phi$ and

$$\theta = \frac{\Phi}{3} - N$$

From Figure 8.51

$$BD = BO - DO = R - R \cos \Phi = R (1 - \cos \Phi)$$

$$= R \left(\frac{\Phi^2}{2!} + \frac{\Phi^4}{4!} + \frac{\Phi^6}{6!} \cdots \right)$$

but $\Phi = L/2R$,

$$\therefore BD = \frac{L^2}{2!2^2 R} + \frac{L^4}{4!2^4 R^3} + \frac{L^6}{6!2^6 R^5}$$

and $Y = \dfrac{L^2}{3 \cdot 2R} - \dfrac{L^4}{7 \cdot 3!2^3 R^3} + \dfrac{L^6}{11 \cdot 5!2^5 R^5}$

$$\therefore \text{Shift} = S = (Y - BD) = \frac{L^2}{24R} - \frac{L^4}{3!7 \cdot 8 \cdot 2^3 R^3} + \frac{L^6}{5!11 \cdot 12 \cdot 2^5 R^5}$$

From *Figure 8.51*

$$Dt_1 = R \sin \phi = R \left(\Phi - \frac{\Phi^3}{3!} + \frac{\Phi^5}{5!} \cdots \right)$$

but $\Phi = L/2R$

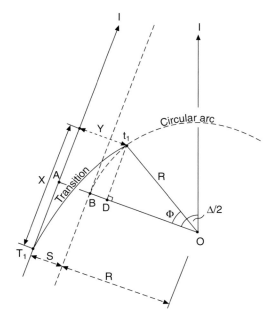

Fig. 8.51

$$\therefore Dt_1 = \frac{L}{2} - \frac{L^3}{3! \, 2^3 \, R^2} + \frac{L^5}{5! \, 2^5 \, R^4} \, \cdots$$

also $\quad X = L - \dfrac{L^3}{5 \cdot 4 \cdot 2! R^2} + \dfrac{L^5}{9 \cdot 4^2 \cdot 4! R^4} \; \cdots$

Tangent length $= T_1 I = (R + S) \, \tan \frac{\Delta}{2} + A T_1 = (R + S) \, \tan \frac{\Delta}{2} + (X - D t_1)$

$$= (R + S) \, \tan \frac{\Delta}{2} + \left(\frac{L}{2} - \frac{L^3}{2! 5 \cdot 6 \cdot 2^2 R^2} + \frac{L^5}{4! 9 \cdot 10 \cdot 2^4 R^4} \right)$$

$$= (R + S) \, \tan \frac{\Delta}{2} + C \tag{8.58}$$

9

Earthworks

Estimation of areas and volumes is basic to most engineering schemes such as route alignment, reservoirs, tunnels, etc. The excavation and hauling of material on such schemes is the most significant and costly aspect of the work, on which profit or loss may depend.

Areas may be required in connection with the purchase or sale of land, with the subdivision of land or with the grading of land.

Earthwork volumes must be estimated to enable route alignment to be located at such lines and levels that cut and fill are balanced as far as practicable; and to enable contract estimates of time and cost to be made for proposed work; and to form the basis of payment for work carried out.

The tedium of earthwork computation has now been removed by the use of micro- and mainframe computers. Digital ground models (DGM)), in which the ground surface is defined mathematically in terms of x, y and z coordinates, are now stored in the computer memory. This data bank may now be used with several alternative design schemes to produce the optimum route in both the horizontal and vertical planes. In addition to all the setting-out data, cross-sections are produced, earthwork volumes supplied and mass-haul diagrams drawn. Quantities may be readily produced for tender calculations and project planning. The data banks may be updated with new survey information at any time and further facilitate the planning and management not only of the existing project but of future ones.

However, before the impact of modern computer technology can be realized, one requires a knowledge of the fundamentals of areas and volumes, not only to produce the software necessary, but to understand the input data required and to be able to interpret and utilize the resultant output properly.

9.1 AREAS

The computation of areas may be based on data scaled from plans or drawings, or direct from the survey field data.

9.1.1 Plotted areas

(1) It may be possible to sub-divide the plotted area into a series of triangles, measures the sides a, b, c, and compute the areas using:

$$\text{Area} = [s(s - a)(s - b)(s - c)]^{\frac{1}{2}} \quad \text{where } s = (a + b + c)/2$$

The accuracy achieved will be dependent upon the scale error of the plan and the accuracy to which the sides are measured.

(2) Where the area is irregular, a sheet of gridded tracing material may be superimposed over it and the number of squares counted. Knowing the scale of the plan and the size of the squares, an estimate of the area can be obtained. Portions of squares cut by the irregular boundaries can be estimated.

(3) Alternatively, irregular boundaries may be reduced to straight lines using *give-and-take lines*, in which the areas 'taken' from the total area balance out with extra areas 'given' (*Figure 9.1*).

(4) If the area is a polygon with straight sides it may be reduced to a triangle of equal area. Consider the polygon *ABCDE* shown in *Figure 9.2*

Take *AE* as the base and extend it as shown, Join *CE* and from *D* draw a line parallel to *CE* on to the base at *F*. Similarly, join *CA* and draw a line parallel from *B* on to the base at *G*. Triangle *GCF* has the same area as the polygon *ABCDE*.

(5) The most common method of measuring areas from plans is to use an instrument called a *planimeter (Figure 9.3(a))*. This comprises two arms, *JF* and *JP*, which are free to move relative to each other through the hinged point at *J* but fixed to the plan by a weighted needle at *F. M* is the graduated measuring wheel and *P* the tracing point. As *P* is moved around the perimeter of the area, the measuring wheel partly rotates and partly slides over the plan with the varying movement of the tracing point (*Figure 9.3(b)*). The measuring wheel is graduated circumferentially into 10 divisions, each of which is further sub-divided by 10 into one-hundredths of a revolution, whilst a vernier enables readings to one thousandths of a revolution. The wheel is connected to a dial which records the numbered revolutions up to 10. On a *fixed-arm planimeter* one revolution of the wheel may

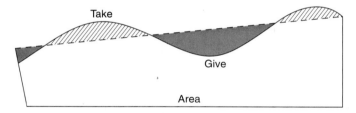

Fig. 9.1

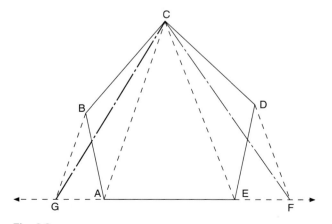

Fig. 9.2

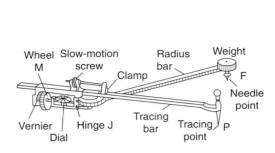

Fig. 9.3(a) *Amsler's polar planimeter*

Fig. 9.3(b)

represent 100 mm^2 on a 1 : 1 basis; thus, knowing the number of revolutions and the scale of the plan, the area is easily computed. In the case of a *sliding-arm planimeter* the sliding arm *JP* may be set to the scale of the plan, thereby facilitating more direct measurement of the area.

In the normal way, needle point *F* is fixed *outside* the area to be measured, the initial reading noted, the tracing point traversed around the area and the final reading noted. The difference of the two readings gives the number of revolutions of the measuring wheel, which is a direct measure of the area. If the area is too large to enable the whole of its boundary to be traversed by the tracing point *P* when the needle point *F* is outside the area, then the area may be sub-divided into smaller more manageable areas, or the needle point can be transposed *inside* the area.

As the latter procedure requires the application of the *zero circle* of the instrument, the former approach is preferred.

The zero circle of a planimeter is that circle described by the tracing point *P*, when the needle point *F* is at the centre of the circle, and the two arms *JF* and *JP* are at right angles to each other. In this situation the measuring wheel is normal to its path of movement and so slides without rotation, thus producing a zero change in reading. The value of the zero circle is supplied with the instrument.

If the area to be measured is greater than the zero circle (*Figure 9.4(a)*) then only the tinted area is measured, and the zero circle value must be added to the difference between the initial and final wheel readings. In such a case the final reading will always be *greater* than the initial reading. If the final reading is *smaller* than the initial reading, then the situation is as shown in *Figure 9.4(b)* and the measured area, shown tinted, must be subtracted from the zero circle value.

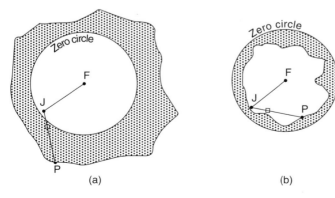

Fig. 9.4

Worked example

Example 9.1

(a) *Forward movement*

Initial reading	2.497
Final reading	6.282
	———
Difference	3.785 revs
Add zero circle	18.546
	———
Area =	22.231 revs

(b) *Backward movement*

Initial reading	2.886
Final reading	1.224
	———
Difference	1.662 revs
Subtract from zero circle	18.546
	———
Area =	16.884 revs

If one revolution corresponds to an area of (A), then on a plan of scale 1 in M, the actual area in

(a) above equals $22.331 \times A \times M^2$.
(b) If the area can be divided into strips then the area can be found using either (a) the trapezoidal rule or (b) Simpson's rule, as follow (*Figure 9.5*).

(a) *Trapezoidal rule*

$$\text{Area of 1st trapezoid } ABCD = \frac{h_1 + h_2}{2} \times w$$

$$\text{Area of 2nd trapezoid } BEFC = \frac{h_2 + h_3}{2} \times w \text{ and so on.}$$

Total area $=$ sum of trapezoids

$$= A = w\left(\frac{h_1 + h_7}{2} + h_2 + h_3 + h_4 + h_5 + h_6\right) \tag{9.1}$$

N.B (i) If the first or last ordinate is zero, it must still be included in the equation.
(ii) The formula represents the area bounded by the broken line under the curving boundary; thus, if the boundary curves outside then the computed area is too small, and *vice versa*.

(b) *Simpson's rule*

$$A = w[(h_1 + h_7) + 4(h_2 + h_4 + h_6) + 2(h_3 + h_5)]/3 \tag{9.2}$$

i.e. one-third the distance between ordinates, multiplied by the sum of the *first* and *last* ordinates, plus four times the sum of the *even* ordinates, plus twice the sum of the *odd* ordinates.

N.B. (i) This rule assumes a curved boundary and is therefore more accurate than the trapezoidal rule. If the boundary was a parabola the formula would be exact.

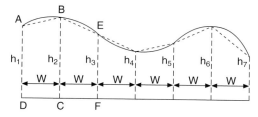

Fig. 9.5 *Trapezoidal and Simpson's rule*

(ii) The equation requires an *odd* number of ordinates and consequently an even number of areas.

The above equations are also useful for calculating areas from chain survey data. The areas enclosed by the chain lines are usually in the form of triangles, whilst the offsets to the irregular boundaries become the ordinates for use with the equations.

9.1.2 Areas by coordinates

Using appropriate field data it may be possible to define the area by its rectangular coordinates. For example:

The area enclosed by the traverse *ABCDA* in *Figure 9.6* can be found by taking the area of the rectangle *a'cDd* and subtracting the surrounding triangles, etc., as follows:

$$
\begin{aligned}
\text{Area of rectangle } a'cDd &= a'c \times a'd \\
&= 263 \times 173 = 45\,499 \text{ m}^2 \\
\text{Area of rectangle } a'bBa &= 77 \times 71 = 5\,467 \text{ m}^2 \\
\text{Area of triangle } AaB &= 71 \times 35.5 = 2\,520.5 \text{ m}^2 \\
\text{Area of triangle } BbC &= 77 \times 46 = 3\,542 \text{ m}^2 \\
\text{Area of triangle } Ccd &= 173 \times 50 = 8\,650 \text{ m}^2 \\
\text{Area of triangle } DdA &= 263 \times 12.5 = 3\,287.5 \text{ m}^2 \\
\hline
\text{Total} &= 23\,467 \text{ m}^2
\end{aligned}
$$

$\therefore$ Area $ABCDA = 45\,499 - 23\,467 = 22\,032$ m$^2 \approx 22\,000$ m^2

The following rule may be used when the *total coordinates* only are given. Multiply the algebraic *sum* of the northing of each station and the one following by the algebraic *difference* of the easting of each station and the one following. The area is half the algebraic sum of the products. Thus, from *Table 9.1* and *Figure 9.6*.

Area $ABCDA \approx 22\,032$ m$^2 \approx 22\,000$ m^2

The value of $22\,000$ m^2 is more correct considering the number of significant figures involved in the computations.

Table 9.1

Stns	E	N	Difference of E	Sum of N	Double area +	−
A	0.0	0.0	−71	71		5041
B	71	71	−92	219		20 148
C	163	148	−100	123		12 300
D	263	−25	263	−25		6575
A	0.0	0.0				—
				Σ		44 064

Area $ABCDA = 22\,032$ m$^2 \approx 22\,000$ m^2

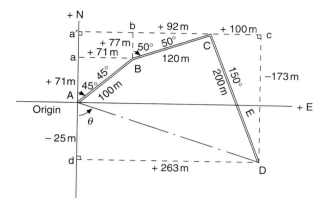

Fig. 9.6

This latter rule is the one most commonly used and is easily remembered if written as follow:

$$(9.3)$$

Thus $A = 0.5[N_A(E_B - E_D) + N_B(E_C - E_A) + N_C(E_D - E_B) + N_D(E_A - E_C)]$
$= 0.5[0 + 71(163) + 148(263 - 71) + -25(0 - 163)]$
$= 0.5[11\ 573 + 28\ 416 + 4\ 075] = 22\ 032\ \text{m}^2$

9.2 PARTITION OF LAND

This task may be carried out by an engineer when sub-dividing land either for large buildings plots or for sale purposes.

9.2.1 To cut off a required area by a line through a given point

With reference to *Figure 9.7*, it is required to find the length and bearing of the line *GH* which divides the area *ABCDEFA* into the given values.

Method

(1) Calculate the total area *ABCDEFA*.
(2) Given point *G*, draw a line *GH* dividing the area approximately into the required portions.
(3) Draw a line from *G* to the station nearest to *H*, namely *F*.
(4) From coordinates of *G* and *F*, calculate the length and bearing of the line *GF*.
(5) Find the area of *GDEFG* and subtract this area from the required area to get the area of triangle *GFH*.
(6) Now area $GFH = 0.5HF \times FG \sin \theta$, difference *FG* is known from (4) above, and θ is the difference of the known bearings *FA* and *FG* and thus length *HF* is calculated.
(7) As the bearing *FH* = bearing *FA* (known), then the coordinates of *H* may be calculated.
(8) From coordinates of *G* and *H*, the length and bearing of *GH* are computed.

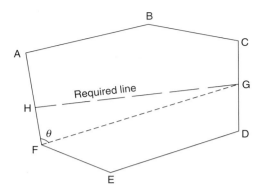

Fig. 9.7

9.2.2 To cut off a required area by a line of given bearing

With reference to *Figure 9.8(a)*, it is required to fix line *HJ* of a given bearing, which divides the area *ABCDEFGA* into the required portions.

Method

(1) From any station set off on the given bearing a trial line that cuts off approximately the required area, say *AX*.
(2) Compute the length and bearing of *AD* from the traverse coordinates.
(3) In triangle *ADX*, length and bearing *AD* are known, bearing *AX* is given and bearing *DX* = bearing *DE*; thus the three angles may be calculated and the area of the triangle found.
(4) From coordinates calculate the area *ABCDA*; thus total area *ABCDXA* is known.
(5) The difference between the above area and the area required to be cut off, is the area to be *added* or *subtracted* by a line *parallel* to the trial line *AX*. Assume this to be the trapezium *AXJHA* whose area is known together with the length and bearing of one side (*AX*) and the bearings of the other sides.

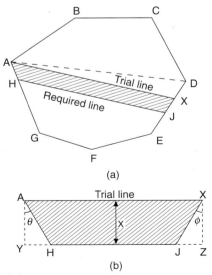

(a)

(b)

Fig. 9.8

(6) With reference to *Figure 9.8(b)*, as the bearings of all the sides are known, the angles θ and ϕ are known. From this $YH = x \tan \theta$ and $JZ = x \tan \phi$; now:

Area of $AXJHA$ = area of rectangle $AXZYA$ − (area of triangle AHY + area of triangle XZJ)

$$= AX \times x - \left(\frac{x}{2} \times x \tan \theta + \frac{x}{2} \times x \tan \phi \right)$$

$$= AX \times x - \left[\frac{x^2}{2} (\tan \theta + \tan \phi) \right] \tag{9.4}$$

from which the value of x may be found.

(7) Thus, knowing x, the distances AH and XJ can easily be calculated and used to set out the required line HJ.

9.3 CROSS-SECTIONS

Finding the areas of cross-sections is the first step in obtaining the volume of earthwork to be handled in route alignment projects (road or railway), or reservoir construction, for example.

In order to illustrate more clearly what is meant by the above statement, let us consider a road construction project. In the first instance an accurate plan is produced on which to design the proposed route. The centre-line of the route, defined in terms of rectangular coordinates at 10 to 30–m intervals, is then set out in the field. Ground levels are obtained along the centre-line and also at right-angles to the line *(Figure 9.9(a))*. The levels at right-angles to the centre-line depict the

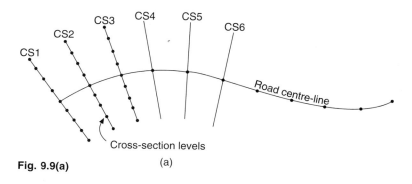

Fig. 9.9(a)

(a)

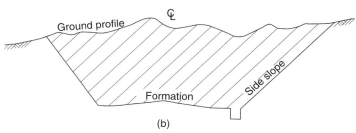

(b)

Fig. 9.9(b) *Cross-sectional area of cutting*

ground profile, as shown in *Figure 9.9(b)*, and if the design template, depicting the formation level, road width, camber, side slopes, etc. is added, then a cross-section is produced whose area can be obtained by planimeter or computation. The shape of the cross-section is defined in terms of vertical heights (levels) at horizontal distances each side of the centre-line; thus no matter how complex the shape, these parameters can be treated as rectangular coordinates and the area computed using the rules given in *Section 9.1.2*. The areas may now be used in various rules (see later) to produce an estimate of the volumes. Levels along, and normal to, the centre-line may be obtained by standard levelling procedures, by optical or electromagnetic tacheometry, or by aerial photogrammetry. The whole computational procedure, including the road design and optimization, would then be carried out on the computer to produce volumes of cut and fill, accumulated volumes, areas and volumes of top-soil strip, side widths, etc. Where plotting facilities are available the program would no doubt include routines to plot the cross-sections for visual inspection.

An example of an earthworks program, written in standard BASIC, is given at the end of the chapter.

Where there are no computer facilities the cross-sections may be approximated to the ground profile to afford easy computation. The particular cross-section adopted would be dependent upon the general shape of the ground. Typical examples are illustrated in *Figure 9.10*.

Whilst equations are available for computing the areas and side widths they tend to be over-complicated and the following method using 'rate of approach' is recommended (*Figure 9.11*).

Given: height x and grades AB and CB in triangle ABC.
Required: to find distance y_1.
Method: *Add* the two grades, using their absolute values, invert them and mutiply by x.

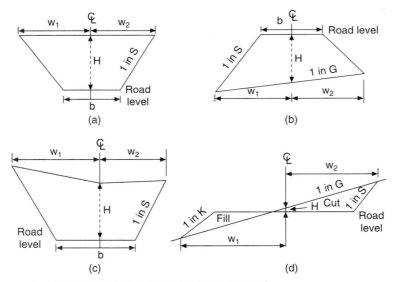

b = Finished road width at road or formation level
H = Centre height
W_1, W_2 = Side widths, measured horizontally from the centre-line and depicting the limits
   of the construction
1 in S = Side slope of 1 vertical to S horizontal
1 in G = Existing ground slope

Fig. 9.10 *(a)* Cuttting, *(b)* embankment, *(c)* cutting and *(d)* hillside section

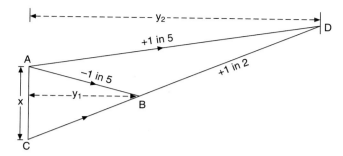

Fig. 9.11 *Rate of approach*

i.e. $(1/5 + 1/2)^{-1}x = 10x/7 = y_1$

Similarly, to find distance y_2 in triangle *ADC*, subtract the two grades, invert them and multiply by x.

e.g. $(1/5 - 1/2)^{-1}x = 10x/3 = y_2$

The rule, therefore is:

(1) When the two grades are running in opposing directions (as in *ABC*), add (signs opposite + −).
(2) When the two grades are running in the same direction (as in *ADC*), subtract (signs same).

N.B. Height x must be *vertical* relative to the grades (*see worked example 9.2*).

Proof

From *Figure 9.12* it is seen that 1 in 5 = 2 in 10 and 1 in 2 = 5 in 10, and thus the two grades diverge from *B* at the rate of 7 in 10. Thus, if *AC* = 7 m then *EB* = 10 m, i.e. $x \times 10/7 = 7 \times 10/7 = 10$ m.
Two examples will now be worked to illustrate the use of the above technique.

Worked examples

Example 9.2. Calculate the *side widths* and *cross-sectional area* of an embankment (*Figure 9.13*) having the following dimensions:

Road width = 20 m   existing ground slope = 1 in 10
Side slopes = 1 in 2   centre height = 10 m

As horizontal distance from centre-line to *AE* is 10 m and the ground slope is 1 in 10, then *AE*

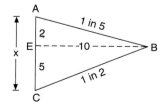

Fig. 9.12

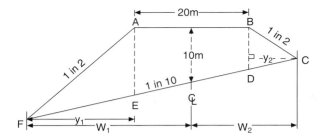

Fig. 9.13

will be 1 m greater than the centre height and *BD* 1 m less. Thus, *AE* = 11 m and *BD* = 9 m, area of *ABDE* = 20 × 10 = 200 m². Now, to find the areas of the remaining triangles *AEF* and *BDC* one needs the perpendicular heights y_1 and y_2, as follows:

(1) $1/2 - 1/10 = 4/10$, then $y_1 = (4/10)^{-1} \times AE = 11 \times 10/4 = 27.5$ m

(2) $1/2 + 1/10 = 6/10$, then $y_2 = (6/10)^{-1} \times BD = 9 \times 10/6 = 15.0$ m

$$\therefore \text{ Area triangle } AEF = \frac{AE}{2} \times y_1 = \frac{11}{2} \times 27.5 = 151.25 \text{ m}^2$$

$$\text{Area triangle } BDC = \frac{BD}{2} \times y_2 = \frac{9}{2} \times 15.0 = 67.50 \text{ m}^2$$

Total area = (200 + 151.25 + 67.5) = 418.75 m²

Side width w_1 = 10 m + y_1 = 37.5 m

Side width w_2 = 10 m + y_2 = 25.0 m

Example 9.3. Calculate the side widths and cross-sectional areas of cut and fill on a hillside section *(Figure 9.14)* having the following dimensions:

Road width = 20 m	existing ground slope = 1 in 5
Side slope in cut = 1 in 1	centre heigh in cut = 1 m
Side slope in fill = 1 in 2.	

As ground slope is 1 in 5 and centre height 1 m, it follows that the horizontal distance from centre-line to *B* is 5 m; therefore, *AB* = 5 m, *BC* = 15 m. From these latter distances it is obvious that *AF* = 1 m and *GC* = 3 m.

Now, $y_1 = (1/2 - 1/5)^{-1} \times AF = \frac{10}{3} \times 1 = 3.3$ m

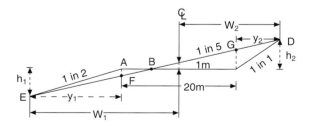

Fig. 9.14

$$y_2 = (1 - 1/5)^{-1} \times GC = \frac{5}{4} \times 3 = 3.75 \text{ m}$$

$$\therefore \text{ Side width } w_1 = 10 \text{ m} + y_1 = 13.3 \text{ m}$$

$$\text{Side width } w_2 = 10 \text{ m} + y_2 = 13.75 \text{ m}$$

Now, as side slope AE is 1 in 2, then $h_1 = y_1/2 = 1.65$ m and as side slope CD is 1 in 1, then $h_2 = y_2 = 3.75$ m

$$\therefore \text{ Area of cut } (BCD) = \frac{BC}{2} \times h_2 = \frac{15}{2} \times 3.75 = 28.1 \text{ m}^2$$

$$\text{Area of fill } (ABE) = \frac{AB}{2} \times h_1 = \frac{5}{2} \times 1.65 = 4.1 \text{ m}^2$$

Example 9.4. Calculate the area and side-widths of the cross-section (*Figure 9.15*):

$$y_1 = \left(\frac{1}{2} - \frac{1}{20}\right)^{-1} \times 10.5 = \frac{20}{9} \times 10.5 = 23.33 \text{ m}$$

$$\therefore \textbf{W}_1 = \textbf{10} + \textbf{y}_1 = \textbf{33.33 m}$$

$$y_2 = \left(\frac{1}{2} + \frac{1}{10}\right)^{-1} \times 9 = \frac{10}{6} \times 9 = 15.00 \text{ m}$$

$$\therefore \textbf{W}_1 = \textbf{10} + \textbf{y}_2 = \textbf{25.00 m}$$

Area I	$= ((10 + 10.5)/2) \ 10 =$	102.50 m²
Area II	$= (\ (10 + 9)/2) \ 10 \ \ \ =$	95.00 m²
Area III	$= (10.5/2) \ 23.33 \ \ \ \ =$	122.50 m²
Area IV	$= (9/2) \ 15.00 \ \ \ \ \ \ \ \ =$	67.50 m²
	Total area $=$	387.50 m²

9.3.1 *Cross-sectional areas by coordinates*

Where the cross-section is complex and the ground profile has been defined by reduced levels at known horizontal distances from the centre-line, the area may be found by coordinates. The horizontal

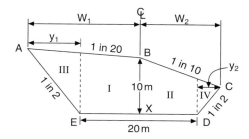

Fig. 9.15

distances may be regarded as Eastings (E) and the elevations (at 90° to the distances) as Northings (N).

Consider the previous example (*Figure 9.15*). Taking X on the centre-line as the origin, XB as the N-axis (reduced levels) and ED as the E-axis, the coordinates of A, B, C, D and E are:

$$\left\{ A = \frac{11.67}{-33.33} \quad B = \frac{10.00}{0} \quad C = \frac{7.50}{25.00} \right\} \quad \text{Ground levels}$$

$$\left\{ E = \frac{0}{-10.00} \quad D = \frac{0}{10.00} \right\} \quad \text{Formation levels}$$

(In the above format the denominator is the E-value.)

Point	E	N
A	−33.33	11.67
B	0	10.00
C	25.00	7.50
D	10.00	0
E	−10.00	0

From Equation (9.3).

$$\text{Area} = \tfrac{1}{2}[N_A (E_B - E_E) + N_B (E_C - E_A) + N_C (E_D - E_B) + N_D (E_E - E_C) + N_E (E_A - E_D)]$$

$$= \tfrac{1}{2}[11.67 \{0 - (-10)\} + 10 \{25 - (-33.33)\} + 7.5 (10 - 0)]$$

(As the N_D and N_E are zero, the terms are ignored.)

$$= \tfrac{1}{2}[116.70 + 583.30 + 75] = 387.50 \text{ m}^2$$

9.4 DIP AND STRIKE

On a tilted plane there is a direction of maximum tilt, such direction being called *the line of full dip*. Any line at right angles to full dip will be a level line and is called a *strike line* (*Figure 9.16(a)*). Any grade between full dip and strike is called *apparent dip*. An understanding of dip and strike is occasionally necessary for some earthwork problems. From *Figure 9.16(a)*:

$$\tan \theta_1 = \frac{ac}{bc} = \frac{de}{bc} = \left(\frac{de}{be} \times \frac{be}{bc} \right) = \tan \theta \cos \phi$$

i.e. tan (apparent dip) = tan (full dip) × cos (included angle)   (9.5)

Worked example

Example 9.5. On a stratum plane, an apparent dip of 1 in 16 bears 170°, whilst the apparent dip in the direction 194° is 1 in 11; calculate the direction and rate of full dip.

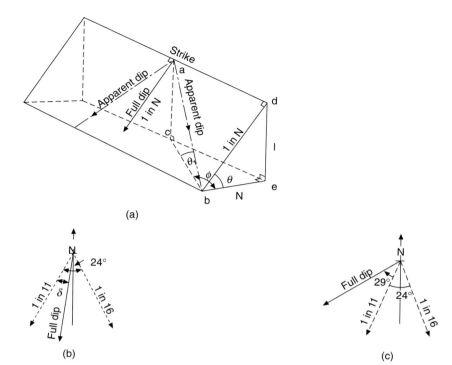

(a)

(b)

(c)

Fig. 9.16

Draw a sketch of the situation (*Figure 9.16(b)*) and assume any position for full dip. Now, using equation (9.5)

$$\tan \theta_1 = \tan \theta \cos \phi$$

$$\frac{1}{16} = \tan \theta \cos (24° - \delta)$$

$$\tan \theta = \frac{1}{16 \cos (24° - \delta)}$$

Similarly, $$\frac{1}{11} = \tan \theta \cos \delta$$

$$\tan \theta = \frac{1}{11 \cos \delta}$$

Equating (a) and (b)

$$16 \cos (24° - \delta) = 11 \cos \delta$$

$$16(\cos 24° \cos \delta + \sin 24° \sin \delta) = 11\cos \delta$$

$$16(0.912 \cos \delta + 0.406 \sin \delta) = 11 \cos \delta$$

$$14.6 \cos \delta + 6.5 \sin \delta = 11 \cos \delta$$

$$3.6 \cos \delta = -6.5 \sin \delta$$

Cross multiply $$\frac{\sin \delta}{\cos \delta} = \tan \delta = -\frac{3.6}{6.5}$$

$$\therefore \delta = -29°$$

N.B. The minus sign indicates that the initial position for full dip in *Figure 9.16(b)* is incorrect, and that it lies *outside* the apparent dip. As the grade is increasing from 1 in 16 to 1 in 11, the full dip must be as in *Figure 9.16(c)*.

∴ Direction of full dip = 223°

Now, a second application of the formula will give the rate of full dip. That is

$$\frac{1}{11} = \frac{1}{x} \cos 29°$$

∴ $x = 11 \cos 29° = 9.6$

∴ Rate of full dip = 1 in 9.6

9.5 VOLUMES

The importance of volume assessment has already been outlined. Many volumes encountered in civil engineering appear, at first glance, to be rather complex in shape. Generally speaking, however, they can be divided into *prisms*, *wedges* or *pyramids*, each of which will now be dealt with in turn.

(1) Prism

The two ends of the prism (*Figure 9.17*) are equal and parallel, the resulting sides thus being parallelograms.

Vol = AL   (9.6)

(2) Wedge

Volume of wedge (*Figure 9.18*) = $\frac{L}{6}$ (sum of parallel edges × vertical height of base)

$$= \frac{L}{6} [(a + b + c) \times h]$$ (9.7a)

when $a = b = c$: $V = AL/2$ (9.7b)

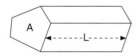

Fig. 9.17 *Prism*

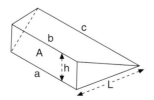

Fig. 9.18 *Wedge*

(3) Pyramid

Volume of pyramid (*Figure 9.19*) = $\dfrac{AL}{3}$ (9.8)

Equations (9.6) to (9.8) can all be expressed as the common equation:

$$V = \frac{L}{6}(A_1 + 4A_m + A_2) \tag{9.9}$$

where A_1 and A_2 are the end areas and A_m is the area of the section situated mid-way between the end areas. It is important to note that A_m is not the arithmetic mean of the end areas, except in the case of a wedge.

To prove the above statement consider:

(1) Prism

In this case $A_1 = A_m = A_2$

$$V = \frac{L}{6}(A + 4A + A) = \frac{L \times 6A}{6} = AL$$

(2) Wedge

In this case A_m is the mean of A_1 and A_2, but $A_2 = 0$. Thus $A_m = A/2$

$$V = \frac{L}{6}\left(A + 4 \times \frac{A}{2} + 0\right) = \frac{L \times 3A}{6} = \frac{AL}{2}$$

(3) Pyramid

In this case $A_m = \dfrac{A}{4}$ and $A_2 = 0$

$$V = \frac{L}{6}\left(A + 4 \times \frac{A}{4} + 0\right) = \frac{L \times 2A}{6} = \frac{AL}{3}$$

Thus, any solid which is a combination of the above three forms and having a common value for *L*, may be solved using equation (9.9). Such a volume is called a *prismoid* and the formula is called the *prismoidal equation*. It is easily deduced by simply substituting areas for ordinates in Simpson's rule. The prismoid differs from the prism in that its parallel ends are not necessarily equal in area; the sides are generated by straight lines from the edges of the end areas (*Figure 9.20*).

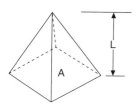

Fig. 9.19 *Pyramid*

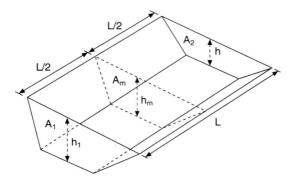

Fig. 9.20

The prismoidal equation is correct when the figure is a true prismoid. In practice it is applied by taking three successive cross-sections. If the mid-section is different from that of a true prismoid, then errors will arise. Thus, in practice, sections should be chosen in order to avoid this fault. Generally, the engineer elects to observe cross-sections at regular intervals assuming compensating errors over a long route distance.

9.5.1 End-area method

Consider *Figure 9.21*, then

$$V = \frac{A_1 + A_2}{2} \times L \tag{9.10}$$

i.e. the mean of the two end areas multiplied by the length between them. This equation is correct

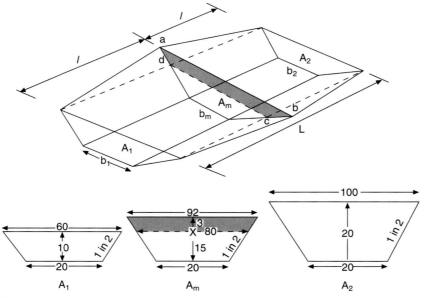

Fig. 9.21

only when the mid-area of the prismoid is the mean of the two end areas. It is correct for wedges but in the case of a pyramid it gives a result which is 50% too great:

$$\text{Vol. of pyramid} = \frac{A + 0}{2} \times L = \frac{AL}{2} \text{ instead of } \frac{AL}{3}$$

Although this method generally over-estimates, it is widely used in practice. The main reasons for this are its simplicity and the fact that the assumptions required for a good result using the prismoidal method are rarely fulfilled in practice. Strictly, however, it should be applied to prismoids comprising prisms and wedges only; such is the case where the height or width of the consecutive sections is approximately equal. It is interesting to note that with consecutive sections, where the height increases as the width decreases, or vice versa, the end-area method gives too small a value.

The difference between the prismoidal and end-area equations is called *prismoidal excess* and may be applied as a correction to the end-area value. It is rarely used in practice.

Summing a series of end areas gives:

$$V = L\left(\frac{A_1 + A_n}{2} + A_2 + A_3 + \cdots + A_{n-1}\right) \tag{9.11}$$

called the *trapezoidal rule* for volumes.

9.5.2 Comparison of end-area and prismoidal equations

In order to compare the methods, the volume of *Figure 9.21*, will be computed as follows: Dimensions of *Figure 9.21*.

Centre heights: $h_1 = 10$ m, $h_2 = 20$ m, $h_m = 18$ m

Road widths: $b_1 = b_2 = b_m = 20$ m

Side slopes: 1 in 2

Horizontal distance between sections: $l = 30$ m, $L = 60$ m

N.B. For a true prismoid h_m would have been the mean of h_1 and h_2, equal to 15 m. The broken line indicates the true prismoid, the excess area of the mid-section is shown tinted.

The true volume is thus a true prismoid plus two wedges, as follows:

(1) $A_1 = \dfrac{60 + 20}{2} \times 10 = 400 \text{ m}^2$

$A_2 = \dfrac{100 + 20}{2} \times 20 = 1200 \text{ m}^2$

$A_m = \dfrac{80 + 20}{2} \times 15 = 750 \text{ m}^2$

Vol. of prismoid $= V_1 = \dfrac{60}{6}(400 + 4 \times 750 + 1200) = 46\,000 \text{ m}^3$

Vol. of wedge 1 $= \dfrac{L}{6}[(a + b + c) \times h] = \dfrac{30}{6}[(92 + 80 + 60) \times 3] = 3480 \text{ m}^3$

Vol. of wedge 2 $= \dfrac{30}{6}[(92 + 80 + 100) \times 3] = 4080 \text{ m}^3$

Total true volume $= 53\,560 \text{ m}^3$

(2) *Volume by prismoidal equation* (A_m will now have a centre height of 18 m)

$$A_m = \frac{92 + 20}{2} \times 18 = 1008 \text{ m}^2$$

$$\text{Vol.} = \frac{60}{6}(400 + 4032 + 1200) = 56\,320 \text{ m}^3$$

Error = 56 320 − 53 560 = +2760 m^2

This error is approximately equal to the area of the excess mid-section multiplied by $\frac{L}{6}$, i.e.

$\frac{\text{Area } abcd \times L}{6}$, and is so for all such circumstances; it would be negative if the mid-area had been smaller.

(3) *Volume by end area*

$$V_1 = \frac{400 + 1008}{2} \times 30 = 21\,120 \text{ m}^3$$

$$V_2 = \frac{1008 + 1200}{2} \times 30 = 33\,120 \text{ m}^3$$

$$\text{Total volume} = 54\,240 \text{ m}^3$$

Error = 54 240 − 53 560 = + 680 m^3

Thus, in this case the end-area method gives a better result than the prismoidal equation. However, if we consider only the true prismoid, the volume by end areas is 46 500 m^3 compared with the volume by prismoidal equation of 46 000 m^3, which, in this case, is the true volume.

Therefore, in practice, it can be seen that neither of these two methods is satisfactory. Unless the ideal geometric conditions exist, which is rare, both methods will give errors. To achieve greater accuracy, the cross-sections should be located in the field, with due regard to the formula to be used. If the cross-sections are approximately equal in size and shape, and the intervening surface roughly a plane, then end areas will give an acceptable result. Should the sections be vastly different in size and shape, with the mid-section contained approximately by straight lines generated between the end sections, then the prismoidal equation will give the better result.

9.5.3 Contours

Volumes may be found from contours using either the end-area or prismoidal method. The areas of the sections are the areas encompassed by the contours. The distance between the sections is the contour interval. This method is commonly used for finding the volume of a reservoir, lake or spoil heap (see *Exercise (9.3)*).

9.5.4 Spot heights

This method is generally used for calculating the volumes of excavations for basements or tanks, i.e. any volume where the sides and base are planes, whilst the surface is broken naturally (*Figure 9.22(a)*). *Figure 9.22(b)* shows the limits of the excavation with surface levels in metres at *A*, *B*, *C* and *D*. The sides are vertical to a formation level of 20 m. If the area *ABCD* was a plane, then the volume of excavation would be:

$$V = \text{plan area } ABCD \times \text{mean height} \tag{9.12}$$

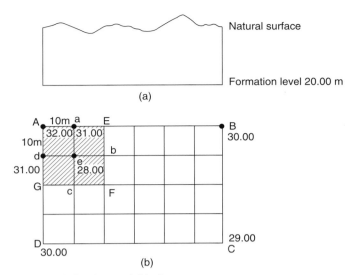

Fig. 9.22 *(a) Section, and (b) plan*

However, as the illustration shows, the surface is very broken and so must be covered with a grid such that the area within each 10-m grid square is approximately a plane. It is therefore the ruggedness of the ground that controls the grid size. If, for instance, the surface *Aaed* was not a plane, it could be split into two triangles by a diagonal (*Ae*) if this would produce better surface planes.

Considering square *Aaed* only:

V = plan area × mean height

$$= 100 \times \frac{1}{4} \, (12 + 11 + 8 + 11) = 1050 \text{ m}^3$$

If the grid squares are all equal in area, then the data is easily tabulated and worked as follows:
 Considering *AEFG* only, instead of taking each grid square separately, one can treat it as a whole.

$$\therefore V = \frac{100}{4} \, [h_A + h_E + h_F + h_G + 2(h_a + h_b + h_c + h_d) + 4h_e]$$

If one took each grid separately it would be seen that the heights of *AEFG* occur only once, whilst the heights of *abcd* occur twice and h_e occurs four times; one still divides by four to get the mean height.

 This approach is adopted by computer packages and by splitting the area into very small triangles or squares (*Figures 1.19* and *9.23*), an extremely accurate assessment of the volume is obtained.

 The above formula is also very useful for any difficult shape consisting entirely of planes, as the following example illustrates (*Figure 9.24*).

Vertical height at *A* and *D* is 10 m.
As *AB* = 40 m and surface slopes at 1 in 10, then vertical heights at *B* and *C* must be 4 m greater, i.e. 14 m.
 Consider splitting the shape into two wedges by a plane connecting *AD* to *HE*.
In *ΔABB′* (*Figure 9.24(a)*):

By rate of approach: $y = \left(1 - \dfrac{1}{10}\right)^{-1} \times 14 = 15.56 \text{ m} = B'H = C'E$

$$\therefore \qquad HE = 20 + 15.56 + 15.56 = 51.12 \text{ m}$$

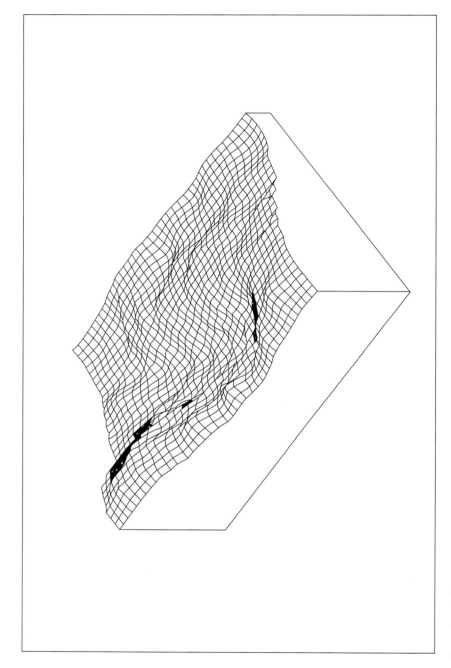

Fig. 9.23 *Ground model*

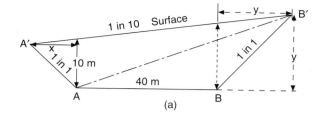

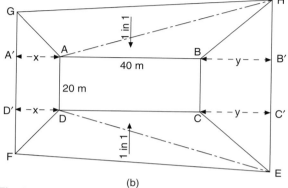

Fig. 9.24

Area of $\triangle ABB'$ normal to AD, BC, HE

$$= \frac{40}{2} \times 15.56 = 311.20 \text{ m}^2$$

$$\therefore \text{Vol} = \text{area} \times \text{mean height} = \frac{311.20}{3} \ (AD + BC + HE)$$

$$= 103.73 \ (20 + 20 + 51.12) = 9452 \ m^3$$

Similarly in $\triangle AA'B'$:

$$x = \left(1 + \frac{1}{10}\right)^{-1} \times 10 = 9.09 \text{ m} = A'G = D'F$$

$$\therefore GF = 20 + 9.09 + 9.09 \times 38.18 \text{ m}$$

Area of $\triangle AA'B'$ normal to AD, GF, HE

$$= \frac{(x + AB + y)}{2} \times 10 = \frac{64.65}{2} \times 10 = 323.25 \text{ m}^3$$

$$\therefore \text{Vol} = \frac{323.25}{2} \ (20 + 38.18 + 51.12) = 11\,777 \text{ m}^3$$

Total vol = 21 229 m³

Check

Wedge $ABB' = \frac{40}{6} \ [(20 + 20 + 51.12) \times 15.56] = 9452 \text{ m}^3$

Wedge $AA'B' = \frac{64.65}{6} \ [(20 + 38.18 + 51.12) \times 10] = 11\,777 \text{ m}^3$

9.5.5 *Effect of curvature on volumes*

The application of the prismoidal and end-area formulae has assumed, up to this point, that the cross-sections are parallel. When the excavation is curved (*Figure 9.25*), the sections are radial and a *curvature correction* must be applied to the formulae.

Pappus's theorem states that the correct volume is where the distance between the cross-sections is taken along the path of the centroid.

Consider the volume between the first two sections of area A_1 and A_2:
Distance between sections measured along centre-line = $X'Y' = D$.
Angle δ subtended at the centre = D/R radians

Now, length along path of centroid = $XY = \delta \times$ mean radius to path of centroid where mean radius

$$= R - (d_1 + d_2)/2 = (R - d)$$

$$\therefore XY = \delta (R - d) = D (R - d)/R$$

$$\text{Vol. by end areas} = \frac{1}{2} (A_1 + A_2) XY = \frac{1}{2} (A_1 + A_2) D (R - d)/R$$

$$= \frac{1}{2} (A_1 + A_2) D (1 - d/R)$$

In other words, one corrects for curvature by multiplying the area A_1 by $(1 - d_1/R)$, and area A_2 by $(1 - d_2/R)$, the corrected areas then being used in either the end-area or prismoidal formulae, in the normal way, with D being the distance measured along the centre-line. If the centroid lay beyond the centre-line, as in section A_3, then the correction is $(1 + d_3/R)$.

This correction for curvature is, again, never applied to earthworks in practice. Indeed, it can be shown that the effect is cancelled out on long earthwork projects. However, it may be significant on small projects or single curved excavations. (Refer *Worked example 9.10*)

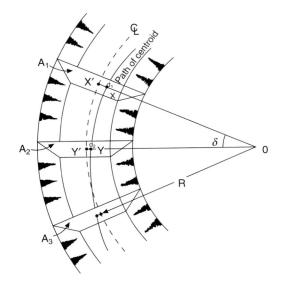

Fig. 9.25

Worked examples

Example 9.6. Figure 9.26 illustrates a section of road construction to a level road width of 20 m, which includes a change from fill to cut. From the data supplied in the following field book extract, calculate the volumes of cut and fill using the end-area method and correcting for prismoidal excess.   (KU)

Chainage	Left	Centre	Right
7500	$-\dfrac{10.0}{36.0}$	$-\dfrac{20.0}{0}$	$-\dfrac{8.8}{22.0}$
7600	$\dfrac{0}{10}$	$-\dfrac{6.0}{0}$	$-\dfrac{14.0}{24.6}$
7650	$\dfrac{16.0}{22.0}$	$\dfrac{4.0}{0}$	$\dfrac{0}{10}$
7750	$\dfrac{13.5}{24.0}$	$\dfrac{22.0}{0}$	$\dfrac{8.6}{26.0}$

N.B. (1) Students should note the method of booking and compare it with the cross-sections in *Figure 9.27*.

(2) The method of splitting the sections into triangles for easy computation should also be noted.

Area of cross-section 75 + 00

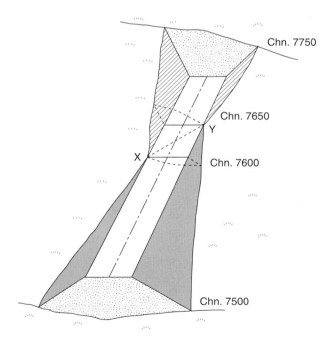

Chn. 7750

Chn. 7650

Y

X

Chn. 7600

Chn. 7500

Fig. 9.26

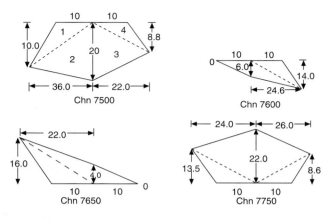

Fig. 9.27

$$\text{area } \Delta 1 = \frac{10 \times 10}{2} = 50 \text{ m}^3$$

$$\text{area } \Delta 2 = \frac{36 \times 20}{2} = 360 \text{ m}^3$$

$$\text{area } \Delta 3 = \frac{22 \times 20}{2} = 220 \text{ m}^3$$

$$\text{area } \Delta 4 = \frac{8.8 \times 10}{2} = 44 \text{ m}^3$$

$$\textit{Total area} = 674 \text{ m}^2$$

Similarly, area of cross-section 76 + 00 = 173.8 m^2

$$\text{Vol. by end area} = \frac{674 + 173.8}{2} \times 100 = 42\,390 \text{ m}^3$$

The equation for prismoidal excess varies with the shape of the cross-section. In this particular instance it equals

$$\frac{L}{12}(H_1 - H_2)(W_1 - W_2)$$

where L = horizontal distance between the two end areas
    H = centre height
    W = the sum of the side widths per section, i.e. $(w_1 + w_2)$

Thus, prismoidal excess = $\dfrac{100}{12}$ $(20 - 6)(58 - 34.6) = 2730 \text{ m}^3$

$$\textit{Corrected volume} = 39\,660 \text{ m}^3$$

Vol. between 76 + 00 and 76 + 50

Line *XY* in *Figure 9.26* shows clearly that the volume of fill in this section forms a pyramid with the cross-section 76 + 00 as its base and 50 m high. It is thus more accurate and quicker to use the equation for a pyramid.

$$\text{Vol} = \frac{AL}{3} = \frac{173.8 \times 50}{3} = 2897 \text{ m}^3$$

$$\therefore \textit{Total vol. of fill} = (39\,660 + 2897) = 42\,557 \text{ m}^3$$

The student should now calculate for himself the volume of cut.

(*Answer:* 39 925 m^3).

Example 9.7. The access to a tunnel has a level formation width of 10 m and runs into a plane hillside, whose natural ground slope is 1 in 10. The intersection line of this formation and the natural ground is perpendicular to the centre-line of the tunnel. The level formation is to run a distance of 360 m into the hillside, terminating at the base of a cutting of slope 1 vertical to 1 horizontal. The side slopes are to be 1 vertical to 1.5 horizontal.

Calculate the amount of excavation in cubic metres. Marks will be deducted if calculations are not clearly related to diagrams. (LU)

Figure 9.28 illustrates the question, which is solved by the methods previously advocated.

Height AB = 36 m as ground slope is 1 in 10

By rate of approach

$$x = \left(1 - \frac{1}{10}\right)^{-1} \times AB = \frac{10 \times 36}{9} = 40 \text{ m} = DD'$$

As the side slopes are 1 in 1.5 and D' is 40 m high, then $D'G = 40 \times 1.5 = 60$ m $= E'H$. Therefore $GH = 130$ m

Area of $\triangle BCD'$ (in section above) normal to GH, DE, CF

$$= \frac{360}{2} \times 40 = 7200 \text{ m}^2$$

$$\therefore \text{Vol} = \frac{7200}{3} (130 + 10 + 10) = 360\,000 \text{ m}^3$$

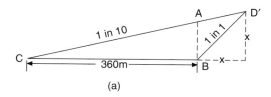

(a)

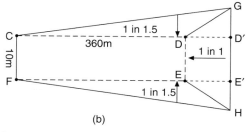

(b)

Fig. 9.28 *(a) Section, and (b) plan*

Check

$$\text{Wedge } BCD' = \frac{360}{6} \left[(130 + 10 + 10) \times 40\right] = 360\,000 \text{ m}^3$$

Example 9.8. A solid pier is to have a level top surface 20 m wide. The sides are to have a batter of 2 vertical in 1 horizontal and the seaward end is to be vertical and perpendicular to the pier axis. It is to be built on a rock stratum with a uniform slope of 1 in 24, the direction of this maximum slope making an angle whose tangent is 0.75 with the direction of the pier. If the maximum height of the pier is to be 20 m above the rock, diminishing to zero at the landward end, calculate the volume of material required. (LU)

Figure 9.29 illustrates the question. Students should note that not only is the slope in the direction of the pier required but also the slope at right angles to the pier.

By dip and strike

tan apparent slope = tan max slope × cos included angle

$$\frac{1}{x} = \frac{1}{24} \cos 36°52' \qquad \text{where } \tan^{-1} 0.75 = 36°52'$$

$$x = 30$$

∴ Grade in direction of pier = 1 in 30, ∴ $AB = 20 \times 30 = 600$ m

Grade at right angles: $\dfrac{1}{y} = \dfrac{1}{24} \cos 53°08'$

$y = 40$. Grade = 1 in 40 as shown on *Figure 9.29(a)*

∴ $DD' = 19.5$ m and $DC = 19.5 \times 30 = 585$ m

From *Figure 9.29(b)*

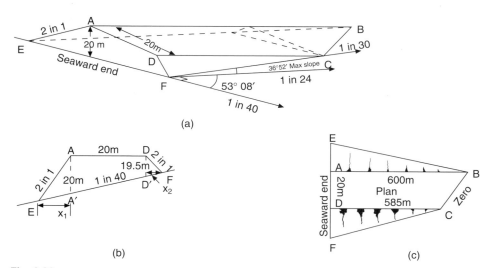

(a)

(b)   (c)

Fig. 9.29

$$x_1 = \left(2 - \frac{1}{40}\right)^{-1} \times 20 = 10.1 \text{ m}$$

$$x_2 = \left(2 + \frac{1}{40}\right)^{-1} \times 19.5 = 9.6 \text{ m}$$

$$\therefore \text{ Area } \Delta EAA' = \frac{20 \times 10.1}{2} = 101 \text{ m}^2$$

$$\text{Area } \Delta DFD' = \frac{19.5 \times 9.6}{2} = 93.6 \text{ m}^2$$

Now, Vol. *ABCD* = plan area × mean height

$$= \left(\frac{600 + 585}{2} \times 20\right) \times \frac{1}{4} (20 + 19.5 + 0 + 0) = 117\,315 \text{ m}^3$$

$$\text{Vol. of pyramid } EAB = \frac{\text{area } EAA' \times AB}{3} = \frac{101 \times 600}{3}$$

$$= 20\,200 \text{ m}^3$$

$$\text{Vol. of pyramid } DFC = \frac{\text{area } DFD' \times DC}{3} = \frac{93.6 \times 585}{3}$$

$$= 18\,252 \text{ m}^3$$

Total vol = (117\,315 + 20\,200 + 18\,252) = 155\,767 m³

Alternatively, finding the area of cross-sections at chainages 0, 585/2 and 585 and applying the prismoidal rule plus treating the volume from chainage 585 to 600 as a pyramid, gives an answer of 155 525 m³

Example 9.9. A 100-m length of earthwork volume for a proposed road has a constant cross-section of cut and fill, in which the cut area equals the fill area. The level formation is 30 m wide, transverse ground slope is 20° and the side slopes in cut-and-fill are $\frac{1}{2}$ horizontal to 1 vertical and 1 horizontal to 1 vertical, respectively. Calculate the volume of excavation in the 100-m length.

(LU)

If the student turns *Figure 9.30* through 90°, then the 1-in-2.75 grade (20°) becomes 2.75 in 1 and the 2-in-1 grade becomes 1 in 2, then by rate of approach:

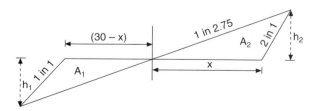

Fig. 9.30

$$h_1 = (2.75 - 1)^{-1}(30 - x) = \frac{30 - x}{1.75}$$

$$h_2 = \left(2.75 - \frac{1}{2}\right)^{-1} x = \frac{x}{2.25}$$

Now, area $\Delta A_1 = \frac{30 - x}{2} \times h_1 = \frac{(30 - x)^2}{3.5}$

area $\Delta A_2 = \frac{x}{2} \times h_2 = \frac{x^2}{4.5}$

But area A_1 = area A_2

$$\frac{(30 - x)^2}{3.5} = \frac{x^2}{4.5}$$

$$(30 - x)^2 = \frac{3.5}{4.5}x^2 = \frac{7}{9}x^2 \text{ from which } x = 16 \text{ m}$$

$$\therefore \text{ Area } A_2 = \frac{16^2}{4.5} = 56.5 \text{ m}^2 = \text{area } A_1$$

$$\therefore \text{ Vol in 100 m length} = 56.5 \times 100 = 5650 \text{ m}^3$$

Example 9.10. A length of existing road of formation width 20 m lies in a cutting having side slopes of 1 vertical to 2 horizontal. The centre-line of the road forms part of a circular curve having a radius of 750 m. For any cross-section along this part of the road the ground surface and formation are horizontal. At chainage 5400 m the depth to formation at the centre-line is 10 m, and at chainage 5500 m the corresponding depth is 18 m.

The formation width is to be increased by 20 m to allow for widening the carriageway and for constructing a parking area. The whole of the widening is to take place on the side of the cross-section remote from the centre of the arc, the new side slope being 1 vertical to 2 horizontal. Using the prismoidal rule, calculate the volume of excavation between the chainages 5400 m and 5500 m. Assume that the depth to formation changes uniformly with distance along the road.   (ICE)

From *Figure 9.31*, it can be seen that the centroid of the increased excavation lies $(20 + x)$ m from the centre-line of the curve. The distance x will vary from section to section but as the side slope is 1 in 2, then:

$$x = 2 \times \frac{h}{2} = h$$

horizontal distance of centroid from centre-line = $(20 + h)$

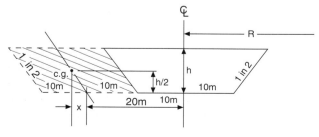

Fig. 9.31

At chainage 5400 m, $h_1 = 10$ m $\therefore (20 + h) = 30$ m $= d_1$
At chainage 5450 m, $h_2 = 14$ m $\therefore (20 + h) = 34$ m $= d_2$
At chainage 5500 m, $h_3 = 18$ m $\therefore (20 + h) = 38$ m $= d_3$
Area of extra excavation at 5400 m $= 10 \times 20 = 200$ m$^2 = A_1$
Area of extra excavation at 5450 m $= 14 \times 20 = 280$ m$^2 = A_2$
Area of extra excavation at 5500 m $= 18 \times 20 = 360$ m$^2 = A_3$

The above areas are now corrected for curvature: $A\left(1 + \dfrac{d}{R}\right)$

At chainage 5400 m $= 200\left(1 + \dfrac{30}{750}\right) = 208$ m^2

At chainage 5450 m $= 280\left(1 + \dfrac{34}{750}\right) = 292.6$ m^2

At chainage 5500 m $= 360\left(1 + \dfrac{38}{750}\right) = 378$ m^2

$\therefore$ Vol $= \dfrac{100}{6}(208 + 4 \times 292.6 + 378) = 29\,273$ m^3

Exercises

(*9.1*) An access road to a quarry is being cut in a plane surface in the direction of strike, the full dip of 1 in 12.86 being to the left of the direction of drive. The road is to be constructed throughout on a formation grade of 1 in 50 dipping, formation width 20 m and level, side slopes 1 in 2 and a zero depth on the centre-line at chainage 0 m.

At chainage 400 m the direction of the road turns abruptly through a clockwise angle of 40°; calculate the volume of excavation between chainages 400 m and 600 m. (KU)

(*Answer:* 169 587 m^3)

(*9.2*) A road is to be constructed on the side of a hill having a cross fall of 1 in 50 at right angles to the centre-line of the road; the side slopes are to be 1 in 2 in cut and 1 in 3 in fill; the formation is 20 m wide and level. Find the position of the centre-line of the road with respect to the point of intersection of the formation and the natural ground.

(a) To give equality of cut and fill.
(b) So that the area of cut shall be 0.8 of the area of fill, in order to allow for bulking. (LU)

(*Answer:* (a) 0.3 m in cut, (b) 0.2 m in fill)

(*9.3*) A reservoir is to be formed in a river valley by building a dam across it. The entire area that will be covered by the reservoir has been contoured and contours drawn at 1.5-m intervals. The lowest point in the reservoir is at a reduced level of 249 m above datum, whilst the top water level will not be above a reduced level of 264.5 m. The area enclosed by each contour and the upstream face of the dam is shown in the table below.

Contour (m)	Area enclosed (m²)
250.0	1 874
251.5	6 355
253.0	11 070
254.5	14 152
256.0	19 310
257.5	22 605
259.0	24 781
260.5	26 349
262.0	29 830
263.5	33 728
265.0	37 800

Estimate by the use of the trapezoidal rule the capacity of the reservoir when full. What will be the reduced level of the water surface if, in a time of drought, this volume is reduced by 25%?

(ICE)

(*Answer*: 294 211 m³; 262.3 m)

(*9.4*) The central heights of the ground above formation at three sections 100 m apart are 10 m, 12 m, 15 m, and the cross-falls at these sections are respectively 1 in 30, 1 in 40 and 1 in 20. If the formation width is 40 m and sides slope 1 vertical to 2 horizontal, calculate the volume of excavation in the 200-m length:

(a) If the centre-line is straight.
(b) If the centre-line is an arc of 400 m radius.   (LU)

(*Answer*: (a) 158 367 m³, (b) 158 367 ± 1070 m³)

9.6 MASS-HAUL DIAGRAMS

Mass-haul diagrams (MHD) are used to compare the economy of various methods of earthwork distribution on road or railway construction schemes. By the combined use of the MHD plotted directly below the longitudinal section of the survey centre-line, one can find.

(1) The distances over which cut and fill will balance.
(2) Quantities of materials to be moved and the direction of movement.
(3) Areas where earth may have to be borrowed or wasted and the amounts involved.
(4) The best policy to adopt to obtain the most economic use of plant.

9.6.1 Definitions

(1) *Haul* refers to the volume of material multiplied by the distance moved, expressed in 'station metres'.

(2) *Station metre* (stn m) is 1 m³ of material moved 100 m

Thus, 20 m³ moved 1500 m is a haul of 20 × 1500/100 = 300 stn m.

(3) *Freehaul* and *overhaul* can best be defined by example. A contractor may offer to haul material a distance of, say, 150 m at 50 p per m³, but thereafter for any distance hauled beyond 150 m the contractor may require an extra 5 p per stn m, i.e. 5 p per m³ moved per 100 m. The distance of 150 m is called the *freehaul distance* and is based on the economical hauling distance of the earthmoving plant used. It may range from 100 m for a bulldozer to 3000 m for self-propelled scrapers. The haul beyond the freehaul distance is termed the *overhaul.*

(4) *Waste* is the material excavated from cuts but not used for embankment fills.

(5) *Borrow* is the material needed for the formation of embankments, secured not from roadway excavation but from elsewhere. It is said to be obtained from a 'borrow pit'.

(6) *Limit of economical haul* is the maximum overhaul distance plus the freehaul distance. When this limit is reached it is more economical to waste and borrow material. For example, assume:

$$\text{Freehaul distance} = 500 \text{ m}$$

$$\text{Overhaul} = 10 \text{ p per stn m (i.e. 10 p per m}^3 \text{ per 100 m)}$$

$$\text{Borrow} = 30 \text{ p per m}^3$$

From these figures it can be seen that to overhaul 1 m³ a distance of 300 m would cost 30 p, equal to the cost of borrow; this then is the maximum overhaul distance. However, before overhaul comes into operation, one may move earth through the freehaul distance of 500 m. Thus the limit of economical haul = (300 + 500) = 800 m.

9.6.2 Bulking and shrinkage

Excavation of material causes it to loosen, and thus its excavated volume will be greater than its *in situ* volume. However, when filled and compacted, it may occupy a less volume than when originally *in situ*. For example, ordinary earth is less by about 10% after filling, whilst rock bulks by some 20% to 30%. To allow for this, a correction factor is generally applied to the cut or fill volumes.

9.6.3 Construction of the MHD

A MHD is a continuous curve, whose vertical ordinates, plotted on the same distance scale as the longitudinal section, represent the algebraic sum of the corrected volumes (cut +, fill −).

9.6.4 Properties of the MHD

Consider *Figure 9.32(a)*) in which the ground *XYZ* is to be levelled off to the grade line *A′B′*. Assuming that the fill volumes, after correction, equal the cut volumes, the MHD would plot as shown in *Figure 9.32(b)*. Thus:

(1) Since the curve of the MHD represents the algebraic sums of the volumes, then any horizontal line drawn parallel to the base *AB* will indicate the volumes which balance. Such a line is called a *balancing line* and may even be represented by *AB* itself, indicating that the total cut equals the total fill.

(2) The rising curve, shown broken, indicates cut (positive), the falling curve indicates fill (negative).

(3) The maximum and minimum points of a MHD occur directly beneath the intersection of the natural ground and the formation grade; such intersections are called *grade points*.

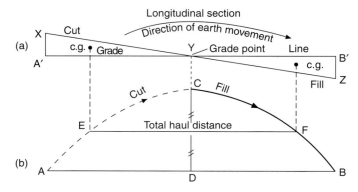

Fig. 9.32   *Mass-haul diagram*

(4) As the curve of the MHD rises above the balance line *AB,* the haul is from left to right. When the curve lies below the balance line, the haul is from right to left.

(5) The total cut volume is represented by the maximum ordinate *CD.*

(6) In moving earth from cut to fill, assume that the first load would be from the cut at *X* to the fill at *Y,* and the last load from the cut at *Y* to the fill at *Z.* Thus the haul distance would appear to be from a point mid-way between *X* and *Y,* to a point mid-way between *Y* and *Z.* However, as the section is representative of volume, not area, the haul distance is from the centroid of the cut volume to the centroid of the fill volume. The horizontal positions of these centroids may be found by bisecting the total volume ordinate *CD* with the horizontal line *EF.*

Now, since haul is volume × distance, the *total haul* in the section is total vol × total haul distance = *CD* × *EF*/100 stn m.

9.6.5 Balancing procedures

In order to illustrate the use of freehaul distance consider *Figure 9.33*:

(1) Assuming a freehaul distance of 100 m; move this scaled distance up and down the MHD, keeping it parallel to the base *A′B′* until it cuts the curve at *E* and *F.*

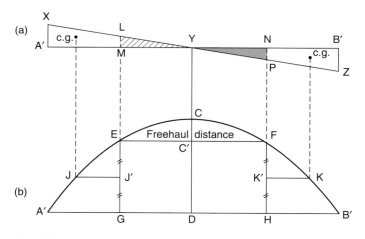

Fig. 9.33

(2) *EF* indicates on the longitudinal section that the cut volume *LMY* equals the fill volume *YNP*. The amount of the volume is *CC'* and it obviously falls within the freehaul distance.

(3) The remaining cut volume *XLMA'* is represented by the ordinate *EG* and is the *overhaul volume*.

(4) The overhaul volume *XLMA'* has now to be filled into *NPZB'*, the average distance being from centroid to centroid. The positions of the centroids are found by bisecting *EC* and *FH*, giving the horizontal distance between centroids *JK*.

(5) Assuming *JK* = 250 m, the overhaul volume has to be moved through this distance. However, the first 100 m of the movement is still within the freehaul contract, and thus the overhaul distance is (250 − 100) = 150 m.

(6) From (5) it is obvious that the total volume (*CC'* + *EG*) = *CD* falls within the freehaul contract.

(7) Thus, the overhaul = overhaul vol × overhaul distance = *EG*(*JK* − *EF*).

Worked examples

Example 9.11. The following notes refer to a 1200-m section of a proposed railway, and the earthwork distribution in this section is to be planned without regard to the adjoining sections. The table below shows the stations and the surface levels along the centre-line, the formation level being at an elevation above datum of 43.5 m at chainage 70 and thence rising uniformly on a gradient of 1.2%. The volumes are recorded in m³, the cuts are plus and fills minus.

(1) Plot the longitudinal section using a horizontal scale of 1:1200 and a vertical scale of 1:240.
(2) Assuming a correction factor of 0.8 applicable to fills, plot the MHD to a vertical scale of 1000 m³ to 20 mm.
(3) Calculate *total haul* in stn m and indicate the haul limits on the curve and section.
(4) State which of the following estimates you would recommend:

    (a) No freehaul at 35 p per m³ for excavating hauling and filling,
    (b) A freehaul distance of 300 m at 30 p per m³ plus 2 p per stn m for overhaul.   (LU)

Chn	Surface level	Vol	Chn	Surface level	Vol	Chn	surface level	Vol
70	52.8		74	44.7		78	49.5	
		+1860			−1080			−237
71	57.3		75	39.7		79	54.3	
		+1525			−2025			+362
72	53.4		76	37.5		80	60.9	
		+547			−2110			+724
73	47.1		77	41.5		81	62.1	
		−238			−1120			+430
74	44.7		78	49.5		82	78.5	

For answers to parts (1) and (2) see *Figure 9.34* and the values in *Table 9.2*.

N.B. (1) The volume at chainage 70 is zero.
    (2) The mass ordinates are always plotted at the station and not between them.
    (3) The mass ordinates are now plotted to the same horizontal scale as the longitudinal section and directly below it.

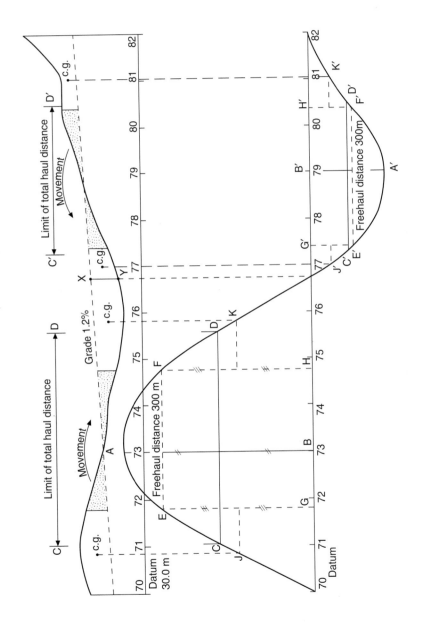

Fig. 9.34

Table 9.2

Chainage	Volume	Mass ordinate (algebraic sum)
70	0	0
71	+ 1860	+ 1860
72	+ 1525	+ 3385
73	+ 547	+ 3932
74	$- 238 \times 0.8 = - 190.4$	+ 3741.6
75	$- 1080 \times 0.8 = - 864$	+ 2877.6
76	$- 2025 \times 0.8 = - 1620$	+ 1257.6
77	$- 2110 \times 0.8 = - 1688$	- 430.4
78	$- 1120 \times 0.8 = - 896$	- 1326.4
79	$- 237 \times 0.8 = -189.6$	- 1516
80	+ 362	- 1154
81	+ 724	- 430
82	+ 430	0

(4) Check that maximum and minimum points on the MHD are directly below grade points on the section.

(5) Using the datum line as a balancing line indicates a balancing out of the volumes from chainage 70 to *XY* and from *XY* to chainage 82.

Total haul (taking each loop separately) = total vol × total haul distance. The total haul distance is from the centroid of the total cut to that of the total fill and is found by bisecting *AB* and *A′B′*, to give the distances *CD* and *C′D′*.

$$\text{Total haul} = \frac{AB \times CD}{100} + \frac{A'B' \times C'D'}{100}$$

$$= \frac{3932 \times 450}{100} + \frac{1516 \times 320}{100} = 22\,545 \text{ stn m}$$

(a) If there is no freehaul, then all the volume is moved regardless of distance for 35 p per m³.

Estimate costs: $(AB + A'B') \times 35 \text{ p} = 5448 \times 35 = 190\,680 \text{ p}$

(b) The purpose of plotting the freehaul distance on the curve is to assess the overhaul.

From MHD:

Cost of freehaul = $(AB + A'B') \times 30$ p per m³ (see *Section 9.6.5(6)*)

$$= 163\,440 \text{ p}$$

Cost of overhaul = $\dfrac{EG\,(JK - EF)}{100} + \dfrac{E'G'(J'K' - E'F')}{100} \times 2 \text{ p}$

$$= 13\,628 \text{ p}$$

Total cost = $163\,440 + 13\,628 = 177\,068$ p

∴ Second estimate is cheaper by $13\,612$ p $= £136.12$

N.B. All the dimensions in the above solution are scaled from the MHD.

Example 9.12. The volumes between sections along a 1200-m length of proposed road are shown below, positive volumes denoting cut, and negative volumes denoting fill:

Chainage (m)	0	100	200	300	400	500	600	700	800	900	1000	1100	1200
Vol. between sections ($m^3 \times 10^3$)		+2.1	+2.8	+1.6	−0.9	−2.0	−4.6	−4.7	−2.4	+1.1	+3.9	+3.5	+2.8

Plot a MHD for this length of road to a suitable scale and determine suitable positions of balancing lines so that there is

(1) A surplus at chainage 1200 but none at chainage 0.
(2) A surplus at chainage 0 but none at chainage 1200.
(3) An equal surplus at chainage 0 and chainage 1200.

Hence, determine the cost of earth removal for each of the above conditions based on the following prices and a freehaul limit of 400 m.

Excavate, cart and fill (freehaul)	60 p/m^3
Excavate, cart and fill (overhaul)	85 p/m^3
Removal of surplus to tip from chainage 0	125 p/m^3
Removal of surplus to tip from chainage 1200	150 p/m^3 (ICE)

For plot of MHD see *Figure 9.35*

Mass ordinates = 0, +2.1, +4.9, +6.5, +5.6 +3.6, −1.0,
  −5.7, −8.1, −7.0, −3.1, +0.4, +3.2
  (obtained by algebraic summation of vols)

(1) Balance line *AB* gives a surplus at chainage 1200 but none at 0.
(2) Balance line *CD* gives a surplus at chainage 0 but none at 1200.
(3) Balance line *EF* situated mid-way between *AB* and *CD* will give equal surpluses at the ends.

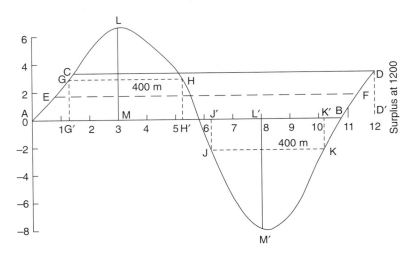

Fig. 9.35

In the second part of the question the prices are quoted in an unusual manner. Excavating, carting and filling within a distance of 400 m, is at 60 p/m^3. Carrying on beyond that distance gives a total price of 85 p/m^3; thus the overhaul is 25 p/m^3.

The question is now tackled in the usual way but it is not necessary to find overhaul distance.

(1) Taking *AB* as base

Freehaul plots at *GH* and *JK*.

Cost of freehaul $= (LM + L'M') \times 60$ p
$\qquad\qquad\quad = (6500 + 8100) \times 60 = 876\,000$ p

Cost of overhaul $= (GG' + JJ') \times 25$ p
$\qquad\qquad\quad\ = (2800 + 2200) \times 25$

Cost of surplus removal $= DD' \times 150$ p
$\qquad\qquad\qquad\qquad\quad = 3200 \times 150$ p

brought forward   876 000 p

$\qquad = 125\,000$ p

$\qquad = 480\,000$ p

Total cost $= 1\,481\,000$ p

$\qquad = £14\,810$

(2)(3) The technique is the same for these situations working to different balancing lines *CD* and *EF* respectively. The student should now attempt this for himself, the answers being (2) £14 130, (3) £143 80.

N.B. As the freehaul lines remain fixed, there is no overhaul on line *CD* in the first loop of the MHD.

Example 9.13. Volumes in m^3 of excavation (+) and fill (−) between successive sections 100 m apart on a 1300-m length of a proposed railway are given.

Section	0	1	2	3	4	5	6	7
Volume (m^3)		−1000	−2200	−1600	−500	+200	+1300	+2100

Section	7	8	9	10	11	12	13
Volume (m^3)		+1800	+1100	+300	−400	−1200	−1900

Draw a MHD for this length. If earth may be borrowed at either end, which alternative would give the least haul? Show on the diagram the forward and backward freehauls if the freehaul limit is 500 m, and give these volumes.

(LU)

Adding the volumes algebraically gives the following mass ordinates

Section	0	1	2	3	4	5	6	7
Volume (m^3)		−1000	−3200	−4800	−5300	−5100	−3800	−1700

Section	7	8	9	10	11	12	13
Volume (m^3)		+100	+1200	+1500	+1100	−100	−2000

These are now plotted to produce the MHD of *Figure 9.36*. Balancing out from the zero end permits borrowing at the 1300 end.

(1) Total haul $= \dfrac{(AB \times CD)}{100} + \dfrac{(A'B' \times C'D')}{100}$ stn m

$\qquad\qquad\quad = \dfrac{(5300 \times 475)}{100} + \dfrac{(1500 \times 282)}{100} = 29\,405$ stn m

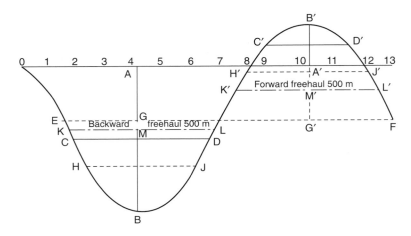

Fig. 9.36

Note: *CD* bisects *AB* and *C'D'* bisects *A'B'*.

(2) Balancing out from the 1300 end (*EF*) permits borrowing at the zero end.

$$\text{Total haul} = \frac{(GB \times HJ)}{100} + \frac{(G'B' \times H'J')}{100}$$

$$= \frac{(3300 \times 385)}{100} + \frac{(3500 \times 430)}{100} = 27\,755 \text{ stn m}$$

Thus, borrowing at the zero end is the least alternative.

Backward freehaul = *MB* = 2980 m³

Forward freehaul = *M'B'* = 2400 m³

9.6.6 Auxiliary balancing lines

A study of the material on MHD plus the worked examples should have given the reader an understanding of the basics. It is now appropriate to illustrate the application of auxiliary balancing lines.

Consider in the first instance a MHD as in *Worked examples 9.11*. In *Figure 9.37*, the balance line is *ABC* and the following data are easily extrapolated:

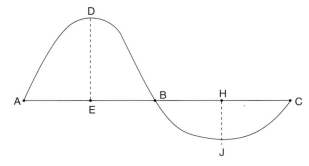

Fig. 9.37

Cut *AD* balances fill *DB*   Vol moved = *DE*
Cut *CJ* balances fill *BJ*   Vol moved = *HJ*

Consider now *Figure 9.38*; the balance line is *AB*, but in order to extrapolate the above data one requires an auxiliary balancing line *CDE* parallel to *AB* and touching the MHD at *D*:

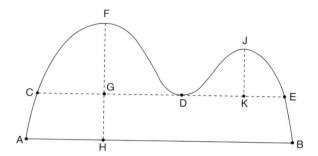

Fig. 9.38

Cut *AC* balances fill *EB*   Vol moved = *GH*
Cut *CF* balances fill *FD*   Vol moved = *FG*
Cut *DJ* balances fill *JE*   Vol moved = *JK*

Thus the total volume moved between *A* and *B* is *FH + JK*

Finally, *Figure 9.39* has a balance from *A* to *B* with auxiliaries at *CDE* and *FGH*, then:

Fill *AC* balances cut *BE*   Vol moved = *JK*
Fill *CF* balances cut *DH*   Vol moved = *KL*
Fill *FM* balances cut *GM*   Vol moved = *LM*
Fill *GO* balances cut *HO*   Vol moved = *NO*
Fill *DQ* balances cut *EQ*   Vol moved = *PQ*

The total volume moved between *A* and *B* is *JM + NO + PQ*.
The above data become apparent only when one introduces the auxiliary balancing lines.

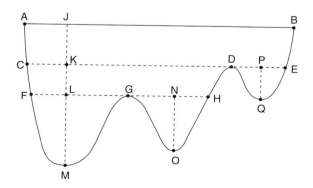

Fig. 9.39

Exercises

(*9.5*) The volumes in m³ between successive sections 100 m apart on a 900-m length of a proposed road are given below (excavation is positive and fill negative):

Section	0	1	2	3	4	5	6	7
Volume (m³)		+1700	−100	−3200	−3400	−1400	+100	+2600

Section	7	8	9
Volume (m³)		+4600	+1100

Determine the maximum haul distance when earth may be wasted only at the 900-m end. Show and evaluate on your diagram the overhaul if the freehaul limit is 300 m.   (LU)

(*Answer*: 558 m, 5500 stn m)

(*9.6*) Volumes of cut (positive) and fill (negative) along a length of proposed road are as follows:

Chainage (m)	0		100		200		300		400		480	
Volume (m³)		+290		+760		+1680		+620		+120		−20
Chainage (m)	500		600		700		800		900		1000	
Volume (m³)		−110		−350		−600		−780		−690		−400
Chainage (m)	1100		1200									
Volume (m³)		−120										

Draw a MHD and, excluding the surplus excavated material along this length, determine the overhaul if the freehaul distance is 300 m.   (ICE)

(*Answer*: 350 stn m)

Earthworks program

```
10   REM: CUT AND FILL CALCULATIONS
20   REM: BY BRIAN MERRONY, SCHOOL OF CIVIL ENGINEERING
30   REM: KINGSTON POLYTECHNIC
40   DIM GY(50) GX(50), RY(50), RX(50), S(2)
50   LS = "————————————————————" \ PRINT LS
60   PRINT"EARTHWORKS" \ PRINT LS
70   REM———— START OF SECTION DATA
80   INPUT"FILL SLOPE"; FS\
90   INPUT "CUT SLOPE"; CS
100  INPUT "NO OF GROUND POINTS"; NG
110  PRINT "ENTER GROUND OFFSETS AND LEVELS (IN PAIRS)"
120  FOR I=0 TO NG-1
130  PRINT "POINT"; I + 1; \ INPUT GX(I), GY (I)
140  NEXT I \ PRINT LS
150  INPUT "NO OF ROAD POINTS"; NR
160  PRINT "ENTER ROAD OFFSETS AND LEVELS (IN PAIRS)" ;
170  FOR I = 1 TO NR
180  PRINT "POINT"; I; \ INPUT RX (I), RY(I)
190  NEXT I \ PRINT LS
200  REM * * * * * CHECK THAT GROUND SPANS ROAD * * * * *
210  IF GX (0) < = RX(1) AND GX (NG-1) > = RX (NR) THEN 230
220  PRINT "GROUND DOES NOT SPAN ROAD" \ GOTO 1110
230  REM * * * * * TREAT LEFT HAND SIDE * * * * *
240  P = O
250  P = P+1 \ IF GX (P) <RX(1) THEN 250
260  A = 1 \ B = 0 \ Q = 1 \ GOSUB 940
270  REM * * * * * TREAT RIGHT HAND SIDE * * * * *
280  P = NG-1
290  P = P-1 \ IF GX (P) > RX(NR) THEN 290
300  A = -1 \ B = NG-1 \ Q = NR \ GOSUB 940
310  REM * * * * * START OF MAIN PASS * * * * *
320  P = 0\ Q = 0 \ F = 1
330  AA = GY (0) \ BB = RY(0) \ HH = ABS (AA-BB)
340  A = 1 \ IF AA < BB THEN A = 2
350  EE = GX(0)
360  REM———— START OF LOOP
370  P = P+1 \ Q = Q+1 \ X1 = EE
380  RX = RX (Q) \ GX = GX (P)
390  IF GX > RX THEN 470
400  IF GX = RX THEN 530
410  REM * * * * * CONDITION GX < RX * * * * *
420  FG = GX \ CC = GY (P)
430  Y1 = BB \ Y2 = RY(Q) \ X2 = RX \ XX = GX
440  DD = (Y2-Y1) * (XX-X1) / (X2-X1) + Y1
450  Q = Q - 1
460  GOTO 560
470  REM * * * * * GX > RX CONDITION * * * * *
480  FG = RX \ DD = RY(Q)
490  Y1 = AA \ Y2 = GY (P) \ X2 = GX \ XX = RX
500  CC = (Y2-Y1) * (XX-X1) / (X2-X1) + Y1
510  P = P-1
520  GOTO 560
530  REM * * * * * GX = RX CONDITION * * * * *
540  FG = GX \ CC = GY (P) \ DD = RY(Q)
550  Y1 = AA \ Y2 = CC \ X2 = GX
560  REM———— EVALUATE AREAS
570  B = 1 \ IF CC < DD THEN B = 2
580  I I = ABS (CC-DD)
590  GG = FG-EE
600  IF A = B THEN 750
610  REM———— TWO TRIANGLES
620  Y = GG*HH / (HH + I I)
630  XX = EE + Y
640  JJ = Y* HH / 2
650  IF F = 1 THEN  F = 0 \ JJ = 0
660  S(A) = S(A) + JJ
670  S3 = S3 + S(1) \ S(1) = 0
680  S4 = S4 + S(2) \ S(2) = 0
690  KK = GG* I I* I I / (2* (HH + I I ))
700  S(B) = S(B) + KK
710  Y = (Y2-Y1)*(XX-X1) / (X2-X1) + Y1
720  PRINT "INTERSECTION"
730  PRINT XX,Y
740  GOTO 770
750  REM ———— TRAPEZIUM
760  IF F = 0 THEN S(A) = S(A) + (HH + I I)* GG / 2
770  REM ———— TEST FOR END OF CROSS-SECTION
780  IF P = NG-1 AND Q = NR + 1 THEN 810
790  AA = CC \ BB = DD \ EE = FG \ HH = II \ A = B
800  GOTO 360
810  REM———— CALCULATE VOLUMES AND UPDATE FOR NEXT
     SECTION
820  PRINT "CUT AREA, FILL AREA" \ PRINT S3, S4
830  S(1) = 0 \ S(2) = 0
840  VI = (S5 + S3) *. 5* LL \ V2 = (S6 + S4)*. 5* LL
850  S5 = S3 \ S6 = S4 \ S3 = 0\ S4 = 0
860  PRINT "CUT VOLS, INCREMENT AND TOTAL"
870  V3 = V3 + V1 \ PRINT VI, V3
880  PRINT "FILL, VOLS, INCREMENT AND TOTAL"
890  V4 = V4 + V2 \ PRINT V2, V4 \PRINT LS
900  PRINT "CHAINAGE INTERVAL (INSERT ZERO TO
     TERMINATE RUN)"
910  INPUT LL
920  IF LL = 0 THEN 1110
930  GOTO 70
940  REM ———— DECIDE CUT OR FILL AND EXTRAPOLATE IF
     NECESSARY
950  XX = RX(Q)
960  Y1 = GY(P-A) \ XI = GX (P-A) \ Y2 = GY (P) \ X2 = GX(P)
970  Y = (Y2-Y1) * (XX-X1) / (X2-X1) + Y1
980  F = 0 \ IF RY(Q) < Y THEN F = 1
990  IF RY(Q) = Y THEN RY(Q) = Y + 0.0001
1000 S = -FS \ IF F = 1 THEN S = CS
1010 Y1 = GY(B) \ X1 = GX (B) \ Y2 = GY (B+A) \ X2 = GX(B+A)
1020 XX = GX(B)
1030 Y = RY(Q)+ABS(RX(Q)-XX)* S
1040 IF (F = 0 AND GY (B) > Y) OR (F = 1 AND GY (B) < Y) THEN
     1100
1050 XX = XX-100* A \ Y = Y + 100* S
1060 GY (B) = (Y2-Y1)* (XX-X1) / (X2-X1) + Y1
1070 GX (B) = XX
1080 IF (F = 0 AND GY (B) > Y) OR (F = 1 AND GY(B) < Y) THEN
     1100
1090 PRINT "EXTRAPOLATION FAILED" \ GOTO 1110
1100 RX (Q-A) = GX(B) \ RY (Q-A) = Y \ RETURN
1110 END
```

EARTHWORKS PROGRAM: BASIC

The program is used for earthwork computation and mass-haul data.

From offsets and levels of existing ground cross-sections and proposed road cross-sections at successive chainage points along the centre-line of a route location; cut and fill volumes are computed, accumulated volumes for mass haul produced and the chainage and level of the intersection of the side slopes and existing ground are found.

Input data (the program is interactive and will request input data)

(1) Side slope in FILL; e.g. for 1 in 2, enter 0.5.
(2) Side slope in CUT; this may be the same as or different from that adopted for fill.
(3) The number of points defining the EXISTING GROUND in the cross-section.
(4) The distance from the centre-line and level of the GROUND POINTS. Distances to the *left* of the centre-line are *negative*, those to the *right* are *positive*. N.B. They are entered into the computer in IN SEQUENCE FROM THE EXTREME LEFT TO RIGHT, of the cross-section.
(5) The number of points defining the proposed ROAD formation.
(6) As in (4), but for road points.
(7) The chainage interval to the next cross-section. If there are no further cross-sections the entry of a zero chainage interval terminates the run. (Refer example overleaf).

Output data

(1) Ground/road intersection points

The distance from the centre-line and level at which the road side-slopes intersect the existing ground; i.e. the top of the cutting or toe of the embankment (the *slope stake* position).

(2) Areas

Areas of the cross-sections.

(3) Volumes

The volume at each cross-section and the accumulated volume. The latter is used for mass haul construction.

An example of the input and output for a hill-side section in cut-and-fill is given overleaf.

EARTHWORKS

INPUT DATA OF CROSS-SECTION LEVELS

NO OF ROAD POINTS? 3
ENTER ROAD OFFSETS AND LEVELS (IN PAIRS)
POINT 1 ? –18, 103.7
POINT 2 ? 0, 104
POINT 3 ? 18, 103.7

INPUT DATA OF PROPOSED ROAD LEVELS

INTERSECTION
–27.5789   98.9105
INTERSECTION
3.61445   103.94
INTERSECTION
21.1111   106.811
CUT AREA, FILL AREA
28.7184   92.2913
CUT VOLS, INCREMENT AND TOTAL
0   0
FILL VOLS, INCREMENT AND TOTAL
0   0

OUTPUT

CHAINAGE INTERVAL (INSERT ZERO TO TERMINATE RUN)
? 30
FILL SLOPE ? 0.5
CUT SLOPE ? 1.0
NO OF GROUND POINTS ? 7
ENTER GROUND OFFSETS AND LEVELS (IN PAIRS)

      etc., etc.... .

10

Setting out (dimensional control)

In engineering the production of an accurate large-scale plan is usually the first step in the planning and design of a construction project. Thereafter the project, as designed on the plan, must be set out on the ground in the correct absolute and relative position and to its correct dimensions. Thus, surveys made in connection with a specific project should be planned with the setting-out process in mind and a system of three-dimensional control stations conveniently sited and adequate in number should be provided to facilitate easy, economical setting out.

It is of prime importance that the establishment and referencing of survey control stations should be carried out at such places and in such a manner that they will survive the construction processes. This entails careful choice of the locations of the control stations and their construction relative to their importance and long- or short-term requirements. For instance, those stations required for the total duration of the project may be established in concrete or masonry pillars with metal plates or bolts set in on which is punched the station position. Less durable are stout wooden pegs set in concrete or driven directly into the ground. A system of numbering the stations is essential, and frequently pegs are painted different colours to denote the particular functions for which they are to be used.

10.1 PROTECTION AND REFERENCING

Most site operatives have little concept of the time, effort and expertise involved in establishing setting-out pegs. For this reason the pegs are frequently treated with disdain and casually destroyed in the construction process. A typical example of this is the centre-line pegs for route location which are the first to be destroyed when earth-moving commences. It is important, therefore, that control stations and BMs should be protected in some way (usually as shown in *Figure 10.1*) and site operatives, particularly earthwork personnel, impressed with the importance of maintaining this protection.

Where destruction of the pegs is inevitable, then referencing procedures should be adopted to re-locate their positions to the original accuracy of fixation. Various configurations of reference pegs are used and the one thing that they have in common is that they must be set well outside the area of construction and have some form of protection, as in *Figure 10.1*.

A commonly-used method of referencing is from four pegs (A, B, C, D) established such that two strings stretched between them intersect to locate the required position (*Figure 10.2*). Distances AB, BC, CD, AD, AC, BD should all be measured as checks on the possible movement of the reference pegs, whilst distances from the reference pegs to the setting-out peg will afford a check on positioning. Intersecting lines of sight from theodolites at, say, A and B may be used where ground conditions make string lining difficult.

Where ground conditions preclude taping, the setting-out peg may be referenced by trisection

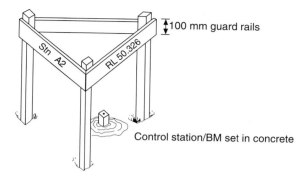

100 mm guard rails

Stn A2

RL 50.326

Control station/BM set in concrete

Fig. 10.1

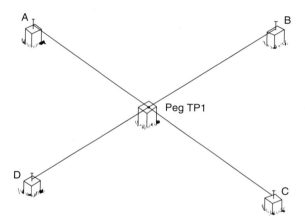

A

B

Peg TP1

D

C

Fig. 10.2

from three reference pegs. The pegs should be established to form well-conditioned triangles of intersection (*Figure 10.3*), the angles being measured and set out on both faces of a 1″ theodolite.

All information relating to the referencing of a point should be recorded on a diagram of the layout involved.

10.2 BASIC SETTING-OUT PROCEDURES USING COORDINATES

Plans are generally produced on a plane rectangular coordinate system, and hence salient points of the design may also be defined in terms of rectangular coordinates on the same system. For instance, the centre-line of a proposed road may be defined in terms of coordinates at, say, 30-m intervals, or alternatively, only the tangent and intersection points may be so defined. The basic methods of locating position when using coordinates is by either polar coordinates or intersection.

10.2.1 By polar coordinates

In *Figure 10.4*, *A*, *B* and *C* are control stations whose coordinates are known. It is required to locate point *IP* whose design coordinates are also known. The computation involved is as follows:

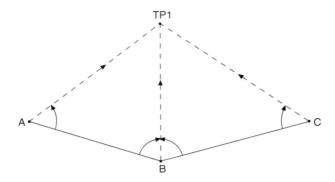

Fig. 10.3

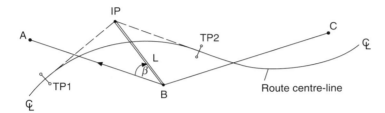

Fig. 10.4

(1) From coordinates compute the bearing *BA* (this bearing may already be known from the initial control survey computations).
(2) From coordinates compute the horizontal length and bearing of *B – IP.*
(3) From the two bearings compute the setting-out angle *AB(IP)*, i.e. β.
(4) Before proceeding into the field, draw a neat sketch of the situation showing all the setting-out data. Check the data from the plan or by independent computation.

The field work involved is as follows:

(1) Set up theodolite at *B* and backsight to *A*, note the horizontal circle reading.
(2) Add the angle β to the circle reading *BA* to obtain the circle reading *B – IP.* Set this reading on the theodolite to establish direction *B – IP* and measure out the horizontal distance *L*.

If this distance is set out by steel tape, careful consideration must be given to all the error sources such as standardization, slope, tension and possibly temperature if the setting-out tolerances are very small. It should also be carefully noted that the sign of the correction is reversed from that applied when measuring a distance. For example, if a 30-m tape was in fact 30.01 m long, when measuring a distance the recorded length would be 30 m for a single tape length, although the actual distance is 30.01 m; hence a *positive* correction of 10 mm is applied to the recorded measurement. However, if it is required to set out 30 m, the actual distance set out would be 30.01 m; thus this length would need to be reduced by 10 mm, i.e. a *negative* correction.

The best field technique when using a steel tape is carefully to align pegs at *X* and *Y* each side of the expected position of *IP* (*Figure 10.5*). Now carefully measure the distance *BX* and subtract it from the known distance to obtain distance *X – IP*, which will be very small, possibly less than one metre. Stretch a fine cord between *X* and *Y* and measure *X – IP* along this direction to fix point *IP.*

Modern EDM equipment displays horizontal distance, so the length *B – IP* may be ranged direct

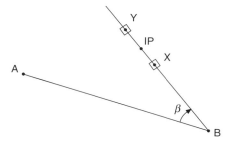

Fig. 10.5

to a reflector fixed to a setting-out pole. The use of short-range EDM equipment has made this method of setting out very popular.

10.2.2 By intersection

This technique, illustrated in *Figure 10.6*, does not require linear measurement; hence, adverse ground conditions are immaterial and one does not have to consider tape corrections.

The computation involved is as follows:

(1) From the coordinates of *A*, *B* and *IP* compute the bearings *AB*, *A* – *IP* and *B* – *IP*.
(2) From the bearings compute the angles α and β.

The relevant field work, assuming two theodolites are available, is as follows:

(1) Set up a theodolite at *A*, backsight to *B* and turn off the angle α.
(2) Set up a theodolite at *B*, backsight to *A* and turn off the angle β.

The intersection of the sight lines *A* – *IP* and *B* – *IP* locates the position of *IP*. The angle δ is measured as a check on the setting out.

If only one theodolite is available then two pegs per sight line are established, as in *Figure 10.5*, and then string lines connecting each opposite pair of pegs locate position *IP*, as in *Figure 10.2*.

10.3 TECHNIQUE FOR SETTING OUT A DIRECTION

It can be seen that both the basic techniques of position fixing require the turning-off of a given angle. To do this efficiently the following approach is recommended:

In *Figure 10.6*, consider turning off the angle β equal to $20°36'20''$ using a Watts No. 1 (20″) theodolite (*Figure 10.7(a)*).

(1) With theodolite set at *B*, backsight to *A* and read the horizontal circle – say, $02°55'20''$.
(2) As the angle β is clockwise of *BA* the required reading on the theodolite will be equal to $(02°55'20'' + 20°36'20'')$, i.e. $23°31'40''$.
(3) As the *minimum* main scale division is equal to 20′ anything less than this will appear on the micrometer (*Figure 10.7(a)*). Thus, set the micrometer to read 11′40″, now release the upper plate clamp and rotate the theodolite until it reads approximately $23°20'$ on the main scale; using the upper plate slow-motion screw, set the main scale to exactly $23°20'$. This process will not alter the micrometer scale and so the total reading is $23°31'40''$, and the instrument has been swung through the angle $\beta = 20°36'20''$.

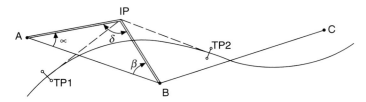

Fig. 10.6

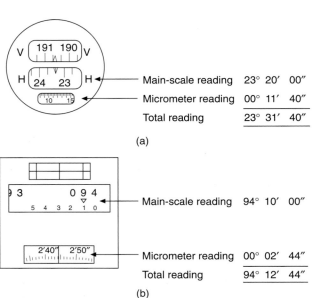

(a)

Main-scale reading	23° 20′	00″
Micrometer reading	00° 11′	40″
Total reading	23° 31′	40″

Main-scale reading	94° 10′	00″
Micrometer reading	00° 02′	44″
Total reading	94° 12′	44″

(b)

Fig. 10.7   *(a) Watts No. – 20″ theodolite, (b) T2 – 1″ theodolite*

If the Wild T.2 (*Figure 10.7(b)*) had been used, an examination of the main scale shows its minimum division is equal to 10′. Thus, to set the reading to 23° 31′ 40″ one would set only 01′ 40″ on the micrometer first before rotating the instrument to read 23° 30′ on the main scale.

Therefore, when setting out directions with any make of theodolite, the observer should examine the reading system to find out its minimum main scale value, anything less than which is put on the micrometer first.

Basically the micrometer works as shown in *Figure 10.8*, and, if applied to the Watts theodolite, is explained as follows:

Assuming the observer's line of sight passes at 90° through the parallel plate glass, the reading is 23° 20′ + S. The parallel plate is rotated using the micrometer screw until an *exact* reading (23° 20′) is obtained on the main scale, as a result of the line of sight being refracted towards the normal and emerging on a parallel path. The distance S, through which the viewer's image was displaced, is recorded on the micrometer scale (11′ 40″) and is a function of the rotation of the plate. Thus it can be seen that rotating the micrometer screw in no way affects the pointing of the theodolite, but back-sets the reading so that rotation of the theodolite is through the total angle of 20° 36′ 20″.

As practically all setting-out work involves the use of the theodolite and/or level, the user should be fully conversant with the various error sources and their effects, as well as the methods of adjustment. Information to this end is available in Chapters 2 and 4.

The use of coordinates is now universally applied to the setting out of pipelines, motorways, general roadworks, power stations, offshore piling and jetty works, housing and high-rise buildings,

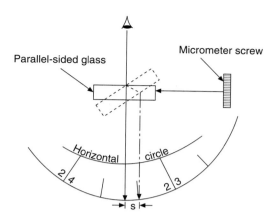

Micrometer screw

Parallel-sided glass

Horizontal   circle

2\4   2\3

s

Fig. 10.8

etc. Thus it can be seen that although the project may vary enormously from site to site the actual setting out is completed using the basic measurements of angle and distance.

There are many advantages to the use of coordinates, the main one being that the engineer can set out any part of the works as an individual item, rather than wait for the overall establishment of a setting-out grid.

10.4 USE OF GRIDS

Many structures in civil engineering consist of steel or reinforced concrete columns supporting floor slabs. As the disposition of these columns is inevitably that they are at right angles to each other, the use of a grid, where the grid intersections define the position of the columns, greatly facilitates setting out. It is possible to define several grids as follows:

(1) *Survey grid*: the rectangular coordinate system on which the original topographic survey is carried out and plotted (*Figure 10.9*).

(2) *Site grid*: defines the position and direction of the main building lines of the project, as shown in *Figure 10.9*. The best position for such a grid can be determined by simply moving a tracing of the site grid over the original plan so that its best position can be located in relation to the orientation of the major units designed thereon.

In order to set out the site grid, it may be convenient to translate the coordinates of the site grid to those of the survey grid using the well-known transformation formula.

$$E = \Delta E + E_1 \cos \theta - N_1 \sin \theta$$
$$N = \Delta N + N_1 \cos \theta + E_1 \sin \theta$$

where $\Delta E, \Delta N$ = difference in easting and northing of the respective grid origins
E_1, N_1 = the coordinates of the point on the site grid
θ = relative rotation of the two grids
E, N = the coordinates of the point transformed to the survey grid

Thus, selected points, say X and Y (*Figure 10.9*) may have their site-grid coordinate values transformed to that of the survey grid and so set-out by polars or intersection from the survey control. Now, using XY as a baseline, the site grid may be set out using theodolite and steel tape,

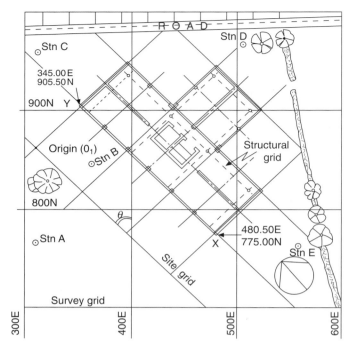

Fig. 10.9

all angles being turned off on both faces and grid intervals carefully fixed using the steel tape under standard tension.

When the site grid has been established, each line of the grid should be carefully referenced to marks fixed clear of the area of work. As an added precaution, these marks could be further referenced to existing control or permanent, stable, on-site detail.

(3) *Structural grid:* used to locate the position of the structural elements within the structure and is physically established usually on the concrete floor slab (*Figure 10.9*).

10.5 SETTING OUT BUILDINGS

For buildings with normal strip foundations the corners of the external walls are established by pegs located direct from the survey control or by measurement from the site grid. As these pegs would be disturbed in the initial excavations their positions are transferred by theodolite on to profile boards set well clear of the area of disturbance (*Figure 10.10*). Prior to this their positions must be checked by measuring the diagonals as shown in *Figure 10.11*.

The profile boards must be set horizontal with their top edge at some predetermined level such as *damp proof course* (DPC) or *finished floor level* (FFL). Wall widths, foundations widths, etc. can be set out along the board with the aid of a steel tape and their positions defined by saw-cuts. They are arranged around the building as shown in *Figure 10.11*. Strings stretched between the appropriate marks clearly define the line of construction.

In the case of buildings constructed with steel or concrete columns, a structural grid must be

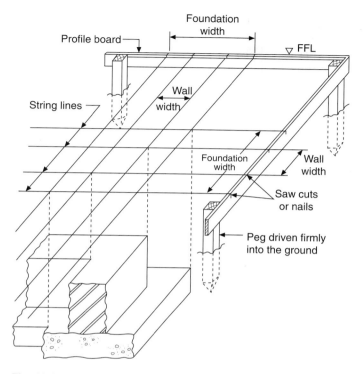

Fig. 10.10

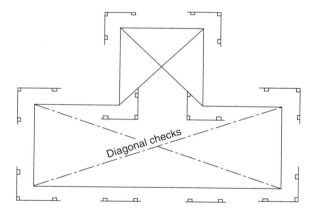

Fig. 10.11

established to an accuracy of about ±2 to 3 mm, the prefabricated beams and steelwork will not fit together without some distortion.

The position of the concrete floor slab may be established in a manner already described. Thereafter the structural grid is physically established by hilty nails or small steel plates set into the concrete. Due to the accuracy required a 1″ theodolite and standardized steel tape corrected for temperature and tension, should be used.

Once the bases for the steel columns have been established, the axes defining the centre of each column should be marked on and, using a template orientated to these axes, the positions of the holding-down bolts defined (*Figure 10.12*). A height mark should be established, using a level, at a set distance (say, 75 mm) below the underside of the base-plate, and this should be constant throughout the structure. It is important that the base-plate starts from a horizontal base to ensure verticality of the column.

10.6 CONTROLLING VERTICALITY

10.6.1 Using a plumb-bob

In low-rise construction a heavy plumb-bob (5 to 10 kg) may be used as shown in *Figure 10.13*. If the external wall was perfectly vertical then, when the plumb-bob coincides with the centre of the peg, distance *d* at the top level would equal the offset distance of the peg at the base. This concept can be used internally as well as externally, provided that holes and openings are available.

10.6.2 Using a theodolite

If two centre-lines at right angles to each other are carried vertically up a structure as it is being built, accurate measurement can be taken off these lines and the structure as a whole will remain vertical. Where site conditions permit, the stations defining the 'base figure' (four per line) are placed in concrete well clear of construction (*Figure 10.14(a)*). Lines stretched between marks fixed from the pegs will allow offset measurements to locate the base of the structure. As the

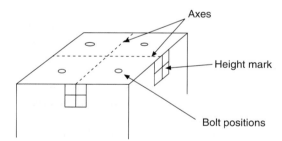

Fig. 10.12

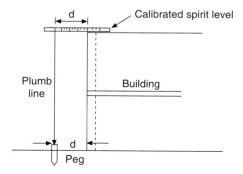

Fig. 10.13

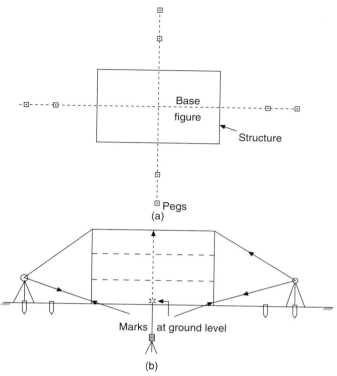

Fig. 10.14 *(a) Plan, and (b) section*

structure rises the marks can be transferred up onto the walls by theodolite, as shown in *Figure 10.14(b)*, and lines stretched between them. It is important that the transfer is carried out on both faces of the instrument.

Where the structure is circular in plan the centre may be established as in *Figure 10.14(a)* and the radius swung out from a pipe fixed vertically at the centre. As the structure rises, the central pipe is extended by adding more lengths. Its verticality is checked by two theodolites (as in *Figure 10.14(b)*) and its rigidity ensured by supports fixed to scaffolding.

The vertical pipe may be replaced by laser beam or autoplumb, but the laser would still need to be checked for verticality by theodolites.

Steel and concrete columns may also be checked for verticality using the theodolite. By string lining through the columns, positions *A – A* and *B – B* may be established for the theodolite (*Figure 10.15*); alternatively, appropriate offsets from the structural grid lines may be used. With instrument set up at *A*, the outside face of all the uprights should be visible. Now cut the outside edge of the upright at ground level with the vertical hair of the theodolite. Repeat at the top of the column. Now depress the telescope back to ground level and make a fine mark; the difference between the mark and the outside edge of the column is the amount by which the column is out of plumb. Repeat on the opposite face of the theodolite. The whole procedure is now carried out at *B*. If the difference exceeds the specified tolerances the column will need to be corrected.

10.6.3 Using optical plumbing

For high-rise building the instrument most commonly used is an autoplumb (*Figure 10.16*). This instrument provides a vertical line of sight to an accuracy of ±1 second of arc (1 mm in 200 m). Any

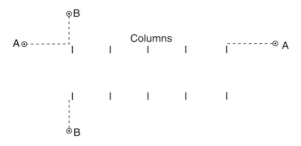

Fig. 10.15

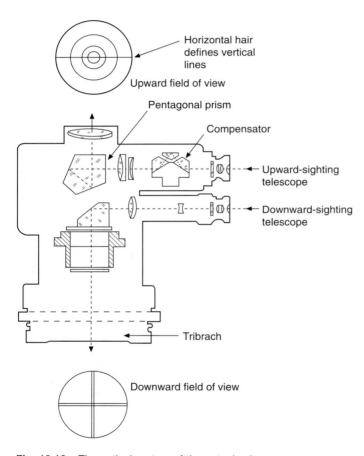

Fig. 10.16   *The optical system of the autoplumb*

deviation from the vertical can be quantified and corrected by rotating the instrument through 90° and observing in all four quadrants; the four marks obtained would give a square, the diagonals of which would intersect at the correct centre point.

A base figure is established at ground level from which fixing measurements may be taken. If this figure is carried vertically up the structure as work proceeds, then identical fixing measurements from the figure at all levels will ensure verticality of the structure (*Figure 10.17*).

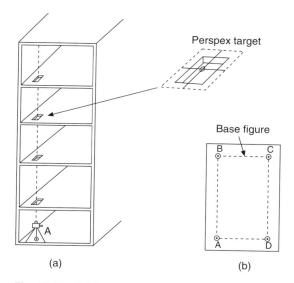

Perspex target

Base figure

B   C

A   D

(a)   (b)

Fig. 10.17 *(a) Elevation, and (b) plan*

To fix any point of the base figure on an upper floor, a Perspex target is set over the opening and the centre point fixed as above. Sometimes these targets have a grid etched on them to facilitate positioning of the marks.

The base figure can be projected as high as the eighth floor, at which stage the finishing trades enter and the openings are closed. In this case the uppermost figure is carefully referenced, the openings filled, and then the base figure re-established and projected upwards as before.

The shape of the base figure will depend upon the plan shape of the building. In the case of a long rectangular structure a simple base line may suffice but *T* shapes and *Y* shapes are also used.

10.7 CONTROLLING GRADING EXCAVATION

This type of setting out generally occurs in drainage schemes where the trench, bedding material and pipes have to be laid to a specified design gradient. Manholes (MH) will need to be set out at every change of direction or at least every 100 m on straight runs. The MH (or inspection chambers) are generally set out first and the drainage courses set out to connect into them.

The centre peg of the MH is established in the usual way and referenced to four pegs, as in *Figure 10.2*. Alternatively, profile boards may be set around the MH and its dimensions marked on them. If the boards are set out at a known height above formation level the depth of excavation can be controlled, as in *Figure 10.18*.

10.7.1 Use of sight rails

Sight rails (SRs) are basically horizontal rails set a specific distance apart and to a specific level such that a line of sight between them is at the required gradient. Thus they are used to control trench excavation and pipe gradient without the need for constant professional supervision.

Figure 10.19 illustrates SRs being used in conjunction with a boning rod (or traveller) to control

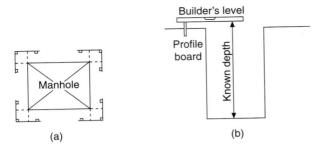

Fig. 10.18

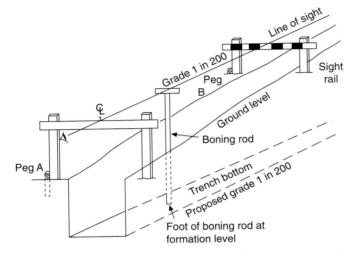

Fig. 10.19

trench excavation to a design gradient of 1 in 200 (rising). Pegs *A* and *B* are offset a known distance from the centre-line of the trench and levelled from a nearby TBM.

Assume that peg *A* has a level of 40 m and the formation level of the trench at this point is to be 38 m. It is decided that a reasonable height for the SR above ground would be 1.5 m, i.e. at a level of 41.5; thus the boning rod must be made (41.5 − 38) = 3.5 m long, as its cross-head must be on level with the SR when its toe is at formation level.

Consider now peg *B*, with a level of 40.8 m at a horizontal distance of 50 m from *A*. The proposed gradient is 1 in 200, which is 0.25 m in 50 m, and thus the formation level at *B* is 38.25 m. If the boning rod is 3.5 m, the SR level at *B* is (38.25 + 3.5) = 41.75 m and is set (41.75 − 40.8) = 0.95 m above peg *B*. The remaining SRs are established in this way and a line of sight or string stretched between them will establish the trench gradient 3.5 m above the required level. Thus, holding the boning rod vertically in the trench will indicate, relative to the sight rails, whether the trench is too high or too low.

Where machine excavation is used, the SRs are as in *Figure 10.20*, and offset to the side of the trench opposite to where the excavated soil is deposited.

Knowing the bedding thickness, the invert pipe level may be calculated and a second cross-head added to the boning rod to control the pipe laying, as shown in *Figure 10.21*.

Due to excessive ground slopes it may be necessary to use double sight rails with various lengths of boning rod as shown in *Figure 10.22*.

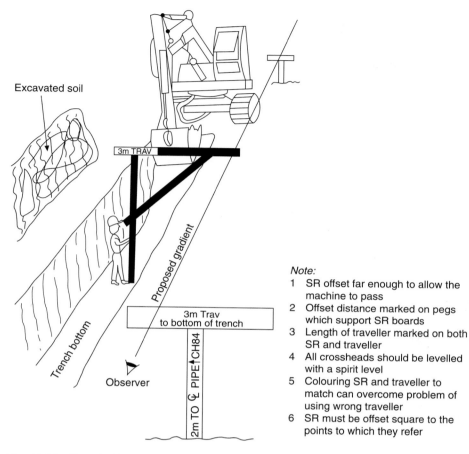

Note:
1 SR offset far enough to allow the machine to pass
2 Offset distance marked on pegs which support SR boards
3 Length of traveller marked on both SR and traveller
4 All crossheads should be levelled with a spirit level
5 Colouring SR and traveller to match can overcome problem of using wrong traveller
6 SR must be offset square to the points to which they refer

Fig. 10.20 *Use of offset sight rails (SR)*

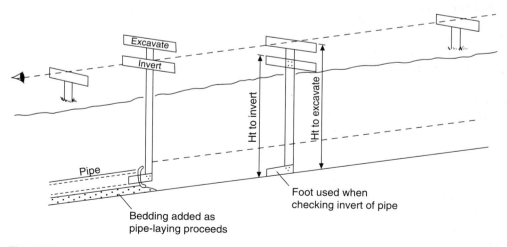

Fig. 10.21

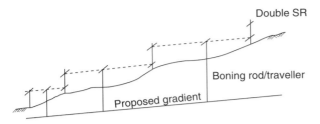

Fig. 10.22

10.7.2 Use of lasers

The word *laser* is an acronym for Light Amplification by Stimulated Emission of Radiation and is the name applied to an intense beam of highly monochromatic, coherent light. Because of its coherence the light can be concentrated into a narrow beam and will not scatter and become diffused like ordinary light.

In controlling trench excavation the laser beam simply replaces the line of sight or string in the SR situation. It can be set up on the centre-line of the trench, over a peg of known level, and its height above the peg measured to obtain the reduced level of the beam. The instrument is then set to the required grade and used in conjunction with an extendable traveller set to the same height as that of the laser above formation level. When the trench is at the correct level, the laser spot will be picked up on the centre of the traveller target, as shown in *Figure 10.23*. A levelling staff could just as easily replace the traveller, the laser spot being picked up on the appropriate staff reading.

Where machine excavation is used the beam can be picked up on a photo-electric cell fixed at the appropriate height on the machine. The information can be relayed to a console within the cabin, which informs the operator whether he is too high or too low (*Figure 10.24*).

At the pipe-laying stage, a target may be fixed in the pipe and the laser installed on the centre-line in the trench. The laser is then orientated in the correct direction (by bringing it on to a centre-line peg, as in *Figures 10.25(a, b)*) and depressed to the correct grade of the pipe. A graduated rod, or appropriately-marked ranging pole, can also be used to control formation and sub-grade level (*Figure 10.25a*). For large-diameter pipes the laser is mounted inside the pipe using horizontal compression bars (*Figure 10.25c*).

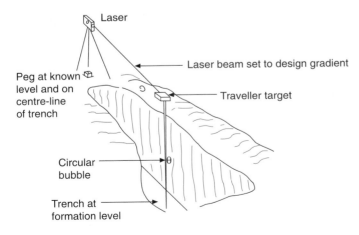

Fig. 10.23

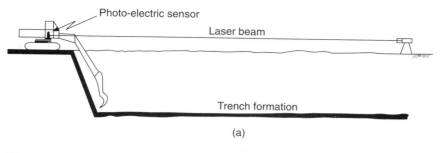

Photo-electric sensor

Laser beam

Trench formation

(a)

(b)

(c)

Fig. 10.24

Where the MH is constructed, the laser can be orientated from within using the system illustrated in *Figure 10.26*. The centre-line direction is transferred down to peg *B* from peg *A* and used to orientate the direction of the laser beam.

10.8 ROTATING LASERS

Rotating lasers (*Figure 10.27(a)*) are instruments which are capable of being rotated in both the horizontal and vertical planes, thereby generating reference planes or datum lines.

When the laser is established in the centre of a site over a peg of known level, and at a known height above the peg, a datum of known reduced level is permanently available throughout the site.

Using a vertical staff fitted with a photo-electric detector, levels at any point on the site may be instantly obtained and set out, both above and below ground, as illustrated in *Figure 10.27(b)*.

Since the laser reference plane covers the whole working area, photo-electric sensors fitted at an

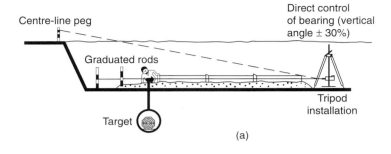

(a)

(b)   *4700 Laser beam aligner*   (c)   *Laser alignment in large-diameter pipes*

Fig. 10.25

appropriate height on earthmoving machinery enable whole areas to be excavated and graded to requirements by the machine operator alone.

Other uses of the rotating laser are illustrated in *Figure 10.28*.

From the above applications it can be seen that basically the laser supplies a reference line at a given height and gradient, and a reference plane similarly disposed. Realizing this, the user may be able to utilize these properties in a wide variety of setting-out situations such as off-shore channel dredging, tunnel guidance, shaft sinking, etc.

10.9 LASER HAZARDS

The potential hazard in the use of lasers is eye damage. There is nothing unique about this form of

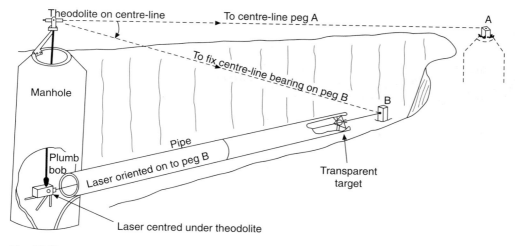

Fig. 10.26

radiation damage; it can also occur from other, non-coherent light emitted, for example, by the sun, arc lamps, projector lamps and other high-intensity sources. If one uses a magnifying lens to focus the sun's rays onto a piece of paper, the heat generated by such concentration will cause the paper to burst into flames. Similarly, a laser produces a concentrated, powerful beam of light which the eye's lens will further concentrate by focusing it on the retina, thus causing an almost microscopic burn or blister which can cause temporary or permanent blindness. When the beam is focused on the macula (critical area of the retina) serious damage can result. Since there is no pain or discomfort with a laser burn, the injury may occur several times before vision is impaired. A further complication in engineering surveying is that the beam may be acutely focused through the lens system of a theodolite or other instrument, or may be viewed off a reflecting or refracting surface. It is thus imperative that a safety code be adopted by all personnel involved with the use of lasers.

Under the Health and Safety at Work Act (1979) most sites will be required to adopt the recommendations of the British Standards Institution guide to the safe use of lasers (BS4803). The BSI classifies five types of laser, but only three of these are relevant to on-site working:

Class 2   A visible radiant power of 1 mW. Eye protection is afforded by the blink-reflex mechanism.
Class 3A   Has a maximum radiant power of 1 to 5 mW, with eye protection afforded by the blink-reflex action
Class 3B   Has a maximum power of 1 to 500 mW. Eye protection is not afforded by blink-reflex. Direct viewing, or viewing of specular reflections, is highly dangerous.

For surveying and setting-out purposes the BSI recommends the use of Classes 2 and 3A only. Class 3B may be used outdoors, if the more stringent safety precautions recommended are observed.

The most significant recommendation of the BSI document is that on sites where lasers are in use there should be a laser safety officer (LSO), who the document defines as 'one who is knowledgeable in the evaluation and control of laser hazards, and has the responsibility for supervision of the control of laser hazards'. Whilst such an individual is not specifically mentioned in conjunction with the Class 2 laser, the legal implications of eye damage might render it advisable to have an LSO present. Such an individual would not only require training in laser safety and law, but would need to be fully conversant with the RICS' Laser Safety Code produced by a working party of certain members of the Royal Institution of Chartered Surveyors.

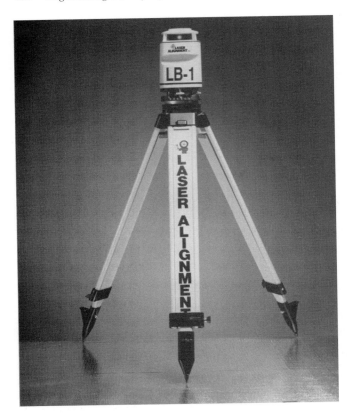

(a)   *LB-1 Rotating laser*

(b)   *1 – Rotating laser; 2 – Laying sub-grade to laser control; 3 – Checking formation level; 4 – Fixing wall levels; 5 – Taking ground levels; 6 – Staff with photoelectric detector fixing foundation levels; 7 – Laser plane of reference*

Fig. 10.27

The RICS code was produced in conjunction with BS4803 and deals specifically with the helium – neon gas laser (He – Ne) as used on site. Whilst the manufacturers of lasers will no doubt comply with the classification laid down, the modifications to a laser by mirrors or telescopes may completely alter such specifications and further increase the hazard potential. The RICS code presents methods and computations for assessing the possible hazards which the user can easily apply to his working

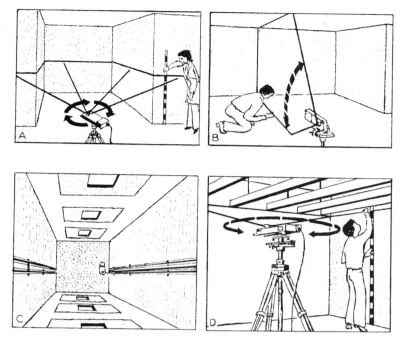

Fig. 10.28 *A – Height control; B – Setting out of dividing walls; C – Use of the vertical beam for control of elevator guide rails, and slip-forming structures; D – Setting out and control of suspended ceiling*

laser system, in both its unmodified and modified states. Recommendations are also made about safety procedures relevant to a particular system from both the legal and technical aspects. The information within the RICS code enables the user to compute such important parameters as:

(1) The safe viewing time at given distances.
(2) The minimum safe distance at which the laser source may be viewed directly, for a given period of time.

Such information is vital to the organization and administration of a 'laser site' from both the health and legal aspects, and should be combined with the following precautions.

(1) Ensure that all personnel, random visitors to the site, and where necessary members of the public, are aware of the presence of lasers and the potential eye damage involved.
(2) Using the above-mentioned computations, erect safety barriers around the laser with a radius greater than the minimum safe viewing distance.
(3) Issue laser safety goggles where appropriate.
(4) Avoid, wherever possible, the need to view the laser through theodolites, levels or binoculars.
(5) Where possible, position the laser either well above or well below head height.

10.10 ROUTE LOCATION

Figure 10.29 shows a stretch of route location for a road or railway. In order to control the construction involved, the pegs and profile boards shown must be set out at intervals of 10 to 30 m along the whole stretch of construction.

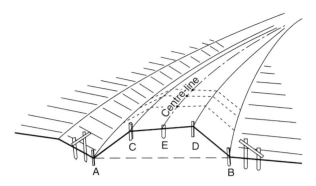

Fig. 10.29

The first pegs located would be those defining the centre-line of the route (peg E), and the methods of locating these on curves have been dealt with in Chapter 8. The straights would be aligned between adjacent tangent points.

The shoulder pegs C and D, defining the road /railway width, can be set out by appropriate offsets at right angles to the centre-line chords.

Pegs A and B, which define the toe of the embankment (fill) or top edge of the cutting, are called *slope stakes*. The side widths from the centre-line are frequently calculated and shown on the design drawings or computer print-outs of setting-out data. This information should be used only as a rough check or guide; the actual location of the slope stake pegs should always be carried out in the field, due to the probable change in ground levels since the information was initially compiled. These pegs are established along with the centre-line pegs and are necessary to define the area of top-soil strip.

10.10.1 Setting-out slopes stakes

In *Figure* 10.30(*a*) and (*b*), points A and B denote the positions of pegs known as *slope stakes* which define the points of intersection of the actual ground and the proposed side slopes of an embankment or cutting. The method of establishing the positions of the stakes is as follows:

(1) Set up the level in a convenient position which will facilitate the setting out of the maximum number of points therefrom.
(2) Obtain the height of the plane of collimation (HPC) of the instrument by backsighting on to the nearest TBM.
(3) Foresight onto the staff held where it is thought point A may be and obtain the ground level there.
(4) Subtract 'ground level' from 'formation level' and multiply the difference by N to give horizontal distance x.
(5) Now tape the horizontal distance $(x + b)$ from the centre-line to the staff. If the *measured distance* to the staff equals the *calculated distance* $(x + b)$, then the staff position is the slope stake position. If not, the operation is repeated with the staff in a different position *until the measured distance agrees with the calculated distance.*

The above 'trial-and-error' approach should always be used on site to avoid errors of scaling the positions from a plan, or accepting, without checking, a computer print-out of the dimensions.

For example, if the side slopes of the proposed embankment are to be 1 vertical to 2 horizontal, the formation level 100 m OD and the ground level at A, say, 90.5 m OD, then $x = 2(100 - 90.5)$

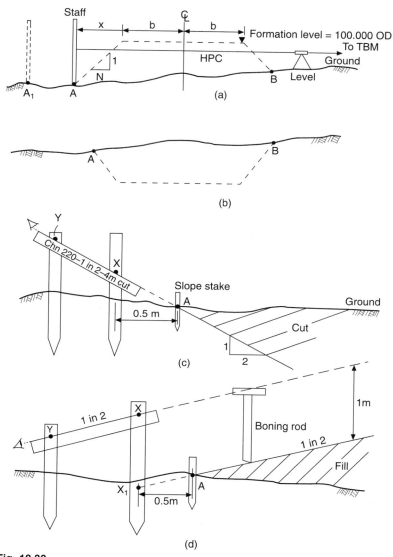

Fig. 10.30

= 19 m, and if the formation width = 20 m, then b = 10 m and $(x + b)$ = 29 m. Had the staff been held at A_1 (which had exactly the same ground level as A) then obviously the calculated distance $(x + b)$ would not agree with the measured distance from centre-line to A_1. They would agree only when the staff arrived at the slope stake position A, as x is dependent upon the level at the toe of the embankment, or top of the cutting.

10.10.2 *Controlling earthworks*

Batter boards, or *slope rails* as they are sometimes called, are used to control the construction of the side slopes of a cutting or embankment (see *Figure 10.30(c)* and (*d*), and *Figure 10.29*).

Consider *Figure 10.30(c)*. If the stake adjacent to the slope stake is set 0.5 m away, then, for a grade of 1 vertical to 2 horizontal, the level of point X will be 0.25 m higher than the ground level at *A*. From *X*, a batter board is fixed at a grade of 1 in 2, using a 1 in 2 template and a spirit level. Stakes *X* and *Y* are usually no more than 1 m apart. Information such as chainage, slope and depth of cut are marked on the batter board.

In the case of an embankment, (*Figure 10.30(d)*) a boning rod is used in the control of the slope. Assuming that a boning rod 1 m high is to be used, then as the near stake is, say, 0.5 m from the slope stake, point x_1 will be 0.25 m lower than the ground level at A, and hence point *x* will be 0.75 m above the ground level of A. The batter board is then fixed from *x* in a similar manner to that already described

The formation and sub-base, which usually have setting-out tolerances in the region of ±25 mm, can be located with sufficient accuracy using profiles and travellers. *Figure 10.31* shows the use of triple profiles for controlling camber, whilst different lengths of traveller will control the thickness required.

Laying of the base course (60 mm) and wearing course (40 mm) calls for much smaller tolerances, and profiles are not sufficiently accurate; the following approach may be used:

Pins or pegs are established at right angles to the centre-line at about 0.5 m beyond the kerb face (*Figure 10.32*). The pins or pegs are accurately levelled from the nearest TBM and a coloured tape placed around them at 100 mm above finished road level; this will be at the same level as the top of the kerb. A cord stretched between the pins will give kerb level, and with a tape the distances to the top of the sub-base, top of the base course and top of the wearing course can be accurately fixed (or dipped).

The distance to the kerb face can also be carefully measured in from the pin in order to establish the kerb line. This line is sometimes defined with further pins and the level of the kerb top marked on.

10.11 UNDERGROUND SURVEYING

The essential problem in underground surveying is that of orientating the underground surveys to the surface surveys, the procedure involved being termed a *correlation*.

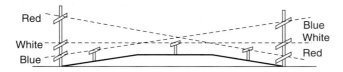

Fig. 10.31

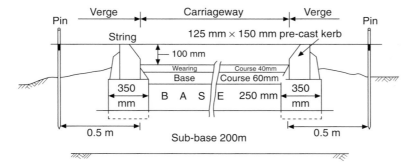

Fig. 10.32

In an underground transport system, for instance, the tunnels are driven to connect inclined or vertical shafts (points of surface entry to the transport system) whose relative locations are established by surface surveys. Thus the underground control networks must be connected and orientated into the same coordinate system as the surface networks. To do this, one must obtain the coordinates of at least one underground control station and the bearing of at least one line of the underground network, relative to the surface network.

Another prime example of orientation is that of the National Coal Board (NCB) of the UK, all of whose underground surveys are required, by law, to be orientated and connected to the Ordnance Survey National Grid (NG) system.

If entry to the underground tunnel system is via an inclined shaft, then the surface survey may simply be extended and continued down that shaft and into the tunnel, usually by the method of traversing. Extra care would be required in the measurement of the horizontal angles due to the steeply inclined sights involved (see Chapter 4) and in the temperature corrections to the taped distances due to the thermal gradients encountered.

If entry is via a vertical shaft, then optical, mechanical or gyroscopic methods of orientation are used.

10.11.1 Optical methods

Where the shaft is shallow and of relatively large diameter the bearing of a surface line may be transferred to the shaft bottom by theodolite (*Figure 10.33*).

The surface station A and B are part of the control system and represent the direction in which the tunnel must proceed. They would usually be established clear of ground movement caused by shaft sinking or other factors. Auxiliary stations c and d are very carefully aligned with A and B using the theodolite on both faces and with due regard to all error sources. The relative bearing of A and B is then transferred to $A'B'$ at the shaft bottom by direct observations. Once again these observations must be carried out on both faces with the extra precautions advocated for steep sights.

If the coordinates of d are known then the coordinates of B' could be fixed by measuring the vertical angle and distance (EDM) to a reflector at B'.

It is important to understand that the accurate orientation bearing of the tunnel is infinitely more critical than the coordinate position. For instance, a standard error of $\pm 1'$ in transferring the bearing down the shaft to $A'B'$ would result in a positional error at the end of 1 km of tunnel drivage of ± 300 mm and would increase to ± 600 mm after 2 km of drivage.

10.11.2 Mechanical methods

Although these methods, which involve the use of wires hanging vertically in a shaft, are rapidly being superseded by gyroscopic methods, they are still widely used and are described herewith.

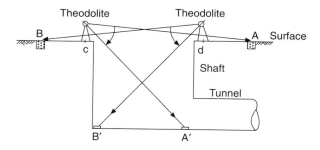

Fig. 10.33

The basic concept is that wires hanging freely in a shaft will occupy the same position underground that they do at the surface, and hence the bearing of the wire plane will remain constant throughout the shaft.

10.11.2.1 Weisbach triangle method

This appears to be the most popular method in civil engineering. Two wires, W_1 and W_2, are suspended vertically in a shaft forming a very small base line (*Figure 10.34*). The principle is to obtain the bearing and coordinates of the wire base relative to the surface base. These values can then be transferred to the underground base.

In order to establish the bearing of the wire base at the surface, it is necessary to compute the angle $W_s W_2 W_1$ in the triangle as follows:

$$\sin \hat{W}_2 = \frac{w_2}{w_s} \sin \hat{W}_s \tag{10.1}$$

As the Weisbach triangle is formed by approximately aligning the Weisbach station W_s with the wires, the angles at W_s and W_2 are very small and equation (10.1) may be written:

$$\hat{W}_2'' = \frac{w_2}{w_s} \hat{W}_s'' \tag{10.2}$$

(The expression is accurate to seven figures when $\hat{W}_s < 18'$ and to six figures when $\hat{W}_s < 45'$.)

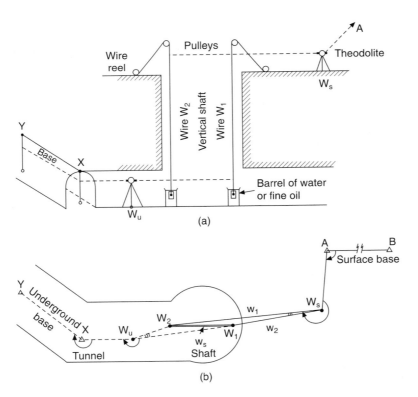

Fig. 10.34 (a) *Section, and* (b) *plan*

From equation (10.2), it can be seen that the observational error in angle W_s will be multiplied by the fraction w_2/w_s.

Its effect will therefore be reduced if w_2/w_s is less than unity. Thus the theodolite at W_s should be as near the front wire W_1 as focusing will permit and preferably at a distance smaller than the wire base W_1W_2.

The following example, using simplified data, will now be worked to illustrate the procedure. With reference to *Figure 10.34(b)*, the following field data is obtained:

(1) Surface observations

Angle BAW_s	$= 90°00'00''$	Distance W_1W_2	$= w_s = 10.000$ m
Angle AW_sW_2	$= 260°00'00''$	Distance W_1W_s	$= w_2 = 5.000$ m
Angle $W_1W_sW_2$	$= 0°01'20''$	Distance W_2W_s	$= w_1 = 15.000$ m

(2) Underground observations

Angle $W_2W_uW_1$	$= 0°01'50''$	Distance W_2W_u	$= y = 4.000$ m
Angle W_1W_uX	$= 200°00'00''$	Distance W_uW_1	$= x = 14.000$ m
W_uXY	$= 240°00'00''$		

Solution of the surface Weisbach triangle:

$$\text{Angle } W_sW_2W_1 = \frac{5}{10} \times 80'' = 40''$$

Similarly, underground

$$\text{Angle } W_2W_1W_u = \frac{4}{10} \times 110'' = 44''$$

The bearing of the underground base XY, relative to the surface base AB is now computed in a manner similar to a traverse:

Assuming
then,

WCB of $AB =$	$89°00'00''$
WCB of $AW_s =$	$179°00'00''$ (using angle BAW_s)
Angle $AW_sW_2 =$	$260°00'00''$

sum $= 439°00'00''$
$-180°$

WCB of $W_sW_2 = 259°00'00''$
Angle $W_sW_2W_1 = +0°00'40''$ } see *Figure 10.35(a)*

WCB $W_1W_2 = 259°00'40''$
Angle $W_2W_1W_u = -0°00'44''$ } see *Figure 10.35(b)*

WCB $W_1W_u = 258°59'56''$
Angle $W_1W_uX = 200°00'00''$

sum $= 458°59'56''$
$-180°$

WCB $W_uX = 278°59'56''$
Angle $W_uXY = 240°00'00''$

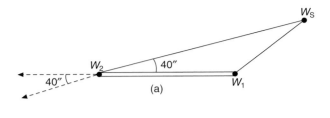

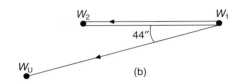

Fig. 10.35

$$\text{sum} = \quad 518°59'56''$$
$$-180°$$

$$\text{WCB } XY = 338°59'56'' \qquad \text{(underground base)}$$

The transfer of bearing is of prime importance; the coordinates can be obtained in the usual way by incorporating all the measured lengths AB, AW_s, W_uX, XY.

10.11.2.2 Shape of the Weisbach triangle

As already indicated, the angles W_2 and W_s in the triangle are as small as possible. The reason for this can be illustrated by considering the effect of accidental observation errors on the computed angle W_2.

From the basic equation: $\sin \hat{W}_2 = \dfrac{w_2}{w_s} \sin \hat{W}_s$

Differentiate with respect to each of the measured quantities in turn:

(1) with respect to W_s

$$\cos W_2 \delta W_2 = \frac{w_2}{w_s} \cos W_s \delta W_s$$

$$\therefore \ \delta W_2 = \frac{w_2 \cos W_s}{w_s \cos W_2} \ \delta W_s$$

(2) with respect to w_2

$$\cos W_2 \delta W_2 = \frac{\sin W_s}{w_s} \ \delta w_2$$

$$\therefore \ \delta W_2 = \frac{\sin W_s}{w_s \cos W_2} \ \delta w_2$$

(3) with respect to w_s

$$\cos W_2 \delta W_2 = \frac{-w_2 \sin W_s}{w_s^2} \delta w_s$$

$$\therefore \delta W_2 = \frac{-w_2 \sin W_s}{w_s^2 \cos W_2} \delta w_s$$

then:

$$\delta W_2 = \pm \left[\frac{w_2^2 \cos^2 W_s}{w_s^2 \cos^2 W_2} \delta W_s^2 + \frac{\sin^2 W_s}{w_s^2 \cos^2 W_2} \delta w_2^2 + \frac{w_2^2 \sin^2 W_s}{w_s^4 \cos^2 W_2} \delta w_s^2 \right]^{\frac{1}{2}}$$

$$= \pm \frac{w_s}{w_s \cos W_2} \left[\cos^2 W_s \delta W_s^2 + \sin^2 W_s \frac{\delta w_2^2}{w_2^2} + \sin^2 W_s \frac{\delta w_s^2}{w_s^2} \right]^{\frac{1}{2}}$$

but $\cos W_s = \dfrac{\sin W_s \cos W_s}{\sin W_s} = \sin W_s \cot W_s$, which on substitution gives:

$$\delta W_2 = \pm \frac{w_2}{w_s \cos W_2} \left[\sin^2 W_s \cot^2 W_s \delta W_s^2 + \sin^2 W_s \frac{\delta w_2^2}{w_2^2} + \sin^2 W_s \frac{\delta w_s^2}{w_s^2} \right]^{\frac{1}{2}}$$

$$= \pm \frac{w_2 \sin W_s}{w_s \cos W_2} \left[\cot^2 W_s \delta W_s^2 + \frac{\delta w_2^2}{w_2^2} + \frac{\delta w_s^2}{w_s^2} \right]^{\frac{1}{2}}$$

by sine rule $\dfrac{w_2 \sin W_s}{w_s} = \sin W_2$, therefore substituting

$$\delta W_2 = \pm \tan W_2 \left[\cot^2 W_s \delta W_s^2 + \left(\frac{\delta w_2}{w_2} \right)^2 + \left(\frac{\delta w_s}{w_s} \right)^2 \right]^{\frac{1}{2}} \tag{10.3}$$

Thus to reduce the standard error (δW_2) to a minimum:

(1) $\tan W_2$ must be a minimum; therefore the angle W_2 should approach $0°$.
(2) As W_2 is very small, W_s will be very small and so $\cot W_s$ will be very large. Its effect will be greatly reduced if δW_s is very small; the angle W_s must therefore be measured with maximum precision.

10.11.2.3 Sources of error

The standard error of the transferred bearing e_B, is the combined effect of:

(1) Errors in connecting the surface base to the wire base, e_s.
(2) Errors in connecting the wire base to the underground base, e_u.
(3) Errors in the determination of the verticality of the wire plane, e_p giving:

$$e_B = \pm (e_s^2 + e_u^2 + e_p^2)^{\frac{1}{2}} \tag{10.4}$$

The errors e_s and e_u can be obtained in the usual way from an examination of the method and type of instruments used. The source of error e_p is vitally important in view of the extremely short length of the wire base.

Given random errors e_1 and e_2, of deflected wires W_1 and W_2 equal to 1 mm, then e_p = 100 sec for a wire base of 2 m. The NCB specifies a value of 2′ 00″ for e_B, then from equation (10.4) assuming $e_s = e_u = e_p$:

$$e_p = \frac{2′\ 00″}{(3)^{\frac{1}{2}}} = 70″,$$

which for the same wire base of 2 m permits a deflection of the wires of only 0.7 mm. These figures serve to indicate the great precision and care needed in plumbing a shaft, in order to minimize orientation errors.

10.11.2.4 Verticality of the wire plane

The factors affecting the verticality of the wires are:

(1)   *Ventilation air currents in the shaft*
All forced ventilation should be shut off and the plumb-bob protected from natural ventilation.
(2)   *Pendulous motion of the shaft plumb*
The motion of the plumb-bob about its suspension point can be reduced by immersing it in a barrel of water or fine oil. When the shaft is deep, complete elimination of motion is impossible and clamping of the wires in their mean swing position may be necessary.
The amplitude of wire vibrations, which induce additional motion to the swing, may be reduced by using a heavy plumb-bob, with its point of suspension close to the centre of its mass, and fitted with large fins.
(3)   *Spiral deformation of the wire*
Storage of the plumb wire on small-diameter reels gives a spiral deformation to the wire. Its effect is reduced by using a plumb-bob of maximum weight. This should be calculated for the particular wire using a reasonable safety factor.

These sources of error are applicable to all wire surveys.

10.11.2.5 Co-planing

The principles of this alternative method are shown in *Figure 10.36*. The triangle of the previous method is eliminated by aligning the theodolite at W_s exactly with the wires W_1 and W_2. This alignment is easily achieved by trial and error, focusing first on the front wire and then on the back. Both wires can still be seen through the telescope even when in line. The instrument should be set up within 3 to 4 m of the nearer wire. Special equipment is available to prevent lateral movement of the theodolite affecting its level, but if this is not used, special care should be taken to ensure that the tripod head is level.

The movement of the focusing lens in this method is quite long. Thus for alignment to be exact, the optical axis of the object lens should coincide exactly with that of the focusing lens in all focusing positions. If any large deviation exists, the instrument should be returned to the manufacturer. The chief feature of this method is its simplicity with little chance of gross errors.

10.11.2.6 Weiss quadrilateral

This method may be adopted when it is impossible to set up the theodolite, even approximately, on the line of the wire base W_1W_2 (*Figure 10.37*). Theodolites are set up at C and D forming a

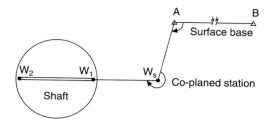

Fig. 10.36

quadrilateral CDW_1W_2. The bearing and coordinates of CD are obtained relative to the surface base, the orientation of the wire base being obtained through the quadrilateral. Angles 1, 2, 3 and 8 are measured direct, and angles 4 and 7 are obtained as follows:

$$\text{Angle } 4 = [180° - (\hat{1} + \hat{2} + \hat{3})]$$

$$\text{Angle } 7 = [180° - (\hat{1} + \hat{2} + \hat{8})]$$

The remaining angles 6 and 5 are then computed from

$$\sin \hat{1} \sin \hat{3} \sin \hat{5} \sin \hat{7} = \sin \hat{2} \sin \hat{4} \sin \hat{6} \sin \hat{8}$$

thus
$$\frac{\sin \hat{5}}{\sin \hat{6}} = \frac{\sin \hat{2} \sin \hat{4} \sin \hat{8}}{\sin \hat{1} \sin \hat{3} \sin \hat{7}} = x \qquad (a)$$

and
$$(\hat{5} + \hat{6}) = (\hat{1} + \hat{2}) = \hat{y}$$

$$\therefore \hat{5} = (\hat{y} - \hat{6}) \qquad (b)$$

from (a) $\sin \hat{5} = x \sin \hat{6}$ $\quad \therefore \sin (\hat{y} - \hat{6}) = x \sin \hat{6}$

and $\sin \hat{y} \cos \hat{6} - \cos \hat{y} \sin \hat{6} = x \sin \hat{6}$

from which $\sin y \cot \hat{6} - \cos \hat{y} = x$

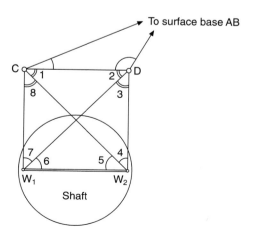

To surface base AB

Fig. 10.37

and
$$\cot \hat{6} = \frac{x + \cos \hat{y}}{\sin \hat{y}}$$
(10.5)

Having found angle 6 from equation (10.5), angle 5 is found by substitution in (*b*).
Error analysis of the observed figure indicates

(1) The best shape for the quadrilateral is square.
(2) Increasing the ratio of the length *CD* to the wire base increases the standard error of orientation.

10.11.2.7 Single wires in two shafts

The above methods have dealt with orientation through a single shaft, which is the general case in civil engineering. Where two shafts are available, orientation can be achieved via a single wire in each shaft. This method gives a longer wire base, and wire deflection errors are much less critical.
   The principles of the method are outlined in *Figure 10.38*. Single wires are suspended in each shaft at *A* and *B* and coordinated into the surface control network, most probably by multiple intersection from as many surface stations as possible. From the coordinates of *A* and *B*, the bearing *AB* is obtained.
   A traverse is now carried out from *A* to *B* via an underground connecting tunnel (*Figure 10.38 (a)*). However, as the angles at *A* and *B* cannot be measured it becomes an open traverse on an *assumed* bearing for *AX*. Thus, if the assumed bearing for *AX* differed from the 'true' (but unknown) bearing by α, then the whole traverse would swing to apparent positions *X'*, *Y'*, *Z'* and *B'* (*Figure 10.38 (b)*).
   The value of α is the difference of the bearings *AB* and *AB'* computed from surface and underground coordinates respectively. Thus if the underground bearings are rotated by the amount α this will swing the traverse almost back to *B*. There will still be a small misclosure due to linear error and this can be corrected by multiplying each length by a scale factor equal to length *AB*/length *AB'*. Now, using the corrected bearings and lengths the corrected coordinates of the traverse fitted to *AB* can be calculated. These coordinates will be relative to the surface coordinate system.
   Alternatively, the corrected coordinates can be obtained directly by mathematical rotation and

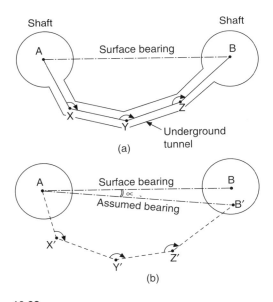

(a)

(b)

Fig. 10.38

translation of the 'assumed' values. Therefore, with A as origin, a rotation equal to α and a translation of AB/AB', the corrected coordinates are obtained from

$$E_i = E_0 + K(E_i' \cos \alpha - N_i' \sin \alpha) \qquad (10.6a)$$

$$N_i = N_0 + K(N_i' \cos \alpha + E_i' \sin \alpha) \qquad (10.6b)$$

where E_0, N_0 = coordinates of the origin (in this case A)
$\quad\quad E_i', N_i'$ = coordinates of the traverse points computed on the assumed bearing
$\quad\quad E_i, N_i$ = transformed coordinates of the underground traverse points
$\quad\quad K$ = scale factor (length AB/length AB')

There is no doubt that this is the most accurate and reliable method of surface-to-underground orientation. The accuracy of the method is dependent upon.

(1) The accuracy of fixing the position of the wires at the surface.
(2) The accuracy of the underground connecting traverse.

The influence of errors in the verticality of the wires, so critical in single-shaft work, is practically negligible owing to the long distance separating the two shafts. Provided that the legs of the underground traverse are long enough, then modern single-second theodolites integrated with EDM equipment would achieve highly accurate surface and underground surveys, resulting in final orientation accuracies of a few seconds. As the whole procedure is under strict control there is no reason why the final accuracy cannot be closely predicted.

10.11.2.8 Alternatives

In all the above methods the wires could be replaced by autoplumbs or lasers.
In the case of the autoplumb, stations at the shaft bottom could be projected vertically up to specially arranged targets at the surface and appropriate observations taken direct onto these points.
Similarly, lasers could be arranged at the surface to project the beam vertically down the shaft to be picked up on optical or electronic targets. The laser spots then become the shaft stations correlated in the normal way.
The major problems encountered by using the above alternatives are:

(1) Ensuring the laser beam is vertical.
(2) Ensuring correct detection of the centre of the beam.
(3) Refraction in the shaft (applies also to autoplumb).

In the first instance a highly sensitive spirit level or automatic compensator could be used; excellent results have been achieved using arrangements involving mercury pools, or photo-electric sensors.
Detecting the centre of the laser is more difficult. Lasers having a divergence of 10″ to 20″ would give a spot of 10 mm and 20 mm diameter respectively at 200 m. This spot also tends to move about due to variations in air density. It may therefore require an arrangement of photocell detectors to solve the problem.
Due to the turbulence of air currents in shafts the problem of refraction has not proved too dangerous.

10.12 GYRO-THEODOLITE

An alternative to the use of wire methods is the gyro-theodolite. This is a north-seeking gyroscope integrated with a theodolite, and can be used to orientate underground base lines relative to true north.

The main type is the suspended gyroscope used by the Wild GAK.1. The essential elements of the suspended gyro-theodolite are shown in *Figure 10.39*.

10.12.1 Theory

The gyroscope is basically a rapidly spinning flywheel with the spin axis horizontal. The gyro spins from west to east, as does the Earth, the horizontal component of the Earth's rotation causing the spin axis to oscillate about the true north position.

Before commencing an explanation of the theory of the north-seeking gyroscope, a revision of Newton's Laws of Motion may prove useful. If the force F increases the velocity of a mass m from V_1 to V_2 in time t then

$$F \propto (mV_2 - mV_1)/t$$

As $(V_2 - V_1)/t =$ acceleration $= a$, then $F \propto ma$.

The constant of proportionality C in the equation $F = Cma$ can be made unity by suitably defining the units of F. Using SI units, C does in fact become unity.

$$\therefore F = (mV_2 - mV_1)/t = \text{rate of change of linear momentum.}$$

Similarly, $T = (I\Omega_2 - I\Omega_1)/t =$ rate of change of angular momentum
thus $T \propto F$ ($I =$ moment of inertia, $\Omega =$ angular velocity of spin)

The theory may be itemized as follow:

(1) *Figure 10.40(a)* indicates a spinning flywheel in which the angular velocity of spin $= \Omega$.

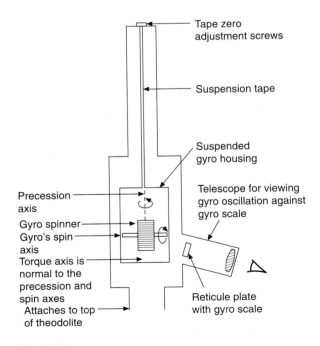

- Tape zero adjustment screws
- Suspension tape
- Suspended gyro housing
- Telescope for viewing gyro oscillation against gyro scale
- Precession axis
- Gyro spinner
- Gyro's spin axis
- Torque axis is normal to the precession and spin axes
- Attaches to top of theodolite
- Reticule plate with gyro scale

Fig. 10.39

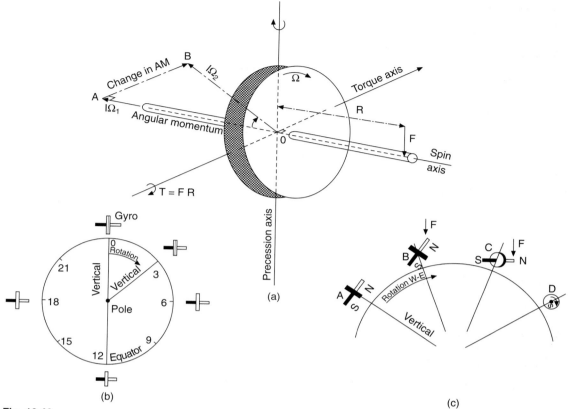

Fig. 10.40

(2) The angular velocity of spin results in an angular momentum vector (AMV) *OA* (similar to that on a RH screw as it enters).

(3) Consider now the AMV changing position to *OB* in the horizontal plane *AOB* during time *t*.

(4) This results in a change in angular momentum of $AB = (I\Omega_2 - I\Omega_1)$, which for a small displacement may be regarded as a vector change at 90° to *OA*.

(5) For the angular momentum vector to change position from *A* to *B*, an additional vector quantity must be superimposed on the system. Such a quantity is the reactive torque effect *T* along the axis parallel to *AB* (torque axis). Thus $T = (I\Omega_2 - I\Omega_1)/t$.

(6) A force *F* acting vertically down on the spin axis will produce the reactive torque effect *T*, i.e. $T = FR$.

Thus to summarize:

The effect of a force *F* acting vertically down on the spin axis of a spinning gyro is to cause the spin axis to *precess* in a horizontal plane about the vertical axis of precession.

Precession will continue, and resistance to the couple likewise, until the plane of the gyro rotor coincides with the plane of the applied couple. Precession then ceases and with it all resistance to the applied couple.

The effect of the Earth's rotation on the gyro fulfils the stated summary as follows:

Consider *Figure 10.40(b)* with the gyro at 0 h having its spin axis E – W. Due to the phenomenon of gyroscopic inertia it will maintain its plane of rotation in space while the Earth's horizon turns in space. Thus, although maintaining its plane, it appears to turn with respect to the Earth at the rate of one revolution per 24 hours.

At A (*Figure 10.40(c)*) consider a weight in the form of a pendulum attached to the gyro axis; it will point to the centre of the Earth, and render the spin axis horizontal. Assume the axis is E – W.

At B, the Earth's rotation causes the axis to show an apparent tilt as described in *Figure 10.40(b)*. The pendulum weight is now no longer evenly supported so that the effect of gravity is felt mainly by the upper end of the axis. The effect of this downward force F is to cause precession as shown at C.

At D, precession has swung the spin axis into the N – S meridian. In this position the direction of the rotor spin is the same as the Earth's and so has no effect on the pendulum weight. Thus, theoretically, the spin axis will point N – S and all movement ceases. In practice, however, the inertia of the system causes the spin axis to over-shoot the N – S meridian, which results in oscillations of the spin axis about the meridian.

From the theory of the spinning wheel, provided that the total angular rotation is small, the motion of the horizontal spin axis about the vertical may be represented by:

$$K_1\ddot{\theta} + K_2\dot{\theta} + K_3\theta = 0 \tag{10.7}$$

where θ = angle between spin axis and true north, and K_1, K_2, K_3 are constants.

Solution of the equation is a damped simple harmonic motion in which θ converges exponentially to zero, with a period of several minutes and an amplitude of about $\pm 2.5°$.

Location of true north, therefore, involves fixing the axis of symmetry of the gyro oscillation as it precesses about true north. However, as it would be time-wasting and uneconomical to allow the spin axis to steady in the direction of true north, various methods have been devised to compute the necessary direction from observations of the damped harmonic oscillation.

10.12.2 Observational techniques

Basically, all the techniques used measure the amplitude or the time of the oscillation about its axis of symmetry.

(1) *Reversal-point method*: In the first instance the instrument is orientated so that the spin axis of the gyroscope is within a few degrees of north. The power is switched on and the spinner brought to full speed, at which point it is gently uncaged; it will then oscillate to and fro about its vertical axis. The oscillation is then tracked, using the slow-motion screws of the theodolite, by keeping the image of the spin axis in the centre of the 'gyro scale' (*Figure 10.41(b)*). The magnitude of the oscillation, which is a damped simple harmonic motion about true north, is then measured on the horizontal scale of the theodolite.

Horizontal circle readings are taken each time the gyro reaches its maximum oscillation east or west of the meridian; these positions (r) are called *reversal points*. The minimum number of readings required is three, the mean of which gives the direction of true north (N).

Various methods exist for finding the mean, the most popular being Schuler's Mean, which is explained by reference to *Figure 10.41(a)*.

$$N_1 = \frac{1}{2}\left(\frac{r_1 + r_3}{2} + r_2\right) = \frac{1}{4}\left(r_1 + 2r_2 + r_3\right) \tag{10.8}$$

An additional observation r_4, enables a second mean to be computed, thus:

$$N_2 = \frac{1}{2}\left(\frac{r_2 + r_4}{2} + r_3\right) = \frac{1}{4}\left(r_2 + 2r_3 + r_4\right) \tag{10.9}$$

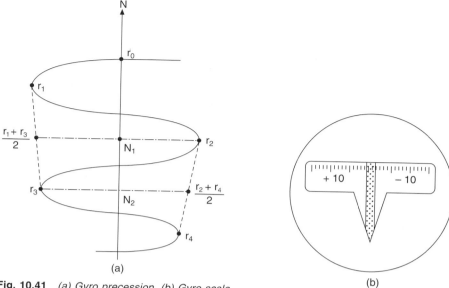

Fig. 10.41 *(a) Gyro precession, (b) Gyro-scale*

The direction of true north (N) is then more accurately obtained from $N = \frac{1}{2}(N_1 + N_2)$.

Dr T.L. Thomas (*Proc., 3rd South African Nat. Survey Conf.*, January 1967) advocates the symmetrical four-point method, using two observations to the east, and two to the west. This formula is simply the mean of ($N_1 + N_2$).

i.e $N = (r_1 + 3r_2 + 3r_3 + r_4)/8$ (10.10)

In this method, as the theodolite moves with the spinner, the suspension tape should not twist and untwist with reference to it, and hence there should be no tape zero error. However, tests have proved that a small tape zero error due to torque in the suspension system does require a correction (see *Section 10.12.3(2)*). Thus, if

N = the horizontal circle reading of gyro north
Z = the tape zero correction, and
N_0 = the horizontal circle reading of true north.

then $N_0 = (N - Z)$ (10.11)

Further corrections also are required for the instrument constant (see *Section 10.12.3(1)*).

(2) *Transit method*: This method, as originally devised for the Wild GAK.1 gyro-theodolite, assumed that the damping effect was zero and that within ±15′ of north (N') the oscillation curve was linear and hence proportional to time.

In this method the theodolite remains clamped and the magnitude of the oscillation is measured against the gyro scale, as shown in *Figure 10.42*. As the moving mark, which depicts the oscillation, passes through the central graduation of the gyro scale, its time is noted by means of a stop watch (with lap-timing facilities).

From *Figure 10.42*, as the moving mark passes the zero of the gyro scale, the time is zero (t_1) and the stop watch is started. When it reaches its western elongation its scale reading is noted (A_w). As it returns to the zero mark, the time (t_2) is noted. Thus $t_2 - t_1 = T_w$, is the time taken for the gyro

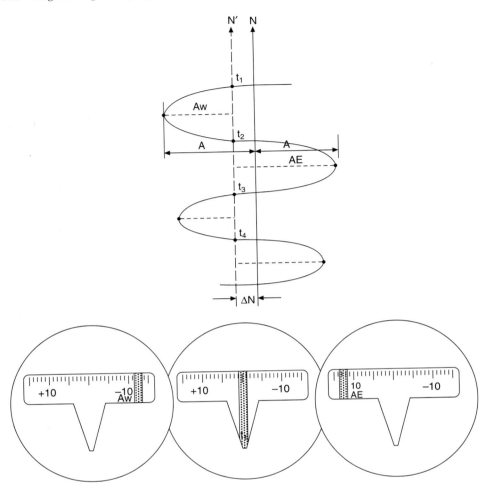

Fig. 10.42

to reach its western reversal point and return and is called the *half oscillation time of the western elongation*. The eastern elongation reading (A_E) is noted and the return transit time t_3, from which $t_3 - t_2 = T_E$.

The correction ΔN, to transform the approximate north reading N' to true north N, is given by

$$\Delta N = CA\Delta T \qquad (10.12)$$

where A = the amplitude = $(A_w + A_E)/2$
C = the proportionality constant, which changes with change in latitude
ΔT = the difference in 'swing' time = algebraic sum of times T_E and T_w (by convention T_E is positive and T_w negative)

It is obvious from the diagrams that if $A_E > A_W$, then N' is west of N and the correction ΔN is positive, and *vice versa*.

The value of C need be obtained only once for each instrument, as follows. With alidade oriented first east and then west of true north, the usual transit observations are taken, giving two equations:

$$N = N_1' + CA_1\Delta T_1$$

$$N = N_2' + CA_2\Delta T_2$$

$$\text{then } C = \frac{N_1' - N_2'}{A_2\Delta T_2 - A_1\Delta T_1} \frac{\text{mins of arc/div}}{\text{of scale/sec of time}}. \tag{10.13}$$

As this method has the theodolite in the clamped mode, the harmonic motion of the spin axis is affected by torque due to precession and torque due to twisting and untwisting of the suspension system.

Also, in the case of the Wild GAK.1, the gyro scale intervals are $10'$ and therefore cannot be read to the accuracy inherent in the gyro itself. The Royal School of Mines (RSM) recommended the introduction of an optical micrometer which enabled readings to 1/100 of the scale readings with an accuracy of about $\pm 4''$.

(3) *Transit method using the Wild modified GAK.1*: This method (*Figure 10.43*) was devised at the RSM and involves timing the moving mark as it passes *any* graduation on the gyro scale, i.e. time t_0 at graduation A_0, then reading the reversal point r_1, and then recording the time t_1 when the mark returns to the selected gyro scale graduation A_0. This process is continued until sufficient readings are available. The value N, of the centre of the oscillations, may be found as follows:

$$T = \text{period of oscillation} = \frac{1}{3}(t_2 - t_0) + (t_3 - t_1) + (t_4 - t_2) \tag{10.14}$$

$$u = \frac{90}{T}(t_1 - t_0) \tag{10.15}$$

$$P = 2\sin^2 u \tag{10.16}$$

$$N = \frac{A_0 + r_1(P-1)}{P} \tag{10.17}$$

Further values of T can be obtained for different values of A_0.

(4) *The amplitude method:* This method (*Figure 10.43*) was also devised at the RSM and uses the modified instrument in the clamped mode within a few degrees of true north. The improved accuracy of reading on the gyro scale enables reversal point readings (r_n) to be made satisfactorily; then

$$N = (r_1 + 3r_2 + 3r_3 + r_4)/8$$

and if $F = $ the fixed reading on the horizontal circle of the theodolite at the period of observation then the horizontal circle reading of true north N_0 is obtained from

$$N_0 = F - sN(1 + C) - Z \tag{10.18}$$

where $C = $ proportionality constant
   $Z = $ tape zero correction (Csr_0) (see *Section 10.12.3(2)*)
   $s = $ the value of 1 div on the gyro scale

It can be seen that methods (3) and (4) could be (and usually are) combined. Other methods of observation exist and may be referred to in 'The Six Methods of Finding North Using a Suspended Gyroscope' by Dr T.L. Thomas (*Survey Review*, Vol. 26, 203, January 1982 and 204, April 1982).

10.12.3 Instrumental errors

(1) Instrument calibration constant

The scale on which the oscillation of the gyroscope is observed, as in the amplitude method, is

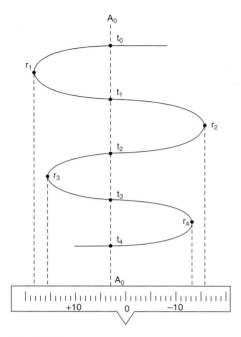

Fig. 10.43

usually not exactly aligned with the north-seeking vector of the gyroscope. Also, the line defined by the gyro scale may not be aligned with the axis of the theodolite. These errors therefore constitute an instrument constant K which can be ascertained only by carrying out observations on a base line of known azimuth.

Then $N = N_G + K$

where N = true or geographical north
        N_G = gyro north (i.e. the apparent N established by the gyroscope)
        K = instrument calibration constant

Thus: Azimuth of known base = $30°25'30''$
        Gyro azimuth of base = $30°28'30''$

$$K\text{-}value = -0°03'00''$$

Tests have shown that the K-value is not a constant but changes slowly over a period of time. Frequent calibration checks are therefore necessary to obtain optimum results when using a gyro-theodolite, and should certainly be carried out just before and immediately after underground observations.

(2) Tape zero error

The tape zero position is defined as *the position of rest of the oscillating system relative to the instrument*, with the gyro in the non-spin mode. In the zero position any torquing forces on the suspension tape holding the gyro-system, due to twist in the tape, are eliminated and the gyro mark coincides with the zero of the gyro scale.

In order to find the tape zero correction (Z), the gyro oscillations (r_n) in the non-spin mode are read against the gyro scale as follows:

$$r_B = (r_1 + 3r_2 + 3r_3 + r_4)/8 \text{ (amplitude readings before spin up)}$$
$$r_A = (r_1' + 3r_2' + 3r_3' + r_4')/8 \text{ (amplitude readings after spin up)}$$

then the weighted mean r_0 is formed from

$$r_0 = (r_B + 3r_A)/4 \tag{10.19}$$

The correction (Z) to normal north-finding observations due to tape zero error is

$$Z = - Csr_0 \tag{10.20}$$

where C = the proportionality constant and is equal to (torque due to tape/torque due to precession) (see *Section 10.12.2.(2)*).

s = value of one gyro scale unit in angular measure

It should be noted that if the tape zero correction Z is not applied to the observations for the instrument calibration constant K, then it will become part of K and a further cause of variation in its value.

(3) Circle drift

This is movement of the horizontal circle caused possibly by vibration of the gyro. Thus the RO of the base line should be observed before and after the gyro observations, and the mean taken.

(4) Change of collimation and eccentricity of collimation

This is difficult to eliminate as one cannot change face with some types of gyro-theodolite. Observational procedures can be adopted to reduce this error.

(5) Circle eccentricity

This form of error has already been discussed in Chapter 4 and can be reduced in gyro work by rotating the circle through 180° relative to the RO between sets of observation.

10.12.4 Observation procedures

Examples of the two basic methods of observation with a gyro-theodolite will now be given to illustrate more clearly the theory already covered.

(1) Reversal point method

The spin axis is brought approximately to north, and the rotor accelerated to full speed and carefully suspended. As the gyro, as indicated on the gyro scale, starts to move away from the centre-line of the scale, it is followed and kept on the centre-line of the scale by rotating the tangent screws of the theodolite. As it reaches, say, its left reversal point (r_1, *Figure 10.41(a)*) movement ceases for a few seconds and the theodolite horizontal circle is read. The gyro is then tracked back, keeping it on the zero of the gyro scale, to its right reversal point r_2 and the theodolite again read. Thus the movement of the gyro is followed by simply keeping it on the centre-line of the gyro index, whilst the amplitude of its movement is measured by the theodolite.

Reversal	Horiz circle reading
r_1 (left)	42° 00′ 31″
r_2 (right)	49° 40′ 32″
r_3 (left)	42° 04′ 02″
r_4 (right)	49° 37′ 21″

Schuler's mean $N_1 = \frac{1}{4}(r_1 + 2r_2 + r_3)$

$$= \frac{1}{4}(42° 00′ 31″ + 99° 21′ 04″ + 42° 04′ 02″)$$

$$= 45° 51′ 24″$$

$$N_2 = \frac{1}{4}(r_2 + 2r_3 + r_4) = 45° 51′ 29″$$

$$\therefore N = (N_1 + N_2)/2 = 45° 51′ 26″ \text{ (horizontal circle reading of gyro north)}$$

Assuming a tape zero correction (Z) of − 10″, then

$$N_0 = N - Z = 45° 51′ 26″ - 10″ = 45° 51′ 16″$$

A check on the computation can be applied by combining N_1 and N_2 to give

$$N = \frac{1}{8}(r_1 + 3r_2 + 3r_3 + r_4)$$

Assuming that the theodolite is now sighted along the base line and the mean horizontal circle reading was, say, 55° 51′ 16″; then the base is obviously 10° clockwise of gyro north and its bearing relative to gyro north is therefore 10°.

The application of the instrument constant K equal to, say, − 03′ 00″ will reduce the bearing of the base line to its correct geographical azimuth, i.e. 09° 57′ 00″ relative to true north.

If the survey is to be connected into the OS National Grid then a correction (δ) for convergence of meridians (i.e. the difference between grid north and true north) must be made and possibly a $(t - T)$ correction (i.e. the difference between an observed bearing and its corresponding grid bearing). In some instance a Laplace correction may need to be applied where the 'deviation of the vertical' is very high; however, in most cases this correction is usually less than 3″ of arc.

In the case of local surveys, of limited extent, on a rectangular Cartesian grid, only the convergence of meridians need to be considered.

Figure 10.44 shows the relationship of the various corrections.

Gyro bearing	$AB = \beta = 10° 00′ 00″$
Instrument constant	$K = -0° 03′ 00″$

Geographical azimuth	$\theta = 09° 57′ 00″$
Convergence of meridians	$\delta = +43′ 09″$ computed from
$(t - T)$	$= +00′ 04″$ geodetic tables

NG bearing of base-line $AB = 10° 40′ 13″ = \phi$

(*N.B.* Z is the tape zero correction $= -Csr_0$.)

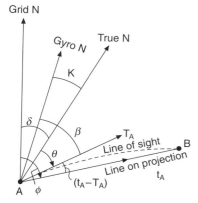

Fig. 10.44

(2) Transit method

In this method the gryoscope is set approximately to north (N') and the whole instrument clamped. Oscillation of the gyro index about the gyro scale is noted and timed. For instance, with the gyro on the centre-line of the scale the reading is zero and the time is zero. As the gyro reaches its left reversal point (A_W) the gyro-scale reading is noted; as it returns to zero the time t_2 is noted (see *Figure 10.42*). Similarly, the gyro-scale reading of the right reversal is noted (A_E) and the time t_3, when the gyro once again returns to the zero of the gyro index. The field data are reduced as follows:

Transit time t	Oscillation time T	Time difference 'ΔT'	Reversal readings A_W/A_E	$\frac{1}{2}(A_W+A_E)$ A	Horizontal circle N'	ΔN
0 m 00.0 s					45° 47′ 00″	
	−3 m 16.1 s		−11.8 (A_W)			
3 m 16.1 s		+7.2 s		12.35		+4.25′
	+3 m 23.3 s		+12.9 (A_E)			
6 m 39.4 s		+7.7 s		12.35		+4.55′
	−3 m 15.6 s		−11.8			
9 m 55.0 s		+7.6 s		12.35		+4.49′
	+3 m 23.2 s		+12.9			
13 m 18.2 s					*Mean*	+4.43′

$$\Delta N = CA\Delta T$$

where C = proportionality constant = 0.0478 min of arc/div of gyro scale × sec of time

$$A = \frac{1}{2}(A_W + A_E) = \frac{1}{2}(11.8 + 12.9) = 12.35$$

ΔT = algebraic sum of times = (−3 m 16.1 s + 3 m 23.3 s)
= + 7.2 s (T_W is − ve, T_E is +ve)

then $\Delta N = 0.0478 \times 12.35 \times 7.2 = 4.25'$

Horizontal circle reading of gyro $N = N' + \Delta N = 45°51'26''$

which is now corrected for Z and/or K, as previously.

The value of C is easily obtained using the above method with the instrument oriented say west of north, then east of north. Thus

$$N = N'_W + CA_1 \Delta T_1 = N'_E + CA_2 \Delta T_2$$

from which

$$C = \frac{N'_W - N'_E}{A_2 \Delta T_2 - A_1 \Delta T_1}$$

The main point to be emphasized about gyro observations is that they can be carried out on a line situated anywhere in the underground workings.

10.13 LINE AND LEVEL

10.13.1 Line

The line of the tunnel, having been established by wire survey or gyro observations, must be fixed in physical form in the tunnel. For instance, in the case of a Weisbach triangle (*Figure 10.45*) the bearing $W_u W_1$ can be computed; then, knowing the design bearing of the tunnel, the angle θ can be computed and turned off to give the design bearing, offset from the true by the distance $X W_u$. This latter distance is easily obtained from right-angled triangle $W_1 X W_u$.

The line is then physically established by carefully lining in three plugs in the roof from which weighted strings may be suspended as shown in *Figure 10.46(a)*. The third string serves to check the other two. These strings may be advanced by eye for short distances but must always be checked by theodolite as soon as possible.

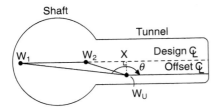

Fig. 10.45   *Plan view*

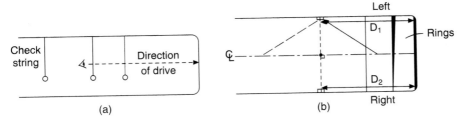

Fig. 10.46   *(a) Section, and (b) plan*

The gradient of the tunnel may be controlled by inverted boning rods suspended from the roof and established by normal levelling techniques.

In addition to the above, 'square marks' are fixed in the tunnel by taping equilateral triangles from the centre-line or, where dimensions of the tunnel permit, turning off 90° with a theodolite, (*Figure 10.46(b)*). Measurements from these marks enable the amount of lead in the rings to be detected. For instance, if $D_1 > D_2$ then the difference is the amount of left-hand lead of the rings. The gap between the rings is termed *creep*. In the vertical plane, if the top of the ring is ahead of the bottom this is termed *overhang,* and the reverse is *look-up.* All this information is necessary to reduce to a minimum the amount of 'wriggle' in the tunnel alignment.

Where tunnel shields are used for the drivage, laser guidance systems may be used for controlling the position and attitude of the shield. A laser beam is established parallel to the axis of the tunnel (i.e. on bearing and gradient) whilst a position-sensing system is mounted on the shield. This latter device contains the electro-optical elements which sense the position and attitude of the shield relative to the laser datum. Immunity to vibrations is achieved by taking 300 readings per second and displaying the average. Near the sensing unit is a monitor which displays the displacements in mm automatically corrected for roll. Additionally, roll, lead and look-up (*Figure 10.47*) are displayed on push-button command along with details of the shield's position projected 5 m ahead. When the shield is precisely on line a green light glows in the centre of the screen. All the above data can be relayed to an engineers' unit several hundred metres away. Automatic print-out of all the data at a given shield position is also available to the engineers. The system briefly described here is the Predictor system, designed and manufactured by ZED Instruments Ltd, London, England.

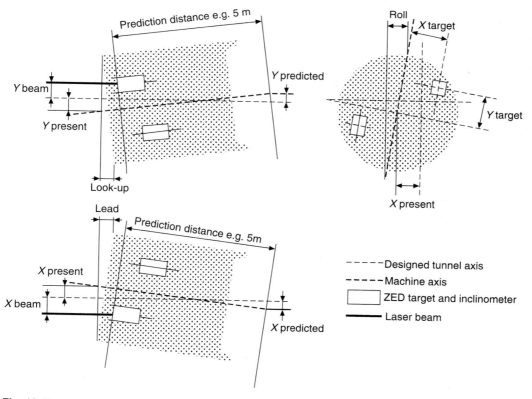

Fig. 10.47

In general the power output of commercial lasers is of the order of 5 mW and the intensity at the centre of a 2-cm diameter beam approximately 13 mW/cm^2. This may be compared with the intensity of sunlight received in the tropics at noon on a clear day, i.e. 100 mW/cm^2. Thus, as with the sun, special precautions should be taken when viewing the laser (see *Section 10.9*).

Practically all the lasers used in tunnelling work are wall- or roof-mounted, and hence their setting is very critical. This is achieved by drilling a circular hole in each of two pieces of plate material, which are then fixed precisely on the tunnel line by conventional theodolite alignment. The laser is then mounted a few metres behind the first hole and adjusted so that the beam passes through the two holes and thereby establishes the tunnel line. Adjustment of the holes relative to each other in the vertical plane would then serve to establish the grade line.

An advantage of the above system is that the beam will be obscured should either the plates or laser move. In this event the surveyor/engineer will need to 'repair' the line, and to facilitate this, check marks should be established in the tunnel from which appropriate measurements can be taken.

In order to avoid excessive refraction when installing the laser, the beam should not graze the wall. Earth curvature and refraction limit the laser line to a maximum of 300 m, after which it needs to be moved forward. To minimize alignment errors, the hole furthest from the laser should be about one-third of the total beam distance from the laser.

10.13.2 Level

In addition to transferring bearing down the shaft, the correlation of the surface level to underground must also be made.

One particular method is to measure down the shaft using a 30-m standardized steel band. The zero of the tape is correlated to the surface BM as shown in *Figure 10.48*, and the other end of the tape precisely located using a bracket fixed to the side of the shaft. This process is continued down the shaft until a level reading can be obtained on the last tape length at B. The standard tension is applied to the tape and tape temperature recorded for each bay. A further correction is made for elongation of the tape under its own weight using:

$$\text{Elongation (m)} = WL/2AE$$

where E = modulus of elasticity of steel (N/mm^2)
L = length of tape (m)

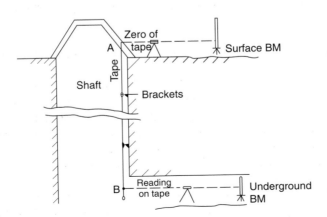

Fig. 10.48

A = cross-sectional area of tape (mm^2)
W = mass of tape (N)

Then the corrected distance AB is used to establish the value of the underground BM relative to the surface BM.

If a special shaft tape (1000 m long) is available the operation may be carried out in one step. The operation should be carried out at least twice and the mean value accepted. Using the 30-m band, accuracies of 1 in 5000 are possible; the shaft tape gives accuracies of 1 in 10 000.

Electromagnetic distance-measuring (EDM) equipment has also been used to measure shaft depths. A special reflecting mirror at the top of the shaft is aligned with the EDM instrument and then rotated in the vertical plane until the measuring beam strikes a reflector at the shaft bottom. In this way the distance from the instrument to the reflector is obtained and subsequently adjusted to give the distance from mirror to reflector. By connecting the mirror and reflector to surface and underground BMs respectively, their values can be correlated. With top-mounted EDM equipment a reflecting mirror is unnecessary and the distance to a reflector at the shaft bottom could be measured direct.

10.14 RESPONSIBILITY ON SITE

Responsibility with regard to setting out is defined in Clause 17 of the ICE Conditions of Contract:

> The contractor shall be responsible for the true and proper setting out of the works, and for the correctness of the position, levels, dimensions, and alignment of all parts of the works, and for the provision of all necessary instruments, appliances, and labour in connection therewith. If, at any time during the progress of the works, any error shall appear or arise in the position, levels, dimensions, or alignment of any part of the works, the contractor, on being required so to do by the engineer, shall, *at his own cost*, rectify such error to the satisfaction of the engineer, unless such error is based on incorrect data supplied in writing by the engineer or the engineer's representative, in which case the cost of rectifying the same shall be borne *by the employer*. The checking of any setting out, or of any line or level, by the engineer or the engineer's representative, shall not, in any way, relieve the contractor of *his* responsibility for the correctness thereof, and the contractor shall carefully protect *and preserve* all bench-marks, sight rails, pegs, and other things used in setting out the works.

The clause specifies three persons involved in the process, namely, the employer, the engineer and the agent, whose roles are as follows:

The *employer*, who may be a government department, local authority or private individual, requires to carry out and finance a particular project. To this end, he commissions an *engineer* to investigate and design the project, and to take responsibility for the initial site investigation, surveys, plans, designs, working drawings, and setting-out data. On satisfactory completion of his work he lets the contract to a contractor whose duty it is to carry out the work.

On site the employer is represented by the engineer or his representative, referred to as the *resident engineer* (RE), and the contractor's representative is called the *agent*.

The engineer has overall responsibility for the project and must protect the employer's interest without bias to the contractor. The agent is responsible for the actual construction of the project.

10.15 RESPONSIBILITY OF THE SETTING-OUT ENGINEER

The setting-out engineer should establish such a system of work on site that will ensure the accurate setting out of the works well in advance of the commencement of construction. To achieve this, the following factors should be considered.

(1) A complete and thorough understanding of the plans, working drawings, setting-out data, tolerances involved and the time scale of operations. Checks on the setting-out data supplied should be immediately implemented.
(2) A complete and thorough knowledge of the site, plant and relevant personnel. Communications between all individuals is vitally important. Field checks on the survey control already established on site, possibly by contract surveyors, should be carried out at the first opportunity.
(3) A complete and thorough knowledge of the survey instrumentation available on site, including the effect of instrumental errors on setting-out observations. At the first opportunity, a base should be established for the calibration of tapes, EDM equipment, levels and theodolites.
(4) A complete and thorough knowledge of the stores available, to ensure an adequate and continuing supply of pegs, pins, chalk, string, paint, timber, etc.
(5) Office procedure should be so organized as to ensure easy access to all necessary information. Plans should be stored flat in plan drawers, and those amended or superseded should be withdrawn from use and stored elsewhere. Field and level books should be carefully referenced and properly filed. All setting-out computations and procedures used should be clearly presented, referenced and filed.
(6) Wherever possible, independent checks of the computation, abstraction and extrapolation of setting-out data and of the actual setting-out procedures should be made.

It can be seen from this brief itinerary of the requirements of a setting-out engineer, that such work should never be allocated, without complete supervision, to junior, inexperienced members of the site team.

All site engineers should also make a careful study of the following British Standards, which were prepared under the direction of the Basic Data and Performance Criteria for Civil Engineering and Building Structures Standards Policy Committee:

(1) BS 5964:1990 (ISO 4463–1, 1989)
    Part 1. Methods of measuring, planning and organization and acceptance criteria.
(2) BS 5606:1990
    Guide to Accuracy in Building.
(3) BS 7307:1990 (ISO 7976:1989)
    Part 1. Methods and instruments.
    Part 2. Position of measuring points.
(4) BS 7308:1990 (ISO 7737:1986)
    Presentation of dimensional accuracy data in building construction.
(5) BS 7734:1990 (ISO 8322:1989)
    Part 1. Methods for determining accuracy in use: Theory.
    Part 2. Methods for determining accuracy in use: Measuring Tapes.
    Part 3. Methods for determining accuracy in use: Optical Levelling Instruments.
    Parts 4 to 8 deal with theodolites and EDM.

These documents supply important information coupled with a wealth of excellent, explanatory diagrams of various setting-out procedures.

For instance BS 5964 provides acceptance criteria for the field data measured, termed permitted deviation (PD), where PD = 2.5 (SD) (Standard Deviation).

If a primary system of control points has been established as stage one of the surveying and setting-out process, its acceptability (or otherwise) is based on the difference between the measured distance and its equivalent, computed from the adjusted coordinates; this difference should not exceed $\pm 0.75\,(L)^{\frac{1}{2}}$ mm, with a minimum of 4 mm, where L is the distance in metres. For angles it is $\pm 0.045\,(L)^{\frac{1}{2}}$ in degrees.

For a secondary system of control points the acceptance criteria is:

For distance, $\pm 1.5\,(L)^{\frac{1}{2}}$ mm, with a minimum of 8 mm.

For angles, $\pm 0.09\,(L)^{\frac{1}{2}}$ degrees.

For levelling between bench marks the general acceptance criterion is ± 5 mm, with slight variations for different situations.

This small sample of the information available in these standards indicates their importance to all concerned with the surveying and setting-out on site.

Worked examples

Example 10.1. The national grid (NG) bearing of an underground base line, *CD* (*Figure 10.49*), is established by co-planning at the surface onto two wires, W_1 and W_2, hanging in a vertical shaft, and then using a Weisbach triangle underground.

The measured field data is as follows:

NG bearing *AB*	74°28′34″
NG coords of *A*	E 304 625 m, N 511 612 m
Horizontal angles:	
BAW_s	284°32′12″
AW_sW_2	102°16′18″
$W_2W_uW_1$	0°03′54″
W_1W_uC	187°51′50″
W_uCD	291°27′48″

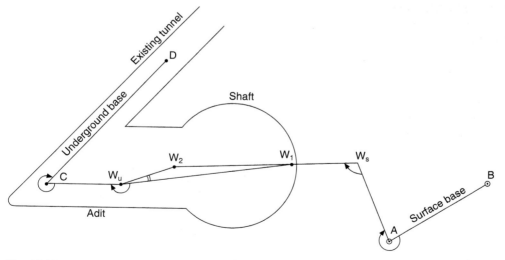

Fig. 10.49

Horizontal distances:

W_1W_2 3.625 m

W_uW_2 2.014 m

In order to check the above correlation, a gyro-theodolite is set up at C and the following horizontal scale theodolite readings to the reversal points of the gyro recorded:

Left reversal point   336° 25′ 18″
Right reversal point   339° 58′ 52″
Left reversal point   336° 02′ 44″

The mean horizontal scale reading on to CD is 20° 51′ 26″

Given: Convergence of meridians   ±0° 17′ 24″
          $(t - T)$ correction   ±0° 00′ 06″
          Instrument constant   −2° 28′ 10″

Compute the difference between the bearings of the underground base as fixed by the wire survey and by the gyro-theodolite.
(KU)

The first step is to calculate the bearing of the wire base using the measured angles at the surface:

Grid bearing of AB =   74° 28′ 34″ (given)
Angle BAW_s =   284° 32′ 12″
                      ———————

Grid bearing of AW_s =   359° 00′ 46″
Angle AW_sW_2 =   102° 16′ 18″
                      ———————

Sum =   461° 17′ 04″
          − 180
                  ———————

Grid bearing W_1W_2 =   281° 17′ 04″ (see *Figure 10.49*)

Now, using the Weisbach triangle calculate the bearing of the underground base from the wire base:

From a solution of the Weisbach triangle

Angle $W_2W_1W_u$ = $\dfrac{234'' \times 2.014}{3.625}$ = 130″ = 0° 02′ 10″

Grid bearing W_1W_2 =   281° 17′ 04″
Angle $W_2W_1W_u$ =   0° 02′ 10″
                      ———————

Grid bearing W_1W_u =   281° 14′ 54″
Angle W_1W_uC =   187° 51′ 50″
                      ———————

Sum =   469° 06′ 44″
          − 180
                  ———————

Grid bearing W_uC =   289° 06′ 44″
Angle W_uCD =   291° 27′ 48″
                      ———————

Sum =   580° 34′ 32″
          −540
                  ———————

Grid bearing CD =   40° 34′ 32″ (underground base)

Compute the grid bearing of *CD* using gyro data:

Horizontal circle reading of gyro north $= \frac{1}{2}\left(\frac{r_1 + r_3}{2} + r_2\right)$

$= \frac{1}{2}[(336°25'18'' + 336°02'44'')/2 + 339°58'52''] = 338°06'26''$

Horizontal circle reading of base *CD* $= 20°51'26''$

$\therefore$ Bearing of *CD* reative to gyro north $= 42°45'00''$

Bearing of *CD* relative to true north $= 42°45'00''$ – instr constant

                                   $= 40°16'50'' = \phi$ (*Figure 10.50*)

With reference to *Figure 10.50*, it can be seen how the 'convergence of meridians' ($\delta\theta$) and $(t - T)$ are applied to transform true N to grid N (see Chapter 5 for details).

$\therefore$ National grid bearing $CD = 40°16'50'' + 0°17'24'' + 0°00'06''$

$$= 40°34'20''$$

Difference in bearings $= 12''$.

Example 10.2. The centre-line of the tunnel *AB* shown in *Figure 10.51* is to be set out to a given bearing. A short section of the main tunnel has been constructed along the approximate line and access is gained to it by means of an adit connected to a shaft. Two wires *C* and *D* are plumbed down the shaft, and readings are taken onto them by a theodolite set up at station *E* slightly off the line *CD* produced. A point *F* is located in the tunnel, and a sighting is taken on to this from station *E*. Finally a further point *G* is located in the tunnel and the angle *EFG* measured.

From the survey initially carried out, the coordinates of *C* and *D* have been calculated and found to be E 375.78 m and N 1119.32 m, and E 375.37 m and N 1115.7 m respectively.

Calculate the coordinates of *F* and *G*. Without making any further calculations describe how the required centre-line could then be set out.

                                                                                 (ICE)

Given data: $CD = 3.64$ m $DE = 4.46$ m

                $EF = 13.12$ m $FG = 57.5$ m

                Angle *DEC* $= 38''$

                Angle *CEF* $= 167°10'20''$

                Angle *EFG* $= 87°23'41''$

Solve Weisbach triangle for angle *ECD*

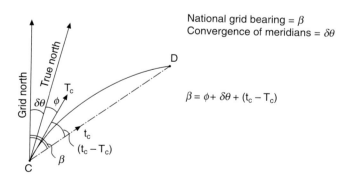

National grid bearing $= \beta$
Convergence of meridians $= \delta\theta$

$\beta = \phi + \delta\theta + (t_c - T_c)$

Fig. 10.50

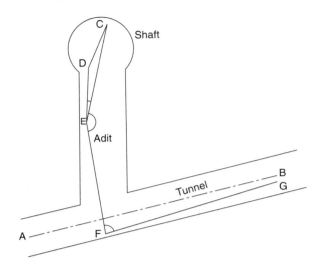

Fig. 10.51

$$\hat{C} = \frac{ED}{DC}\hat{E} = \frac{4.46}{3.64} \times 38'' = 47''$$

By coordinates

Bearing of wire base $CD = \tan^{-1}\dfrac{-0.41}{-3.62} = 186°27'19''$

$$\therefore \text{WCB of } CE = 186°27'42'' - 47'' = 186°26'55''$$
$$\text{WCB of } CE = 186°26'55''$$
$$\text{Angle } CEF = 167°10'20''$$

$$\text{WCB of } EF = 173°37'15''$$
$$\text{Angle } EFG = 87°23'41''$$

$$\text{WCB of } FG = 81°00'56''$$

Line	Length (m)	WCB	Coordinates		Total coordinates		
			ΔE	ΔN	E	N	Station
					375.78	1119.32	C
CE	8.10	186°26'55''	−0.91	−8.05	374.87	1111.27	E
EF	13.12	173°37'15''	1.46	−13.04	376.33	1098.23	F
FG	57.50	81°00'56''	56.79	8.99	433.12	1107.22	G

Several methods could be employed to set out the centre-line; however, since bearing rather than coordinate position is critical, the following approach would probably give the best results.

Set up at G, the bearing of GF being known, the necessary angle can be turned off from GF to give the centre-line. This is obviously not on centre but is the correct line; centre positions can now be fixed at any position by offsets.

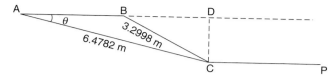

Fig. 10.52

Example 10.3. Two vertical wires A and B hang in a shaft, the bearing of AB being $55°\,10'\,30''$ *(Figure 10.52)*. A theodolite at C, to the right of the line AB produced, measured the angle ACB as $20'\,25''$. The distances AC and BC were 6.4782 m and 3.2998 m respectively.

Calculate the perpendicular distance from C to AB produced, the bearing of CA and the angle to set off from BC to establish CP parallel to AB produced.

Describe how you would transfer a line AB above ground to the bottom of a shaft. (LU)

$AB \approx AC - BC = 3.1784$ m

Angle $BAC = \dfrac{3.2998}{3.1784} \times 1225'' = 1272'' = 21'\,12'' = \theta$

By radians CD $\quad = AC \times \theta$ rad $= \dfrac{6.4782 \times 1272}{206\,265} = 0.0399$ m

$$\begin{aligned}
\text{Bearing } AB &= 55°\,10'\,30'' \\
\text{Angle } BAC &= 21'\,12'' \\[-2pt]
&\rule{4cm}{0.4pt} \\
\text{Bearing } AC &= 55°\,31'\,42'' \\
\therefore \text{ Bearing } CA &= 235°\,31'\,42''
\end{aligned}$$

$$\begin{aligned}
\text{Angle to be set off from } BC = ABC &= 180° - (21'\,12'' + 20'\,25'') \\
&= 179°\,18'\,23''
\end{aligned}$$

Exercises

(10.1) (a) Describe fully the surveying operations which have to be undertaken in transferring a given surface alignment down a shaft in order to align the construction work of a new tunnel.

(b) A method involving the use of the three-point resection is often employed in fixing the position of the boat during offshore sounding work.

Describe in detail the survey work involved when this method is used and discuss any precautions which should be observed in order to ensure that the required positions are accurately fixed.

(ICE)

(10.2) Describe how you would transfer a surface bearing down a shaft and set out a line underground in the same direction.

Two plumb lines A and B in a shaft are 8.24 m apart and it is required to extend the bearing AB along a tunnel. A theodolite can only be set up at C 19.75 m from B and a few millimetres off the line AB produced.

If the angle BCA is $09'\,54''$ what is the offset distance of C from AB produced? (ICE)

*(Answer:*195 mm)

Index